Conservation Agriculture

Environment, Farmers Experiences, Innovations, Socio-economy, Policy

Edited by

Luis García-Torres

Institute for Sustainable Agriculture, CSIC, Cordoba, Spain, and European Conservation Agriculture Federation (ECAF), Brussels, Belgium

José Benites

Conservation Agriculture Working Group, Agriculture Department, Food and Agriculture Organization (FAO), Rome, Italy

Armando Martínez-Vilela

European Conservation Agriculture Federation (ECAF), Brussels, Belgium

and

Antonio Holgado-Cabrera

European Conservation Agriculture Federation (ECAF), Brussels, Belgium

KLUWER ACADEMIC PUBLISHERS
DORDRECHT / BOSTON / LONDON

A C.I.P. Catalogue record for this book is available from the Library of Congress.

ISBN 1-4020-1106-7

Published by Kluwer Academic Publishers,
P.O. Box 17, 3300 AA Dordrecht, The Netherlands.

Sold and distributed in North, Central and South America
by Kluwer Academic Publishers,
101 Philip Drive, Norwell, MA 02061, U.S.A.

In all other countries, sold and distributed
by Kluwer Academic Publishers,
P.O. Box 322, 3300 AH Dordrecht, The Netherlands.

Printed on acid-free paper

The content of this report does not necessarily reflect the official opinion of the Food and Agricultural Organisation of the United Nations, LIFE Programme of EU Commission DG-Environment and European Conservation Agriculture Federation. Neither these entities nor any persons acting on their behalf is responsible for the use that may be made of the information in this book.
Trademark notice: Product or corporate names may be trademarks or registered trademarks, and are used only for identification and explanation, without intent to infringe.

Printed in the Netherlands.

PREFACE

In large parts of the developed and developing worlds soil tillage by plough or hoe is the main cause of land degradation leading to stagnating or even declining production levels and increasing production cost. It causes the soil to become more dense and compacted, the organic matter content to be reduced and water runoff and soil erosion to increase. It also leads to droughts becoming more severe and the soil becoming less fertile and less responsive to fertiliser.

During the last 50 years there has been a profusion of agricultural development programmes throughout the world trying to arrest erosion, which is perceived as one of the main causes of land degradation. Sadly, though not surprisingly, the have failed to remedy the situation. To date, it has proved difficult to achieve conservation of soil and water productive potentials using physical erosion control methods which, though widely demonstrated in temperate and tropical areas, have not found favour with either the farmers on whose lands they were implemented nor with the soils they were supposed to protect.

There is a growing number of experiences in conditions of both mechanised and un-mechanised agriculture, on small and large farms in both temperate and tropical zones that further and significant improvements in conservation-effective agriculture are indeed possible, and acceptable to farmers, in addressing these varied concerns now strongly indicate that sustainable production systems can be achieved when the basic principles of good farming practice are applied. The terminology being adopted for such systems by FAO, ECAF and other organisations is "Conservation Agriculture". This implies conformity with all three of the following general principles: no mechanical soil disturbance, direct seeding or planting; and permanent soil cover, making particular use of crop residues and cover crops; judicious choice of crop rotations.

Globally, conservation agriculture (CA) is being practised on about 80 million hectares, mostly in South and North America and Australia. The rate of adoption of these practices is growing exponentially on small and large farms in South America, stimulated by both economic and environmental pressures.. The benefits have been markedly positive in both agricultural, environmental, economic and social terms. For example: crops suffer less from drought; yields improve; production costs reduce; floods and soil erosion are minimized, and streamflows improve; farm-family satisfaction rises, and their activities diversify. The principles are applicable to both minute, small and large farms; the practical applications vary from situation to situation.

In countries where conservation agriculture has not yet been developed, the first step toward CA adoption would be to sensitize stakeholders to land issues and CA opportunities so as to create awareness and willingness to change. Thereafter, activities would initiate the change process, primarily to identify pathways of change, pilot CA farming, establish support for knowledge and information systems, and build capacity of local institutions and producer organizations.

At global level, the formation of groups with a common interest concerning the concepts and practices of conservation agriculture have now led to the establishment of an interdisciplinary Conservation Agriculture Working Group (CAWG), within the Agriculture Department of FAO and the European Conservation Agricultural Federation (ECAF). This unites national CA associations in the Belgium, Denmark, France, Germany, Great Britain, Italy, Portugal, Spain, Switzerland and others. FAO and ECAF have become very active developing field projects, creating awareness, networking and advising on policy. They promote the dissemination of information through workshops and international meetings, the I World Congress on Conservation Agriculture (Madrid, October 2001) being a current high-profile example to spread CA technologies further. But its benefits will spread rapidly and widely only where Government policies, services and infrastructure facilitate the conversion to conservation agriculture.

The referred World Congress called upon politicians, international institutions, environmentalists, farmers and private industry to support further develop and embrace the concept of conservation agriculture as the means of ensuring the continuity of the land's ongoing capacities to yield food, other agricultural products and water in perpetuity. Its main objective was to consider and promote the world-wide adoption of the principles and locally-suitable practices of conservation agriculture. This book brings together the key notes lectures and other outstanding contributions of the I World Congress on Conservation Agriculture, and provides an updated view of the environment and economic advantages of CA and of its implementation in diferent areas of the World.

The Editors

CONTENTS

CONTRIBUTORS

AHMED, M.; Scientific Officer Wheat Research Centre, Bangladesh Agriculture Research Institute, Bangladesh

AMADO, T.J.C., Professor Federal University of Santa Maria, RS, Brazil. CNPq Researcher. e-mail: *tamado@creta.ccr.ufsm.br*

AMIN, M.R.; Scientific Officer Wheat Research Centre, Bangladesh Agriculture Research Institute, Bangladesh

ASHBURNER, J. E.; FAO Regional Office for Africa, PO Box 1628, Accra, Ghana

BACHMAN, T.; Food and Agricultural Organization (FAO), Viale delle Terme di Caracalla; 00100 Rome, Italy

BAKER, C.J.; Centre for International No-Tillage Research and Engineering, 50 Nannestad Line; RD 5; Feilding 5600; New Zealand, e-mail: *bakerbnt@inspire.net.nz*

BAKSH, E.; Senior Scientific Officer, Wheat Research Centre, Bangladesh Agriculture Research Institute, Bangladesh

BALOTA, E. L.; IAPAR; Area de Microbiologia do Solo, Cx. Postal 481, 86.001-970; Londrina, PR, Brazil, e-mail: *diva@pr.gov.br*

BASSO, B.; Dipartimento di Produzione Vegetale; Universita' degli Studi della Basilicata; Via N.Sauro, 85, Potenza, Italy, e-mail: *basso@unibas.it*

BENITES, J.; AGLS; Food and Agricultural Organization (FAO), Viale delle Terme di Caracalla; 00100 Rome, Italy, e-mail: *Jose.Benites@FAO.ORG*

BODRUZZAMAN, M.; Senior Scientific Officer, Wheat Research Centre, Bangladesh Agriculture Research Institute, Bangladesh

BÖHRNSEN, A.; Institute of Agricultural Engineering, Justus-Liebig-University, 35390 Giessen,, DE

BOT, A.J., Freelance consultant FAO. e-mail: *alexandra.bot@12move.nl*

BOUCHE, M.B.; INRA, Laboratoire de Zooécologie du Sol, 2, Place Pierre Viala, F-34060 Montpellier cedex 1

BOUZINAC, S.; Centre de Cooperation Internationale en Recherche Agronomique por le Développement CIRAD-CA-GEC, a/c Tasso de Castro, BP 504, Agencia Central CEP, 74000-970 Goiânia Go, Brazil

BRUNOTTE, J.; Production Engineering and Building Research, Federal Agricultural Research Centre, Braunschweig, Germany

BWALYA, M.; ACT Secretariat, c/o IES/University of Zimbabwe, P O Box MP, 167 Harare, e-mail: *ACTsecre@africaonline.co.zw*

CALEGARI, A.; Agronomic Institute (IAPAR); Rodovia Celso Garcia; Cid, KM 375CEP-86001-870; Londrina, Parana – BRASIL, e-mail: *asoiapar@pr.gov.br*

CALISSAYA, A.; PROMETA/CIFEMA, Casilla 831, Cochabamba, Bolivia

CARTER, A.D.; ADAS Consulting Limited, Rosemaund, Preston Wynne, Hereford HR 1 3PG, UK

CHOUDHARY, M. A.; Western Australia Department of Agriculture, Northam, WA, Australia

CIOLKOSZ, E. J.; The Pennsylvania State University, 116 ASI Building, University Park, PA 16802-3504, USA

CLANCY, S.H.; Conservation Agriculture P.O. Box 477 Devils Lake, North Dakota. 58301 USA

COLOZZI FILHO, A.; IAPAR; Área de Microbiologia do Solo, Cx. Postal 481, 86.001-970, Londrina, PR, Brazil, e-mail: *diva@pr.gov.br*

COLLINS, R. M.; Western Australia Department of Agriculture, Northam, WA, Australia

CROVETTO, C.; Fundo Chequen; Las Heras 2095; Casillaa 1626; Concepción – Chile, e-mail: *crovetto@entelchile.net*

DA VEIGA, M.; EECN/EPAGRI; BR 282, km. 342; Caixa Postal 116; 89620-000 Campos Novos; Santa Catarina (BRASIL), e-mail: *milveiga@epagri.rct-sc.br*

DAUPHIN, F.; FAO; Via delle Terme di Caracalla; 00100-ROME – ITALY, e-mail: Francois.*Dauphin@fao.org*

DE FREITAS, P. L.; Researcher Embrapa Soils Centre, Rio de Janeiro, RJ, Brazil. E-mail: *pedro.freitas@terra.com.br*

DE GRAAFF, J.; Erosion and Soil and Water Conservation Group, Department of Environmental Sciences, Wageningen University, Nieuwe Kanaal 11, 6709 PA Wageningen, The Netherlands

DE SOUZA ANDRADE, D.; IAPAR; Área de Microbiologia do Solo, Cx. Postal 481; 86.001-970; Londrina; PR; Brasil, e-Mail:*diva@pr.gov.br*

DE WRACHIEN, D.; EurAgEng Incoming President and Coordinator; of the Field of Interest on Soil and Water; Head of the Institute of Hydraulics Agricultural; State University of Milan, Italy

DERPSCH, R.; GTZ; Casilla de Correo 1859 – (Paraguay), e-mail: *rderpsch@quarta.com.py*

DHILLON, S.S.; Agronomist, Punjab Agricultural University: College of Agriculture: Department of Plant Breeding, Ludhiana 141 004, India

DO PRADO WILDNER, L.; RELACO Secretario 1999-2001, EPAGRI, Santa Catarina, Brasil, e-mail: *lpwild@epagri.rct-sc.br*

DUIKER, S.W.; The Pennsylvania State University, 116 ASI Building, University Park, PA 16802-3504, USA

EKBOIR, J.; CIMMYT, Apdo. Postal 6-641, 06600 Mexico, D.F., Mexico, e-mail: *jekboir@cgiar.org*

EL TITI, A.; State Institute for Plant Protection; Reinsburgstraße 107; D-70197 Stuttgart (Germany), e-mail: *eltiti@lfpst.bwl.de*

EPPERLEIN, J.; Gesellschaft für konservierende Bodenbearbeitung (GKB) e.V. Dorfstraße 9, D - 13051 Berlin

FAROOQ, N.; 1022-C, Canal View Housing Society; Thokar Niaz Baig; Lahore, (Pakistan), e-mail: *nasreenayub@hotmail.com*

FLESKENS, L.; Erosion and Soil and Water Conservation Group, Department of Environmental Sciences, Wageningen University, Nieuwe Kanaal 11, 6709 PA Wageningen, The Netherlands

FRESCO, L.O.; FAO (Agriculture Department); Via delle Terme di Caracalla; 00100-Rome – Italy

GAMEZ, P.; PROMETA/CIFEMA, Casilla 831, Cochabamba, Bolivia

GARCÍA TORRES, L.; Profesor, IAS/ CSICS; Córdoba; Spain; and ECAF, Brussels, Belgium, e-mail: *ml1gatol@uco.es; conservation.agriculture@ecaf.org*

GILL, M. A.; Director General Agriculture (Water Management) Punjab; 21-Sir Agha Khan Soyyem Road; Lahore, (Pakistan), e-mail: *mushtaqgill@hotmail.com*

GRANVAL, P.; Direction de l'Evaluation et de la Prospective (ONCFS), Domaine de Saint-Benoist F-78610 Auffargis.

GREGOIRE, T.; North Dakota State University Extension Service, P.O. Box 477 Devils Lake , North Dakota. 58301 USA

GUPTA, R.; Facilitator of the RWC.

HAMELL, M.; Environment Quality and Natural Resources, Environment Directorate-General, European Commission. Rue de la Loi, B-1049 Brussels (Belgium)

HAMILTON- MANNS, M.; Grassworks Consultancy, Feilding, NEW ZEALAND

HARRINGTON, L.; Director, Natural Resources – CIMMYT; Apdo. Postal 6-641; 06600 Méjico, D.F., (México), e-mail: *cimmyt@cgiar.org L.Harrington@cgiar.org*

HOBBS, P. R.; CIMMYT regional representative for South Asia and co-Facilitator of the rice-wheat consortium for the Indo-Gangetic Plains of South Asia, e-mail: *p.hobbs@cgiar.org*

HOSSAIN, A.B.S.; Director Wheat Research Centre Bangladesh Agriculture Research Institute, Bangladesh

HUNGRIA, M.; Empresa Brasileira de Pesquisa Agropecuária (Embrapa Soja), Cx- Postal 231, 86001-970, Londrina, PR, Brazil

HUNTER, J.M.; The Pennsylvania State University, 116 ASI Building, University Park, PA 16802-3504, USA

ISENSEE, E.; Institute for Agricultural Engineering, University Kiel, Germany

KARAJEH, F.; ICARDA, PO Box 5466, Aleppo, Syria

KAUMBUTHO, P.; KENDAT, Kenya Network for Draught Animal Technology, PO Box 30197, Nairobi, Kenya. e-mail: *kendat@ken.healthnet.org*

KOOHAFHAN, P.; FAO (Agriculture Department); Via delle Terme di Caracalla; 00100-Rome – Italy, e-mail: *parviz.koohafkan@fao.org*

KUBIAK, R.; State Education and Research Center Neustadt, Department Ecology. Breitenweg 71, Neustadt an der Weinstraße, 67435-DE, e-mail: *sreuter.slfa-nw@agrarinfo.rlp.de*

KURGANOVA, I.; Institute of Physicochemical and Biological Problems of Soil Science, RAS, Pushchino, Moscow region, 142290 – Russia

KURISHBAYEV, A.; Kazakh Research Institute of Cereals; Shortandy-1; 474070 Akmoliskaya oblast. Kazakhstan, e-mail: *niizern@kaznet.kz*

LANDERS, J. N.; Brazilian National Science Council Scholar and Executive Secretary, Association for Zero Tillage in the Cerrado, MUDB Conjunto 9, Lote 5, postcode 71680-090, Lago Sul, Brasília, DF, Brazil, e-mail: *john.landers@apis.com.br*

LEAKE, A.; Head of Project; The Allerton Research and Educational Trust; Loddington House; Main Street. Loddington. Leicestershire; LE7 9XE, e-mail: *aleake@allertontrust.org.uk*, Website *www.allertontrust.org.uk*

LECONTE, D.; Domaine Expérimental Fourrager du Vieux Pin, F-61310 Le Pin au Haras

LOPES DE GERENYU, V.O.; Institute of Physicochemical and Biological Problems of Soil Science, RAS, Pushchino, Moscow region, 142290 - Russia

MALIK, R.K.; Agronomist/weed scientist at Haryana State University, Hisar, India.

MANFRINATO, W. A.; Graduate student, Biogeochemistry, Centro de Energia Nuclear na Agricultura, Piracicaba, SP, Brazil. e-mail: *warwick@merconet.com.br*

MARTÍNEZ-VILELA, A.; ECAF Executive Director, Brussels, Belgium. e-mail: *conservation.agriculture@ecaf.org; www.ecaf.org*

MATTANA SATURNINO, H.;*President, , Association for Zero Tillage in the Cerrado, SCRLN 712, Bloco C, Loja 18, Asa Norte, Brasília, DF, Brazil, postcode 70.760-533.* E-mail: *helvecio@gcsnet.com.br*

McGARRY, D.; Department of Natural resources; Queensland Government; RSK, DNR, Block "B"; Meiers Road Indooroopilly, Qld 4068. Australia, e-mail: *mcgarrd@dnr.qld.gov.au*

MEISNER, C.A.; CIMMYT Agronomist in Bangladesh and Cornell University Adjunct Professor; P.O. Box 6057 Gulshan Dhaka – 1212 Bangladesh; e-mail: *c.meirner@cgiar.org*

MIELNICZUK, J.; Professor Federal University Rio Grande do Sul , Porto Alegre, RS, Brazil.

MILLER, D.A.; The Pennsylvania State University, 116 ASI Building, University Park, PA 16802-3504, USA

MORRISON, J. E.; USDA, Agricultural Research Service, Temple, Texas, 76502 USA. E-mail: *morrison@brc.tamus.edu*

MUELLER, J. P.; Crop Science Dept., Box 7620; North Carolina State University, Raleigh, NC 27695-7620 (U.S.A.), e-mail: *paul_mueller@ncsu.edu*

MUYS, B.; Domaine Expérimental Fourrager du Vieux Pin, F-61310 Le Pin au Haras

MWALLEY, P.; SCAPA, Ministry of Agric, PO Box 3163, Arusha, Tanzania. e-mail: *scapa@habari.co.tz*

NALEWAJA, J.; Department of Plant Sciences, Lofsgard Hall,P.O. Box 5051, FARGO, North Dakota 58105-5051, USA, e-mail: *John_Nalewaja@ndsu.nodak.edu*

O'NEILL, D.H.; Silsoe Research Institute, Wrest Park, Silsoe Bedford, MK45 4HS, UK

PALA, M.; ICARDA, P.O.Box 5466, Aleppo, Syria

PEIRETTI, R.; AAPRESID; Paraguay 777, 8vo piso, oficina 4; 2000 Rosario, Pcia de Santa Fe (Argentina), e-mail: *aapresid@bcr.com.ar*

PEREIRA, M. H.; Confederación de Asociaciones Americanas para la Agricultura Sustentable (CAAPAS); Rua 7 de Setembro, 800; 84035 Ponta Grossa; Paraná. (Brasil), e-mail: *mhp@convoy.com.br*

PEZO, D. A.; Crop Science Dept., Box 7620; North Carolina State University, Raleigh, NC 27695-7620 (U.S.A.), e-mail: *paul_Mueller@ncsu.edu*

PIERI, C.; Senior Agro-Ecologist, Rural Development Department, Room MC5-771; The World Bank, 1818 H Street, N.W., Washington , DC 20433, USA, e-mail: *cpieri@worldbank.org*

REICOSKY, D. C.; United States Department of Agriculture (USDA), North Iowa- Ave. Morris; 56267 Minnesota; U.S.A., e-mail: *dreicosky@mail.mrsars.usda.gov, Reicosky@morris.ars.usda.gov*

REUTER, S.; State Education and Research Center Neustadt, Department Ecology. Breitenweg 71, Neustadt an der Weinstraße, 67435-DE, e-mail: *sreuter.slfa-nw@agrarinfo.rlp.de*

RIBEIRO, M. F.; Agronomic Institute (IAPAR); Rodovia Celso Garcia Cid, Km. 375; CEP-86001-970; Londrina, Paraná; BRASIL, e-mail: *asoiapar@pr.gov.br*

RITCHIE, W. R.; Centre for International No-Tillage Research and Engineering (CINTRE)

ROCKSTROM, J.; IHE-Delft, PO Box 3015 DA Delft, The Netherlands. e-mail: *jro@ihe.nl*

ROSE, S. C.; ADAS Consulting Limited, Gleadthorpe, Meden Vale, Mansfield, Notts NG20 9PF, UK, e-mail: *steve.rose@adas.co.uk*

SADAT, M.A.; Scientific Officer Wheat Research Centre, Bangladesh Agriculture Research Institute, Bangladesh

SANT'ANNA DE BARROS, G.; Professor, Socio-Economics Department, Escola Superior de Agricultura Luiz de Queiroz, Piracicaba, SP, Brazil. E-mail: *gscbarro@esalq.usp.br*

SAPRONOV, D.V.; Institute of Physicochemical and Biological Problems of Soil Science, RAS, Pushchino, Moscow region, 142290 - Russia

SAXTON, K.; Greensbridge, Winterborne, Kingston; Dorset DT11 9BJ, U.K., e-mail: *f.shaxson@mcmail.com*

SCHLAEPFER, N. P.; The Department for International Development (DFID), UK.

SCHWARZ, R.; Swiss Soil Conservation Association, Oberdorf 7, CH-2514 Ligerz

SEGUY, L.; Centre de Coopertion Internationale en Recherché Agronomique pour le Dévelopment (CIRAD), e-mail: *lucien.seguy@cirad.fr*

SERRANO DE NOREÑA, F.; Director Manager, Environment Directorate, Community of Madrid, Madrid, Spain. E-mail: *feres@iies.es*

SHARP, G.; Sinclair Knight Mertz, 49 Annand Street, Toowoomba, Queensland 4350, Australia

SHAXSON, F.; Greensbridge, Winterborne, Kingston; Dorset DT11 9BJ, U.K., e-mail: *f.shaxson@mcmail.com*

SIMS, B.; PROMETA/CIFEMA, Casilla 831, Cochabamba, Bolivia.

SIMS, B.G.; Silsoe Research Institute, Wrest Park, Silsoe Bedford, MK45 4HS, UK

SOLORZANO, N.; FAO, Ministry of Agriculture and Husbandry; Apdo. 8198-1000, San José, Costa Rica.

SOMMER, C.; Production Engineering and Building Research, Federal Agricultural Research Centre, Braunschweig, Germany

STEINER, K.; Deutsche Gesellschaft für Technische Zusammenarbeit (GTZ) GmbH Postfach 5180; 65726 Eschborn, e-mail: *kurt.steiner@gtz.de*

STURNY, W.G.; Bernese Soil Protection Service, Ruetti, CH-3052 Zollikofen, e-mail: *sturny@no-till.ch*

SULEIMENOV. M.; ICARDA, P.O. Box 4564, 700000 Tashkent, Uzbekistan.

TAPIA, A.; Farmer and ABULAC; President; C/ Plaza, 86; 09410-Quintanarraya (Burgos), e-mail: *tapiape@teleline.es*

TEBRÜGGE, F.; Professor, Institut für Landtechnik; Justus-Liebig-Universität GieBen; Braugasse 7; 35390 GieBen; Germany, e-mail: *friedrich.tebruegge@agrar.uni-giessen.de*

TEMESGEN, M.; National Agrcultural Mechanisation Research Center, PO Box 954, Nazareth, Etiopía. e-mail: *narc@telecom.net.et*

THEOTO ROCHA, M.; Graduate student, Socio-Economics Department, Escola Superior de Agricultura Luiz de Queiroz, Piracicaba, SP, Brazil, e-mail: *matrocha@carpa .ciagri.usp.br*

TRECENTI, R.; Training manager, Association for Zero Tillage in the Cerrado, SCRLN 712, Bloco C, Loja 18, Asa Norte, Brasília, DF, Brazil, postcode 70.760-53, E-mail: *apdc@tba.com.br*

TRUCCO, V.; Asociación Argentina de Productores en Siembra Directa (AAPRESID), Paraguay 777, 8vo piso, oficina 4; 2000 Rosario, Pcia de Santa Fe (Argentina), e-mail: *vtrucco.aapresid@bcr.com.ar*

WALL, P.C.; Centro Internacional para la Mejora del Maiz y el Trigo (CIMMYT), CC 2305, Santa Cruz, bolivia

WALTMAN, W. J.; Nebraska Research Initiative, Computer Science & Engineering, University of Nebraska, 113 Nebraska Hall, Lincoln NE 68588-0517, USA

WEISS, J.; Professor, Sustainable Development Center, University of Brasilia. E-mail: *weiss@loreno.net*

WEISSBAC, M.; Institute for Agricultural Engineering, University Kiel, Germany

YOUNG, N.; President, Western Australian No-Tillage Farmers Association (Inc)

ZAMBRANA, L.; PROMETA/CIFEMA, Casilla 831, Cochabamba, Bolivia

I. CONSERVATION AGRICULTURE GLOBAL IMPROVEMENTS

D.C. REICOSKY

CONSERVATION AGRICULTURE: GLOBAL ENVIRONMENTAL BENEFITS OF SOIL CARBON MANAGEMENT

USDA-Agricultural Research Service, North Central Soil Conservation Research Lab, 803 Iowa Ave. Morris, MN 56267 USA

Agricultural carbon (C) sequestration may be one of the most cost effective ways to slow processes of global warming. Numerous environmental benefits may result from agricultural activities that sequester soil C and contribute to environmental security. As part of no-regret strategies, practices that sequester soil C help reduce soil erosion and improve water quality and are consistent with more sustainable and less chemically dependent agriculture. While we learn more about soil C storage and its central role in direct environmental benefits, we must understand the secondary environmental benefits and what they mean to production agriculture. Increasing soil C storage can increase infiltration, increase fertility and nutrient cycling, decrease wind and water erosion, minimize compaction, enhance water quality, decrease C emissions, impede pesticide movement and generally enhance environmental quality. The sum of each individual benefit adds to a total package with major significance on a global scale. Incorporating C storage in conservation planning demonstrates concern for our global resources and presents a positive role for soil C that will have a major impact on our future quality of life.
Key Words: soil organic matter, soil quality, environmental quality, conservation tillage, zero tillage, direct seeding, carbon sequestration.

INTRODUCTION

Conservation agriculture aims to conserve, improve and make more efficient use of natural resources through integrated management of available soil, water and biological resources combined with external inputs. Conservation agriculture contributes to global environmental conservation as well as to enhanced and sustained agricultural production and can play a central role in global agricultural policy. Food security and sustainability are important for all citizens. Agriculture, the major industry for food and fiber production, is known to cause emission and storage of greenhouse gases. Intensification of agricultural production has been an important factor influencing greenhouse gas emission. Agricultural activities contribute to carbon dioxide (CO_2) emissions to the atmosphere through the combustion of fossil fuel, soil organic matter (SOM) decomposition, and biomass burning. Improved conservation agricultural

L. García-Torres et al. (eds.), Conservation Agriculture, 3–12.

practices have great potential to increase soil carbon (C) sequestration and decrease net emissions of CO_2 and other greenhouse gases that contribute to global environmental security.

World soils are an important pool of active C and play a major role in the global C cycle and have contributed to changes in the concentration of greenhouse gases in the atmosphere. Agriculture is believed to cause some environmental problems, especially related to water contamination, soil erosion, and greenhouse effect (Houghton, Hackler & Lawrence, 1999; Schlesinger 1985; Davidson & Ackerman, 1993). The soil contains two to three times as much C as the atmosphere. In the last 20 years, intensive agriculture has caused a C loss between 30 and 50 %. By minimizing the increase in ambient CO_2 concentration through soil C management, we minimize the production of greenhouse gases and minimize potential for climate change. Recent results suggest scientific agriculture can also lessen environmental problems and mitigate the greenhouse effect. In fact, agricultural practices have the potential to store more C in the soil than farming emits through land use change and fossil fuel combustion (Lal et al., 1998).

Soil quality is the fundamental foundation of environmental quality. Soil quality is largely governed by SOM content, which is dynamic and responds effectively to changes in soil management, primarily tillage and C input. This review will primarily address soil C and its associated environmental benefits. Other recent reviews on the role of C sequestration in conservation agriculture were presented by Robert (2001), Uri (1999), Tebrugge & Guring (1999), Lal et al. (1998) and Lal (2000). Agriculture has an opportunity to offset some CO_2 emissions and will be a small, but significant player in sequestering C.

KEY ROLE OF SOIL ORGANIC MATTER

Soil organic C represents a key indicator for soil quality, both for agricultural functions (production and economy) and for environmental functions (C sequestration and air quality). Soil organic matter is the main determinant of biological activity because it is the primary energy source. The amount, diversity and activity of soil fauna and microorganisms are directly related to SOM content and quality. Organic matter and the biological activity that it generates, have a major influence on the physical and chemical properties of the soils. Soil aggregation and stability of soil structure increases with increasing organic C. These factors in turn increase the infiltration rate and available water holding capacity of the soil as well as resistance against erosion by wind and water. Soil organic matter also improves the dynamics and bio-availability of main plant nutrient elements.

Soils contain relatively small amounts of C that could be considered analogous to a catalyst for biological activity where a small amount has a big impact. Farmers are the primary soil managers who each have a tremendous responsibility to maintain SOM for environmental benefit of the global population. Thus, farmers who use conservation agriculture or direct seeding techniques are providing ecosystem services and helping to maintain environmental quality for all of society. Quality food production and economic and environmentally-friendly management practices that are socially acceptable will lead to sustainable production and be mutually beneficial to farmers and all of society. It is important, therefore, that C loss from the soil system through historical land use of farming practices be restored to its natural potential using direct seeding and conservation tillage methods for sustainable production.

SOURCES AND SINKS IN AGRICULTURAL SYSTEMS

Agricultural systems contribute to C emissions through several mechanisms including direct use of fossil fuels in farm operations, indirect use of energy inputs for manufacturing chemicals (typically fertilizers), irrigation and grain drying and through intensive tillage of soils resulting in the loss of SOM. With conservation agriculture techniques, soils can accumulate C to offset other C losses. Thus, the soil can be converted from a «source» of C to a «sink» for C with improved soil and crop management.

Preliminary assessments indicate that soil C sequestration can be a tool to offset C emissions from burning fossil fuels. We in agriculture play a significant role because of the large amount of soil C in the C cycle within agricultural production systems. The limited use of crop rotations combined with intensive tillage decreases soil quality and soil organic matter. Any operation that removes or incorporates crop residue contributes to the decline of soil C through increased biological oxidation. The drive to maximize profit in food and fiber production has created environmental problems that have slowly crept up on conventional agriculture and now requires new knowledge, research and innovation to overcome these concerns.

A CASE FOR CONSERVATION AGRICULTURE AND ZERO TILLAGE

Tillage or soil preparation has been an integral part of traditional agricultural production. Tillage is also a principle agent resulting in soil perturbation and subsequent modification of the soil structure with soil degradation. Intensive tillage loosens soil, enhances the release of soil nutrients for crop growth, kills the weeds that compete with crop plants for water and nutrients and modifies the circulation of water and air within the soil. Intensive tillage can adversely affect soil structure and cause excessive break down of aggregates leading to potential soil movement via erosion. Intensive tillage causes soil degradation through C loss and tillage-induced greenhouse gas emissions that impact productive capacity and environmental quality.

Recent studies involving a dynamic chamber, various tillage methods and associated incorporation of residue in the field indicated major C losses immediately following intensive tillage (Reicosky & Lindstrom, 1993, 1995). The moldboard plow had the roughest soil surface, the highest initial CO_2 flux and maintained the highest flux throughout the 19-day study. High initial CO_2 fluxes were more closely related to the depth of soil disturbance that resulted in a rougher surface and larger voids than to residue incorporation. Lower CO_2 fluxes were caused by tillage associated with low soil disturbance and small voids with no-till having the least amount of CO_2 loss during 19 days. The large gaseous losses of soil C following moldboard plowing compared to relatively small losses with direct seeding (no-till) have shown why crop production systems using moldboard plowing have decreased SOM and why no-till or direct seeding crop production systems are stopping or reversing that trend. The short-term cumulative CO_2 loss was related to the soil volume disturbed by the tillage tools. This concept was explored when Reicosky (1998) determined the impact of strip tillage methods on CO_2 loss after five different strip tillage tools and no-till. The highest CO_2 fluxes were from the moldboard plow and subsoil shank tillage. Fluxes from both slowly declined as the soil dried. The least CO_2 flux was measured from the no-till treatment. The other forms of strip tillage were intermediate with only a small amount of CO_2 detected immediately after the tillage operation. These results suggested that the CO_2 fluxes appeared to be directly and linearly related to the volume of soil disturbed. Intensive tillage fractured a larger depth and volume of soil and increased aggregate surface area available for gas exchange that contributed to the vertical gas flux. The narrower and shallower soil disturbance caused less CO_2 loss suggests that the volume of soil disturbed must be minimized to reduce C loss and impact on soil and air quality. The results suggest environmental benefits and C storage of strip tillage over broad area tillage that needs to be considered in soil management decisions.

Reicosky (1997) reported that average short-term C loss from four conservation tillage tools was 31 % of the CO_2 from the moldboard plow. The moldboard plow lost 13.8 times more CO_2 as the soil not tilled while conservation tillage tools averaged about 4.3 times more CO_2 loss. The smaller CO_2 loss from conservation tillage tools was significant and suggests progress in equipment development for enhanced soil C management. Conservation tillage reduces the extent, frequency and magnitude of mechanical disturbance caused by the moldboard plow and reduces the large air-filled soil pores to slow the rate of gas exchange and C oxidation.

Carbon loss associated with intensive tillage is also associated with soil erosion and degradation that can lead to increased soil variability and yield decline. Tillage erosion or tillage-induced translocation, the net movement of soil down slope through the action of mechanical implements and gravity forces acting on the loosened soil has been observed for many years. Papendick, McCool, & Krauss (1983) reported original topsoil on most hilltops had been removed by tillage erosion in the Paulouse region of the Pacific Northwest of the US. The moldboard plow was identified as the primary cause, but all tillage implements will contribute to this problem (Grovers et al., 1994; Lobb & Kachanoski, 1999). Soil translocation from moldboard plow tillage can be greater than soil loss tolerance levels (Lindstrom, Nelson & Schumacher, 1992; Grovers et al., 1994; Lobb, Kachanoski & Miller, 1995; Poesen et al., 1997). Soil is not directly lost from the fields by tillage translocation, rather it is moved away from the convex slopes and deposited on concave slope positions. Lindstrom et al. (1992) showed that soil movement on a convex slope in southwestern Minnesota, USA could result in a sustained soil loss level of approximately 30 t ha^{-1} yr^{-1} from annual moldboard plowing. Lobb et al. (1995) estimated soil loss in southwestern Ontario, Canada from a shoulder position to be 54 t ha^{-1} $yr^{-1}$from a tillage sequence of moldboard plowing, tandem disk and a C-tine cultivator. In this case, tillage erosion, as estimated through resident Cesium137, accounted for at least 70 % of the total soil loss. The net effect of soil translocation from the combined effects of tillage and water erosion is an increase in spatial variability of crop yield and a likely decline in soil C related to lower soil productivity (Schumacher et al., 1999).

ENVIRONMENTAL BENEFITS OF SOIL CARBON

The main direct benefit of conservation agriculture or direct seeding is the immediate impact on SOM and soil C interactions. Soil organic matter is so valuable for what it does in soil, it can be referred to as «black gold» because of its vital role in physical, chemical and biological properties and processes within the soil system. Agricultural policies are needed to encourage farmers to improve soil quality by storing C that will also lead to enhanced air quality, water quality and increased productivity as well as to help mitigate the greenhouse effect. Soil C is one of our most valuable resources and may serve as a «second crop» if global C trading systems become a reality. While technical discussions related to C trading are continuing, there are several other secondary benefits of soil C impacting environmental quality that should be considered to maintain a balance between economic and environmental factors.

Soil C is so important that it can be compared to the central hub of a wheel as shown in Fig 1. The wheel represents a circle, which is a symbol of strength, unity

Fig 1. Environmental sustainability wheel with benefits emanating from the soil C hub.

and progress. The «spokes» of this wagon wheel represent incremental links to soil C that lead to the environmental improvement that supports total soil resource sustainability. Many spokes make a stronger wheel. Each of the secondary benefits that emanate from soil C contributes to environmental enhancement through improved soil C management. Soane (1990) discussed several practical aspects of soil C important in soil management. Some of the «spokes» of the environmental sustainability wheel are described in following paragraphs.

Increased SOM has a tremendous effect on soil water management because it increases infiltration and the water holding capacity. The primary role of SOM in reducing soil erodibility is by stabilizing the surface aggregates through reduced crust formation and surface sealing, which increases infiltration (Le Bissonnais, 1990). Enhanced soil water-holding capacity is a result of increased SOM that more readily absorbs water and releases it s1lowly over the season to minimize the impacts of short-term drought. In fact, certain types of SOM can hold up to 20 times its weight in water. Hudson (1994) showed that for each one % increase in SOM, the available water holding capacity in the soil increased by 3.7 % of the soil volume. The extra SOM prevents drying and improves water retention properties of sandy soils. In all texture groups, as SOM content increased from 0.5 to 3 %, available water capacity of the soil more than doubled. Other factors being equal, soils containing more organic matter can retain more water from each rainfall event and make more of it available to plants. This result plus the increased infiltration with higher organic matter and the decreased evaporation with crop residues on the soil surface all contribute to improve water use efficiency.

Ion adsorption or exchange is one of the most significant nutrient cycling functions of soils. Cation exchange capacity (CEC) is the amount of exchange sites that can absorb and release nutrient cations. Soil organic matter can increase CEC of the soil from 20 to 70 % over that of the clay minerals and metal oxides present. In fact, Crovetto (1996) showed that the contribution of the organic matter to the cation exchange capacity exceeded that of the kaolinite clay mineral in the surface 5 cm of his soils. Robert (2001) showed a strong linear relationship between organic C and CEC of his experimental soil. The CEC increased four-fold with an organic C increase from 1 to 4 %. The toxicity of other elements can be inhibited by SOM which has the ability to adsorb soluble chemicals. The adsorption by clay minerals and SOM is an important means by which plant nutrients are retained in crop rooting zones.

Soils relatively high in C, particularly with crop residues on the soil surface, are very effective in increasing SOM and in reducing soil erosion loss. Reducing or eliminating runoff that carries sediment from fields to rivers and streams will enhance environmental quality. Under these situations, the crop residue acts as tiny dams that slow down the water runoff from the field allowing the water more time to soak into the soil. Worm channels, macropores and plant root holes left intact increase infiltration (Edwards, Shipitalo & Norton, 1988). Water infiltration is two to ten times faster in soils with earthworms than in soils without earthworms (Lee, 1985). Soil organic matter contributes to soil particle aggregation that makes it easier for the water to move through the soil and enables the plants to use less energy to establish to root systems (Chaney & Swift, 1984). Intensive tillage breaks up soil aggregates and results in a dense soil making it more difficult for the plants to get nutrients and water required for their growth and production.

The reduction in soil erosion leads to enhanced surface and ground water quality, another secondary benefit of higher SOM (Uri, 1999). Crop residues on the surface help hold soil particles in place and keep associated nutrients and pesticides on the field. The surface layer of organic matter minimizes herbicide runoff, and with conservation tillage, herbicide leaching can be reduced as much as half (Braverman et al., 1990). The enhancements of surface and ground water quality are accrued through the use of conservation tillage and by increasing SOM. Increasing SOM and maintaining crop residues on the surface reduces wind erosion (Skidmore, Kumar & Larson, 1979). Depending on the amount of crop residues left on the soil surface, soil erosion can be reduced to nearly nothing as compared to the unprotected, intensively tilled field.

Another key factor is SOM that can decrease soil compaction (Angers & Simard 1986; Avnimelech & Cohen, 1988). Soane (1990) presented different mechanisms where soil «compactibility» can be decreased by increased SOM content: 1) improved internal and external binding of soil aggregates; 2) increased soil elasticity and rebounding capabilities; 3) dilution effect of reduced bulk density due to mixing organic residues with the soil matrix; 4) temporary or permanent existence of root networks; 5) localized change electrical charge of soil particles surfaces, 6) change in soil internal friction. While most soil compaction occurs during the first vehicle trip over the tilled field, reduced weight and horsepower requirements associated with forms of conservation tillage can also help minimize compaction. Additional field traffic required by intensive tillage compounds the problem by breaking down soil structure. The combined physical and biological benefits of SOM can minimize the affect of traffic compaction and result in improved soil tilth.

Maintenance of SOM contributes to the formation and stabilization of soil structure. Another spoke in the wagon wheel of environmental quality is improved soil tilth, structure and aggregate stability that enhances the gas exchange properties and aeration required for nutrient cycling (Chaney & Swift, 1975). Critical management of soil airflow with improved soil tilth and structure is required for optimum plant function and nutrient cycling. It is the combination of many little factors rather than one single factor that results in comprehensive environmental benefits from SOM management. The many attributes suggest new concepts on how we should manage the soil for the long-term aggregate stability and sustainability.

A secondary benefit of less tillage and increasing SOM is reduced air pollution. CO_2 is the final decomposition product of SOM and is released to the atmosphere. Research has shown that intensive tillage, particularly the moldboard plow, releases large amounts of CO_2 as a result of physical release and enhanced biological oxidation (Reicosky et al., 1995). With conservation tillage, crop residues are left more naturally on the surface to protect the soil and control the conversion of plant C to SOM and humus. Intensive tillage releases soil C to the atmosphere as CO_2 where it can combi-

Reicosky, D.C., Kemper, W.D., Langdale, G.W., Douglas, Jr C.W. and Rasmussen, P.E. (1995) Soil organic matter changes resulting from tillage and biomass production. *Journal of Soil and Water Conservation* **50**:253-261.

Reicosky, D.C. and Lindstrom, M.J. (1993) Fall tillage method: effect on short-term carbon dioxide flux from soil. *Agronomy Journal* **85**: 1237-1243.

Reicosky, D.C. and Lindstrom, M.J. (1995) Impact of fall tillage and short-term carbon dioxide flux. In *Soil and Global Change.* pp. 177-187. Ed R Lal Chelsea, MI: Lewis Publishers.

Robert, M. (1996) Aluminum toxicity a major stress for microbes in the environment. In: *Environmental Impacts,* pp. 227-242. Eds P M Huang et al., Vol. **2,** Soil component interactions. CRC press.

Robert, M. (2001) Carbon Sequestration in Soils: Proposals for land management. AGLL, FAO, United Nations, Rome. Report No. XXX. 69 pp.

Schlesinger, W.H. (1985) Changes in soil carbon storage and associated properties with disturbance and recovery. In *The changing carbon cycle: A global analysis.* pp. 194-220. Eds J R Trabalha and D E Reichle. New York: Springer Verlag.

Schumacher, T.E., Lindstrom, M.J., Schumacher, J.A. and Lemme, G.D. (1999) Modelling spatial variation and productivity due to tillage and water erosion. *Soil Tillage Research.* **51**: 331-339.

Skidmore, E.L., Kumar, M. and Larson, W.E. (1979) Crop residue management for wind erosion control in the Great Plains. *Journal of Soil and Water Conservation* **34**: 90-94.

Soane, B.D. (1990) The role of organic matter in the soil compactibility: A review of some practical aspects. *Soil Tillage Research* **16**: 179-202.

Tebregge, F. and Guring, R.A. (1999) Reducing tillage intensity – a review of results from a long-term study in Germany. *Soil Tillage Research* **53**: 15-28.

Uri, N.D. (1999) Conservation Tillage in U.S. Agriculture. In *Environmental, Economic, and Policy Issues,* p 130. Binghamton, NY: The Haworth Press, Inc.

C.J. BAKER, R.M. COLLINS & M.A. CHOUDHARY

FACTORS AFFECTING THE UPTAKE OF NO-TILLAGE IN AUSTRALIA, ASIA AND NEW ZEALAND

Centre for International No-Tillage Research and Engineering, Feilding, NZ.

Western Australia Department of Agriculture, Northam, WA, Australia.

Massey University, Palmerston North, NZ

In Australia and Asia there are compelling reasons for undertaking no-tillage, and adoption has followed expected patterns with regional differences dictated largely by climate, soil type, and sociological factors. The advantages of no-tillage in these regions have been sufficiently strong to mask other possible weak links in the technical chain. By contrast, in New Zealand there are less compelling reasons to undertake no-tillage because of that country's favourable agricultural climate and integrated crop and animal systems. New Zealand farmers have therefore been free to respond positively or negatively to the level of fail-safeness that the no-tillage technique has offered over the years. After responding negatively in the 1970's to the early 1990's, a recent upswing in response in New Zealand has followed the release of a new design of no-tillage opener (Cross Slot™) that has attained a field fail-safeness rating of 99%.

Key words: No-tillage, adoption, incentives, openers, risk, fail-safe, Cross Slot™

INTRODUCTION

Nothing affects the voluntary uptake of no-tillage more than crop yields and the net returns earned by practicing farmers, regardless of government or other incentives. This is clearly illustrated in the USA where legislation and strong government incentives have failed to increase adoption rates above Canada and most Latin American countries with more limited incentives. Nonetheless, since 69% of the world's food is from annual cereals (Borlaug 1994) substitution of no-tillage for tillage has the potential to profoundly affect the sustainability of world food production from annual crops.

Many cropping farmers across the globe (rightly or wrongly) still view no-tillage as increasing their business risks compared with tillage and are reluctant to accept the increased short-term risks for the long-term gains of their nations. While no-tillage is undoubtedly gaining ground, it has yet to become a mainstream technique on a

L. García-Torres et al. (eds.), Conservation Agriculture, 13–20.

widespread basis. The challenge is to persuade all farmers that no-tillage can be more risk-free than tillage and should be the preferred crop establishment technique.

The purpose of this paper is to examine how farmers have perceived these risks in three contrasting regions displaying different national and regional incentives for no-tillage (Australia, Asia and New Zealand) and how this has affected adoption rates. In one region (New Zealand) increased risk of failure from no-tillage had a negative effect. However development of almost fail-safe no-tillage drill technology later reversed this trend. The lessons learnt in New Zealand have far-reaching ramifications for other countries and should be an important model for policy makers and legislators.

To understand the relevance of the New Zealand experience it is appropriate to firstly outline the experiences of Australia and Asia since those two regions had strong but different incentives that are similar to those in many other countries.

AUSTRALIA

In Australia the pressure to adopt no tillage has been driven by soil and water conservation and reduced costs. A number of state and federal incentives have maintained no-tillage at the forefront of preferred techniques. However gains have been somewhat regional, ranging from 20% to 50% in different states. Western Australia is the state with greatest adoption. Figure 1 illustrates the reasons for adoption of no-tillage in that state (Wallwork, 2000).

The issues of adoption in Western Australia are unique to that state. Western Australia has a winter dominant-rainfall pattern. In favourable years, newly-establishing crops grow into cooling and declining water stress conditions. No-tillage results in yield benefits from early seeding without penalty to subsequent machinery operations (weed spraying and additional nitrogenous fertiliser spreading) compared with «full-cut» «direct drill» operations where all the surface soil is disturbed.

Western Australian soils are old and weathered. «Sand over clay» «duplex» soils have the worst and best features of both soils. At the surface, soil is free flowing and never sticky, although the light texture can often be a negative factor in marginal moisture conditions. The underlying clay can lead to waterlogging conditions and often restricts root penetration. No-tillage gives a benefit of leaving the surface with stronger structure and more quickly accessible after wet weather. Acidic «deep sands» may be several metres deep, with a small clay content and a wide range of sand particle sizes, resulting in root-impeding compaction from rainfall, livestock and machinery. «Heavy soils» tend to be in the lower part of the landscape and set very hard in the dry season. Another group is the «grey clays» that alternate between concrete hard and smeary when fully wet. While cultivation may be needed to break up hard soil to form an adequate seedbed, organic matter accumulation from no-tilling over several years allows these soils to become more «mellow» and more easily sown. The sand soils do not accumulate organic matter in the same way. The long hot dry summer generally mitigates against organic matter accumulation, with anything above 2% being unusual.

The combination of climatic and soil conditions has meant that sandy soils near the south coast are extremely erodible from strong wind events, leading to early adoption of no-tillage, particularly of disc seeders. Maintaining stubble cover has been important throughout the state to reduce both wind and water erosion. Repetitive reliance on herbicides and limited rotations have produced herbicide resistance so serious that a special research centre has been established to develop new integrated weed control programmes.

Western Australia's city of Perth is the most isolated city in the world with a population of 1m or more. Declining rural populations and increasing farm sizes for economic survival have maintained a self-reliant attitude to life by the farming community including finding lower cost options to overcome the inherent problems imposed by soils and climate.

ASIA

Considerable progress has been made in the 14 million hectares that make up the Indo-Gangetic region of South Asia. Conventional tillage practices have impeded sustainable agriculture, causing soil compaction, deteriorating soil physical properties and biological degradation. High productivity rice-wheat systems are fundamental to employment and the livelihoods of hundreds of millions of rural and urban poor. Even although crop yields continue to increase slowly, productivity has declined in some areas. This has been due to declining soil fertility, an increase in problem weeds, and groundwater depletion and salinity, leading to food security problems.

Timeliness of crop establishment is important. Significant yield penalties occur if winter crops are planted late. Delays in sowing wheat after the photosensitive aromatic rice varieties reduce yields by up to 35 kg/ha/day (1%) and reduce the efficiency of fertilizer and water-use (Figure 2). Other factors include, quality of seed, soil condition, fertilizer rates, weed control, and the timing and quality of irrigation (Choudhary, Gill and Pulatov, 1999). Water use efficiency is particularly important in the region as about 90% of crop production is on irrigated land. Deterioration of the vast irrigation and drainage systems through improper management of watershed areas and deforestation has become a serious threat to the sustainability of agricultural growth. Low delivery efficiency, inequitable distribution, and insufficient cost-recovery have been the main problems.

No-tillage of crops following 'puddled' rice and cotton crops has become a viable alternative, allowing conservation and utilisation of antecedent soil moisture, time saving, reduced dependency on weather for suitable working days, and minimised soil structural breakdown. No-tillage (also known as «zero tillage» and «resource conservation tillage» in South Asia) was introduced into rice-wheat systems in the Indo-Gangetic Plains through a modified version of a New Zealand no-tillage drill capable of sowing wheat into standing rice stubble. These machines have advanced sowing dates, improved yields, reduced costs, raised water and fertilizer-use-efficiency, reduced weed germination, and fostered crop diversity. Farmers adopting no-tillage report using 75% less diesel fuel and 10-30% less water, while at the same time improving yields. Widespread adoption of no-tillage practices could have important environmental benefits. If adoption spreads to less than half of the rice-wheat area, five billion cubic meters of water could be saved each year together with 0.5 billion litres of diesel fuel, representing an annual reduction of 1.3 million tons of CO_2

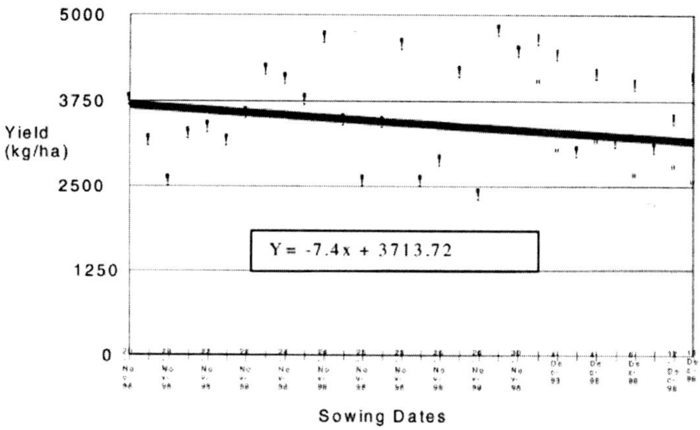

Figure 1. Trend of wheat yield penalties with conventional tillage sown after 20 November.

emissions. If no-tillage leads to reduced burning of crop residues, annual CO_2 emissions could be reduced by a further 17 million tonnes.

In the past three years the annual area sown to wheat by no-tillage in rotation with paddy rice in South Asia, has grown to 100,000 hectares, driven largely by reductions in cost. The area under no-tillage is expected to more than double each year in the short term, depending on how quickly private machine shops can manufacture new drills, although improved openers are now needed that can sow wheat and simultaneously apply fertilizer into heavy rice residues. Such openers would allow farmers to stop burning crop residues entirely.

NEW ZEALAND

It is estimated (C.J.Baker unpublished data, 2001) that about 3% of New Zealand's tractorable land is currently no tilled annually, represented by 2.5% (150,000 hectares) of pastureland and 6% (20,000 hectares) of cropped land. New Zealand was one of the first countries in the world to experiment with no-tillage (Cross 1957). However that country deflected away from the technique in later years and has only recently returned to taking it seriously. It is the reasons for the deflection that other countries could learn from, together with the reasons (which now appear more sustainable than ever) for its recent return.

Because of an unusually favourable climate, strong integration of animal and plant agricultural systems and extensive rotations of crop varieties (compared with more monoculture arable cropping in many other countries) there are few compelling reasons to no-till in New Zealand, even although all the usual advantages from doing so exist. Three distinct classes of crop (arable, forage and pastures) are regularly established, often in rotation. Double cropping is considered the norm. Successful no-tillage machines are expected to cope with all three crop types and their associated residues.

They therefore experience a wider variety of conditions, residue levels and seed types than in most other countries. In addition, many of these crops are grown on sloping land. Under these conditions some New Zealand no-tillage machines operate for up to nine months of the year and any one machine might travel up to 8,000 kilometres per year in drilling mode.

New Zealand has only regional soil erosion problems, is under little environmental pressure to ban burning of residues, has only regional droughts and floods, and disguises the decline in soil carbon and soil structure brought about by tillage by regularly introducing pastures and integrated animal grazing into cropping rotations. Therefore, to be attractive to New Zealand farmers, no tillage had to offer significant advantages, and become virtually fail-safe, since the consequences of continuing to till the soil have so far not been catastrophic to individual farmers in that particular nation. Furthermore, until 2000 (when a modest sustainable farming fund was created for approved projects) there were no government incentives for no-tillage of any nature.

New Zealand was targeted for early introduction of no-tillage by Imperial Chemical Industries Ltd when paraquat was first introduced in the 1960's. Although some successes were achieved in the field despite the (later-identified) limitations of triple disc openers, failures were also common. In any case paraquat was soon superseded by the broader spectrum translocated herbicide, glyphosate from Monsanto. While glyphosate heralded the real birth of no-tillage in many countries in the 1970's, ironically after an initial burst of interest and adoption in New Zealand, it eventually coincided with the virtual demise of no-tillage in that country. The fault did not lie with the herbicide. World (and New Zealand) no-tillage machinery design had not advanced significantly since the 1970's (Baker 1981). Failures continued. In the absence of any compelling reason to use no-tillage, many farmers reverted to tillage, which had stood the test of time as far as they were concerned.

No amount of favourable publicity from other countries (which had greater incentives to persist despite similar periodic failures) nor promotion by commercial interests in New Zealand altered the overall decline in interest in no-tillage during the 1980's and early 1990's. However, during that period a significant new no-tillage opener and drill technology (Cross Slot™) was developed in New Zealand and tested in the USA and Australia (Baker *et al*, 1979; Baker, Saxton and Bligh, 1995; Saxton & Baker, 1990). Its mode of action was different from any other known no-tillage opener and resulted from a 28-year scientific study that had isolated and eliminated the causes of previous failures that continued to limit the biological performance of most other designs of no-tillage opener (Ritchie, Baker and Hamilton-Manns, 2001). The new technology was released onto the New Zealand market in 1995 and is currently being licensed internationally.

It took 3-5 years for sceptical New Zealand farmers to become convinced that Cross Slot™ openers were not just older technologies re-packaged, and the superior crop results they observed were not just chance events. As time advanced however, and similar results were reported with the new technology from the USA and Australia, it became obvious that a higher level of fail-safeness with no-tillage machinery

Table 1. Performance of crops and pastures no-tilled with Cross Slot™ drills and planters in New Zealand over a four-year period.

Year	*Total area sown by Cross Slot drills and planters (ha)*	*Total No. of fields covered by Cross Slot drills and planters*	*Crops assessed to be at or above district average Yield (%)*	*Crops assessed to be below district average yield (%)*	*Impairment of crops caused by management factors (weeds, pests etc) (%)*	*Impairment of crops caused by machine limitations (%)*
96/97	5,083	812	92	8	6	2
97/98	12,337	1,970	89	11	9.8	1.2
98/99	10,438	1,378	90	10	9.3	0.7
99/00	11,267	2,146	88	12	11	1
Total/av.	**39,125**	**6,306**	**89.8**	**10.2**	**9**	**1.2**

had been achieved. Drill technology was at last matching herbicide and pesticide technologies in terms of sophistication and reliability (evidenced by crop yields) whereas until then drill technology everywhere in the world had lagged behind the other critical inputs.

Commercial owners of Cross Slot™ machines in New Zealand were surveyed over a 4-year period (1996-2000). About half of the respondents were contractors that no-tilled on multiple farms. The other half were farmer-owners. Some were both. The survey covered 39,000 hectares and involved assessments of some 6,000 separate fields sown by 20 operators involving a very wide range of crop species, soils, surface residues, management systems and topographies. The summarized data are shown in Table 1. In the belief that eliminating failures is at least as important in reducing risk as recording successes, the surveys also sought to identify the causes of impaired crop performance. Nonetheless positive results ranged from the equal highest milling wheat yield for the country in 2000 (9.3 tonne/ha), through consistently top regional pea yields and outstanding arable, forage and pasture crops, to those that did no more than consistently equal the district averages.

The consistently high crop yields recorded in the USA and Australia (some up to 11 years) with Cross Slot™ machines were not included in the survey (Baker, Saxton and Ritchie, 1996; Grabski, Schafer and Desborough, 1995; Saxton and Baker, 1990).

Causes of impaired crop performance attributable to machine design were reduced to approximately 1% after 1996/97. However, a further 9-11% of sown crops had continued to suffer from inadequate management. While it is doubtful if machine-causes can be entirely eliminated, there continues to be room for improvement of management aspects. Although experiments in controlled climatic conditions had highlighted differences in the biological performances of no-tillage opener designs as early as 1976 and subsequently (Baker 1976; Baker *et al*, 1996; Choudhary and Baker, 1981; Ritchie, Baker and Hamilton-Manns, 2000) the authors are not aware of comparable surveys of biological reliability of no-tillage field drills and planters in other countries (or indeed in New Zealand with other machines).

CONCLUSIONS

When the rejection by New Zealand farmers of inadequate no-tillage opener designs is balanced against their more recent recognition of superior designs, the most obvious lesson to be learnt is that unless no-tillage systems become more fail-safe as a whole, they may not be voluntarily chosen as the mainstream methods of food production anywhere. The New Zealand experience shows that without the influence of compelling reasons to overlook failures (such as exist in Australia and Asia) farmers will quickly revert to old habits. But if incentives have to be continuously balanced against inconsistent field results this will create a very fragile basis for the universal advancement of no-tillage. Conversely the study also shows that if the weakest link in the chain (in this case opener design) can be corrected, positive and enduring responses can result regardless of «outside» incentives. If world no-tillage has progressed this far using strong incentives but inferior opener designs that are far from biologically fail-safe, how far can it go with the same incentives and better opener designs (such as Cross Slot™) that are more tolerant of sub-optimal conditions than tillage itself?

REFERENCES

Baker, C.J. (1976) Experiments relating to the techniques of direct drilling of seeds into undisturbed sprayed pasture. *Journal of Agricultural Engineering Research,* **21** (**2**): 133-145.

Baker, C.J. (1981) How direct drilling equipment developments in Europe and North America relate to New Zealand. *Proceedings of the Conservation Tillage Technical Seminar, Christchurch, New Zealand.* pp. 105-121.

Baker, C.J., McDonald, J.H., Seebeck, K., Rix, C.S. and Griffiths, P.M. (1979) Developments with seed drill coulters for direct drilling: III An improved chisel coulter with trash handling and fertilizer placement capabilities. *New Zealand Journal of Experimental Agriculture* **7**, 189-196.

Baker, C.J., Saxton, K.E. and Bligh, K. (1995) Cross Slot no-tillage technologies. *Proceedings of the National No-Tillage Conference, Indianapolis, Indiana, USA.* pp. 165-184.

Baker, C.J., Saxton, K.E. and Ritchie, W.R. (1996) *No-Tillage Seeding: Science and Practice.* 258 pages. Published by CABI, Oxford, England.

Borlaug, N.E. (1994) Feeding a human population that increasingly crowds a fragile planet. *Supplement to Transactions of the 15th World Congress of Soil Science, Acapulco, Mexico.*

Choudhary, M.A. and Baker, C.J. (1981) Physical effects of direct drilling equipment on undisturbed soils: III Wheat seedling performance and in-groove micro-environment in a dry soil. *New Zealand Journal of Agricultural Science,* **24**: 189-195.

Choudhary, M.A., Gill, M.A. and Pulatov, A. (1999) Prospects of no-till wheat establishment in Rotation with rice or cotton in Central and South Asia. *Proceedings of the International Workshop on Conservation Tillage: A viable option for Sustainable Agriculture in Eurasia, Shortandy, Kazakistan.* pp. 125-133.

Cross, M.W. (1957) Overdrilling machinery and methods. *Proceedings of the Massey Agricultural College Dairy Farmers' Conference, Massey University, New Zealand.* pp. 77-84.

Grabski, A.S., Schafer, B.M. and Desborough, P.J. (1995) A comparison of the impact of 14 years of conventional and no-till cultivation on physical properties and crop yields of a loam soil at Grafton, NSW. *Proceedings of the National Controlled Traffic Conference, Rockhampton, Queensland, Australia.* pp. 97-102.

Ritchie, W.R., Baker, C.J. and Hamilton-Manns, M. (2000) *Successful No-tillage in Crop and Pasture Establishment.* 102 pages. Published by Monsanto New Zealand Ltd.

Ritchie, W.R., Baker, C.J. and Hamilton-Manns, M. (2001) The development and transfer of a new no-tillage technology in New Zealand. *Proceedings World Congress on Conservation Agriculture, Madrid, Spain.* (in press).

Saxton, K.E. and Baker, C.J. (1990) The Cross Slot drill opener for conservation tillage.Proceedings of the Great Plains Conservation Tillage Symposium, Bismarck, North Dakota, USA. pp. 65-72.

Wallwork, S. (2000) State-by-State adoption gives min till the edge. In Min Till Drill: A guide to minimum tillage cropping systems, Kondinin Group publication, Perth, Western Australia.

N. FAROOQ

CONSERVATION AGRICULTURE AND THE RURAL WOMEN. BREAKING BARRIERS FOR A NEW HORIZON

Chairperson, Community Welfare Society, Chak No. 26/UCC, District Sheikhupura, Pakistan.

INTRODUCTION

I have had the experience of working as a lady farmer on a project at DFID village in Murdike Shiekupura district Pakistan. Community welfare society took up the challenge of transferring conservation agriculture technologies to the poor farmers and the rural women of the area. We have worked on this project with community participation in field schools and by holding seminars for the farmers. Here farmers experimented with researchers, extension workers and agronomists on new technologies. With community participation in research fields the farmers themselves access the impact of these new technologies. My experience with conservation agriculture for the past four years has shown fruitful results and 20-30% farmers of the area have switched from traditional and conventional ways to conservation agriculture.

Rural women in developing countries play a vital role both in communities and national economics as producers of food and caretakers of their families. The contribution of women to economic has increased in the agriculture sector especially in areas where there is male migration to urban areas. So the women folk comprised the major part of farm labour, but sad to say that this major participant is hardly ever appreciated. The male domination, the uneducated mullahs, and the feudal lords have tried their best to keep them in the background, but it would be appropriate now to bring them in the forefront if we really want a breakthrough. Improving their lot would definitely have a positive effect on the whole in the agriculture sector.

For many years we have seen to missed on the most vital issue of the millennium (the gender issue), but we must understand now that the absence of the literate, healthy, contented mother, sister and daughter amounts to a blind man directing traffic on a busy road intersection. Being a lady farmer and worked with rural women in closed

L. García-Torres et al. (eds.), Conservation Agriculture, 21–30.

quarters for the past 20 years I could understand their problems. So my initial step was to include them in all the training programmes on conservation agriculture on the farm. The Pakistani women's work is in actual fact a triple burden; it is productive, reproductive and domestic. But since their economic productivity is not officially recognized, society and government at large have ignored their basic problems. The benefits and facilities for workingwomen especially in the agriculture sector are non-existent. They are drawing less wages then men; they have no social security and health coverage, and despite the efforts of women organizations, successive government pledges, the vast majority of women have seen no improvement in their status.

Talking of the role of women in agriculture, the rural women role as food providers and food producers link them directly with conservation agriculture. We must give them recognition as partners in decision-making and as beneficiaries of development activities in response to the needs for sustainable development. Rural women in developing countries hold a key to future agricultural systems and to livelihood security. They carry out multiple jobs in farm related activities for land preparation, to food processing and in marketing activities, tending to animals, looking after the kitchen, feeding the family members and hauling fuel and water. But in the agriculture field the rural women's role is under valued and their contribution to rural development is hardly appreciated. Undoubtedly they face greater difficulties then men in every field.

One of the strength of conservation agriculture is economics, with economics and crop production, it is a matter of time before people are going to be looking across the field and recognizing that this is the way they will have to go. At present we have still have good conventional farmers who are doing a good job, but they are doing it in old traditions. As the land starts to change hands into the new generations, economic is going to drive people to conservation agriculture as the same as it drove people to continuous cropping to increase income and rural development

Much ink has been split in order to inform the problems of women especially the lower middle class and the rural women, but the fact is that we only use flowery words or adorn our stories but always miss the chance when the iron is hot. It is high time that in keeping the teaching the Islam and the Holy prophet (pbuh) some meaningful and solid work should be done with joint hands to achieve the objective. The affluent class should extent necessary financial assistance and well reputed volunteer agencies and government support should launch this emancipation campaign

COMMUNITY WELFARE SOCIETY CHAK 26 UCC

Chak 26 UCC is a village of Sheikhupura district in the Punjab province of Pakistan. It is situated about 30 km from historic city of Lahore. Socio-economic conditions of its people are similar to those of other Pakistani villages. A Community Welfare Society (CWS) was established in the village during 1985 as an NGO with, interalia, following objectives.

Organization of Water Users Associations

Remodeling/reconstruction of tertiary irrigation network (watercourses) to enhance conveyance efficiencies

Promotion of Irrigation Agronomy activities and Conservation Tillage Practices i.e. Precision Land Leveling, Zero Tillage Technology, Mechanical Rice Transplanting, Furrow-Bed planting.

Training and education of Water Users/Farmers in On Farm Water Management techniques by holding regular training and refresher courses

In addition, the CWS operates an agro-based Vocational Training Center (VTC) for the rural women and small farmers of the area. Its main aim is to transfer Conservation Agriculture Technologies to the poor farmers and rural women. The society acts as a Network for researchers, machinery manufactures, extension workers, and farmers to share information of new technologies as well as to provide feed back to the scientists on performance of new innovations. Following key technological interventions have been selected to address major production constraints by VTC.

a) Provide skill training to woman folk in all agricultural related activities
b) Transfer industrial know how for women from non-farming families
c) Help arrange credit at local level
d) Facilitate family planning services
e) Promote literacy for school drop outs
f) Operate nursery classes and Day-care services for children

PROMOTION OF CONSERVATION AGRICULTURE PRACTICES AND THEIR BENEFITS

The achievements made by Community Welfare Society in promotion and adoptions of Conservation Agriculture practices are very encouraging during past years. Some of these are discussed below:

1. Watercourse Improvement

Tertiary irrigation conveyance channels in Pakistan are called watercourses. These are operated and maintained by the shareholders being provided water through channels. A watercourse commands an area of about 400 acres and is shared by 40 to 50 farm families. Studies have indicated that about 40% of irrigation water is lost during its conveyance through watercourses because of their aging and deteriorated conditions.

Improvement of the watercourses consists of complete demolishing of community channel and its rebuilding/re-aligning according to the engineering design with clean compacted soil. Parts of reconstructed channel are lined and necessary water control structures are installed to improve conveyance of the canal as well as tubewell water. All these works are carried out through active participation of the beneficiary farmers who contribute entire skilled and unskilled labour in addition to 30% of the material costs. First watercourse in the Community Watercourses area was constructed during 1988 with the support of On Farm Water Management staff. So far, 20 watercourses have been renovated in the adjoining villages.

Studies and previous experience of watercourse improvement has shown that, on an average, annual water saving in an improved watercourse are about 240 Acre Feet. Accordingly, improvement of above said 20 watercourses area resulting in overall annual saving of about 4,800 Acre Feet of water besides other socio-economic benefits. Because of enhanced water supplies, farmers are able to increase cropping intensity and cropped area at their farms ultimately leading to enhanced crop production from their lands.

2. LASER Land Levelling

The Community Welfare Society provides free technical assistance to the farmers for precision land leveling and farm planning in addition to training the farmers and tractor operators in its operation. Resultantly, a big compact block of the area has been levelled around the village. LASER technology has been proved to be beneficial as it minimizes the cost of operation, ensures better degree of accuracy in much lesser time, saves irrigation water, ascertains uniform seed germination, increases fertilizer use efficiency, and resultantly enhances crop yields. Impact studies on LASER Land Levelling reveal that it

- Curtails the irrigation application losses up to the extent of 25%
- Reduces labour requirements for irrigation by about 35%
- Enhances the irrigated area by about 2% by brining the number as well as length of field ditches and dikes to a minimum.
- Increases the crop yields by about 20%

Up till now, an area of about 400 acres owned/operator by members of CWS have been precisely levelled with LASERs technology.

3. Zero Tillage Technology for Growing Wheat

Zero Tillage is a special technique of establishing crops without seedbed preparations. The implement used for this purpose is known as Zero Tillage Drill, which is capable to seeding the crop in untilled soil through existing residue cover. The technique has been found useful specially for planting wheat in the rice-harvested fields where long duration rice varieties, uncertain rains, and excessive soil moisture do not permit its timely sowing. Beside this, the technology also minimizes the use of water, saves expenditure involved in seedbed preparations, and enhances the productivity. Post-sowing weed control in Zero Tillage technique requires less use of herbicides as compared to the conventional sowing system. Because of reduced uses of diesel/fuel, it decreases air pollution and green house gas emission.

An area of about 200 acres was planted with zero tillage drill on unplowed land between standing rice stubble by 25 farmers in and around the village during 2000-01. Wheat yield obtained under zero tillage is either better that wheat planted after tillage operation or quite comparable with it. The main benefit of use of zero-tillage technique include saving the cost of tillage operations which range from 17.4% to 35.7% of total wheat production costs, i.e. saving of Rs. 400 to 1,000 per acre and ensure timely sowing of wheat crop. In addition, this technique saves 30-50% irrigation water in first irrigation after sowing and 15-20% in subsequent events. It reduces weed infestation and improves soil fertility and microbial activities. It also enhances water and fertilizer use efficiency. This all translate into 15-20% increase in wheat productivity, if properly implemented.

4. Bed Planting of Wheat

Technology of raising row crops on beds and furrows is gaining popularity amongst the progressive farmers. Main advantages associated with furrow-bed-irrigation technology of crop production are:

(i) savings of about 30 percent irrigation water;
(ii) reduced chances of plant submergence due to excessive rain or over irrigation;
(iii) lesser crusting of soil around plants and, therefore, more suitable for saline and sodic soils;
(iv) adaptable for various crops without changing basic design/layout of farm;
(v) enhanced fertilizer use efficiency due to local application; and
(vi) minimizes the chances of lodging of crops

CIMMYT-Mexico has sponsored trials of growing wheat also on raised beds on two different sites in the area served by Community Welfare Society community center. The results noticed in this regard are quite encouraging. It is found that this technique ensures better crop stand and yield with less water and other inputs like seed, fertilizer etc. The technology is being tested for many other crops like vegetables, corn and rice. Its results indicate that it is a potential technology of the future

5. Mechanization of Rice Cultivation

Rice is Pakistan's important food, industrial, and export crop. Nature has been blessed the country with climate to grow several types of rice not only to meet local food requirements but also to get significant share in international market. Pakistan is perhaps the only country in the world that has the largest acreage under fine-grained aromatic basmati rice. Rice export contribution to nation' economy is very significant.

Despite a lot of technological achievements in paddy cultivation, its yields are very low due to a number of constraints which include poor water management, inadequate agronomic practices etc. Optimum plant density and timeliness of transplanting operation in paddy is considered essential for getting higher yields. These problems can only be overcome if dependence on manual labour is minimized. All these factors have encouraged the agricultural experts to introduce appropriate machine methods for rice transplanting.

I. Mechanical Rice Transplanters

Pakistan Agricultural Research Council and On Farm Water Management Program, Punjab coordinated for promotion of use of mechanical rice transplanters amongst the farmers. One rice transplanter was provided to the Vocational Training Center of Community Welfare Society through which an area of seven acres was transplanted at three different sites during 2000-01. It helped in establishing 80 to 100% more plant population per acre as compared to manual transplanting. The yields obtained from manually transplanted and machine transplanted plots were, however, found almost similar. Reasons for the same are being studied and CWS is coordinating further research in this area between farmers and scientists.

II. Parachute Rice Transplanting Technology

Technique of parachute transplanting of rice technology was conceived in China. Rice nursery is raised on plastic sheets specially designed for the purpose. These sheets have depressions in them and two pre-soaked seeds are put per depression. These depressions are filled with mixture of soil and farmyard manure above the seeds. Irrigation water is applied regularly according to the requirement of seedlings, which are allowed to grow for 25-30 days.

The rice nursery raised with above technique has rice seedlings with a soil ball at their roots. The uprooted rice seedlings are put on a plastic sheet attached with air-emitting pipe of a power blower and broadcasted in the puddled field. The seedlings are pushed and droped in a projectile way with air. In this way rice seedlings are transplanted in the field as soil ball helps to put the roots in mud of the puddled field and their erected establishment. Parachute transplanting of rice technology is being promoted amongst members of Community Welfare Society. It is gaining popularity and farmers are willing to adopt it for broadcasting of rice seedlings.

The rice transplanting is carried out in scorching heat of summer, and it is probably the tough most laborious job. Moreover, the activity is carried out in standing water, which makes it a great threat for human health. The women, who perform most of the job, are more exposed to these adverse effects. Mechanized rice transplanting and parachute broadcasting of rice seedlings seem to be key promising alternatives. Major advantages of parachute technology have been noted as follow.

- It requires less labour as one person alone can transplant rice in one acre/day against five persons required for manual transplanting.
- Growing of nursery requires no special skill and expertise.
- Plastic sheets can be used for many years to grow nursery.
- It is much economical method compared to manual transplanting.
- Facilitates early establishment of rice seedlings due to no damage to plant roots during uprooting.

With efforts of Community Welfare Society, 20% farmers of the area have become familiarized to new technologies. They are being benefited and enjoying the outcome of this change in the form of increased farm income. The CWS is trying for adoption/

promotion of zero tillage and bed planting technologies for various crops, and mechanical transplanting of rice and parachute broad casting of rice seedlings. The main aim of various demonstrations and field days organized by CWS for the farmers is to introduce them with those technologies which offer conservation of water, reduction in cost of production, and obtain increased yields.

IMPACT OF RCT ADOPTION ON WOMEN WELFARE

The main objective of Conservation Agriculture Technologies is to conserve land, water, and energy resources, which lead to reduction in cost of production. Their adoption, therefore, helps in increasing overall income of the farmers and their families. It has direct effect on women welfare as increased farm and family income facilitates her to manage the family budget more easily as a house manager. The increase in farm income is ultimately spent on family welfare.

The rural women, who are the major participants of the farm labor, find Conservation Agricultural Technologies extremely beneficial and time saving. Previously, they had no awareness and knowledge about such innovations. Rice transplanting is probably most laborious work at the farm, which is mostly done by women in standing hot water under the burning sun. Conservation agriculture technologies for the rice transplanting like mechanical transplanting and parachute broadcasting of rice seedlings would facilitate to reduce this drudgery. They will know be able to plant an acre of rice with broadcasting by power blower in 2-3 hours as compared to traditional method which takes a group of 8-10 women half a day to transplant rice seedlings on same area. Conservation Agriculture Technologies would release women from labour intensive and less productive jobs like rice transplanting to devote more time and attention for important tasks of their family e.g. children welfare and other household activities. Lesser drudgery will result in the improvement of health and quality of life of rural women. These technologies will reduce working hours at the farm leading to allocate more time for women in the house and their children for schooling and education instead of carrying out menial jobs.

The issue of economics fundamental rights to life, liberty and security of person shall be the foremost concern for women movement in Pakistan. It is true that a substantial change in institutional responses to violence against women is necessary to create an environment in which women can truly enjoy their fundamental rights guaranteed in the constitution without distinction on basis of gender. We have seen that the problems related to women are multi-dimensional starting at home with their parents and in-laws, emotional, health, educational and economic problems as well. Violence against them has also to be dealt with. Looking into their constraints their needs, their ignorance, their hard work, we have to find for them remedies and cures.

We must look in to the development policies of the rural women. Financial empowerment is one remedy but government policies and plans for rural development are directed towards men and exclude women. Policy makers consider the women views secondary. However rural women's efforts to expand their activities are constrained among other things to their limited access to financial services due to lack of collateral and due to illiteracy and no awareness. They have limited mobility due to domestic and field duties. Extension services and technical assistance are usually provided to male farmers and non-towards women's empowerment and income generating programmes. The women should have ready and easy access to credit facilities next to home. This credit system should be accompanied by technical training advice, on management, production and marketing. So what is most important now is to give women economic empowerment. We must remember that their gender equality is strongly related to human poverty, so the constraints to expand their economic activities should be easily overcome as by providing easy credit facilities and soft loans with a minimum amount of interest rates.

Rural women have been forgotten, un-noticed and sidelined. It is of prime importance that their role is explicitly and fully taken into account .A policy formulated with women's need and interest in mind can help to reduce sexual and class discrimination, injustice and inequality. The policy for women should encompass and focus on the legal economic, social and religious spheres. Law leaves no option and leaves no choices, so improvement and revision of legislation benefiting women can play an important role in the elevation of women's status. Lots of problems arise when women are not given proper recognition as farmers in the own right, status as head of families and work units. If the law ensures proper recognition of women's rights to inherit, they're right to use and own land and their right to access to financial resources. Rural women can bring out remarkable positive change in their own small world

A lot can be achieved by eliminating the financial constraints on rural women. The existing economic condition of women inhabits them from fully using their energies and enhancing their work units. Efforts are needed to provide them with new and latest equipment machinery and tools, including production, harvesting and processing storage and transport. We also need to set up training centers where they can polish their existing and learn new technical skills. Broadening research and experimentation free of cost within women's activities is vital to acquaint them with the growing world. One of the most important steps taken for the economic uplift of women is the provide financial schemes such as micro credit. Women need finances to improve their lifestyle and environment but limitations like high interest rates, no collateral and lack of knowledge and access to financial services hold them back from progressing.

Positive changes in the social sphere must be achieved. Better living condition and quality life must be ensured. Health, balance diet, birth control and planned parenthood is as important but what is most essential is to educate them, practical literacy schemes are essential for their uplift. Women should play a role in decision making at home, in communities and at professional levels. Islam as a religion has given women a very high status, as mothers, sisters and daughters. So let us give them their due rights which they deserve honestly and truly.

RECOMMENDATIONS

1. Establishment of a Network of conservation agriculture is urgently needed to coordinate interaction amongst all stakeholders, which will help accelerated promotion of resource conserving technology.
2. Arrangements are required to be made to provide training to literate rural women who will in turn train and educate other village women about Conservation Agriculture practices.
3. Involvement of NGOs should be encouraged in transfer of technologies to the farming community.
4. Alternate job opportunities may be created in the villages for more effective and efficient utilization of women labour to be released from less productive agricultural activities.
5. Provision of micro credit for the women, those live in rural area may be accelerated enabling them to run their own small-scale entrepreneurs for increasing family income.

L.O. FRESCO

AGRICULTURE AND NATURAL RESOURCE MANAGEMENT. THE ROLE OF CONSERVATION AGRICULTURE

Agriculture Department, Food and Agriculture Organization of the United Nations, Viale delle Terme di Caracalla, 00100 Rome/Italy

FOOD DEMAND AND AGRICULTURAL PRODUCTION IN THE NEXT DECADES

Globally, growth in agricultural production over the last few decades has been more than sufficient to meet the growing and changing demands for agricultural products. However, eight hundred million people continue to be under-nourished because they lack the means to access these products.

As world population and food production has increased, the area of agricultural land per caput has decreased, and at the same time there has been a shift to higher quality and more diversity of products. Substantial investments in agricultural research by both the private and public sector have made a major contribution to the achievement of the gains in agricultural productivity. However, the rate of increase in both yield and total production of two of the main staple food crops, rice and wheat, has been slowing down over the past decade. At the same time, an increasing demand for fish, fruit, vegetables, meat, eggs and milk from burgeoning urban consumers in many developing countries, along with increasing wealth, is contributing to the development of a more diverse agricultural production scenario world-wide.

According to FAO projections, crop production in developing countries will be 70 percent higher in 2030 than in the 1995/97 period. Most of this increase (80 percent) will be accounted for by a further intensification of crop production and the remainder by the expansion of arable land. These increases in crop production will not only have to meet the direct food requirements of an expanding population but will also need to satisfy the increasing demand for feed in livestock and aquaculture production.

At the same time, new demands are being placed on the agricultural sector for greater protection of biodiversity and watersheds; its contribution to the mitigation of greenhouse

L. García Torres et al. (eds.). Conservation Agriculture, 31–36.

gas emissions and its role in sustaining rural livelihoods. Consumers are paying more attention to the quality and the safety of agricultural products; a trend which is leading to increased regulation.

WHY SHOULD WE BE CONCERNED?

While it is likely that on a global basis the projected gains in productivity will meet the growth in demand for agricultural products, at least in the medium term, there are still some areas of real concern. In many developing countries, growth in agricultural production has not kept pace with the nutritional requirements of their populations and a net trade deficit in major commodities is developing. In the long run, growth in demand in some countries is forecast to be greater than that of supply, and the deficit is projected to grow. It is estimated that, for example, in sub-Saharan Africa, a growth rate in production of about 4% per year is needed to keep up with population growth. This contrasts with the average productivity growth rate of only about 2% that has been achieved over the past four decades. Within the context of limited growth of arable land, a requirement for more international trade in food commodities is the likely consequence, with industrialised countries expanding their agricultural output. However, there is doubt as to whether some developing countries will generate sufficient foreign exchange to import increasing quantities of food.

With regard to production, critical thresholds such as yield ceilings and availability of water resources for irrigation start to become limiting factors. There is also a growing concern about damage to the natural resource base as intensity of production continues to increase.

In many developing countries the percentage of the population living in urban areas is increasing at a much faster rate than in the rural areas. Although not generally recognized, labour is already a major seasonal constraint which may severely limit agricultural production. Furthermore much of the urban migration is by the young «economically active» section of society, which, combined with the major scourge of HIV/AIDS in Africa, puts additional pressure on rural labour. The sheer daily grind, the demands placed on limited family labour, the lack of services, the prevalence of hunger and disease, and the inability to afford technologies which might improve livelihoods is driving people from the rural to the urban areas - to a perceived better life.

The future impact on natural resources of the way in which agriculture is practiced will be shaped by two counteracting forces: continuing pressure from the requirement to expand and intensify crop and livestock production; and declining pressure because of technological change and institutional responses to environmental degradation caused by agriculture. Agro-environmental impacts will be largely a continuation or acceleration of recent trends, such as calls to slow down the rate of deforestation and rangeland clearance for crop production. It is thus anticipated that the main impact on the environment will come from the intensification of crop production on land already under cultivation. However, the increasing use of precision farming and other advanced technologies, and a reduced use of pesticides and mineral fertilisers brought about by regulatory measures, may mitigate some of these negative effects.

The challenge of producing sufficient food and other agricultural products from a dwindling natural resource base in order to satisfy greater and more diverse demands from a growing population is not just a question of resource-efficient production technology. Increasingly, food safety legislation is stimulating farms and firms to develop new technologies that respond to improved standards. Of equal importance is that fact that agriculture will have to continue to meet social objectives by sustaining and improving the livelihoods of the 1.3 billion poor, many of whom live in rural areas and depend on agriculture in one way or another. Future agricultural research and development will be driven by complex environmental, public health and social objectives. Here, trade-offs will have to be appropriately balanced, and international action will be required in order to progress towards meeting development targets and aspirations.

THE SHRINKING NATURAL RESOURCE BASE

In developing countries, some 2.8 billion ha of land are potentially available and suitable for rainfed agriculture. Of these, some 960 million ha are already cultivated. However, much of the remaining potentially available land (1.84 billion ha) is forest or classified as reserves, or is marginal land. In South Asia, the Near East and North Africa there are many countries where there are no significant areas of uncultivated land available which is suitable for agricultural development.

Evidence is now emerging that a major contributing factor to the slow-down of the yield increase is land degradation and soil impoverishment. Much of the new land that has been recently brought into production through population pressures is in fragile agro-ecological zones. This is most apparent in semi arid, sub-humid and humid tropical and sub-tropical areas characterised by highly dynamic biophysical processes combined with high and growing human population densities. The move to continuous cultivation caused by population pressures has led to substantial alterations in soil properties (physical, chemical and biological). Soil degradation and the costs of mitigating measures along with the decline in productivity contribute to a reduction in farm incomes, part of a downward spiral leading to abandoned farms and migration to the cities.

In areas where crop production relies totally on rainfall, there is a growing competition between irrigated agriculture and other demands for water. It is estimated that by 2025, the demand for domestic and industrial use and for irrigated agriculture will at least equal and may exceed available water from rivers, streams or underground resources. Only more efficient water use will ensure that sufficient water is available to maintain regular flows in rivers and to mitigate the effects of drought.

Significant increases in food output can thus only realistically be achieved by intensifying the use of those land and water resources already in production. In many areas, this may not be achievable using conventional cultivation methods and presently available labour resources.

WHAT CAN CONSERVATION AGRICULTURE DO?

For agriculture to be sustainable, economically attractive and socially acceptable, it must successfully exploit the productive potential of those crop and animal genetic resources which are best adapted to the local environment. This is achieved by effectively and efficiently using available natural resources but without depleting them. Important ingredients in building such forms of agriculture are the reduction and eventual avoidance of run-off, and high and sustained water and nutrient use efficiencies. These are enabled by interventions which continuously enhance the soil physical and biological properties, as well as ensuring effective nutrient cycling procedures, e.g. by integrating crop and livestock production.

Conservation agriculture, using cover crops and crop residues in conjunction with zero-tillage and direct seeding, meets these requirements. Substantial environmental benefits are being experienced in fragile environments where soils are already degraded or threatened by degradation. Farming which relies on manual or low mechanization technology, benefits through the elimination of conventional land preparation and tillage which are the most arduous and time consuming tasks for agricultural labour. In tractor based, fully mechanized systems, significant decreases in fuel consumption have been recorded.

However, farmers have to go through a considerable change in attitudes and convictions in order to accept the new principles. In fully mechanized farming systems in particular, the acceptance and adoption of conservation agriculture may be constrained by the relatively high investment costs required for the technological change and the suppressed yields sometimes experienced in the initial period of adaptation. Institutional and land tenure aspects may limit prospects for technology uptake where there are, for example, customary laws allowing livestock to graze off crop residues. In mixed crop-livestock systems, practices such as stall-feeding or controlled grazing will need to replace free grazing on harvested fields. Furthermore, there may be other agro-ecological limitations such as rainfall requirements of particular species of cover crops limiting the applicability of conservation agriculture in semi-arid areas.

Conservation agriculture practices are generally more complex than conventional ones, in that they make use of the benefits that one crop may bring to the next in the sequence, and call for coherent management practices extending over more than one or two crop seasons. Farmers may need to adapt some of the new technologies to suit their specific needs and conditions, but may also need to develop more complex management skills to balance crop rotations with market requirements. Conservation agriculture technology is thus knowledge-intensive and the increased technical skill and managerial requirements may make adoption difficult to achieve in some countries without a major effort by all concerned.

The adoption of conservation agriculture, as is the case with all environmentally balanced agricultural technologies, will be much enhanced if appropriate incentive schemes are devised; schemes which will allow the farmer to reap some of the environmental benefits which will occur, through, for example, reduced run off from catchment areas and the consequent reduction in downstream siltation. The trend towards rewarding farmers for adopting environmentally beneficial technologies, or conversely, of establishing penalties for consequential environmental damage occurring off-farm, within an enabling institutional framework over the longer term, is likely to further stimulate the uptake of conservation agriculture principles and practices. Similarly, the promotion of "good farming practices" and perhaps some certification of production processes, may help improve markets for the produce through information to consumers and a consequent change in purchasing preferences.

CONCLUSIONS

Conservation agriculture is a term which embraces a complex package of technologies which is demonstrating that intensive agricultural production can be compatible with the maintenance of the natural resource base across a range of production systems. It is increasingly being practised in Latin America, North America, and Australia and is being actively promoted and adopted in other regions. In many situations it will need to be combined with a wide range of options that advanced technology offers, from sophisticated plant breeding and controlled release of mineral fertiliser to precision farming. Most of these technologies are compatible with conservation agriculture even though specific adaptation of varying degrees may be required. Its successful adoption will also require inputs of technical expertize as well as political support. Although it may not be a universal panacea for solving the problem of sustainably increasing food supplies, conservation agriculture may help in restoring the credibility of the agricultural sector at a time when public attention is focused on its negative effects and when its economic importance is declining in many countries.

II. FARMER EXPERIENCES WITH CONSERVATION AGRICULTURE

C. CROVETTO

NO TILL, THE STUBBLE AND THE SOIL NUTRITION

Soil Conservation Society of Chile, Las Heras 2095 Casilla 162
Concepción Chile

INTRODUCTION

No till has been introduced in most American countries as an agronomic system in soil management with high conservation practices to overcome the serious erosion problems within the region.

On Chequen farm, located in the coastal range of central Chile, after 23 years sowing without plowing, significant changes have been observed in the structure and fertility in old eroded soils. This change has occurred due to the increased organic matter within the soil profile. The organic matter increase in the Alfisols soil on Chequén farm is due to the parallel yields increase in the different: wheat, triticale, corn, lupine and pasture rotation. It is this rotation which affects the stubble amount over the soil.

The irrigated corn and dryland wheat rotation will annually leave 12 t/ha of stubble over the soil when this one is properly managed. The organic matter increase at 0.2% annually, which is initially on soil surface and later in underlying horizons, has generated an appropriate nutrition of this layer and has favored the microbiology and endemic mesofauna within this zone. This larger amount of biological activity has increased the organic carbon and with it, its humic coumpounds.

In the corn-wheat rotation, the largest carbon amount fixed was achieved in the first 5 cm while in 5-10 cm- and 10-30 cm-depths it was smaller. This is compared with a lupine-wheat and lupine-prairie rotation. The carbon balance in the corn-wheat rotation showed humic compounds, like fulvic and humic acids plus humins, were higher in the 5 cm depth compared with the other two rotations. However, in 5-10 cm- and 10-30 cm-depths, the contents of these compounds were larger than the horizon at 5 cm. The corn-wheat rotation indicates a larger carbon content that is superior to the wheat-lupine and pasture rotation.

THE CARBON CYCLE

The carbon cycle is vital in the soil's life. Carbon is the essential part of all cellular

L. García-Torres et al. (eds.), Conservation Agriculture, 39–48.

structure of much animal biomass and vegetative biomass within the soil. The photosynthetic process is responsible for the plants´s carbon synthesis by absorbing the carbon dioxide (CO_2) from the atmosphere. The plants receive the solar light energy through the chloroplasts that contain chlorophyll, a green factory rich in carbohydrates, lipids and protein, that form the foundation of life on the planet. Under those conditions the plants absorb CO_2 and release molecular oxygen (O_2) to the atmosphere, maintaining the gas balance. The soil organic carbon is of great importance as it constitutes the essential organic matter part and it is a basic compound for carbohydrates made up of carbon (C), hydrogen (H) and oxygen (O).

Any remaining stubble left on the soil will begin decomposition as soon as the water content and temperature are adequate. If the stubble is buried and the soil is moist and contains adequate oxygen levels, a rapid oxidation of incorporated organic matter will take place because the microorganisms will be nurtured excessively from the stubble. This phenomenon implies quick carbon loss that escapes to the atmosphere as carbon dioxide (CO_2). Another form of carbon loss can also be generated under anaerobic conditions which creates methane gas (CH_4). The release of these greenhouse gases may be linked to the global warming.

The organic matter oxidation starts when the crop stubble is initially attacked by cellulolitic fungi and later by all kinds of microorganisms that include soil mesofauna. In these circumstances, the rapid organic matter decomposition from plowing the soil surface layer, decomposes the soluble carbohydrates and low molecular weight humic compounds which are not very stable. This rapid decomposition and release of CO_2 is what prevents the build up of organic carbon within the soil profile.

At the recent World Congress of the Environment, which took place in June of 1998 in Bonn, Germany, the level of CO_2 in the atmosphere in 1850 was reported as 280 ppm and that in 1997 it was 365 ppm. At the present time the increase of CO_2 in the atmosphere is 1.5 ppm annually. This CO2 increase may be partially responsible for the devastating effects of the El-Nino and La-Nina weather patterns.

The improper handling of forest and permanent pasture coupled with traditional farming practices throughout the World, which have continually oxidized the organic matter and release CO_2 into the atmosphere, have made large contributions to global warming of the Earth. The carbon excess in the atmosphere is responsible for 50% of the greenhouse effect compared with other gases. (D. Reicosky, personal communication)

These practices, which have occurred throughout the years, have degraded the soil. Forests, permanent pastures and agricultural soils have been depleted from organic reserves. These depleted reserves have left the soil vulnerable to soil erosion and decreased fertility for growing crops.

THE SOIL NUTRITION

The soil biology is composed of a great variety of microorganisms, fungi and mesofauna which have daily food and energy requirements that must be met. These small organisms are directly responsible for formation of soil rich in humic substances and consequently result in higher levels plants fertility.

A soil, which has high biological activity, has the advantage of building up humic compounds. This occurs because of abundant biological excretion generated in metabolic cycles of soil organisms. Keeping stubble on the soil surface enables continuous renovation and proliferation of a favorable environment with no disturbance to cause a change in soil structure. Anytime you change this condition by tillage, you could harm soil biology and along with its native fertility.

Organic carbon is essential for the survivability all the soil's biology. To guarantee that the soil maintains its biological activity, **farmers must leave the crops stubble on the soil surface.**

The farmer that sows without plowing can harvest twice, both the grain and straw. It is of vital importance that the farmer remember «**the grain is for man, the stubble is for the soil**». Returning the stubble to the soil helps recharge the carbon and is the payment to the soil for the grain that it produced. Fertilizers by themselves only feed the plants and do not nurture the soil (Crovetto, 1996).

Today the fertility plants is based on different exogenous contributions that farmers give to the soil. In fact, in a large part of the world, farmers must provide nearly all the nutrition needed by the plants for grain production. This is necessary, as the soils have been depleted their natural nutrients because the farmers have been feeding only the plants and not the soil.

Apropriate soil nutrition will increase the content of organic carbon and consequently of humus.

The humus, while very small particle (less than 0.002 mm), has a strong negative electric charge which increases the cation exchange capacity (CEC) of the soil. This natural phenomenon, improves the soil fertility since the CEC increases retention of useful cations for plant nutrition, like calcium (Ca^{++}), magnesium (Mg^{++}), potassium (K^+), sodium (Na^+) and ammonium (NH_4^+). This diminishes the risk of loosing these basic cations through leaching.

The stubble and soil nutrition, are the basis of a permanent agriculture (Crovetto, 1997).

The humus and calcium can be combined in the soil making this chemical of great importance to increase soil nutrition. Calcium (Ca^{++}), in acid soils, can displace the hydrogen ion (H^+) on the exchange complex. This improves soil pH and available phosphate for better plant nutrition. (Figure 1). The humus increases the negative charge

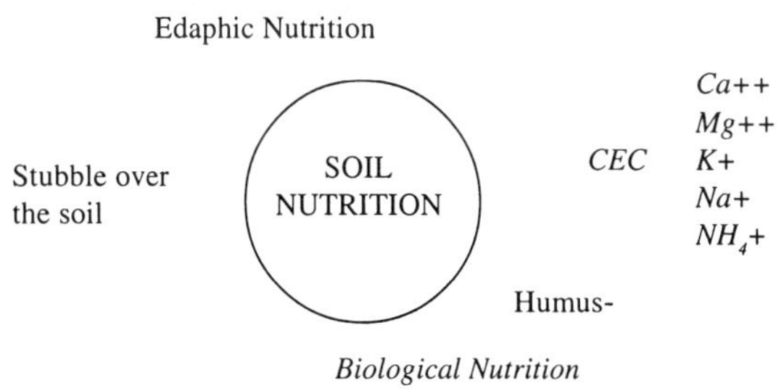

Fig 1. Environmental sustainability wheel with benefits emanating from the soil C hub.

of the soil retaining the useful exchange cations. Without humus or calcium in the soil, farming cannot be productive (García and García, 1982).

The quantity of stubble left on the soil is important, as well as the quality of that stubble that affects the soil biology. Not all stubble has the same value for soil nutrition. It is important to consider the lignin content of the stubble along with the quantity of that stubble. The lignin is directly responsible for the formation of humins. The stubble types which contribute most to the formation of humins are those of small grains (cereals) like triticale, rye, wheat, oat, rice and forest residues.

The composition of different stubbles harvested on Chequén farm highlight the high content of lignin of soybean stubble (11.9%). However the total contribution of this stubble is low due to the small amount of plant biomass. At 9.3% lignin, wheat stubble also provides an important contribution of lignin. (table l).

The corn-wheat rotation on Chequén is the most effective in producing humic compounds, what is coincident with higher soil fertility.

No-till allows the stubble to remain over the soil which increases biological activity producing both micro and macro pores within the soil profile. These pores will provide the soil with the proper aeration required for plant growth. The plant roots also generate a large amount of channels within the profile that will increase water infiltration while also improving the field capacity or availability of water for the plants. The channels or conduits generated by the biology of the soil in no till farming, are generally rich in humus, which can absorb 15 times their weight in water.

The stubble is a vital source of carbon and consequently of humus. The consumption of humus varies with the soil texture. Soils which contain larger soil particles such as sand require a much larger quantity of stubble to generate humic compounds than do the finer clay soils. According to Fuentes (1994). the mineralization of the humus in a clay-loam soil has an annual rate of 1.3%. This would indicate that the humic replacement should be of 620 kg of humic compounds per hectare or equivalent to 2,550 kg. of stubble per hectare since the isohumic coefficient for the wheat straw is 0.22 (defined as the humus quantity formed from a unit in weight of crop stubble added). The reference quantity maintains the necessities of the active soil biology. This means that if we want to increase the soil organic matter content, we should apply a quantity larger than 2,550 kg straw /ha.

Table 1. Stubble composition on Chequen farm. Concepcion, Chile*

Identification	*Carbohydrates Soluble (%)*	*Cellulose (%)*	*Hemicellulose (%)*	*Lignin (%)*
Corn stover with leaves without ears	*5.1*	*43.2*	*26.7*	*8.4*
Grain sorghum straw Without seeds	*1.5*	*35.6*	*25.0*	*4.6*
Soybean straw with leaves and pods without grains	*2.2*	*39.2*	*12.6*	*11.9*
Wheat straw with leaves and spike without grains	*4.1*	*49.1*	*23.9*	*9.3*
Triticale straw with leaves and spikes without grains	*7.0*	*45.5*	*24.2*	*8.6*

(*) Base dry matter

Not all the stubble and organic compounds have the same isohumic coefficient. The stubbles that contributes more humus to the soil are those that have wide carbon to nitrogen ratios. The lower nitrogen in the stubble, the greater is its ability to contribute to increase the humic compounds. Nevertheless, the addition of nitrogen to the stubble with high relationship C:N favors this contribution. In this sense, the lignin content of the stubble will contribute to a better quality humus, expressed in humins.

Cereals stubble have great importance in the formation of humus in the soil as they can contribute up to 20% of their dry weight. However, the manures of domestic animals only contribute 10% of their dry weight in humic materials. The manures are less stable and, consequently, decompose faster (Fuentes. 1994). We have treated the soil as an open mine instead of a real digestive animal rumen.

The small contribution of humic materials in the manure is due to the digestive system of the ruminant animal. One can degrade a large part of the carbon content, especially stubble, biologically through the ruminal liquids and its complex anaerobic enzymatic-digestive system. For this reason, the contribution of humic materials of the manure is notably inferior. This refers basically to soil nutrition and not plant nutrition.

The manure of domestic animals should be applied on the soil surface as soon as possible after having been excreted. The sooner that manure is applied the greater the availability of the nutrients for the soil and or the plants. The carbon losses, nitrogen and other useful components by leaching or gaseous conversion, will be directly proportional to the time which has elapsed since the manure has been excreted until its surface application. The manure should be applied on the soil surface, preferably before a rain or irrigation and most favorably with low temperatures. The losses from gasification of the ammonia (NH_3) content in the manure when it is applied on soil surface is insignificant compared to the losses of carbon and damage to the soil when it is incorporated by tillage. The direct addition to the fresh manure of 100 to 200 k of lime (CaCo3) per ton of manure highly can avoid the ammonia loss.

MAIN COMPONENTS OF ORGANIC MATTER

The soil organic matter is made up of many different compounds whose nitrogen content, carbohydrates, fiber and diverse minerals give origin to cellulose, hemicellulose and lignin. The order of decomposition of the organic components in the soil is: cellulose, hemicellulose and lignin. This decomposition order does not necessarily compare with that of the mammals, especially ruminant animals. (Table 2).

CELLULOSE

The main compound for generating humus in soil is cellulose which can constitute about 50-70% of the dry matter of stubble. Cellulose is formed by a long chain of glucose molecules ($C_6H_{12}O_6$) and is gradually hydrolyzed under natural environmental conditions (Rosell, 1997. personal communication).

HEMICELLULOSE

The hemicellulose is a polysaccharide that is hydrolyzed slowly in an acid reaction where the cellulose produces pentosans which spread and accumulate in the soil. (Rosell, 1997. personal communication).

LIGNIN

The lignin is a precursor of the more resistant humic material to biodegradation like humins. The humin is formed by molecules of substituted fenilpropane. This compound generates a recalcitrant three dimensional structure to the natural and biological physical agents. The enzymes and other depolymerizatión mechanisms (hydrolysis, oxidation-reduction, etc.) favor the decomposition of the lignin and their humification (Rosell. 1997. personal communication).

SOIL HUMIC COMPOUNDS

Humus is a generic name for selected forms of carbon and is present in the soil in the forms of: fulvic acid, himatomelanic acid, humic acid and humins (Labrador, 1996). Humus has as fundamental characteristic of small size in all its forms. This compound possesses a net negative electric charge that is greater when humus is of finer particles. The humic particles cannot always be classified according to the chemical composition of the original organic material, but can be classified by molecular weight and their reaction to acid, alkaline compounds, and alcohol.

FULVIC ACID

Fulvic acids are characterized by their relatively low molecular weight and can be associated with the polysaccharides. They are soluble in alkaline and acid

Table 2. Division of the Humic Substances

FRACTION	*ALKALI*	*ACID*	*ALCOHOL*
Fulvic Acid	*Soluble*	*Soluble*	———
Himatomelanic Acid	*Soluble*	*Nonsoluble*	*Soluble*
Humic Acid	*Soluble*	*Nonsoluble*	*Nonsoluble*
Humin	*Nonsoluble*	*Nonsoluble*	*Nonsoluble*

Modified after Labrador, 1996.

reactions. For their larger comparative content of carboxylic acid, they possess a great capacity to dissociate native minerals in the soil. This characteristic in influenced strongly in their genesis. The anionic colloids can form stable complexes with polyvalent cations like Fe+++, Al+++, Cu++, etc. (Labrador, 1996). This remarkable edafic function decreases the phosphate fixation on the aluminum sesquioxides or iron complexes.

HIMATOMELANIC ACID

A complex mixture of humic compounds like fulvic and humic acid soluble in alkalis and alcohol, produce himatomelanic acid (Labrador. 1996).

HUMIC ACID

Humic acids are the major humic compounds in the soil. They are insoluble to the acids and alcohol and they have a medium molecular weight. They are thin and flat particles bonded to each other, those that form a reticular spongy material (Labrador, 1996). This can be one of the most remarkable physical-chemical characteristics of

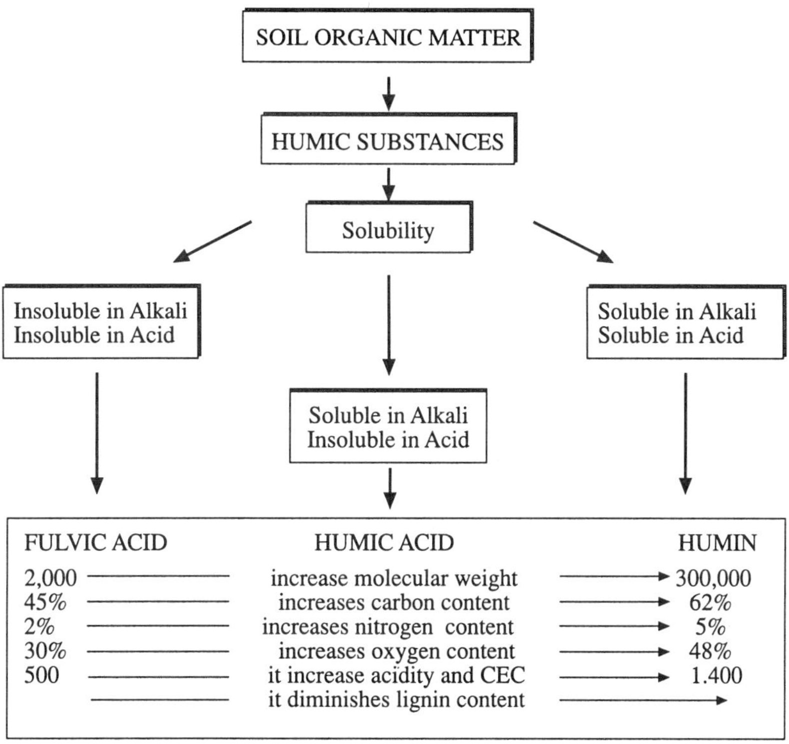

Collins *et al.*, 1997

Figure 2. Characterization of the Soil Organic Carbon relative to their stability and transformation

the humic acid, because it enables a great capacity for water retention and a strong anionic charge that improves CEC noticeably. The humic acid regulates the process of oxididation-reduction of the edafic system, providing some oxygen to the plant roots.

Similar to the fulvic acids, humic acids can form complex substances, especially through metallic ions. According to Schnitzer (mentioned by Labrador), the humic acid is the main factor in genesis of soil and the formation of a good soil structure and in the availability and mobility of certain plant nutrients. It is also important in the agrochemicals controlled persistence and degradation in the soil profile (Labrador, 1996).

Our experience in the no till soils of Chequen farm, herbicides show a very little residual effect. Herbicides are degraded by the active soil physiology and, without a doubt, are affected by the presence of humic compounds in general.

HUMIN

Humin is the more stable component in the soil because of its high molecular weight. Humins are insoluble to the degrading chemical agents and remain strongly bonded to the finest mineral colloids in the soil (clay particles). This characteristic probably allows it to remain in the soil profile for a long time when the soil in not disturbed.

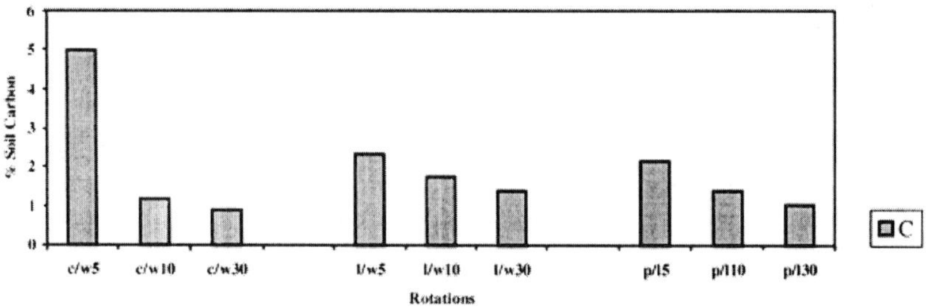

Figure 3. Total carbon in Chequen soils

	c/w5	c/w10	c/w30	l/w5	l/w10	l/w30	p/l5	p/l 10	p/l 30
C%	5	1.19	0.9	2.34	1.75	1.4	2.16	1.41	1.05

Meaning of abbreviations: c/w = corn-wheat; l/w = lupine/wheat; p/l = pasture/lupine; C = carbon 5-10-30 = cm depth

The humin structure is basically derived from lignin content of the organic matter that results in a unique structural relationship.

Figure 2 explains the stability and behavior of humic compounds in soil. Humic compounds are prominent in the humins, the largest molecular weight material that increases cation exchange capacity (CEC) that contributes to better soil quality (Collins et al., 1997). This favorable condition has been observed in the Chequen soils after 20 years of sowing without plowing or harrowing.

The soil CEC has increased from 11 meq/100 g to 26 meq/100 g during this period.

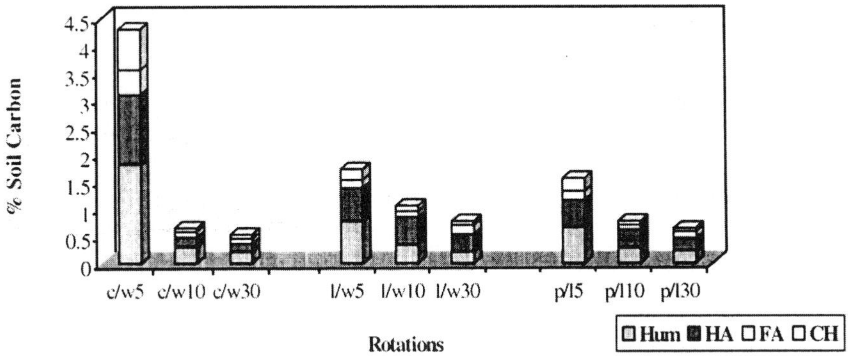

Figure 4. Carbon balance in Chequen soils

Humins are very desirable and their persistence will depend on how the farmer manages his soil. Any soils under conventional tillage will stimulate the oxididation-reduction processes and, consequently, will affect the stability of all the humic compounds.

No till or direct seed farming stimulates the formation of all the humic compounds when permanently adding organic matter to the soil. The carbon balance is significantly higher in the corn-wheat rotation, compared with the wheat-lupine rotation and prairie-lupine, especially the humins content. The carbon balance in Chequen soils explains that the humin content, humic and fulvic acids are superior in the first 5cm depth in the corn-wheat rotation, compared to the other rotations. This is due to the large amount of stubble (carbon) input that this rotation contributes to the soil organic carbon. The corn-wheat rotation contributed the biggest content of total carbon to the soil, although only in the first 5 cm of the profile as the figures 3 and 4 explaine.

		Hum	HA	FA	CH
c/w = corn/wheat	5cm	1.82	1.28	0.46	0.74
c/w = corn/wheat	10cm	0.30	0.18	0.10	0.07
c/w = corn/wheat	30cm	0.21	0.15	0.09	0.08
l/w = lupine/wheat	5cm	0.77	0.62	0.14	0.20
l/w = lupine/wheat	10cm	0.34	0.51	0.11	0.10
l/w = lupine/wheat	30cm	0.21	0.33	0.16	0.07
p/l = pasture/lupine	5cm	0.66	0.50	0.17	0.23
p/l = pasture/lupine	10cm	0.28	0.34	0.08	0.07
p/l = pasture/lupine	30cm	0.23	0.23	0.12	0.06

Meaning of abbreviations: CH = carbohydrates; FA = fulvic acid; HA = humic acid; Hum = humins

CONCLUSIONS

The direct seeding or no till farming system improves soil nutrition, it stimulates soil biological activity, increases the soil organic matter content and enables the formation of humic substances which improves soil physical, chemical, and biological conditions.

The soil nutrition is based on the stubble that the farmer leaves on the soil surface. This is the price the farmer has to pay to use the soil resource. Feeding the soil should be a daily process, which implies the stubble presence should be permanent.

The surface stubble decomposes to form different humic compounds, basically depending on their C:N relationship and lignin content. The lignin is responsible for the humin formation and at the same time provides the soil more stable humic materials that last longer.

The farmer should not only be concerned about the quantity of stubble that leaves on the soil, but also of stubble quality. Those with high carbon/nitrogen relationship are the most beneficial to the soil.

REFERENCES

Collins, H.P., Paul, E.A., Paustian, K. and Elliot, E.T. (1997) Characterization of Soil Organic matter relative to its stability and turnover. CRC Press. Inc. New York. pp. 52-54.

Covetto, C. (1996) Stubble over the soil; the vital role of plant residue in soil management to improve soil quality. *American Society of Agronomy.* Inc. Madison. pp. 245.

Crovetto, C. (1997) La cero labranza y la nutrición del suelo. 5ª National congress of AAPRESID. Publishing Víctor Trucco. Argentina. pp. 73-78.

Fuentes, J.L. (1994) El suelo y los fertilizantes. Mundi Presses. Madrid. pp. 55-71.

García, J. and Garcia, R. (1982) Edafología y fertilizacion agrícola. Aedos. Barcelona. pp. 24-44.

Labrador, J. (1996) La materia orgánica en los agrosistemas. Mundi Presses. Madrid. pp. 24-28

V. H. TRUCCO

ARGENTINE AGRICULTURE: AN INNOVATIVE EXPERIENCE

President of AAPRESID - Argentine Association of No Till Farmers
Paraguay 777, 8vo piso, oficina 4, 2000 Rosario, Pcia de Santa Fe (Argentina)

INTRODUCTION

Change and the speed of change characterize the beginning of the 21st century.

As agricultural entrepreneurs we are perplexed at the amount of variables to be considered.

While new regulatory frameworks have resulted from the 1992 Rio Summit, the advent of Sustainable Development and the Desertification, Climate Change and Biosafety Protocols, their intentions are good, but they are of doubtful application.

Processes develop at the WTO within the framework of Globalization and Free Trade. The Codex Alimentarium and the regulatory frameworks of countries are prone to defend the stance of the economic blocks rather than to achieve a rational, sustainable and productive agriculture to meet the feeding needs of the people and to preserve natural resources.

Agriculture has no doubt enabled the development of today's sophisticated world, given that those who produce food are but a few, while the rest just need the time to consume it.

However, the fact that the Planet has fed mankind for thousands of years has taken its toll, particularly in recent years. Deserts, shortage of water, contamination, the greenhouse effect, are consequences known by all.

Feeding over 6 billion people has been the feat of the farmers of the world with the help of science and agricultural technology. By themselves, they are not enough, but without them, it would have been impossible.

I think the time has come for us to do things better with the help of science and common sense. I am persuaded it is possible and that we are already proving it.

Amongst those who give their opinion we should distinguish between those who are not involved in production and those who must put things into practice.

We are aware that some groups will not have us use agrochemicals or fertilizers to prevent contamination, nor allow us to use biotechnology «just in case». That only leaves us with low productivity agriculture.

L. García-Torres et al. (eds.), Conservation Agriculture, 49–58.

Farmers are expected to stay in their farms as part of the landscape, like in olden times. Though for the romantic and nostalgic few that could be desirable, it is totally unrealistic, impossible and unnecessary nowadays, since it is not the only way of having a sustainable agriculture, nor the only reason for those wishing to reside away from the big cities, doing productive work.

GLOBALIZATION AND SUSTAINABILITY

The world evolves as do relations among countries and their economies, having an impact on individuals and society. New information technologies, with the appearance of Internet, its speed and the possibility of reaching the uttermost corners of the world at the same time, has imposed «Globalization» which means «the continuity of space».

Science, technology, the markets, a competitive edge, may become threats or opportunities for individuals and society.

Globalization is not a theory, it is reality resulting from new circumstances. Globalization is the result of economic, technical, scientific and political development of our times.

However, there is another aspect of economic development, an increased per capita consumption in a growing population that has an impact on the earth's vital system and its natural resources. The «time» variable has come into the picture. The economics of Globalization does not take into account the deterioration of natural resources and the environment, it is not concerned with whether current productive processes will allow future generations to obtain from Nature the resources needed for their subsistence. For this reason we must introduce the concept of «the continuity of time» in economics, and this is what «Sustainability» is all about.

Who will worry as to whether business ventures are sustainable or not when environmental damages are not included in the costs and when profits resulting from sustainable practices do not represent an income for those that practice it?

Business must be profitable to be able to meet current needs. However, profitability and competition must result from preservation and even amelioration of the environment and of natural resources.

Let us think for a moment in farming as it is done today, within the scenario of Globalization, i.e., the economic relations between the individual, his business and the markets. Profitability is a measure of success, no matter what has happened to the soil, the water or the air.

People are still not aware of the situation. The US and Europe still do not understand that subsidies –a resource of society- are used to sustain and increase soil degradation, higher CO_2 levels in the air creating the greenhouse effect, and contamination of the water. Prices fall because production increases in spite of the fact that more fertilizers and agrochemicals are required. Isn't this contradictory?

Can we think of our future as farmers along these lines?

It is necessary to harmonize social and economic development with the preservation of natural resources and the environment. Sustainable development is required, not in words, but in deeds.

In a world ruled by money, markets and competition, this objective calls for the valuing of environmental assets, assigning them a significant economic value in terms of business profitability. Pure air, drinking water, fertile soil, climate, biodiversity, must be valued not only in words, but should have their own weight in the costs and earnings of the business.

Introducing such a change of paradigm is a great challenge; however, there will be no hope if benefits cannot be reaped in an inhabitable world.

To introduce «the continuity of time» in economics is not easy, but it is necessary, and agriculture can be the starting point of such a change in paradigm.

THE FARMER'S PARTICIPATION

We must think of Sustainability, because our acts not only have a consequence in space and in «the present»; our deeds have a consequence in time, in the same way that what our grandparents did has a consequence in our present scenario. Man has been responsible for many deserts, as well as pesticides and fertilizers in water and the warming up of the earth.

While communication technology has led to Globalization, the responsibility of the international community should lead to Sustainability.

Globalization is not simple and has not been equally beneficial to all, but it is inevitable. Sustainability, likewise, will cause problems to some in the short term, but will no doubt prove beneficial to all in the long term.

We, agricultural entrepreneurs, are also to be held responsible. I understand the natural reluctance towards change, particularly when our needs are met, as may be the case of a European farmer.

But we must understand that change is necessary and inexorable. Change happens either through reason or through desolation.

We must anticipate events. There is another way of producing and of organizing ourselves. It is not my intention to instruct anybody about the way things should be done; I simply bring you an experience that has changed our way of envisaging the future. We no longer see it as a threat, we see it as an opportunity.

I do not believe that for one farmer to be successful, another farmer must fail. There are too many things to be done for anybody to be left out. There will always be room for productive and creative work that is no longer an economic need but a human requirement for a long existence.

Being a pioneer is no longer a virtue, it is a requirement. Paradigms no longer last a generation, they last a short time in our lives, so that we must always have a young attitude towards life.

CHANGE OF PARADIGMS: GENETICS AND THE ENVIRONMENT; BIOTECHNOLOGY AND NO TILL

We started the new century with two new concepts, two real changes in paradigm. One is the revolution of biotechnology leading to unthought of advances in plant genetics.

We shall have plants that can grow under impossible conditions until now, for example in saline soils. Others will have no dehiscence, a great advantage to crops. Some fruits –very different from what they are today- will have more nutritional properties, a higher content of aminoacids or essential oils, to replace what the human body cannot synthesize.

Plants will be used for the manufacture of nutraceuticals, that is, medicinal molecules that are currently obtained from chemical synthesis.

Likewise, new molecules will be synthesized through crops for industrial applications, such as biodegradable plastics, color fibers, etc.

Another fundamental change is the development of zero tillage.

Land can no longer be classified into arable and non arable.

Agriculture no longer carries the costs brought about by tillage.

This is because we have found the way to improve the soil's «environment» that will determine the conditions for the seed to express its potential.

Today we know that production is the result of seed genetics and the environment. If we sow a seed of high potential in a good environment, in fertile soil under adequate conditions of humidity and so on, the crop will have a good yield.

If the seed is not good, good environmental conditions will not suffice, and if we sow a good seed on arid soil, the outcome will still not be satisfactory.

The 21st century has started with great changes of paradigm, that of genetics through biotechnology, and that of soil environment, through the development of no till systems.

THE BUSINESS OF PRODUCING

Another issue to be borne in mind is that the world has changed. Not only must it be possible to be productive, it is mandatory to be profitable.

Farmers have always considered agriculture as a way of life; a family activity in the fields, that has always had its good times and its bad times, but never bad enough to threaten its continuity.

54,000 farmers have abandoned their activity in the last seven years.

Official banks do much rescheduling and a large part of the best land is mortgaged or about to be sold in auction.

Agrochemical companies have admitted they are facing the worst period of defaults in history, and their funding is becoming more and more selective.

Many agrochemical and seed dealers have had to close down in view of their difficulties to collect from their customers, high overheads, etc.

For the same reason, many cooperatives have gone bankrupt.

The sowing pools, i.e. investment funds created to promote the farming business by leasing 40,000 to 100,000 hectares of land to grow crops through contract firms in charge of sowing, spraying, etc., have disappeared because they were not profitable and even lost money.

In spite of so much pessimism, production increased by 40%, the greatest rise in the whole of Argentine history of agriculture.

Some made it, others evidently did not.

Many economic changes have taken place in the last few years: inflation, hyperinflation and stability. We have gone from interest rates eaten up by inflation to high rates; from a closed to an open economy with deregulation; from very high prices to very low prices for commodities.

What we have gained from these last few years is experience, and experience is telling us that farming in the future will have to be «very professional». By this I do not mean that it will be an activity reserved to engineers, but rather to those who understand how agriculture works, how economics works; to those who are aware of strengths and weaknesses, who are able to discover opportunities and fight against prejudice.

Farming will no longer be possible without understanding agriculture and business.

Attention must be paid to the change in paradigm.

THE NECESSARY CHANGES

We must change; circumstances so demand it.

The first thing we must change is our mindset; the paradigms of agriculture must change.

I honestly believe that the Argentine experience proves it is possible to produce in a different way. The meeting of farmers in CAAPAS, the Confederation of American Associations for Sustainable Agriculture, shows that it is possible to produce without destroying the soil, without causing erosion, improving the organic matter of the soil through carbon sequestration, avoiding contamination and being profitable at the same time.

«The challenge is to innovate», to change the paradigms, to spot opportunities and increase the strengths. AAPRESID is a change of paradigm in itself as far as organizing farmers is concerned. They do not meet for economic reasons as is the case of cooperatives, nor union motivations as is the case of the rural societies, but rather to create a «network and share innovation».

Farms are no longer «Mom & Pop's» they have become a «network of integrated businesses.»

ASAP: HIGHLY PRODUCTIVE SUSTAINABLE AGRICULTURE

Ideas are good when feasible. This is why I shall talk about what we do in Argentina and how I envisage farming for this 21st century that has just begun in order to meet the demands of the day, that is, being *sustainable, productive and profitable.*

For some years now in AAPRESID we refer to ASAP (Highly Productive Sustainable Agriculture).

In order to achieve ASAP we require changes of paradigm; we can no longer solve problems the way we used to.

The efficiency of ASAP leads us to think that agriculture is not only food, but that it can have other functions such as the use of biofuel or carbon sequestration, with the consequent positive impact on the greenhouse effect. Based on the change in paradigm of a Highly Productive Sustainable ASAP Agriculture, innovation will take place through the introduction of adequate technologies and new organizational structures.

Agriculture has become a complex and professional activity that supersedes the concept of «farmer». We must now think of an organization segmented into functions and firms that provide services and fulfill a role, strengthened by its very professional stakeholders.

THE OUTCOME

The change in paradigm in Argentine agriculture has led to increased acreage under no till. Figure 1 shows very few hectares under no till in the 80's to over 11 million hectares at the end of the 20th century, accounting for 46% of Argentine farming.

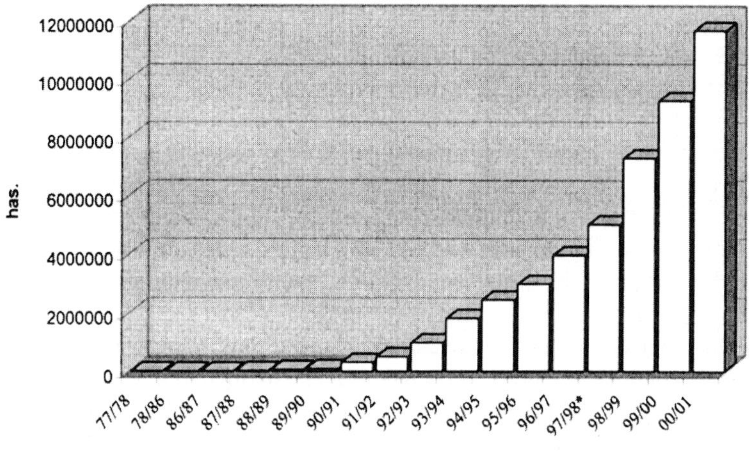

Figure 1. Evolution

Another change in paradigm is that of biotechnology, as shown in Figure 2 which shows the evolution of land cultivated with soybean in the 90's reaching a plateau at 6 million hectares.

Following the 96/97 campaign and the appearance of the first biotechnological event, Roundup Ready (RR) -glyphosate resistant soybean- started with 150 thousand hectares the first year to end up covering around 7 million hectares during the 99/2000 campaign, over a total of 9 million hectares. Production went from 12 million tons to over 20 million tons. Results speak by themselves as to the impact of biotechnology in Argentina.

Biotechnology has also had a significant impact in the plant health market. Until 1997 the market of agrochemicals had been growing steadily up to 925 million dollars that year. With the introduction of RR soybean, the sale of agrochemicals fell drastically to just over 600 million dollars, saving farmers around 300 million dollars. See Figure 3

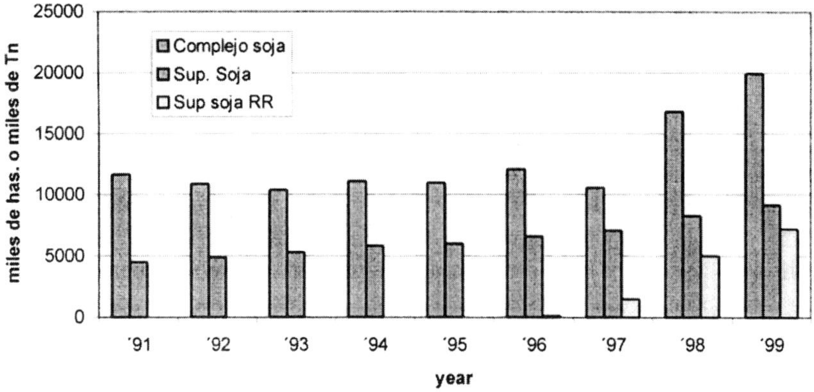

Figure 2. Evolution of total surface area for soybean, for RR, and soya exports in Argentina.

This has not only been good in terms of our own economy, but it has been beneficial to the environment since the change in the use of agrochemicals has favoured glyphosate, a molecule that fixes quickly to the soil and is later degraded by soil bacteria.

So we are confronted with a real win-win situation vis-à-vis the farmer and Nature.

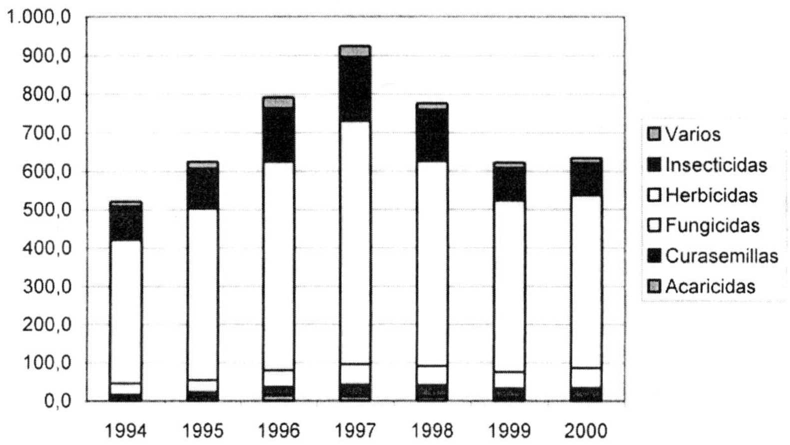

Figure 3. Evolution of the Argentine Plant Health Market.

AGRICULTURE AS A «SERVICE TO THE ECOSYSTEM»

If we hope to preserve the planet -the environment- with productive soils, pure air, pure water and predictable climate, we must stop using the soil as a mine or a dumping field. We must stop using rivers as the sewage disposal for contaminated fluids, or even the air as a warehouse.

In a community where nobody pays for eroding the soil, contaminating the waters or the air, it is quite natural that nobody will wake up in the morning concerned about the soil, the air or the water. Only the press will be concerned, the press that writes interesting articles for people to buy their papers on weekends while they comment that things cannot continue the way they are.

This is why a change in paradigm is required whereby "service to the ecosystem" provided by agriculture takes place through the preservation of productive soils, clean waters and carbon sequestration.

This can only be achieved when farming is practiced adequately, with zero tillage, crop rotation and fertilization of crops, not water.

Thus we will not only obtain high yields necessary to feed the world, but we shall be leaving the stubble to feed the living organisms that keep our soils fertile and enrich them with organic matter that will contribute to mitigate the greenhouse effect.

This is not achieved with conventional agriculture that has fed over 6 billion people at a "war cost".

We need to develop "Peace agriculture", that harmonizes production with the vital processes of Nature.

CONCLUSIONS

We are not required to choose between agriculture and the environment, productivity and Nature, rural and urban.

If we practice agriculture with a different mindset, we shall not bring about the consequences we attribute to agriculture today. The problem is not producing from soil, the problem is how we are producing.

I believe this new possibility you call Conservation Agriculture is not a recipe, it is an evolving concept that will no doubt adapt to each circumstance supported by science and technology.

I am concerned that good intentions might be misinterpreted, that Sustainable Development, Biosafety, the Greenhouse effect, might become mere political discussions. I am concerned that these fora be filled with speeches, many quite distant from reality. I am concerned that people are reluctant to accept mitigations such as the sequestration of carbon through no till, or that biotechnology may be caught in non-tariff barriers.

I am concerned that science may not be accepted to overcome current difficulties. I am also concerned that farmers are not represented at other fora.

I am concerned that farmers be forced to beg for their earnings.

But I am happy about this Congress. I am happy that the European Federation of Conservation Agriculture and FAO convened this Congress.

I am happy that we, South American farmers, have been asked to tell you our story.

I am happy that important officials responsible for the drawing up of agricultural policies for the European Union are participating at this Congress.

I am happy that Europe has begun to recognize that certain farming methods have a strong impact on the environment.

I hope that my story may contribute to change the concept of this type of agriculture.

I hope that my participation may contribute to confraternity among farmers of the world.

I hope that my participation may contribute to the organization of the world entrepreneurs, to become a projection of AAPRESID in Argentina and CAAPAS in America.

ACKNOWLEDGEMENTS

Opinions stated herewith are the result of my personal experience and permanent exchange at AAPRESID and CAAPAS particularly enriched by discussions with Donald Reicovsky and Prof. Francesco di Castri, even if I am solely responsible for whatever has been stated here.

A. TAPIA PEÑALBA

THE SOIL, AGRICULTURE AND I

Calle Plaza, nº 86; Quintarraya. Burgos-Spain

In spite of variety improvements, of chemical fertilizers and of technical means with which progress has provided us, the soil has been, is and will always be the basic resource of all agricultural activity. If no fertile soil existed, there would be no agriculture and, if this disappeared, the negative effective would be felt by mankind, starting with professionals like me who earn a living from it. And, in an opposite sense, the effect of our work could also be the cause of the loss of the soil used for our activity.

Starting from this prior reflexion, I should like to expound my ideas on a form of agriculture which compatibilizes conservationist criteria with productive and economic aspects, both of which are undoubtedly in need of being updated in our sector. I shall do this from a practical point of view, based on the experience I have acquired in the eighteen years I have been carrying out Conservation Agriculture.

With all those years of practice, I am able to affirm that this technique, when well applied, is the best **agronomic alternative**, is an **environmental necessity**, is a **social obligation,** it should be a **political priority**, and perhaps a **philosophy of life.**

In the seventies, when I began as a farmer, this country was undergoing great changes and one of them was a determining factor in my becoming a farmer; the phenomenon of the emigration of the rural population to urban areas meant that land became available for me to extend the surface of my farm. My ambition was to become an ATP (a Farmer with Tenure) for which I owned the land but did not have the necessary mechanical and financial means.

In addition, in those days, agriculture was set up with ancestral structures and was socially margined, politically despised and professionally resigned to. I rebelled against all that and conceived it as a productive activity capable of giving me professional satisfaction, economic benefits and a good standard of living, at the same time as reivindicating its social and political recognition.

I refused to accept that in order to be an agricultural *empresario* I had to fulfill the old Spanish saying : «Plough deep down, throw away the rubbish and don't take the books on agriculture seriously». I was of the opinion that this cliché responded to the cultural circumstances and technical limitations of an era now gone by.

L. García-Torres et al. (eds.), Conservation Agriculture, 59–68.

It was precisely by reading books that I discovered an alternative type of sowing to the conventional one, thus reducing soil tilling, machinery investments and the consequent savings in the final output and which was called Direct Sowing. The discovery of this technique was decisive in making my expectations begin to come true.

It was at that point that, when comparing the conventional ways of sowing with this new technique, I made a mental calculation and began to arrive at a series of conclusions «by spending less I shall reduce risks and even by harvesting less I shall still earn the same».

The years went by and I found that even «spending less, I harvested the same amount and therefore earned more».

I began to realize that not so much traditional tilling but maintaining the natural structure of the soil was a determining factor in production. I noticed that by providing stubble to the surface, fertile soil was generated, resulting in better harvests and «spending less, I harvested more and earned much more». I therefore left off being a tiller-farmer and became a producer-farmer.

Since then, I not only look up at the sky asking for rain but I look down more frequently at the earth to get to know it better and to help it to supply me with its crops. Gradually, I sought a more reasoned explanation for this favourable evolution of my land and began to notice how, by means of the corresponding analyses, the availability of water in my crops increased, the levels of organic matter rose, rising from 0.5% to 1.6% in 7 years, this being generated from the stubble and vegetable residues.

EQUIVALENCE
1 Tm. of hay = 60 Kg. of fertilizer of the 12-12-24
1 Tm. of hay = 7u. N + 7u. P + 14 u. K

I have changed the colour of the soil precisely because of its greater content in humus. Its smell is different too, probably because of the bacterial activity in decomposing vegetable matter. Its volume has increased, 1 litre of tilled soil weighs 1,740 g, whilst the same litre of soil with 9 years of Direct Sowing weighs 1,410 g, there generally being a direct relationship between the density of the soil and its fertility. This weight-volume relation also demonstrates that the plant earth content in the soil is higher, which therefore prevents its clumping.

The contribution of crop remains has increased its biological activity and serves as a food for it.

I have been able to observe a proliferation of animal species, including mice and small moles. These rodents are, in principle, negative elements because they partly destroy my seeds and, in some years, have been a real problem. At the same time, I have noticed an increase in the number of birds, of prey, which regulate the population of these mammals.

There is also an increase in migratory birds and in cynegetic species such as the wild boar, which on some occasions causes some nuisance in my crops. However, I am convinced that on a short or medium term, nature will recover the balance it has lost.

I have also observed an increase in all types of fauna in the subsoil. Especially significant is the number of worms; in a count made in 1 m^2 of earth in a profile of 12 cm, tilled in the conventional manner, I found two, while the same count in my adjoining plot with six consecutive years of Direct Sowing gave forty-two. As my master Carlos Crovetto says in his book: « the weight of worm excreta can reach 50 Tm/ha. Annually, or more». In this regard, I believe that in the future the soil will be valuated on the basis of the number of worms in it».I do not need to till the soil, they do that; I do not need to put any fertilizer on it, they fertilize it; I do not need to meteorize the soil, they oxygenate it; my efforts are devoted to looking after them and providing food and a natural habitat for them to go on increasing the productive capacity of my soil.

The climate of this country limits, to a great extent, its agricultural possibilities; but, above all, the scant fertility of its land is the most influential factor in its low productivity. It is here that we farmers, particularly, have the ability to change this precious resource; with a positive effect, by having a responsible, conservationist attitude, or negatively, by being egoistical and over-exploiting the land.. Intensive tilling has contributed to this impoverishment and has accelerated the process of erosion and desertification of our lands.

Several studies have demonstrated that in certain areas soil losses of up to 80Tm per ha and year have been measured, with the resulting diminution in its fertility. This process is being endured to a greater or lesser extent by 70% of agricultural land in Spain. This phenomenon has been perceived for a longer time than one generation so that, although it is a real problem, it may go on being unnoticed by people in general.

I have proved that, in practice, this factor is, to a great extent, avoidable if the soil is kept protected.

Let us imagine that the increase in organic matter of 1.1% occurring on my land were to be produced in the 19 million ha of the agricultural land in my country, and that its biodiversity had increased and its erosive process stopped, its production potential would undoubtedly be much greater and the agricultural and environmental panorama would be different.

At present, I farm 462 has. of dry land and 18 of irrigated land , made up of several farms located at a radius of 150 km to the south west and north of the province of Burgos. The physical structure of the soil is highly varied as is its production capacity. With a relatively basic pH of between 7.6 and 8.4 , with very low levels of organic matter. The annual rainfall ranges between 440 mm in the south west and 580 mm in the north, at heights of 1,020 m and 830 m, respectively, with annual temperatures of between –10ºC and 36ºC and daily variations reaching 24ºC. The crops extensively sown are, in alternation: cereals (barley, wheat), rapeseed, cereals, legumes (vetch, chickpeas), cereals, proteaginous vegetables (peas), cereals, sunflower, cereals; and more experimental crops of fibre and, for industrial use, Kenaf, hemp, etc..... In part of my land I sow, alternately, rapeseed, black oats and vetch as a plant cover on which I sow the annual crop.

It is very important to successfully manage both the abundance of stubble and these plant covers since they can cause the soil to become excessively cold and some other drawbacks, as will be seen later.

I have also had some experience with multiannual crops such as Cynara Cardunculus, which is highly indicated for soil protection with a permanent cover and, if necessary, to be used as biomass. In my opinion, this thistle species, apart from its industrial use, is useful for sowing in lands which our Common European Agrarian Policy destine for the removal of food crops and for fallow land. These plots are usually the marginal ones with a greater erosion risk so that by establishing this crop during its 5-8 year cycle the soil could be improved and recovered for agriculture, provided that Direct Sowing is used and that its structure is never again destroyed by traditional tilling.

Both the productions of cereals and of legumes and proteaginous vegetables are multiplications of basic seeds, in collaboration with different producing and commercializing companies, which comply with the characteristics of quality, salubriousness and purity required by the European regulations.

My machinery includes a 130 hp tractor, a loading shovel, irrigation equipment, a trailer, a fertilizing machine, pesticide treatment equipment, two machines specifically for direct sowing (ploughshare and discs), a small , a computer, several calculators and many ballpoint pens,the latter being work objects in continuous use and just as important, or more so, as the tractor and the sower. Other highly profitable investments are those in subscriptions to specialized agriculture journals,and in expenditure in telephones, fax,Internet, etc.In this way, I take an interest in all the agrarian regulations conditioning my business and I am highly aware that I need all these tools in order to plan, negotiate and commercialize my production. I should mention here that I am the only direct worker on my farm, with the irreplaceable help of Rosa, who is the person who ministers to and supports me.

From an economic and agricultural competitiveness point of view, one of the cost variables influencing all farms is the investment and depreciation in machinery.

In this country, there are 11,104,475 tractors and mechanical cultivators and 11,131,969 working equipment for these tractors, subfloorers, mouldboard or disc ploughs, harrows, cultivators, rotocultivators, etc. (sowers are not included).

Tractors , motorcultivators	*1,104.475*
Working equipment	*1,131.969*

Each liter of fuel consumed by these vehicles equals 4 kg of wheat, which multiplied by 74.2 liters consumed per ha is 297 kg., and in order to acquire some of these tools sales of 110,000 kg are needed. In other words,taking into account that the average production in dry land cereals in Spain is 2,800 kg/ha, the whole gross production of 39 ha would be necessary to buy just one of these machines; in many areas in Spain, in years like this past one, the production has been based on more than 100 has.

EQUIVALENCES

1 liter of diesel oil = 4 Kg of wheat

74.2 l. Consumed per Ha. = 297 Kg. of wheat

1 moldboard five furrow plough = 110,000 Kg. of wheat =harvest of 39 Ha.

I consider that, in many cases, these tools are not only unnecessary but that they make agriculture economically unfeasible and are agressive weapons which destroy the soil, produce «lumps» and are used by what I call «agricultural terrorists».

Another measurement index determining production costs is the number of horse power used per ha cultivated. At present, this index in Spain is 2.6 HP/Ha and in my case it is lower, i.e. 0.70 HP/ha. The tractor I use to do all the tasks has an HP of 130, it is six years old and has kept its original tyres (having clocked up 4,550 hours of work).

On my farm (480 has.) the savings in mechanical work time per hectare compared with conventional agriculture is 3.20 hours/Ha. (1,536 hours per agricultural season), the reduction in the costs of fuel being 39 l/ha., (18,720 liters per season, which means 11,591,200 pesetas = 9,563.30).

SAVINS IN DIRECT SOWING COMPARED TO CONVENTIONAL SOWING

SAVING / HA.	*SAVING /SEASON*	*SAVING / NATIONAL*
3.20 hours of work	*1,536 hours*	*60,800.000 hours*
39 liters of diesel oil	*18,720 liters*	*741,000.000 liters*
30 u.of P	*14,400 u. of P.*	*570,000.000 de u. de P.*
15 u. of K.	*7,200 u. of K.*	*285,000.000 de u. de K.*
15´03 _ in fertilizers	*5 _*	*285,500.000 _*

Apart from this savings in fixed expenditure, in the plots with over eight years of direct sowing, I have managed to drastically reduce the supply of fertilizers. In some cases, no potassium has been necessary for four consecutive years, two without using phosphorous with no production differences being noted. I consider that I therefore save 50% in this fertilizer, a total of 2,500 pesetas/Ha. (1,200.000 Ptas./ season) or, equal to 45 units of phosphorous plus potassium/ha.

I should not like to be interpreted as suggesting that in Direct Sowing it is not necessary to use chemical or mineral fertilizers. It is just one more fact related to my farm, with the specific conditions in my land and with my personal way to cultivate regarding alternatives, plant covers, etc. In any event, if fertilizers are reduced, this is done on the basis of the corresponding soil analysis or the comparative results of harvests.

Let us again suppose that all Spanish farms opted for the Conservation Agriculture technique and that these data were transferable, the savings originated would be the following: 73.07% of HP, 61 million working hours, 741 million liters of fuel, 855 million units of phosphorous and potassium, with a total of 47,500 million pesetas.

These very significant macrofigures are the result of applying data from the 480 has. of my farm to the 19 million has. of cultivated land in this country. I do not therefore claim to put any more worth on them than the statistical one,but there is no doubt that, even with all the imaginable nuances and variables, these figures should be taken into account so that on our farms we will apply savings criteria and can increase their profitability. For me, there is another interpretation, which is that many manufacturers and suppliers whose business is based on selling products to farmers would not be able to do so if this manner of sowing were to become generalized. These probable vested economic interests which are against Direct Sowing are preventing its spread, in the same way as there may be some interests in favour of it; both, of course, are equally legitimate. However, I should like to see both those against and those in favour sitting down with us to discuss the objective data of Conservation Agriculture results.

I must begin by admitting that not always are the desired effects obtained. I have had some failures and these have taught me what not to do in order to obtain good harvests. I also know some farmers who have failed for different reasons, in some cases because they have not applied the Direct Sowing technique but simply tossed the seed onto the soil without any technical or professional criterion but with the unique aim of receiving agricultural subsidies. In other cases, some colleagues, due to their ignorance of some of the technique's basic aspects, have applied it in the wrong way, or some whose soil is a very difficult one will have to wait for results on a longer term. Similarly, some servicing firms have done the work hurriedly without paying the necessary attention required by this manner of sowing. It is my belief, therefore, that these failures are not attributable to Direct Sowing as a technique but, probably, to their erroneous application.

Direct Sowing is not a panacea which can be explained by an instruction handbook guaranteeing success in all and every one of the farms where it is applied; among other things, because there is an infinity of differentiating factors between these farms, making it impossible to apply literally the same standard in all of them. What undoubtedly can be of use to all the farmers in the world is one same rule, that of understanding and assuming the «Conservation Agriculture concept», which would the first, indispensable condition for applying the technique satisfactorily.

And in order to do so, we have to believe in it and, in the event of a failure, we have to analyze where the error was and try to correct it in order to reach perfection. This is only achieved by putting some enthusiasm into one's work and with time and practice.

Farmers who, through being bad professionals have failed with conventional agriculture, will have difficulties in obtaining positive results. So will those farmers whose only and exclusive objective is to reduce costs without any other considerations, as the correct interpretation of Conservation Agriculture demands being generous with the soil, at least covering it with hay, and not selling it or using it for cattle, and if necessary sowing plant covers for it. As one of the world's leading experts in this technique, Manoel Pereira, has said: «hay for the soil, grain for mankind».

From my experience I consider, however, that Conservation Agriculture should be contemplated within a plan for the integrated improvement of our farms and perhaps as a starting point for Precision Agriculture, which implies rationalizing all the aspects having an influence on a more logical and futuristic agriculture.

One more fault in traditional agriculture has been the non rotation of crops. In this country there has been a generalization of the consecutive sowing of cereals in dry lands but this single crop, in addition to hampering the control of weeds, has caused insect species which were originally created to facilitate crop diversity to become pests. The fact of having abandoned having an appropriate alternation of crops has hindered the proliferation of certain species, both plant and animal, with negative consequences on the biodiversity in general and for the availability of varied vegetable proteins for the manufacture of feeds, which have been replaced by the feeds of animal origin that are causing so many problems in cattle health.

Moreover, appropriate crop rotation means the diversification of risks and is, in my opinion, an advisable practice for any farm. In the case of Direct Sowing, it acquires special relevance and has great added advantages, which are so important that they may determine the sucess or failure of the technique, for various reasons. One of them is that it facilitates the control of some biennial weeds and of certain gramineae by being able to treat them with low herbicide doses. I suspect that the consecutive sowing of one same crop may accentuate the risk of pests and endemic cryptogramic diseases, especially in land with a profuse plant cover. By alternating crops we will reduce these risks because a certain biological control is produced and, at any rate, as the problems are different in different crops it is easier to solve them.

At the same time, the extraction of nutrients and water is different and at different levels so that these reserves in the soil can be maintained in equilibrium. Crops with a greater root volume create a fabric in the soil, which improves its physical structure, those with tap roots break up the tilling earth and transfer to lower levels the fertilizers applied to the surface, and the legumes fix nitrogen in the soil. The plant residues contributed by the different species leave a more heterogeneous and beneficial plant cover in our fields because of its distinct content in nutrients which restitute our soil and which, due to their different carbon-nitrogen relation, take a shorter or longer time to transform. These changes in stubble fields make sowing tasks easier, but they basically save us from having the problems that may arise from the risk of generating allelophathies, which is a question of vital importance, especially when sowing is done on plant covers or stubble of the same species as the crop being established next. In any case, to prevent these allelochemical effects, the decomposition of these plant residues should be aerobic and should be prevented, as far as possible, from coming into contact with the soil. For this, reaping should be done by leaving the stubble as high as possible, scattering grain remains and hay homogeneously all over the plot, preferably unthreshed. As I have said, I have heard about some negative effect of this very important aspect,but this would probably need more technical explanations than I am capable of giving.

Another aspect is that the fact that in Spain only one yearly harvest is obtained, with exceptions, means that the profit margins obtained per hectare are minimal, which obliges us to sow over a large surface in order to obtain sufficient profits. Alternating crops with different vegetative cycles enables us to extend the sowing season and adapt each crop to the appropriate time for it, with a much longer real sowing time available.This added advantage of this cropping plan can only be carried out with the Direct Sowing technique, which enables us to use up all the working time available for sowing in taking advantage of the favourable conditions of the soil and not waste time in the prior tilling required by conventional agriculture.

A good sowing plan as a complement to Conservation Agriculture has permitted me personally to devote time to attending congresses and participating in courses and lectures in order to receive more professional training.

In short, I cannot conceive any feasible agriculture based on monoculture and not considering any adequate alternative. Crop rotation may be different in every area and every farm but it is, in all cases, advisable in spite of the difficulties which may arise,both in marketing and in its adaptation to certain climates, latitudes, etc.

FROM AN ENVIRONMENTAL POINT OF VIEW

I believe that the bad productive habits in industry have mainly been to blame for the deterioration of our ecosystem, but I keep wondering to what extent modern agriculture is also responsible for this degradation process.

As already mentioned, Conservation Agriculture prevents, to a great extent, erosion problems, permitting a lower consumption of fuel and of chemical fertilizers.

Erosion is not only a problem of fertile soil loss but it also has pollutant effects caused by runoffs from rivers and streams,transporting chemical waste. Those of us living in the country remember with some nostalgia when we used to catch now non existent fish and crabs in our rivers.

Any reduction in fuel and fertilizers is not only an economic question but signifies a parallel diminution of their impact on the environment.

In Direct Sowing, the use of herbicides to replace the tilling prior to sowing becomes necessary. There are many not always harmless or degradable products on the market for this use. The use of these chemicals should be rationalized in order for us to respect our conservationist values.It is therefore fundamental to have information on their effects and to select those whose formulation controls weeds without interfering with the soil by using minimal doses. For instance, mixing products to use a smaller amount of active material will give better synergisms.

Contrary to what might be thought, I maintain that in Conservation Agriculture, the medium-long term use of herbicides is lesser than in conventional agriculture. With appropriate crop alternation, with the main pre-sowing treatment at the right time and by using the most suitable product, the infestation of weeds, both annual and biennial, is inferior; in my case, even, in some of the plots, I have not needed to use herbicides prior to sowing.

The fact that the soil is not disturbed by mechanical tilling signifies not burying any seeds which might be on the surface. With intensive tiling, it is possible to arrive at the removal of the emergent weeds but the seeds lying in a latent state in the subsoil simply change places and are situated in conditions which encourage their germination.

Conservation Agriculture is based, as has also been mentioned, on maintaining stubble and plant covers in the soil, resulting in several environmental advantages such as, for instance, the absorption of carbon from the atmosphere.

Stubble incineration

The *burning of stubble,* an unfortunately generalized activity in this country, is incompatible with this technique. Harvest residues should not be removed in this way, which ignores the harm being done to a patrimony belonging to all of us. I am guilty of having done so in the past, following orders from my father, and I am left with the sensation of having committed a crime. I am sure that my farmer colleagues who use this vile practice are unaware that they are destroying the soil, annihilating living beings and polluting the air as well as, in time, bringing about their own ruin and that of future generations. This activity has the added risk of causing other more serious disasters such as fires in forests, woods, etc.

I have been saying for some time that in lands with intensive tilling and in which the hay remains are burnt, there is an excess of iron from the wearing out of the ploughs, and a deficit in organic matter and nitrogen, very little potassium and the only phosphorous remaining is that used by the farmer to burn the stubble.

It may be that *society* finds itself excluded from the profitability offered by Conservation Agriculture to the agricultural sector, but it has every right to demand that this activity should affect the common environment as little as possible. For that reason, it is not easy to defend our claims for agricultural subsidies if we do not assume our responsibility for the influence exercised by us in a large part of the territory. We farmers should, in all respects, be worthy of receiving agricultural subsidies, politicians should justify them and society should consent to them.

In recent times, those responsible for *designing agrarian policies* have become more sensitive to environmental topics, but in actual fact there are no incentives which would be necessary for this sector's professionals to contemplate practising Sustainable Agriculture.

In Spain some norms contemplate the possibility of obtaining certain agroenvironmental aids but these have hardly ever been applied.

On the contrary, the Traditional White Fallow Lands Regulation and the Environmental White Fallow Lands contradict the idea of this agricultural system.

But there is, above all, a norm in the form of a law, which, because of its origin and ancestral connotations, is somewhat incomprehensible nowadays. I refer to the so-called «Law for Use of Agricultural Resources» by which owners of agricultural lands are obliged to give them up for cattle grazing. This grazing limits the possibilities of adequately carrying out Direct Sowing on our lands for many reasons, such as it clumps the soil, flattens stubble, may transport pests and weeds, reduces the effects of herbicide application, etc. The authors of this Law have tried to justify it, by its title, because they presume that the plant residues and plant covers on our land are resources which are only fit for cattle.

They apparently are unaware of the fact that the best use and activity we can give to our stubble and plant covers is for our own soil, so that it can be physically and biologically regenerated.

I believe that all agricultural legislation is a determining factor in the evolution, positively or negatively, of agriculture and I would therefore ask legislators to assume their responsibility too.

In my introduction I said that we farmers carrying out Conservation Agriculture have a particular *philosophy of life*; I do not really know why but Conservation Agriculture creates some kind of addiction, it gets us hooked and, above all, it builds up our hopes and has generated a permanent urge to improve which is carried over to our everyday lives, coinciding with what Jenofonte said: «*Agriculture is at one and the same time a source of satisfaction, a means to increase one's wealth, a means to accustom one's own body to all that is appropriate for a free man*».

One consequence of this way of understanding life and relationships with our farmer colleagues has been the creation of the first non-profitmaking Direct Sowing association in Europe, A.B.U.L.A.C. (Burgos Association of Conservation Tilling), which serves to transfer and acquire experiences from farmers worldwide, and with which we can improve our work, increase our social relations and increase our quality of life.

I should like to take this opportunity to thank the FAO and all the organizers and sponsors who have made this event possible for having had the honour of expressing my preoccupations here;and offering our collaboration and unconditional aid to agriculturalists everywhere.

F. TEBRÜGGE & A. BÖHRNSEN

FARMERS' AND EXPERTS' OPINION ON NO-TILLAGE IN WESTERN EUROPE AND NEBRASKA (USA)

Justus-Liebig-University
Institute of Agricultural Engineering, Braugasse 7, 35390 Giessen, Germany, (phone: +49-6419937202; fax +49-6419937209,

INTRODUCTION

The cycle of economics is increasingly influenced by an extensive globalisation and the same is accurate for the European agriculture. In the sense of the competence of competition drastic measures are necessary for the decrease of costs of production on farms with regard to constant fertility and yield. In Europe, at the moment, the no-tillage system is spreading slowly in spite of many varied scientific results, about positive effects on the ecosystem of the soil and with regard to the income, is this in USA, Canada and South-America already on more than 30 Mio. ha with increasing tendency practised.

METHOD

Among scientists, advisers and farmers were experiences and results of experiments concerning the applicability of no-tillage crop production on European level reported and discussed in connection with an EU-financed concerted action (AIR 3-CT 93-1464).

A main emphasis of the concerted action was to carry out a common questionnaire with the participation of 6 members of the EU and Switzerland. The aim was to register and to analyse the experiences of farmers (n: 111) as applicators of the no-tillage system (total about 53.000 ha arable land) as well as the attitudes of experts (n: 176) from 9 EU-countries and Switzerland, working in the fields of research, advice and industry (tab. 1 and 2).

Questionnaires on farmers were carried out before on a national level in France, Spain and Italy (EC-Workshop Proceedings I, III, IV). The structure of the answers is largely identical with those of the colleagues from the participated 6 EU-countries here. In connection with a research in Nebraska (USA), the possibility was additionally

L. García-Torres et al. (eds.), Conservation Agriculture, 69–78.

offered, to register the opinions and experiences of 50 no-till farmers (with total 20.000 ha arable land) and 18 experts about the application of no-tillage under the conditions there, under the same questionnaire, in order to compare the results in main topics with those reached in the EU-countries.

Table 1. Basis dates of no-tillage-farms (n: EU 111, USA-NB 50)

	Cultivation structure of crops in % of the arable land					
	AF-ha	*% NT*	*Cereals*	*Maize*	*Leguminosae*	*other (mostly)*
CH	20	33	46	16		38 (sugar beets/rape/potatoes)
D	394	65	60	3	5	32 (sugar beets / rape /alfalfa)
DK	409	44	73		2	25 (rape /grass/alfalfa)
GB	519	23	65		9	25 (rape)
I	646	15	23	20	15	42 (sugar beets /sunflowers)
NL	14	25		(75)		25 (potatoes)
P	380	30	63	8		30 (sunflowers)
USA-NB	397	69	8	50	34	13 (Sorghum/alfalfa)
ø EC	340	34	58	12	8	(sugar beets or rape ca.20%)

Table 2. Basis dates of the experts in % (n: EU 176, USA-NB 18)

Experts' fields of work

Advice	science	Industry	Other
52	26	7	15

Experts' areas of competence

soil tillage	crop production	Engineering	Fertilisation	plant protection	Economy
57	47	42	29	22	21

Sources of experts' experience

Literature	own experience	from farmers	from colleagues	Own scientific research
73	59	54	36	35

RESULTS

A comparison with the answers from EU-farmers with a farm size of 340 ha in average (14–646), who apply the no-tillage system on 34% of the arable land and those of the experts shows, that concerning the important arguments for the no-tillage system, these are mainly agronomically. The percentage of answers shows, that the effects of reducing working time (98%), costs (98%), fuel consumption (86%), horse power requirement (79%) and a higher trafficability (88%) are considered as very important (fig.1).

Motives concerning soil and environmental effects of no-tillage are of less importance for the farmers, according to a frequency of 50% each for arguments as less soil erosion, higher soil biological activity, more earthworms, water conservation, less nitrate leaching and a higher infiltration rate. As the percentage of answers (between of 61% and 88%) indicates, these motives are estimated much higher by the experts.

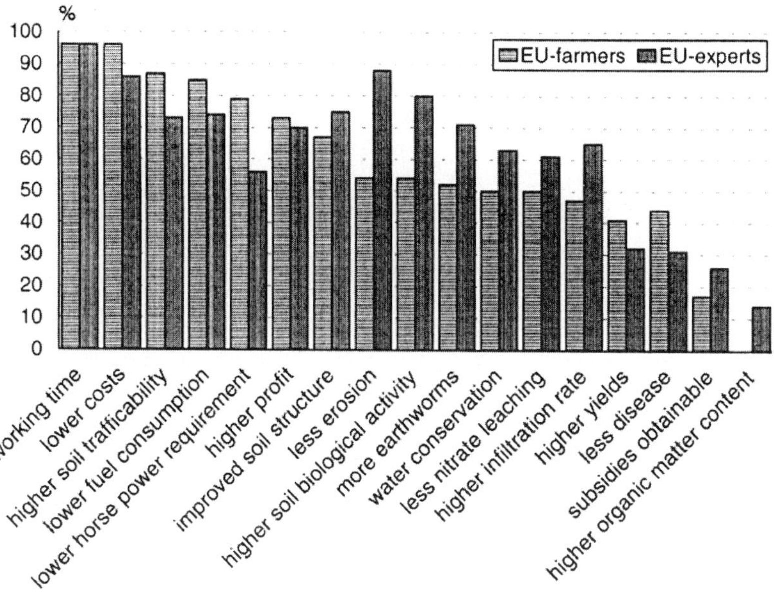

Fig. 1. Structure of motivation of no-tillage farmers and experts in the EU concerning application of no-tillage

This is nearly the same level which was given to agronomic motives by the experts, here as well the effect of reducing work time is a valued argument (97%).

The structure of motivation of the NB farmers is in contrast to the EU farmers much more emphasised on soil and environmental criterions, as water conservation, improved soil structure, less erosion, less nitrate leaching, and higher organic matter content. The NB experts' answers concerning this subject are largely identical to those of the NB farmers as well as to the EU experts (fig.2). Higher profit as a motive is, in contrast to both asked groups from the EU, as well as to the NB experts for NB farmers with a frequency of 93%, a decisive criteria for the application of the no-tillage system.

The experiences and opinions concerning the influence of no-tillage on the change of inputs of fertiliser, herbicides, fungicides and other plant protection (slugs, mice) compared with conventional tillage differ very much, especially in case of herbicides. Whereas the majority of the farmers (64%) estimate a herbicide input of the same and decreased level, 70% of all experts give an increasing input.

The level for all other pesticides as well as nitrogen fertiliser is given with a high frequency of 65% on average of the EU-experts and 80% of the EU-farmers similar to the input of the conventional tillage. Whereas 2% to 18% of the farmers' naming apportion on an increase and 8% on a decrease of agrochemicals with the exception of herbicides, in dependency of the products 43% to 27% of the EU-experts are of the opinion, that no-tillage is in comparison to conventional tillage connected with an increase of agrochemicals.

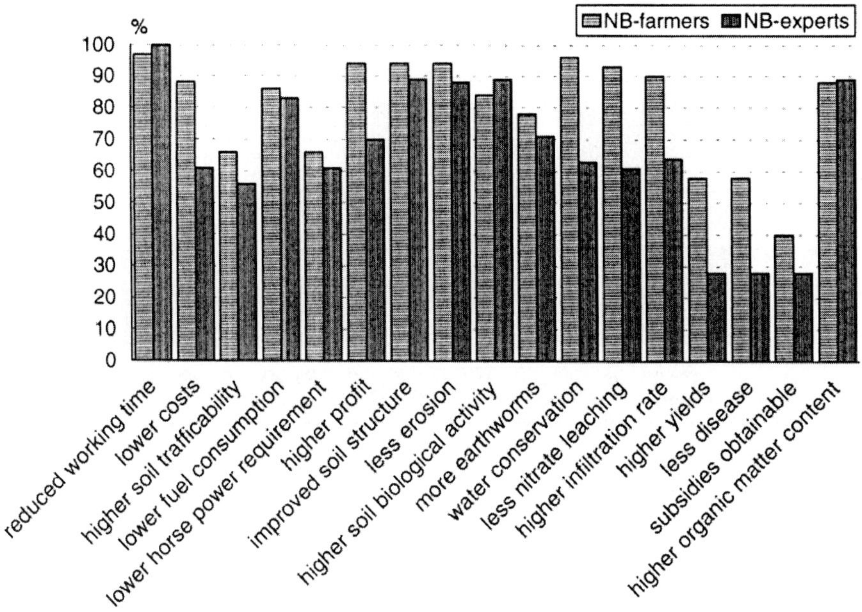

Fig. 2. Structure of motivation of no-tillage farmers and experts in Nebraska (NB)-USA concerning application of no-tillage

A comparison of the answers of the EU-farmers with those of the EU-colleagues from Nebraska leads to a diametrical picture concerning the input of herbicides. Whereas 56% of the EU-farmers give an unchanged input, the no-till farmers from Nebraska answer contrary. Only 34% give an unchanged, 6% a decreased, but 57% an increased input of herbicides.

Concerning the experiences and opinions to the yields in case of no-tillage, there are as well visible differences in the assessment. 54,2% of the farmers estimate the yield, added up about all crops, at an unchanged, 21,1% at a decreased and 21,9% at an increased level. In contrast to this 36,6% of the experts are of the opinion, that the yield level is the same, 12,8% increased, but 47,2% a decreased in case of a long-term application of the no-tillage system (tab.3).

Referring to the crops 62% respectively 69% of the experts expect the yields of cereals respectively leguminosae to be at the same and increasing level under no-tillage conditions in comparison to conventional tillage. However, decreased outputs of yields up to 17% are expected by 74% of the experts in case of sugar beets. 13% yield decrease in case of rape and 15% in case of maize are expected by 58% respectively 36% of the experts.

This topic is investigated by the Justus-Liebig-University (JLU) in Giessen since 18 years on five pedogenetic different locations. A comparison of the experts' generally estimation of yields with the results from literature on European level and with experiments of the JLU reveals, that the experts view the influence of no-tillage on yields very pessimistically. This view is not congruent with most of the no-tillage farmers'

Table 3. Farmers' experience and experts' opinion in EU countries about yield effected by no and conventional tillage

	Farmer	*Expert*	*Farmer*	*expert*	*Farmer*	*expert*
Yields	***Decrease***		***at the same level***		***increase***	
Cereals						
% of answers	*18,7*	*37,9*	*69,2*	*49,7*	*12,1*	*12,4*
yield %	*12,9*	*11,2*	*+/- 0*	*+/- 0*	*12,4*	*10,7*
Maize						
% of answers	*27,0*	*35,7*	*32,4*	*35,7*	*40,5*	*11,9*
yield %	*20,7*	*14,7*	*+/- 0*	*+/- 0*	*19,6*	*13,2*
winter-rape						
% of answers	*31,8*	*57,8*	*50,0*	*35,9*	*18,2*	*6,3*
yield %	*13,9*	*13,3*	*+/- 0*	*+/- 0*	*13,3*	*13,3*
Leguminosae						
% of answers	*16,2*	*30,8*	*65,2*	*46,0*	*16,7*	*23,1*
yield %	*18,6*	*14,5*	*+/- 0*	*+/- 0*	*12,5*	*11,7*
sugar beets						
% of answers	*10,0*	*74,0*	*60,0*	*15,5*	*30,0*	*10,3*
2ôield %	*13,9*	*17,2*	*+/- 0*	*+/- 0*	*10,0*	*11,0*
TOTAL						
% of answers	*21,2*	*47,2*	*54,2*	*36,6*	*23,5*	*12,8*
yield %	*14,2*	*14,2*	*+/- 0*	*+/- 0*	*13,6*	*12,0*

experiences. They give even in case of crops, which are supposed to be less suitable for no-tillage, as sugar beets, maize and rape with a frequency of 90%, 73% respectively 68%, unchanged respectively increased yields in comparison to conventional tillage.

The results of the question about the reasons for the relatively low acceptance of the no-tillage system in the practice, indicate, that the applicants regard the motives of their colleagues especially in insufficient advice (73%), insufficient experience (65%), and high purchase costs (61%). With a decreasing frequency (56-50%) they consider fear of less yields, problems with weeds, higher management requirement and a lack of analysis of the production costs as reasons for the farmers. With 40% each are insufficient scientific results and not a useful drill technique, given as counterarguments. A less importance (32-23%) is attached to the reputation of neighbours, problems with pests and the consent of the landlord (fig.3).

In contrast to this, the experts consider the motives of the farmers for the relatively low acceptance of the no-tillage system especially in the fear to have less yields (93%) (confer the experts' estimation of yields). The experts assume tradition, insufficient knowledge by the farmers (88% each), and that they are less prepared to take high risks (85%) as arguments for the farmers not to use no-tillage. Less important arguments are insufficient analysis of the production costs (76%) and high purchase costs (73%), insufficient technique (68%), insufficient scientific results (65%) and the lack of economic necessity (62%). Only a secondary importance is given to the reputation of neighbours (46%) and finally the objection by the landlord (25%) (fig.4).

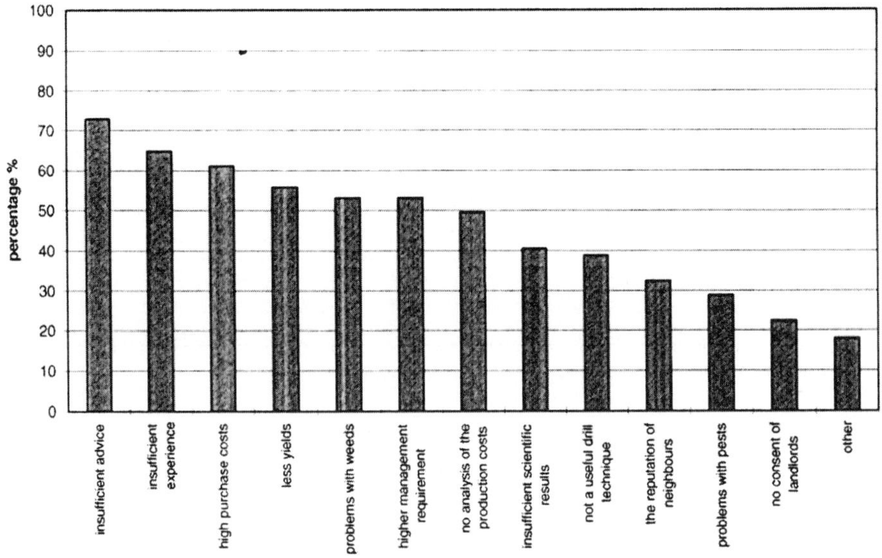

Fig. 3. Important arguments for other farmers not to use no-tillage. Farmers' answers, percentage [%] of all farmers' replies

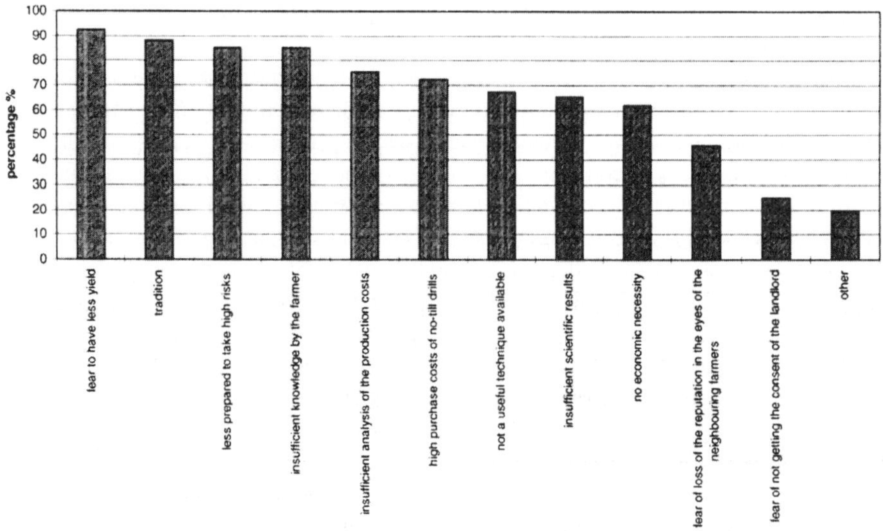

Fig. 4. Important reasons for the relatively low acceptance of no-tillage. Experts' opinions, percentage [%] of all experts' replies

From the experts' answers from Nebraska reveals a different evaluation in the structure of motivation. Priority is given to arguments as tradition, fear to have less yields and high purchase costs (72% on average). In the middle ranks the reputation

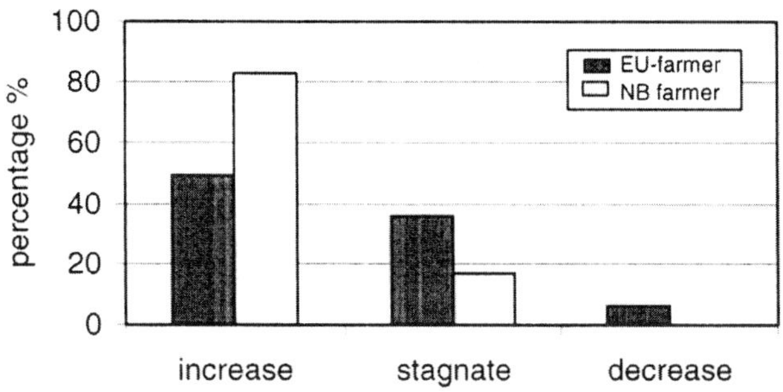

Fig. 5: The income will .... in case of long-term application of no-tillage. Farmers' answers percentage [%] of all farmers' replies

of neighbours, the acceptance by the landlord (56%). The lack of economic necessity and insufficient analysis of the production costs are selected with a frequency of 44% each. Insufficient scientific results (33%) as well as insufficient techniques (22%) play in comparison to the EU-experts (68%) only a secondary role.

The concluding question about the long-term effects of the no-tillage system on the income (fig.5) is answered by 54% of the EU-farmers with an increase, 39% with the same level. 7% of the no-till farmers are of the opinion, that the income decreases in comparison with the conventional tillage. Their colleagues from Nebraska, with a much more experience in case of no-tillage, assess the influence on the income more optimistically. 83% assume an increasing and 17% an unchanged income.

The EU-experts estimate 34% of the arable land in their countries, approx. 23 Mio. ha, suitable for no-tillage crop production. The USA-NE-experts estimate with 75% of the arable land, approx. 6 Mio. ha (tab.4) a much higher area in their country which is suitable for no-tillage.

Table 4: Percentage of arable land suitable for no-tillage crop production, assumed by experts.

Nationality	*Arable land[1] (1.000 ha)*	*Arable land suitable for no-tillage (%)*	*Arable land suitable for no-tillage (1.000 ha)*
CH		49,0	
D	11.805	37,1	4.381
DK	2.510	44,0	1.104
E	13.954	45,5	6.349
F	18.302	31,8	5.820
GB	5.949	17,8	1.058
GR	2.250	37,0	833
I	9.030	22,5	2.032
NL	899	26,3	236
P	2.326	45,0	1.047
Total/average EU	67.025	34,1	22.859
Nebraska USA	***8.000***	***75,0***	***6.000***

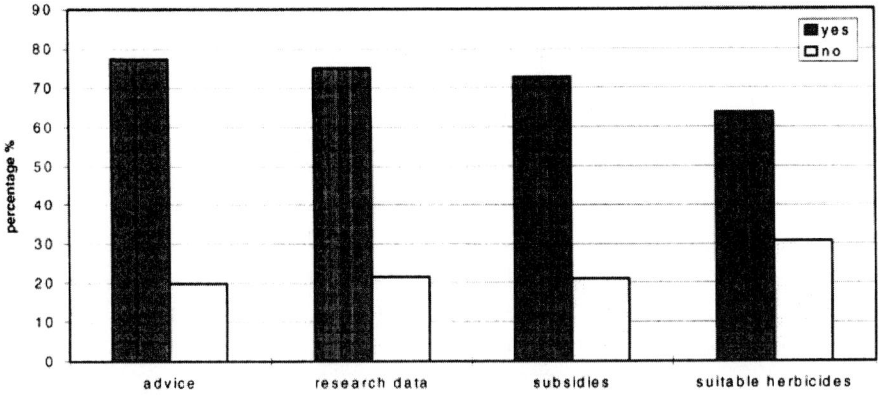

Fig. 6. More farmers would use no-tillage if ... are available. Experts' opinions percentage [%] of all experts' replies

The question, if more farmers would use no-tillage in case of specific advice for the farmers, more available research data on no-tillage, available subsidies and or more suitable herbicides, is answered - in average of the named premises (fig.6) - with 23% negatively.

72% of the EU-experts and 87 % of the NE-experts from the USA are of the opinion, that such an availability, especially in the field of advice , would increase the application of no-tillage.

SUMMARY

An Europe-wide survey as well as in Nebraska (USA) amongst no-till farmers reveals that the most important factors for adoption of no-tillage are the reduction of costs and working hours on the one hand and the awareness of a sustainable utilisation of the soil and the environment on the other. Obstacles for a widespread adoption of no-till are considered the higher management requirements, insufficient advice and experience with the new technique, higher cost for seeding equipment and the fear of lower yields.

Experts working in the field of rural extension and research have a much more negative opinion about no-till as compared to no-till farmers. Although they agree on the ecological and economical benefits of no-till. The so-called experts expect increased problems with weeds, diseases and pests, soil compaction and lower yields. For them the main reasons for the low acceptance of no-till by the farmers are the fear of lower yields, maintenance of traditional methods, insufficient technical knowledge and analysis of the real production costs.

The contradiction between research results and the opinion of "experts" and farmers indicates that it is necessary to overcome the apparent impediments for the adoption of no-till in order to achieve the goal for a reduction of environmental pollution and increase of the production efficiency.

Therefore some of the reasons for the foundation of the German Society for Conservation Tillage (GKB) and his membership in the European Conservation Agriculture Federation (ECAF) are the results of the 4 no-tillage EC-workshops as well as the contradiction of the replies of the questionnaire by farmers and experts in Europe and Nebraska and last not least the German Government law for soil protection in 1999.

ACKNOWLEDGEMENT

This report is a part of an EC-Concerted Action (AIR 3-CT 93-1464 and has been supported by the Commission of the EC, Directorate-General for Agriculture DG VI-F-II 3.

REFERENCES

Tebrügge, F., Böhrnsen, A. (eds.) (1994, 1995, 1996, 1997, 1998). Experience with the Applicability of No-tillage Crop Production in the West-European Countries. Proceedings of the EC-Workshop I, II, III, IV, Wissenschaftlicher Fachverlag Dr. Fleck Giessen ISBN-Nr.: 3-930600-16-1 (1994); 3-930600-46-3 (1995); 3-930600-69-2 (1996); 3-930600-95-1 (1997).

Tebrügge, F., Böhrnsen, A. (eds) (1998) Final Report of Concerted Action No AIR 3 CT 93-1464, Review Papers, Summaries & Conclusions, Fachverlag Köhler Gießen ISBN-Nr.: 3-922306-30-6 (1998).

CD-ROM-NO-TILL Databank (1998) Experience with the Applicability of No-tillage Crop Production in the West European Countries, Institut für Landtechnik, Braugasse 7, 35390 Giessen, Germany.

J.N. LANDERS, H. MATTANA SATURNINO, P.L. DE FREITAS & R. TRECENTI

EXPERIENCES WITH FARMER CLUBS IN DISSEMINATION OF ZERO TILLAGE IN TROPICAL BRAZIL

Brazilian National Science Council Scholar and Executive Secretary, Association for Zero Tillage in the Cerrado, MUDB Conjunto 9, Lote 5, postcode 71680-090, Lago Sul, Brasília, DF, Brazil

President, Association for Zero Tillage in the Cerrado, SCRLN 712, Bloco C, Loja 18, Asa Norte, Brasília, DF, Brazil, postcode 70.760-533

Researcher Embrapa Soils Centre, Rio de Janeiro, RJ, Brazil.

Training manager, Association for Zero Tillage in the Cerrado, SCRLN 712, Bloco C, Loja 18, Asa Norte, Brasília, DF, Brazil, postcode 70.760-533

Since 1992, a network of 42 «Clubes Amigos da Terra»(CATs) and similar farmer organizations has been developed in the Cerrado tropical wet/dry savannah region of Brazil. The essential characteristics of these are that they are solely for the promotion of Zero Tillage as a sustainable system of agriculture. They are apolitical, non-commercial and farmer-run. The operational basis of the CATs is farmer-to-farmer exchanges of experiences on a regular basis later on-farm research and technical training. There is an effective NGO network for ZT promotion and development in Brazil, with backward and forward linkages.

INTRODUCTION

The first farmer club to promote ZT in Brazil, «The Earthworm Club», was constituted in Ponta Grossa, Paraná state (PR) in 1979 (Borges 1993). The same author cites that in 1982, this model was followed in Rio Grande do Sul state (RS), where today there is a state network of some 40 «Clubes Amigos da Terra»(CATs). In the «Cerrado» region (wet/dry tropical savannah) the first CAT was formed in Jataí. Goiás state (GO) in 1992, the same year that the Association for Zero Tillage Farmers in the Cerrado (APDC)was founded. Since that year, 20 clubs have been founded in 5 tropical states and are affiliated to the APDC; another ten are in the initial stage of informality. Twelve other farmer

L. García-Torres et al. (eds.), Conservation Agriculture, 79–84.

foundations and associations also interact with APDC,. which acts as a regional support agency. and is affiliated in turn to the National Federation for Zero Tillage in Crop Residues (FEBRAPDP). This mode of farmer-run NGO activity has been extremely effective in promoting ZT in the tropical region, which now has an estimated 5 million ha under ZT, over 95% of which was adopted since APDC's foundation.

ESSENTIAL CHARACTERISTICS OF THE CLUBES AMIGOS DA TERRA

For the CATs, ZT represents the high road to agricultural sustainability at high production levels (Hernani et al., 1996 and Muzilli et al., 1993), combined with environmental conservation (Landers, 1996). The principal objectives of the CATs are the promotion and development of the Zero Tillage system, but they also encompass the principles of sustainable farming and environmental responsibility. The officers of a CAT have a majority of practising farmers, but the CATs are open to anyone interested in ZT agriculture. They are usually restricted to one or two municipalities, are not-for-profit, non-commercial and apolitical. The presidents of the CATs constitute the governing council of the APDC, giving effective control of the association's priorities to the farmers.

The adoption of ZT requires a certain amount of courage, first to risk implementing a technology with a totally different logic from the conventional paradigms and second to absorb the criticisms of the neighbours. Once the farmer has passed through the portal of ZT adoption, his whole set of paradigms changes and he begins to see the soil as a living entity and that farming in harmony with nature is a lot easier than fighting it with implements imported from other climes. This improves profitability, which then engenders and empowers environmental responsibility. The whole set of new experiences to be faced generates a positive climate for collaboration amongst ZT farmers, to which the CATs are a response. The philosophy of regarding problems which arise in ZT as obstacles to be overcome through persistence and creativity permeates the thinking of the ZT farmer, who has come to understand how much he would lose with a return to Conventional Tillage as a supposed corrective measure for diseases, pests or some other motive. The increased efficiency brought by ZT has been the motivating force for the increasing professionalization of Brazil's farmers.

MODUS OPERANDI

The first clubs in the Cerrado region were founded by «Gaucho» farmers who had migrated to the region when it was opening up, 20 years ago, bringing with them knowledge of the CAT principles from RS. APDC promotes CAT formation through telephone conversations with interested parties in far-flung parts of the region, informal meetings, lectures and PRA activities. From the initial contact, some clubs take five years to achieve the critical mass to begin regular activities. The advent of a leader or core group of leaders, prepared to devote some of their time to organising the CAT's activities is the crucial factor. This has to occur naturally for the group to have continuity. Commercial interests at this point will tend to subvert the long-term success of the CAT, although agribusiness support with no strings attached is always welcome and useful. Good companionship is a byword for the relations between members of a CAT, often cemented through communal «Churrascos» (barbecues) or other activities, which also involve family members. There is also a natural tendency to reduce chemical inputs and encourage biological controls, such that the ZT system merits being viewed as halfway house to organic farming and not its antithesis.

APDC publicises the activities of affiliated CATs in its quarterly bulletin, also making available a practical publication on the principles of organization and operation of the clubs with the title «CAT-Kit». This also includes instructions for formal registration of the articles of association required to become a legal entity. The foundation of a CAT usually happens with a few interested farmers deciding to discuss mutual problems in ZT adoption. After several meetings, the group grows and meetings become regular. At this point, to obtain support for more ambitious activities, a formal organisation becomes necessary. However, when CATs are founded precipitately, in euphoria, without the initial informal phase of mutual assessment, the chances of nominating ineffective officers is high.

The CATS meet regularly, usually once or twice a month at the same place and time, except for field activities. One of the main principles is that of horizontal communication, farmer-to-farmer, to transmit different experiences, giving equal emphasis to mistakes, successes and problems. This means of technology transfer is the most cost-effective since it occurs automatically at zero cost. The principle of pooled experience leads to faster technological development in the group than would be the case for the individual. There is also a *de facto* free-in-free-out principle in technology generated, which is treated as a common good. Activities progress from the initial adoption phase, where all is a novelty, to a consolidation phase of technology adjustment and a third phase with group and individual development of new technology and application of more advanced management methods. Table 1. Illustrates the evolution of different activities over time.

The CAT acts as a local mobiliser of support for farmers, through collaborative activities with agribusiness, local government, farmer organizations (co-operatives, farmers' unions etc.). and government research, training and extension agencies. Getting the right leadership is crucial to success. For this reason an informal initial phase of activities is important for the members to see who fits in which position. Some CATs have attempted to set up buying pools, but this practice was either discontinued or provoked a heavy setback in CAT activity. Admitting commercial activity in a CAT makes one or two firms happy and the rest angry. The principle used by CAT-Uberaba of a broad platform of 12 supporting firms for an on-farm research programme, with no sales opportunities within CAT activities, appears to be a reasonable compromise. In Uberaba, firms were chosen on a first-come-firsst-served basis, which was transparent.

Table 1. Evolution of the activities of a CAT

Adoption Phase	***Consolidation Phase***	***Mature Phase***
Motivation events	*Technical seminars*	*Participative rural appraisal*
Interchanges of experience guest speakers	*Farmer presentations and*	*Guest speakers and farmer presentations*
Farm visits for members	*Field days for all comers*	*Organization of regional ZT events*
Machinery clincs for beginners analysis	*Performance tests for planters, drills, sprayers, etc.*	*Performance indicators and financail*
On-farm demonstrations-liaison with extension	*On-farm testing of new technology – liaison with research*	*On-farm testing in CAT networks-liaison with research*
Technical training for managers	*Specialization courses for managers*	*Advanced management groups*
Operator training	*Operator training*	*Professional CAT management*
		Leadership courses
		Interchanges between CATs

FORWARD AND BACKWARD LINKS

Deriving from its integrating capacity, the ZT movement has been a trigger to technology development and transfer. In the Cerrado region, APDC has acted as a facilitator between CATs, universities, government research and extension services, other government entities and all sectors of agribusiness (suppliers and buyers). Training courses have been executed in conjunction with the Ministry of Agriculture and the state government of São Paulo (450 trainees in 3 years, principally extensionists). The Ministry of the Environment and Amazônia has supported APDC with a program to foster the creation new CATs and support to the whole CAT network. The presidents of the CATs constitute the governing council of APDC. When Brazil's new water law is implemented, on a river basin basis, CAT participation through Municipal Development Committees may be useful, but party politics need to be avoided.

The CATs offer an effective network for on-farm technology development and dissemination of new technology. An example of this is the mobilisation by APDC of the Embrapa Cerrados Centre and a network of CATs and farmer foundations to carry out on-farm testing of over 40 alternative second crop options.

REFERENCES

Borges, G. de O., (1993) Resumo Histórico do Plantio Direto no Brasil. In: *Plantio Direto no Brasil*, Ed. Embrapa-CNPT, FUNDACEP e Fundação ABC. Editora Aldeia Norte, Passo Fundo, RS, Brazil.

Hernani, L.C., Salton, J.C., Fabrício, A.C., Dedecek, R., Alves Júnior, M. (1996) Perdas por erosão e rendimentos de soja e de trigo sob diferentes sistemas de preparo de um latossolo roxo de Dourados (MS). *Congresso Latino-Americano de Ciência do Solo, 13o*. Águas de Lindóia, SBCS. [Proceedings on CD ROM].

Landers, J.N. (1996) O Plantio Direto na agricultura: O caso do Cerrado. In *Gestão Ambiental no Brasil: experiência e sucesso.* Rio de Janeiro: Fundação Getúlio Vargas. pp. 03-33

Muzilli, O., Vieira, M.J., Oliveira, E.L. (1994) Avaliação dos sistemas de Plantio Direto e Convencional em diferentes sucessões de culturas.. IAPAR-Programa Recursos Naturais. Relatório Técnico Anual. Londrina.

ANNEX

A case study – on-farm research programme, CAT Uberaba, Minas Gerais state (MG)

This CAT was formed in 1996, motivated by participation of several farmers from the «Triangle» region of MG in FEBRAPDP's 5th National Meeting of Zero Tillage in Crop Residues, organized by APDC and held for the first time in the Cerrado region. Uberaba is in a region where ZT adoption occurred quite late, amongst relatively advanced farmers. The initial adoption phase was short and, in 1998, the club held a participative rural appraisal to determine priorities for an on-farm experimentation programme addressed to local needs. In this exercise, four sub-groups of 8-10 persons each listed and ranked problems independently followed by an exposé of each group's results and a final consolidated ranking. The sub-groups ensured greater participation of each member. The farmers placed as a condition that all trials be carried out on-farm with their own machinery. At the same time, the researchers required replicated trials in order to further their careers through scientific publications. These two requirements were compatibilized The details of how this project evolved are indicated below.

Objetives : (i) technology refinement for specific local conditions, (ii) on-farm validation of small plot results, (iii) integration between all actors in technology development and transfer, (iv) improved availability of technology for the farmer and (v) support for research and extension activities.

Methodology : (i) Participatory rural appraisal with CAT members and technicians, (ii) synthesis of the limiting factors to better performancce, (iii) development of a research plan by the technicians, (iv) discussion of the plan and re-formulation according to farmers' implementation capacity and required outputs, (v) selection of co-operating farmers.(19 farms, 10 trials on 24 sites), (vi) presentation of the budgeted plan to a meeting of agribusiness firms, (vii) selection of 12 participating firms by bidding process (conditions included rights to participate in field days, presentation of short commercial seminars and divulgation of the firm's participation and non-interference with the choice of research topics).

Implementation: (i) trials were planted with the farmer's own equipment on large plots designed for mechanical harvesting, (ii) weighing of yields was carried out on a truck scale with one ton bulk bags or harvesters with transducer equipment, borrowed from a seed firm, (iii) crops involved were soybeans, maize, cotton and cover crops,

(iv) the field work was supervised by an independent local consultant agronomist paid for this service, (v) the experimental data were collected by research scientists from EPAMIG (Minas Gerais state agricultural research service) and Embrapa Maize and Sorghum centre, (vi) incremental operating costs of the researchers were covered by CAT-Uberaba, (vii) the Triangle farmer foundation located in Uberaba MG acted as treasurer for the programme.

Results: (i) an anual meeting of the CAT evaluates the results and a report is prepared for CAT members and the supporting agribusiness firms, (ii) with three years' results, recommendations are formalised, (iii) demands for additional trials are evaluated.and (iv) field days are held to demonstrate progress and results.

Discussion: (i) the 60 CAT members in 1998 represented a planted area of 43000 ha – today membership has grown to 100 members, showing considerable interest in the CAT's activities, (ii) the principle of the results being freely available is attractive to the sponsors, (iii) using the farmer's planting equipment , means that it is already calibrated for the local soil and surface residue conditions, (iv) the very large samples ensure accurate results, (v) no data have yet been published, but the farmers are following results closely, making their own interim judgements and immediately applying the knowledge obtained, (vi) although this CAT's members are predominantly medium and large mechanized farmers, the results are made available to the state extension service that serves the small farmer and can adapt most of the results for this situation, and (vii) the local mayor has requested the CAT to become involved in a municipal program for small farmers.

Conclusions : (i) this methodology ensures high cost-effectiveness in research by targetting the most pressing problems of the farmers, (ii) for this reason, the private sector has given considerable support, (iii) the mechanism of technology transfer through farmer-to-farmer contacts is the fastest and cheapest and (iv) the social responsibility of the ZT farmer is emphasized by the free availability of information and readiness to collaborate with the dissemination of ZT technology to the small farmer, extending an activity in which APDC has been developing pilot projects with other CATs and pther farmer organisations.

III. INTERNATIONAL NETWORKS FOR CONSERVATION AGRICULTURE

*L. GARCÍA-TORRES, *A. MARTÍNEZ-VILELA
& **F. SERRANO DE NOREÑA[1]

CONSERVATION AGRICULTURE IN EUROPE: CURRENT STATUS AND PERSPECTIVES

* *ECAF, Brussels, Belgium*

** *Environment Directorate, Community of Madrid*

This paper briefly outlines some environmental problems caused by conventional agriculture in Europe and the perspectives for the adoption of conservation agriculture, as a consequence of environmental and economic concerns. Information on the European Conservation Agriculture Federation and its national associations is also given.
Keywords: conventional agriculture, agri- environment, policy, networks

ENVIRONMENTAL PROBLEMS CAUSED BY CONVENTIONAL AGRICULTURE IN EUROPE

There is an obvious interdependence between agriculture and environment. Indeed, in the EU 50.5% of the total territory is agricultural land. In Europe, *conventional agricultural* is still by far the predominant. This is mainly characterised by straw burning and/or removing and aggressive tillage operations (inverting the soil profile by ploughing and disking). Consequently, this type of agriculture has consistent negative effects on soil, water and air quality, global climate, biodiversity and landscape.

The *environmental problems* caused by conventional agriculture are well described in many sources. For example, in Europe, *soil degradation* due to erosion and compaction processes seriously affect nearly 157 million hectares (16% of Europe, nearly 3 times as large as France), with an average erosion rate of 17 ton ha^{-1} yr^{-1}. In Southern Europe this problem is even more severe: 50% to 70% of agricultural land has a moderate to high risk of erosion (average erosion rates of 40-60 ton ha^{-1} yr^{-1}). This has contributed substantially to increase the risk of *desertification* in the most vulnerable areas. The erosion problem has a strong economic impact on the affected agricultural land (53 EUR ha^{-1} yr^{-1}), and off-site on the surrounding civil public

L. García-Torres et al. (eds.), Conservation Agriculture, 87–92.

infrastructure (32 EUR ha^{-1} yr^{-1}). *Water quality* is seriously impaired by conventional agriculture. Soil sediment from eroded agricultural land is by far the most important contaminant of surface water. Pesticides are also largely transported by the run-off adsorbed to eroded soil particles. Straw burning and/ or mouldboard ploughing produce extra *carbon dioxide (CO_2) emissions* to the atmosphere and reduce the potential CO_2 sink effect of the soil, thereby decreasing the organic matter content of the soil (over 50% in the first 20-30 years of cultivation), and thus contributing to global warming. *Biodiversity* is also drastically reduced in conventional agriculture since the tilled soil profile and the bare soil for a long period of time does not provide food and shelter for wildlife at critical times. In a few words, our current agricultural soil management is really unsustainable.

CONSERVATION AGRICULTURE FOR EUROPE IS NEEDED

The adoption of conservation agriculture practices, such as direct sowing or cover crop greatly counteracts the negative environmental effect previously described. In principles, conservationist techniques protect permanently the soil from erosion and degradation leaving the crops residues (stubble) or cover crops (mulching) over the soil. The composition, structure and natural biodiversity of the soil profile is altered as little as possible. Furthermore, less contamination of the surface water occurs, the emissions of CO_2 to the atmosphere are reduced and the biodiversity consistently increases. Generally, the use of herbicides, as part of the conservationist techniques, is much more environmental friendly than in conventional agriculture.

Furthermore, others «new» agricultural modalities, such as organic agricultural or integrated agriculture, will fail to solve the environmental problems caused by conventional agriculture, unless clearly adopt the principals and operations of conservation agriculture in order to protect the soil from erosion and compaction.

The energy saving of conservation agricultural techniques is another important factor to be considered. As an example, in olive crop, an important crop in Southern Europe, the estimated fuel saving is about 60 to 80 litres per hectare, as comparison to conventional tillage. Generally, conservation agriculture reduce the energy consumption and work rate of farming operations in the range of 15%-50% and increase the energetic productivity –i.e. the yield output per energy input- by 25%-100%. In summary, a strong body of scientific and technological research supporting the environmental benefits and agronomic performance of conservation agriculture has been developed world wide, and also in Europe, in the past few decades.

SHIFTING SITUATION IN EUROPE

In the 1990´s, the widespread adoption of conservation agriculture has been consistently increasing in several countries (USA, Canada, Brazil, Argentine, among others) but notably not in Europe. In the EU this can be largely attributed, as compared to the US, to lack of institutional support (environmental subsidies), and in comparison with

Argentine and Brazil to the lack of pressure for saving costs. In both cases the need for technology transfer should also be underlined.

Currently the level of adoption of CA in Europe is low and can be estimated in about 10-15 % of the agriculture surface in Spain and Portugal and of 5- 10 % in France. However, the expectations of adopting conservation agriculture in Europe are very consistent. Several reasons support this statement. EU administration through Agenda 2000 and recent Council Regulations such as 1259/99 (cross compliance) and/ others (1257/99 & 1750/ 1999) have set up the framework to implement at national level more effective agri-environmental measures to protect the soil. For example, in Spain early 2001 a new law (RD 4/ 2001) regulate for the first time state wide subsidies for conservationist techniques such us direct drilling, minimum tillage and cover crops. Others UE national government such as France and Portugal also took similar actions.

THE EUROPEAN CONSERVATION AGRICULTURE FEDERATION (ECAF)

ECAF brings together eleven national associations, which promote among Europe's farmers the soil management «best practice» aspects of conservation agriculture. With member associations in Belgium, Denmark, France, Germany, Greece, Italy, Portugal, Slovakia, Spain, Switzerland and the United Kingdom (see Table 1), ECAF represents the interests of the majority of the European Union's cropped farmland.

ECAF was constituted in Brussels on 14th January 1999, as a non-profit making association, subjected to the Belgium laws. It was conceived to encourage any issue focused on maintaining the agrarian soil and its biodiversity in the context of sustainable agriculture. ECAF is not involved in any commercial product, equipment and/ or trademark.

At national level ECAF's member organisations aim to:

- improve technology transfer to farms;
- promote agricultural and environmental policies supportive of sustainable soil management;
- improve information exchange in the research, policy and practitioner communities; and
- research, develop, evaluate and promote soil management systems to improve crop production and protection of the environment.

To reinforce the aims of its members at European level ECAF operates as:

- The principal point of contact for discussions on conservation agriculture with European policy makers.
- A clearinghouse at European level to collect and spread information to farmers and agrarian technicians about conservation agriculture.
- A focus for encouraging the investigation, development, and teaching of all aspects of conservation agriculture.
- A collaborator with other international and national organisations that have related and complementary objectives.

The Spanish Association for Conservation Agriculture, promoter and founder member of ECAF, and ECAF has been supported by two EU- LIFE Environment Projects (96-E-·338 and 99-E- 308, respectively).

PERSPECTIVES

In summary, conservation agriculture in Europe is still in an introductory phase. However, recent strong environmental concern and the support which is receiving from the EU and national administrations through the agri- environmental measures, envisages a rapid trend of adoption in the present decade. It should also remind that the VI$^{th.}$ Environmental Action Program of the EU (2001-2010) points out the preservation of natural resources, emphasising the protection of agricultural soil against erosion (*http://www.europa.int/comm/ environment/ newprg/index.htm*, 2001).

REFERENCES

EEA, European Environment Agency, (1998). Europe's Environment, The Second Assessment (Dobris Report), Elsevier Science LTD, p. 293.

García Torres, L. (editor) (1999). Conservation Agriculture in Europe: Environmental, Economic and EU Policy Perspective, European Conservation Agriculture Federation, Brussels, Belgium, p. 23.

García Torres, L.(editor) (2000). *Agricultura de Conservación en el Olivar: Cubiertas Vegetales*, Edita: Asociación Española Agricultura de Conservación/ Suelos Vivos, 36 páginas.

http://www.europa.int/comm/environment/newprg/index.htm (January, 2001)

NOTES

[1] First ECAF Executive Director (1999)

Table 1. ECAF National Associations

Foreningen for reduceret jordbearbejdning i Danmark (FRDK) DENMARK
Raedersgade 5, 8700 Horsens Denmark
Tel: +45.7015.4000 • Fax: +45.7562.2566 E-mail: BeA@lro.dk

Association pour la Promotion d´une Agriculture Durable (APAD) FRANCE
6, rue Roger Bacon 75017 Paris (France)
Tel/ Fax: +33 1 45 741467 E-mail: apadorg@compuserve.com Web: www.apad.asso.fr

Gesellschaft für Konservierende Bodenbearbeitung (GKB) GERMANY
Dorfstrasse, 9 13051 Berlin (Germany)
Tel: +49 30 96275512 • Fax +49 30 96275550 E mail: jana.epperlein@rz.hu-berlin.de
Web: www.gkb-ev.de

Associazione Italiana per la Gestione Agronomica e Conservativa del Suolo (AIGACoS) ITALY
COTIR SS 16 Nord, 240 Località Zimarino 66054 Vasto (Italy)
Tel.: +39 0873 310059 • Fax: +39 0873 310307 E-mail: pisante@cotir.it Web: www.aigacos.it

Associação Portuguesa de Mobilização de Conservação do Solo (APOSOLO) PORTUGAL
Rua Diana de Liz Apt 152 7002-502 Évora (Portugal)
Tel.: +351 66 771685 • Fax: +351 66 769381 E-mail: aposolo@netc.pt Web: www.aposolo.pt

---

Asociación Española Agricultura de Conservación. Suelos Vivos (AEAC.SV) SPAIN
CIFA Alameda del Obispo, s/n Apartado 3092 14080 Córdoba (Spain)
Tel: +34 957 422099 • Fax: +34 957 422168 E mail: nolabrar@arrakis.es
Web: www.aeac-sv.org

UK Soil Management Initiative (SMI) UNITED KINGDOM
Baytree Farm Wartnaby Rd Ab Kettleby Melton Mowbray LE14 3JJ England
Phone/fax +44 (0) 1664 823440 E-mail: smi@smi.org.uk Web: www.smi.org.uk

---

Belgian Association in Research Application on Conservation Agriculture (BARACA) BELGIUM
2, Passage des Déportés, 5030 Gembloux
Fax: 081/62 21 81; E-Mail: stephane.loyen@aigx.be

Finland Consercation Agriculture FINLAND
E-mail: yrjo.salo@mtt.fi

Hellenic Association for promotion of Conservation Agriculture (HACA) GREECE
Aristotle University of Thessaloniki. School of Agriculture. Lab. of Agricultural Engineering. 540 06 Thessaloniki – Greece
Phone: +3031-998744/ 998748 • Fax: +30 31-998729 e-mail: tsatsare@agro.auth.gr

---

Talajkímélö Mezögazdászok Magyarországi Egyesülete (Hungarian Conservation Agriculture Association) (TMME) HUNGARY
1023 Budapest Vérhalom utca 12-16; Fax 36-1-345 06 79; E-mail: pupika@axelero.hu

---

Conservation Agriculture Ireland (CAIR) IRELAND
Contact person: John Geraghty

Zdruzenie pre rozvoj bezorbovych technologii v rastlinnej vyrobe (Association for development of no-tillage technologies in plant production) SLOVAK NO-TILL CLUB (SNTC) SLOVAK REPUBLIC
VURV, Bratislavska cesta 122, 921 68 Piestany, Slovak Republic
Tel.: ++ 421 838/7722326, 7722 327 • Fax: ++ 421 838/7726 306, E-mail: lehocka@vurv.sk

---

Swiss Soil Conservation Association (Swiss No-Till) SWITZERLAND
E.mail: info@no-till.ch; Web: www.no-till.ch

L. SEGUY & S. BOUZINAC

DIRECT SEEDING ON PLANT COVER: SUSTAINABLE CULTIVATION OF OUR PLANET'S SOILS

Centre de Coopération Internationale en Recherche Agronomique pour le Développement CIRAD-CA-GEC, a/c Tasso de Castro, BP 504, Agencia Central CEP, 74000-970 Goiânia GO, Brazil

Savannas *(cerrados)* in the humid tropics of Brazil cover an area of 200 million ha, a quarter of which is under intensive agriculture. Tillage techniques used in the *cerrados* have resulted in serious failures and highly degraded soils. CIRAD and its public and private Brazilian partners have developed direct seeding systems over permanent mulch or plant cover. Seven million ha are currently being cropped with direct seeding strategies in the *cerrados* as an upshot of this research. Three systems have been developed: continuous grain production with two annual crops, followed by temporary dry-season grazing; production of grain crops and sward for mulching and temporary grazing; and an annual sequence of grain crops on live forage cover. These systems cost-effectively produce high stable crop yields, function like natural ecosystems, and require that farmers be constantly innovative and flexible.
Key words: Cover crops, Direct seeding, Cost-effective, Cropping system, Organic matter, Brazil

INTRODUCTION

In an agricultural setting, direct seeding refers to a strategy whereby seeds are sown directly in untilled soil, i.e. in small furrows or holes drilled with special tools. The soil therefore hostsboth the crop and a mulch substrate-which can be straw residue from the previous crop or material of a plant intercropped with the main crop. This mulch and second crop serve to protect the soil permanently from weather hazards, to promote biological activity that will churn up the subsoil, to recycle nutrients and to supply the soil with essential organic matter. Environment-friendly herbicides are applied prior to seeding and during the cropping period to control weeds.

FROM ANCESTRAL TECHNIQUES TO MODERN AGRICULTURE

Direct seeding is not a novel principle. Ancient Egyptian and South American Inca farmers used a simple planting stick to make holes in the soil in which they hand-

L. García-Torres et al. (eds.), Conservation Agriculture, 93–100.

planted seeds that were subsequently covered by kicking earth into the hole. Nowadays, in humid tropical forest regions of Latin America, Africa and Asia, hundreds of thousands of hectares are direct seeded by farmers under shifting slash-and-burn agriculture conditions. In Central America and Mexico, beans *(Tapado)* and maize are traditionally seeded directly in soil with a volubilate plant *(Mucuna)* cover.

In countries of the North, direct seeding without any preliminary cultivation was first used in a modern agriculture context in USA in the late 1940s, during a period when the Great Plains were ravaged by wind erosion (the famous Dust Bowl). This trend gained momentum in the 1960s with the development of paraquat-based nonselective herbicides, propulsed by the work of Harry and Lawrence Young on their Kentucky farm. In 1966, Allis Chalmers, the farm machinery manufacturer, designed the first zero-till seeder. Soybean could thus be direct seeded into wheat stubble mulch. Shiney Phillips, research scientist at the University of Lexington (Kentucky), focused on the extension of direct seeding systems throughout USA and Latin America. The agricultural area under direct seeding systems thus increased from 2.2 million ha in 1973 to more than 20 million ha in 1998, which represents 16% of the total crop area.

In Brazil, direct seeding was first implemented in Rio Grande do Sul state in 1969. These systems were then developed and used on a large scale in Paraná state –where soils were severely eroded– under the initiative of Herbert Bartz, with the collaboration of IEAME (an agropastoral research institute of the South, based in Londrinas) and GTZ (Gesellschaft fur Technische Zusammenarbeit Eschborn, Germany), who imported the Chalmers zero-till seeder in 1972. From 1973 to 1981, research carried out by IAPAR (the agricultural research institute of Paraná), with the cooperation of ICI and GTZ, developed suitable crop rotations. Simultaneously, the work of pioneer farmers Franck Dijksitra and Manoel Henrique Pereira (President of the FEBRAPDF, the Brazilian direct-seeding federation, from 1992 to 1998) led to quick adoption of direct seeding techniques throughout Paraná and southern states in Brazil. In the 1990s, the greatest extension of direct seeding took place in the *cerrados* of central and western Brazil, owing to the work of CIRAD, John Landers (Executive Secretary of APDC, the *cerrados* direct seeding association) and APDC, in collaboration with farmer partners in southwestern Goias state. In this region, within a decade direct seeding systems were introduced on an area of some 3 million ha. In Brazil, the main crops grown under direct seeding are soybean, maize, wheat, barley, sorghum, sunflower, irrigated rice, and more recently rainfed rice, cotton and forage species.

Since 1970, 20 million ha were switched to direct seeding in Latin America, including 10 million in Brazil. This represents a major agricultural revolution in this tropical and subtropical region. The present paper focuses chiefly on direct-seeding applications in humid *cerrados* areas of Brazil, south of the Amazon Basin.

BRAZILIAN *CERRADOS*, THE LAST ARABLE LAND BEFORE THE GREAT FOREST

The *cerrados* region accounts for around half of all arable land in Latin America, i.e. 243 million ha, mainly in Brazil, Colombia and Venezuela (which represents twice the crop area of USA). In Brazil, the *cerrados* cover a 200 million ha area (23% of all land in Brazil), 50 million ha of which could potentially be used for intensive mechanized agriculture. Brazilian agricultural development specialists agree that sustainable development of this area could lead to the production of 150 million t of grain, 9 millions t of meat and 300 million m^3 of wood, while retaining 20% of this area for environmental conservation.

Cerrados*: an agricultural frontier threatened by destructive farming techniques*

In Brazil, *cerrados* in the hot humid tropical region were first cropped in the late 1970s, with the colonization of frontier areas in the western and central-western states by farmers from the southern states. Annual rainfall levels range from 2 000 to 3 000 mm, most of which falls from October to April, often at intensities above 100 mm/h. The ferrallitic soils are acidic, with low nutrient content (nitrogen, phosphorus, potassium calcium, magnesium and zinc), and soon deprived of organic matter after the natural vegetation is cleared. The agriculture was initially oriented towards industrial-scale soybean monocultures for export. The exclusive use of disc ploughs –because of their rapidity and multiuse potential– caused soil compaction and hence catastrophic wind and water erosion. Soil productivity inevitably declined despite massive use of pesticides and chemical fertilizers. Within a few years, spectacular regional financial collapses left the land empty and desolate. The collapses were aggravated by the fact that these frontier areas are located far from processing and marketing centres, and that overpriced inputs cut farm-gate prices by 20-30% more than in southern Brazilian states.

Research involvement: geared towards collaborating with and assisting farmers in their environment

CIRAD and its Brazilian partners became involved in frontier agricultural areas in Mato Grosso state, where 1 million ha are currently being cropped. This development-oriented research was first carried out in *cerrados* areas until 1994 and then in forested areas, with the aim of developing innovative «forest-friendly» farming systems to propose to future farmers colonizing this area. In all cases, two objectives had to be fulfilled: sustainable soil use and economic risk management. This meant replacing industrial-scale soybean monocultures with farming systems based on crop diversification and implementing radically different cultivation techniques, while promoting product quality to achieve maximum added value.

A first step was to replace disc ploughs with mouldboard ploughs and tined cultivators and to set up 2-year crop rotations (rainfed rice/soybean or maize/soybean) thus in some ways applying traditional agricultural techniques of countries of the North. Crop yields, as well as erosion, weed and pest control were improved. The upshot of these initiatives was very clear within 5 years, i.e. insufficient erosion control and loss of half of the organic matter in cropped soils. It was thus impossible to continue using these techniques without substantially increasing expensive input supplies. Alternative ways to cultivate these tropical soils had to be found.

Recherch came up with a farming concept adapted to humid tropical constraints, i.e. based on rainforest function. The forest is a generally stable system with complex efficient activity capable of producing high quantities of green matter, recycling mineral nutrients and completely protecting the soil, without nutrient loss or export. To achieve this type of result, in addition to the main crops, new systems therefore have to include permanent live or dead plant cover that will fulfil these three roles. The system could be summarized as follows (Fig. 1): no tillage; crop residue is conserved and the supplementary plant matter produced by one or several specifically grown plants provide permanent soil cover during the rainy and dry seasons; the top 5-cm soil layer is nutrient rich with very high biological activity, resembling that found in the forest. Three direct seeding cropping systems were tested: grain production with two-crop annual sequences, followed by temporary dry-season grazings, or 3-4 year rotations with grazing for livestock.

Direct seeded grain production: how can a suitable soil cover be obtained?

Under subtropical or temperate climates, with the cold season, crops can be directly seeded into mulch from the previous crop because they decompose slowly and efficiently cover the soil. However, under hot humid tropical climatic conditons, harvest residue is quickly decomposed into mineral nutrients which, when it rains, are leached to deep soil layers –especially nitrates, potassium and calcium–. There is also no soil erosion control.

Every year, a cash crop is also grown sequentially with one or several support crops that can –with minimal inputs– produce a thick mulch layer on the soil surface and an extensive root system. For instance, millet or *guinea* sorghum are sown and treated with a nonselective herbicide (glyphosate, etc.) to create a mulch layer in which the main crop is subsequently sown. The agronomic functions of this mulch crop are complementary. It hampers soil erosion and acts as a biological pump, i.e. its powerful root system recycles nutrients and restructures the soil, thus the soil is worked by biological processes rather than mechanically with ploughs, etc. It is a source of organic matter (carbon) for the soil and nutrient elements for the main crop. This plant matter also buffers climatic fluctuations (temperature and soil moisture), promoting the development of beneficial organisms (earthworms, etc.) in the top soil layer. The mulch layer also hampers the growth of most weeds and limits crop pest infestations by blocking light penetration to the soil surface, while having some allelopathic effects (release of substances that inhibit or stimulate the growth of subsequent plants or others growing in the vicinity). There are, of course, many possible combinations. For instance, the action of millet and sorghum can be reinforced by mixed sowing with a forage species (*Brachiaria* sp.) which remains green throughout the year: it can be grazed while also hindering field fires; *Brachiaria* roots remain active during the dry season, thus enhancing the recycling work of the sorghum stubble mulch cover.

In humid tropical regions (Brazilian *cerrados*), two cash crops a year can be grown (soybean, maize, rainfed rice, cotton), but generally a single cash crop is simply intercropped with the following cover crops:

- before sowing the cash crop: millet, sorghum, *Eleusine coracana*;
- at crop harvest: millet, sorghum, associated or not with forage species, sunflower; millet or sorghum associated with *Bracharia* sp., producing 7-13 t/ ha of above-ground dry matter and more than 4 t/ha of roots in the top 50-cm soil layer.

The easiest systems to implement, such as soybean in a sequences with millet or sorghum, were disseminated and readily adopted between 1992 and 1998 –they now prevail on 4.5 million ha in the Brazilian *cerrados*–.

Systems with grain crops in rotation with grazings

In the above systems, the forage species (*Bracharia* sp.) is only grown for a few months a year concomitantly with millet or sorghum for the purposes of enhancing the surface biomass. In mixed grain crop livestock production systems, forage crops are grown for several years in rotations with grain crops.

The pasture crop is direct seeded without fertilizers in the mulch layer formed by the residue from the last cash crop. *Bracharia brizantha* and *Panicum maximum* produce high quantities of biomass until the onset of the dry season; these grazings can support 1.7 to 2.2 head of cattle per ha for an average daily gain (ADG) of around 450 g over the 100-120 days of the dry season. The pasture crop is subsequently wilted with a nonselective herbicide and the cash crop can be direct seeded 15-25 days later. Recommended rotation periods range from 2 to 5 years.

Cropping on live forage cover

In these systems, perennial plants with vegetative reproductive organs (*Cynodon, Paspalum, Arachis, Pueraria* and *Calopogonium* spp.) are grown to provide permanent soil cover. This cover is sown on a yearly basis after the direct seeded cash crop is harvested –the forage species is then treated with nonselective herbicides– (paraquat, diquat) to limit competition with the main crop. Herbicides selective for the crop are then applied at very low dose until the crop cover is complete. Note that nonselective herbicides do not pollute the soil because they are totally intercepted by the targeted green biomass and then trapped in the thick mulch. After harvest, the forage cover is quickly established and can be grazed. These systems have not yet been widely adopted by farmers, especially because of the high cost of *Arachis* sp. seed and *Cynodon* sp. cuttings.

Ten years of agronomic and economic analyses in the cerrados

The systems that we have just described have been compared with conventional practices for more than 10 years on farms in the *cerrados* region. The comparison of newly cleared forest lands or fields with a long cultivation history (e.g. 18 years of soybean monoculture) revealed that biological factors are essential to cost-effectively achieve high sustainable yields: in all cases, direct seeding systems on plant cover generated more income than systems with conventional techniques, even those involving mouldboard ploughing. Yields remained stable for a sowing period of 60 days from the onset of the rainy season, whereas they declined by 30-60% with cropping systems involving tillage. Fertilizer consumption was substantially lower. On farms run by farmers skilled at direct seeding procedures, for soybean and rainfed rice based systems, production costs were reduced to $US300-550/ha, depending on prevailing prices, and net margins reached $US150 to more than $US600/ha. Direct seeding systems have also made it possible to reclaim fields infested with *Cyperus rotundus* weeds that are impossible to control with conventional farming procedures.

The crop sanitary status was enhanced: marked reduction in fungal diseases on cereal crops, in bacterial blight on cotton crops, and in nematodes on all crops. The persistent mulch layer was found to promote the spread of *Nomurea* sp. fungi, which in turn biologically (partially) controlled most defoliating caterpillar pests of cotton and soybean. However, we noted an increase in soil-borne fungi such as *Rhyzoctonia* and *Fusarium* spp., a problem that can be overcome by carefully choosing the plant species used for stubble mulch and applying suitable seed treatments.

The use of these systems improved the soil characteristics. Carbon recycling is a constant function of these systems, but this feature varied according to the type of species grown: in 5 years, the carbon content in the [0-20] horizon increased by 0.5 % for sequences of soybean on millet or sorghum (associated with *Brachiaria ruzizenis*

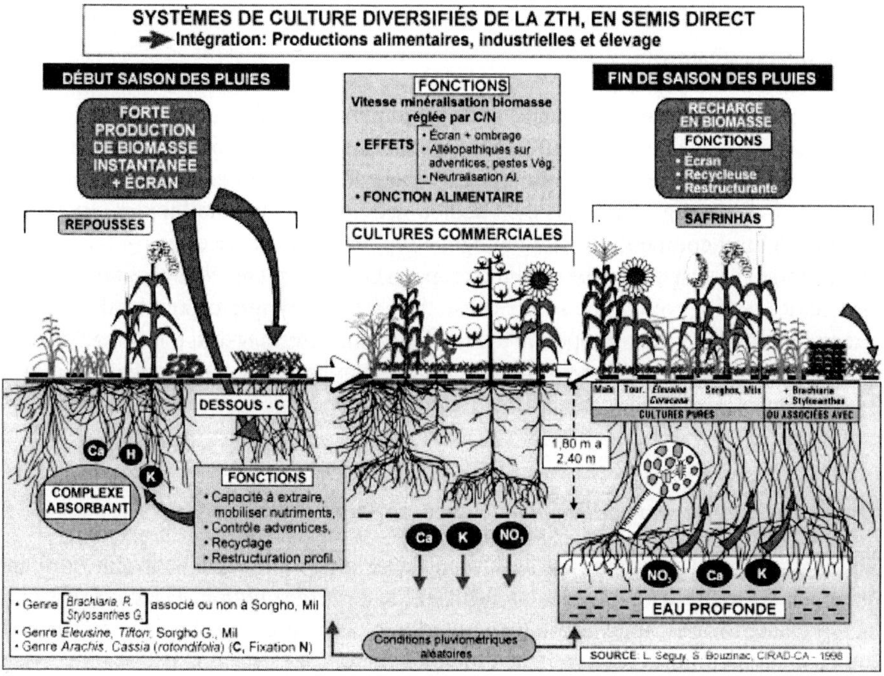

or *Eleusine* sp.), or sequences of maize on *Arachis* sp.; it increased by 1% with systems involving 5-year rotations of grain crops and grazings. Similarly, the cation exchange capacity (CEC) rose from 1 to more than 3 meq/100 g and the exchangeable base saturation rate increased from 15 to 30%, depending on the cropping system.

CONCLUSION

In hot humid tropical regions of Brazil, direct seeding systems have been introduced to create sustainable agricultural scenarios with the aim of enhancing and reclaiming the natural biodiversity. The features of these systems –establishement of a permanent plant or mulch cover to protect the soil and promote soil fauna and all other biological activity, crop rotations, and crop-livestock production combinations– make agrosystems resemble natural ecosystems. Direct seeding systems on plant cover efficiently reduce the emission of greenhouse gases resulting from conventional agricultural activities. Farmers who adopt these techniques actually break with the tillage tradition and are «retrained» on natural biological functioning. Managing agrosystems through the implementation of direct seeding on plant cover is a highly efficient tool to enhance the continuous professionalization of farmers. These techniques can be adapted to the socioeconomic conditions of tropical and temperate countries. In temperate areas, their implementation can substantially reduce nitrate and pesticide pollution of groundwater, while stalling soil erosion and «reconstructing» the biology of the soils.

M. H. PEREIRA

SURVIVAL FARMING

CAAPAS, Rua 7 de Setembro, 800, 84035 Ponta Grossa, Paraná – Brasil

Agriculture, as a profitable activity in South American countries, has been the focus of constant attention from leading political figures and technicians for whom the government has shown very little interest.

This activity has always been related to a national plan of production to be established in order to obtain relative control over regulating stocks, avoiding not only a shortage of food, but also imports, which can only benefit suppliers and intermediaries.

This rather complex situation has a highly negative effect on the annual planning of a productive farm making the farmer uncertain as to the reaction of the market especially when it comes to the perspective of making any profits. The farmer already knows that his plans might have to be altered at any moment by decisions totally beyond his control and which can result in losses.

The most common example of this are the imports which occur at the same time as the harvest and which push market prices below their production costs.

So far nothing has been mentioned about financial support for the farmer, a financial aid which is well-known to be impossible in underdeveloped countries. However, what really has disastrous consequences are the economic measures to control the inflation rate which usually penalizes the productive sector.

Farmers in Brazil, Argentina, Paraguay, Bolivia and other neighbouring countries have no control over their costs, no perspectives of a stable market and the very little financial support that can be had is usually unavailable when it is most needed. As a consequence, all farmers seem to have been contaminated by the same virus and suffer from the same disease, showing the same symptoms of demotivation and lack of support in an activity of high risk. The repetition of some climatic conditions, for example, may very easily make the farmer go out of business even after decades of working in this area.

This situation shows no sign of changing mainly due to few political representatives in this sector who very timidly manifest themselves in defence of the farmer, not to mention the chronic behaviour of economists when it comes to the interests of primary production.

L. García-Torres et al. (eds.), Conservation Agriculture, 101–104.

What one means by farming produce is the enormous diversity which can be extracted from the soil, plants and animals. Lack of support towards other activities such as dairy farming and its by-products, results in a low-profit activity. The result of this is that every day breeders are having to shut down their work, especially in Brazil.

From this we conclude that there are vast continental expanses to be explored with suitable technology adapted to the soil and climatic conditions. These expanses guarantee the rural population conditions for subsistence. Having had to endure such dire situations for so many years, the rural population has become very insecure regarding their income.

Today, in some areas of Latin American countries, a family may be carrying all their belongings on a wagon or truck and may be moving to the the suburbs of a city looking for a job and a salary that, even being low, might be more secure and give them a better standard of living. This is generally only an illusion. Nowadays most urban centres suffer from uncontrolled overpopulation resulting in an increase in poverty, violence and prostitution.

HOW LONG IS THIS SOCIETY GOING TO LIVE WITH RISING LEVELS OF POVERTY IN THE LOWER CLASSES?

The government cannot or does not want to solve the problem. On the contrary, what usually happens with rules of the economy is that there is an increasing tendency not to allow wealth to reach the productvie sector of agriculture and cattle breeding.

What has been observed in the past is that every time when the market surprised the most optimistic analysts with compensating prices, somebody in the economy area of the government would come up with an even bigger surprise. Creating a new index rate, which was usually called a contingency, exportation quota, confiscation and others and with these measures obtain part of the profit of a good crop. The government has never shown solidarity with the farmers at times of losses during the bad years and has never been present when having to check the reasons for losses. Without doubt, the government has been an extremely bad partner only showing up whenever it has meant to obtain a share of the profit.

This behaviour has made it impossible for the farmer to make enough capital and become self-sufficient from good crops. This would enable him to work on his own resources. The few who have managed this heroic feat are the ones who have progressed the most and today are free from the high interest rates charged by the banks when loaning money for agriculture. Unfortunately, I repeat, these producers are very few. *«If the government cannot help, it should at least not get in the way!»*

Now, if we refer to the present moment we notice that in spite of the farmer being discouraged, he still continues with the same activity in the same way in which he continues to sow in the hope, or better, in the certainty, that he will harvest. He has also developed ways of productivity with better and greater results. Better because today he can plant without interfering with the natural conditions of soil, air and water therefore contributing effectively to the improvement of the environment. For this reason, as a retribution, he receives a chart with lower costs in terms of fuel consumption, replacement of machine parts, longer use of his machinery, fertilizers and herbicides, resulting in an increase in the average production.

Nowadays we can understand the relevance of this successful programme called *«No-tillage system (planting on the straw) - Harmony between Man and Nature»*. This is the anonimous result of farmers, technicians, researchers, herbicide and farming equipment manufacturers, as well as companies that produce seeds and fertilizers. They have all invested time and money for more than 2 decades to turn this proposition into the most viable one that has come to farming at the end of the 20th century.

It is common to hear from a farmer, no matter what size his farm is, that it is due to the no-tillage system that his family are still in this activity. This system has also made it possible for the children to go to school and university. This fact is in accordance with what the farmer and cattle breeder, Frank Djikstra, who pioneered this technique, wrote in his book, *«Porque utilizo o plantio direto» («The reasons why I use the no-tillage system»):*

> *«The no-tillage system is not only a different technique, but it is also a matter of survival.»*

Nowadays there is a tool with medium or large animal traction which has been given to the small farmer, and which has yielded excellent results, totally different from the ones obtained with the intensive use of machinery used for nearly half a century in tropical and sub-tropical climates.

It has been reported by research that during the years in which the so-called conventional system was used, an average of 20 tons of productive soil per hectare were lost during a season as opposed to the productivity of 2 tons of grains. It must also be considered that today these results would never result in a favourable profit. The need for change was a matter of survival.

The most striking fact about all this is that groups of farmers, technicians and sometimes researchers from developed countries have been visiting the farms that have shown positive results in soil conservation, generating their stability or in increased productivity with less use of pesticides, herbicides and fertilizers.

The surprise is that these new characters come from countries where farmers are given generous support receiving subsidies for their crops, generating a paternalist and artificial attitude which results in bad habits difficult to be changed. Compared to the penalties applied to farmers in underdeveloped countries, they are totally out of our reality, which is what really occurs. These are different worlds!

WHAT IS THERE IN THE FUTURE FOR US?

We are sure that if we have equal conditions, we will be able to progress because we have the technology, soil, climate and human resources who are conscious of their responsibility. But if the differences between the benefits to farmers in underdeveloped countries and the ones in developed countries are kept, our position in the commodities market will always be at a disadvantage.

We know that old areas used for farming, cattle and sheep breeding have lost most of their fertile soil. These areas are found in rich countries and are kept active by financial support.

For this reason we believe that the no-tillage system without the use of machinery, can have the same effect as a blood transfusion to a dying person, injecting the plasma of stray into him (straw, harvest remains) for the minimum percentual of organic matter to be reborn which is something hard to be seen nowadays in the chemical analysis of the soil. This is currently occurring in many European countries, in an activity which has been carried out for centuries feeding only the people but not the soil.

When we see a 200-year-old tree with uncovered roots, we understand the madness of the farmer, highlighted by Edward Faulkner in his book «Plowman's Folly», edited in the middle of the last century. At that time he was taken for a madman and a visionary in his predictions and claims, but no-one really followed his instructions or listened to his concerns. Is there still enough time left?

From the little I know of European agriculture, what I have seen so far has not made me happy, neither in mechanical engineering to attack the soil, nor in plant behaviour or the quality of the farming areas which can no longer stand any more aggression.

No-tillage is the last chance to make this agonizing soil react. This soil is certainly expecting a better treatment for its weaknesses for then be able to recover its levels of sustainability.

There is something at the end of these considerations which is always in my mind and that must be mentioned: *«Poor soil, rich farmer!»* «It doesn't make sense!» is what the most common man involved in farming would say.

M. DA VEIGA, L. DO PRADO WILDNER & J.R. BENITES

LATIN AMERICAN CONSERVATION AGRICULTURE NETWORK – RELACO

RELACO President 1999-2001, EPAGRI, Santa Catarina, Brasil
RELACO Secretary 1999-2001, EPAGRI, Santa Catarina, Brasil
FAO/RELACO Coordinador 1987-2001, AGSL/FAO, Roma, Itália

The Latin American Conservation Agriculture Network (RELACO) is a professional, non-profit making association that brings together persons who are interested in and capable of contributing efficiently to the development of conservation agriculture science, technology and production in Latin America and the Caribbean. It was founded in 1987 and amongst its activities, it has organised five Latin American Seminars and published the proceedings. It has also promoted the exchange of advisors amongst the participating member countries, delivered training courses on the theme and it has also published a wide range of technical documents and a manual concerning conservation tillage systems for Latin America. RELACO is mainly supported by national institutions from the member countries together with FAO, but it is open to other funding sources with the objective of increasing the number of activities of the Network. Biennial meetings are held with the aim of interchanging conservation agriculture experiences amongst the member countries.

INTRODUCTION

The main areas for both present-day and future development of rainfed agriculture in Latin America lie in the semi-arid and sub-humid zones. It is in these regions that the lowest levels of agricultural production and of land productivity are characteristic. There are increasing difficulties to even maintain these levels in a profitable manner, more so in those areas only recently put into agricultural production. At the same time in many countries, one observes a continuous migration of the active rural population towards urban centres, increasing poverty levels, malnutrition of numerous sectors of the population and an increasing dependence on food importation to satisfy requirements.

One of the main reasons for this situation can be traced back to land degradation caused mainly by inadequate practices of use and management under various combinations of effects due to climate, soils and the existing socio-economic conditions in the countries of the region. Deforestation together with the exploitation of fragile marginal lands on mixed farms is accelerating the process of degradation. Other contributing factors include

L. García-Torres et al. (eds.), Conservation Agriculture, 105–110.

land holdings and tenancy, increasingly costly agricultural production inputs and the lack of or inconsistency of land use policies, strategies or plans. To this list should be added a generally poor understanding of appropriate systems of land management under different agro-ecological conditions and the recent economic problems that have negatively affected the agricultural sectors of almost all the countries in the region.

Clear symptoms of soil degradation are to be observed in the majority of Latin American countries, provoked by intensification and gradual mechanisation of agricultural and livestock production. They thus derive mainly from the land use and management practices and are noticeable in a variety of ways such as erosion due to water and wind, compaction, capping and crusting, loss of organic matter and nutrients, salinity and alkalinity, etc. Together they represent the major limitations for the expansion and intensification of agriculture in Latin America and thus, for the food production needs for their resident population both at present and in the future.

Faced with this situation, there is a need to initiate, develop, strengthen and/or reorient according to each case, studies and projects that can lead towards efficient systems of conservation land management. At the same time, there is a need to diffuse knowledge so as to contribute to a better understanding of the causes of soil degradation and to find solutions satisfying farmer needs. In this sense it seems that the most vital factor directly related to the degradation process within land management systems, is the intensity of soil tillage.

A lack of long-term policies for rational exploitation of natural resources is generally characteristic of many countries within the region. Coupled with this, there is a lack of stability or the necessary political vision to guarantee economic support to already established conservation programmes. Failure to optimise activities and to interact, examples of unnecessary duplication and gaps in information in certain areas, coupled with a general low efficiency in the use of both human and material resources, are all common situations. These can be traced to the general lack of long-term integrated programmes and projects that focus on, co-ordinate and contribute towards bringing together individual institutions and organisations.

It is within this context that networks for research and technology transfer can play a vital role. They are an effective means for promoting the efficient use of scientific resources and the interchange and training of technical personnel of the participating countries. Attempts to use networks to transfer technologies require a minimum need for local expertise and an organisational framework that is able to conduct and support a systematic evaluation of experimental results. It is with this objective of facilitating the exchange of experiences between member countries that FAO supports technical networks that promote meetings and specialist workshops. These will normally be aimed at defining criteria, evaluating the extent and severity of present-day and potential land degradation, selecting pilot areas and formulating development and investment projects. All of these efforts are focussed land conservation through practices of conservation agriculture.

A BRIEF HISTORY OF RELACO

In response to the situation described above, the Latin American Conservation Tillage Network (RELACO) was created, oriented towards diffusing knowledge related to conservation tillage systems. Its activities started in 1986 under the auspices of FAO with the initial objective of assisting national research institutions from the Chaco region (Argentina, Bolivia and Paraguay). Venezuela joined the network in 1988 and in 1991 the existing network arrived at a membership including the additional inscriptions from nine other countries (Argentina, Bolivia, Brazil, Costa Rica, Nicaragua, Paraguay, Peru, Dominican Republic and Venezuela). The 3rd Biennial Seminar held in San José, Costa Rica resulted in the additional membership of Chile, Colombia, Cuba, Ecuador, El Salvador, Honduras and Mexico. The 4th Biennial Seminar held in Morélia, Mexico saw a change in the breath of the thematic elements within the network and the General Assembly decided to change the name to the Latin American Conservation Agriculture Network, although the acronym was to stay the same (RELACO). During the 5th Biennial Seminar held at Santa Catarina, Brazil it was formally decided to open up the network to all countries within Latin America.

According to its statutes, RELACO is established as a non-profit making organisation that links together persons interested and able to efficiently contribute to the development of conservation agriculture science, technology and production in Latin America and the Caribbean. Its major objectives are as follows:

a) Stimulate research, studies and teaching programmes that promote conservation agriculture production systems
b) Diffuse results of advances in research and studies concerning conservation agriculture practices, particularly those undertaken in Latin American countries
c) Promote the development of conservation agriculture production systems in Latin America
d) Co-operate in the promotion of ideas, efforts and organised groups that coincide with the principles of conservation agriculture as proposed by RELACO
e) Stimulate the activities of National Networks and similar conservation agriculture programmes to promote together with RELACO concepts of conservation agriculture within their countries
f) Maintain close relationships with state and private institutions, organisations and companies that are following similar or the same aims both at national and international levels.

BIENNIAL SEMINARS

The 1st Biennial RELACO Seminar and a training course on tillage systems was held from 18-27 November 1991 at the Sáenz Peña Agricultural Experiment Station of the National Institute for Agricultural and Livestock Production (INTA). 28 participants from 9 countries were in attendance. The meeting was organised jointly by the FAO Land and Water Development Division and INTA. During the meeting, the sessions stimulated the participants to reflect on the need to organise a Tillage Network within the region whose structure would constitute an extension to the network already established in 1986 and which comprised 4 member countries. All the training material used during the course was collated and in November 1992, it was published as the Manual of Tillage Systems for Latin America (the FAO Soils Bulletin No. 66). It was also agreed that the Seminar Co-ordinator should also co-ordinate operation of the extended network over the next two-year period, at which time he would be replaced by the co-ordinator of the next biennial meeting. Ing. Roberto Casa, Director of the Soils Institute of the INTA Natural Resources Research Centre, was chosen to be the Regional Co-ordinator of RELACO for the period 1992-93.

The 2nd Biennial Seminar was held in Venezuela from the 14 – 19 November 1993 at the Centre for Agricultural and Livestock Research of FONAIAP in Acarigua/ Guanare, Portugesa State. The event was sponsored jointly by the Venezuelan Fund for Research in Agriculture and Livestock (FONAIAP), FAO, the Llanos Ezequiel Zamora National Experimental University of Venezuela (UNELLEZ) and the International Society for Soil Science (ISSS). A workshop was held during the event titled «Effects of tillage systems on the degradation and productivity of soils», which was attended by 27 professionals from 9 countries. The workshop saw the drawing up of a series of recommendations based on the experiences of the participants, which later served as the basis for the conclusions and recommendations of the seminar. The regional co-ordinating of RELACO for the period 1994-95 was assigned to FONAIAP, the Regional Co-ordinator being Dr. Francisco A. Ovalles.

The 3rd Biennial Seminar was held in the Faculty of Agronomy of the University of Cosat Rica (UCR) at San Pedro de Montes de Oca and was co-ordinated by the Ministry of Agriculture and Livestock (MAG). It was supported by FAO, the Costa Rica Association of Soil Science and the Agronomy Research Centre of UCR. The theme was «Sustainable use of soil on hillsides – the essential role of conservation tillage systems» and was attended by 64 professionals from 34 national institutions in 17 countries. The main recommendation of the seminar was to focus on participatory planning to promote soil conservation management practices, leading towards increased agricultural production and as a basis for sustainable development and good conservation of soil and water. It was also recommended to use micro-catchment areas as units for planning and executing activities. The Regional Co-ordination of RELACO for the period 1995-97 was assumed by Ing. Nils Solórzano from the Costa Rica Ministry of Agriculture and Livestock.

The 4th Biennial Seminar was organised in Mexico at Morélia/Michoacán under the auspices of the National Institute for Research in Forestry, Agriculture and Livestock (INIFAP), the National Centre for Sustainable Agricultural Production (CENAPROS) and FAO. Numerous relevant papers were presented by the 128 RELACO members present, including various case studies and particular keynote speeches. Amongst the recommendations it was highlighted that crop rotations and inter-cropping, together with the use of green manure and compost were of great importance in conservation agriculture systems. The General Assembly of RELACO was held during the meeting and the network changed its name to the Latin American Conservation Agriculture Network, although its acronym was unchanged as RELACO. Dr. Ramón Claverán Alonso from CENAPROS was elected as the RELACO Co-ordinator for the period 1997-99.

The 5th and most recent Biennial RELACO seminar took place in Brazil at Florianópolis/ Santa Catarina from the 3-7 October 1999 and was jointly organised by the Santa Catarina Enterprise for Agricultural and Livestock Research and Rural

Extension (EPAGRI) and FAO. The theme of this fifth meeting was «Direct drilling: a tool for conservation agriculture» for which there were various keynote speeches and then discussions in 6 working groups with the 102 technicians from the 11 participating countries. It was during this meeting that RELACO started to become associated with various producers' organisations such as the Confederation of American Producers for Sustainable Agriculture (CAAPAS) and the Brazilian Federation for Direct Drilling through Residues (FEBRAPDP). The General Assembly approved the proposal that all countries of Latin America should be incorporated as members of RELACO. Co-ordination of the network for the period 1999-2001 was assigned to EPAGRI, Ing. Milton da Veiga being elected to occupy the post.

It is important to highlight that RELACO, throughout its long history, has received both technical and financial assistance from FAO through its Land and Water Development Division, together with support from national institutions. This has enabled both the organisation of the biennial seminars and operation of the network.

Network activities and results

The main activities developed by the network in order to achieve its objectives are summarised below:

- Biennial seminars, held alternately in countries in the North and South of Latin America.
- Printing and distribution of scientific, technical and extension publications
- A Web-site describing the Network and including information relating to the the most recent biennial meeting (http//www.relaco.cjb.net), together with an e-mail address to facilitate information exchange (relaco@epagri.rct-sc.br)
- Circulation of a newsletter three times per year containing relevant information about co-ordination of the network and other matters of interest concerning conservation agriculture
- Continuous interchange of information amongst the members
- Communication with the Ministries of Agriculture and other authorities in Latin American countries
- Organisation of training courses and workshops
- Establishment of contacts with enterprises producing machines, equipment and other agricultural inputs
- Formal and informal exchanges between specialists
- Consultancy and advisory services at different levels
- Communication and interchange with Associations for Conservation Agriculture in Latin America and other parts of the world.

The main achievements of the network may be summarised as follows:

- Organisation of a total of 5 Biennial Seminars over the last 13 years, funded by FAO and and national institutions.

- Publication and distribution of the Proceedings of these seminars, together with approximately 40 additional technical documents.
- Establishment of the formal constitution and designation of the legal personnel necessary to guarantee the full and continuing functioning of the Network.
- Presentation and discussion during the biennial meetings of some of the most significant Latin American case studies, publishing and distributing these afterwards
- RELACO has created a solid sense of conscience and ethics regarding the conservation of natural resources amongst its members and also with those others with access to the information generated.
- RELACO has now gained a certain prestige and has been represented in numerous international meetings not only on Latin America but also in other parts of the world.
- One particular document published by RELACO concerning land management systems in Latin America and written by members from Argentina, has been widely distributed and is still being used by both technicians and producers
- The results and experiences of RELACO are now being transferred to similar organisations in other parts of the world
- RELACO has also introduced to Latin America, knowledge and experiences from other parts of the world through inviting selected participants to one or other of its biennial seminars
- RELACO has assisted in the organisation of various national networks focussed on practices of conservation agriculture in many Latin American countries
- Presently, RELACO is working alongside producer organisations and trying to focus their attention on conservation agriculture both at national and regional levels.

REFERENCES

Alonso, R.C., da Veiga, M. and Benites, J.R. Latin American Conservation Agriculture Network (RELACO). In: Global Forum for Agricultural Research. Dresden, Germany, (2000) *Proceedings*.

Benites, J.R. Red Latinoamericana de Labranza Conservacionista (RELACO): Los diez primeros años de actividades (1987-1997). In: *Reunión Bienal de la RELACO*, **5**. *Memorias*. Florianópolis/SC, Brasil, (1990) EPAGRI/RELACO: CD ROM, 2000. 53p.

Benites, J.R., Ovalles, F.A. y de Bisbal, E.C. *Reunión Bienal de La Red Latinoamericana de Labranza Conservacionista*, **2**. Guarare/Acarigua, Venezuela, (1993) RELACO, publicación especial, **1** (1994), 25p.

RELACO. III Reunión Bienal de la Red Latinoamericana de Labranza Conservacionista. El uso sostenible del suelo en zonas de ladera: El papel esencial de los sistemas de labranza conservacionista. San José, Costa Rica, 1995. *Memorias*. Ed. por Floria Bertsch y Carlos Monreal. San José, C.R.: accs, (1996) 307P.

RELACO. IV Reunión Bienal de la Red Latinoamericana de Agricultura Conservacionista. Estudios de caso en América Latina. Morelia/Michoacán, México, (1997) *Memorias*. Ed. por Ramón C. Alonso y Fernando O. R. Vilchis. 361p.

RELACO. V Reunión Bienal de la Red Latinoamericana de Agricultura Conservacionista. Siembra directa: una herramienta para la Agricultura Conservacionista. Florianópolis/Santa Catarina, Brasil, (1999) *Memorias* en CD ROM. Ed. por Milton da Veiga.

K. STEINER & M. BWALYA

THE AFRICAN CONSERVATION TILLAGE NETWORK ENHANCING CONSERVATION TILLAGE IN AFRICA

Deutsche Gesellschaft für Technische Zusammenarbeit (GTZ) GmbH. Department Rural Development, P O Box 5180, 65760 Eschborn.

ACT Secretariat, c/o IES/University of Zimbabwe, P O Box MP, 167 Harare

Soil tillage by hoe or plough was identified as one of the principal causes of soil degradation in vast regions of Africa. Inspired by the impressing spread of no-tillage practices in Latin America GTZ conducted a study on the potential of no-tillage practices in sub-Saharan Africa. Limited access to information and lacking exchange between research and development institutions within and across countries was identified as one reason for the limited advance of these new technologies throughout Africa. Participants of an international workshop conducted in Zimbabwe proposed the founding of a regional conservation tillage network. With financial and technical from GTZ and FAO, the objectives and structure of such a network were defined, and a permanent secretariat opened early 2000. The network is seen as a network of country networks. Purpose of the network is to enhance the dissemination of conservation tillage in smallholder agriculture. The core functions were defined as, support to the formation/strengthening of national networks; dissemination, mobilisation, provision and inventory of information; documentation and dissemination of the evidence of conservation tillage benefits; organisation of learning through exposure/seeing; support to lobbying at national level; support to the development of dissemination material.

Key words: conservation tillage, networking, dissemination of conservation tillage, information material, lobbying

INTRODUCTION

The degradation and irreversible destruction of soils have reached an alarming scale. Especially in the semi-arid and arid areas of Africa soil degradation is threatening the livelihood of millions of farm households. Main causes of land degradation are the increasing number of livestock causing overgrazing of pasturelands and intensive cropping with conventional tilling. Conventional tilling where the soil is inverted using various forms of ploughing leads to formation of compaction (plough pans) (Douglas *et al.*, 1999) and accelerated decomposition (loss) of organic matter. Moreover, crop residues are removed or burned that leave the soil unprotected to

L. García-Torres et al. (eds.), Conservation Agriculture, 111–122.

climatic hazards such as rains, wind and sun. Feeding livestock on crop residues, leaving behind a bare surface is another reason for accelerating land degradation in semi-arid and arid lands in Africa. In sub-humid and humid areas increasing demand for arable land from a rapidly growing population has led to ever-shorter fallow periods, which no longer enable the restoration of soil fertility. Declining soil fertility and increasing weed pressure increase the workload of farmers, mainly women farmers, while yields persistently decline. In Sub-Saharan Africa yield levels of staple food crops such as maize, sorghum and millet are down to an average of 1 ton grain per hectare in semi-arid rainfed farming systems, less then 25 % of the on-farm potential (Scherr, 1999).

There is an urgent need for the introduction of sustainable soil management practices. The efforts of research and development organisations in the past decades did not lead to the required breakthrough. Most of the technologies developed, such as soil and water conservation practices, planted fallows or alley cropping, increase the labour demand and thus the production costs while yields increase, if at all, only with a time delay.

The development of conservation tillage methods and especially direct planting in North and South America and the rapid spread of these alternatives in countries like Brazil have inspired researchers and developers in African countries in the last years to adapt these methods to the environmental and socio-economic conditions of the mainly small holder farmers. These efforts, however, did not lead to a break-through, as they were conducted by individuals in an isolated manner. The exchange between institutions within and between countries was lacking. Apart from some commercial farmers in southeastern Africa, the search for more sustainable soil management practices was not led by farmers. This lack of participatory approaches to development of such innovative technologies as conservation tillage is a major explanation to the limited adoption among farmers and the difficulty in making conservation tillage a valid alternative by government services and development organisations.

STEPS TOWARDS NETWORKING

The impressing spread of direct planting systems in countries like Brazil or Paraguay, where the German technical assistance was involved in technology development from the early begin, induced the German Ministry for Development Cooperation to look for possibilities of transferring the knowledge from Latin America to African partner countries. As a first step, a study on the potential of direct planting in Africa was conducted in 1997, taking into consideration the ecological and socio-economic frame conditions (Steiner, 1998). Main results of the study were, that (1) most favourable conditions for the application of no-tillage practices prevail in the sub-humid and humid regions, mainly because of the high biomass production. The widespread cultivation of root and tuber crops, however, limits the application of no-tillage practices. (2) In the semi-arid regions, which are in fact the main agricultural production zones, the relative short growing season and the competition for crop residues from livestock husbandry, are constraining the adoption of direct planting. In these regions, second-best methods like minimum tillage (ripping) may have, however, a good potential. (3) In most rural areas the socio-economic prerequisites for a change of the production systems are lacking. The rural infrastructure is underdeveloped, in particular access to inputs such as implements and agrochemicals (fertilisers and herbicides), credits, information and sales outlets. More-over these techniques require increased management inputs, this applies particularly to weed control and ground cover (crop residues management, green manures/covercrops). Perhaps the major limiting factor is, however, the traditional communal land tenure system, which limits the right of land use to the growing season. After crop harvest the fields are open for common grazing. This practice makes a sound crop residue management as well as the use of covercrops, preconditions for no-tillage systems, impossible. The search for a solution to this constraint is aggravated by the fact, that migrating pastoralists use graze their cattle in the dry season in farming areas. (4) The study concluded that despite these constraints no- and minimum tillage practices have the potential to contribute to a more sustainable soil management and consequently an increased food security in the main agricultural zones of the African continent. All persons interviewed expressed the need for a better exchange of information, for more collaboration across borders, and for networking. It is expected that by sharing of information and experiences being generated from within the continent and by more co-operation between all actors in the field of land management (farmers (small holders and commercial farmers, government services, development organisations and the private sector, mainly implement manufacturers and chemical companies) the rate and extent of development and adoption of conservation tillage technologies (minimum and no-tillage) will accelerate.

In a second step the German development co-operation (GTZ), in co-operation with FAO, the Agricultural Research Council (ARC) of South Africa and the Zimbabwe farmers Union (ZFU) organised an international workshop entitled «Conservation tillage for sustainable agriculture» in 1998 in Harare, Zimbabwe. The term «conservation tillage«, favoured in southeastern Africa embraces all practices that minimise soil disturbance and contribute to a sustained use of agricultural soils. Workshop participants recommended, amongst others, the initiation of a regional network. There was a long debate as to whether the term «conservation farming» or «conservation tillage» should be applied. The term «conservation tillage» was finally chosen in order to narrow the focus of the network. A working group was mandated to work out the terms of references for such a network. The main interested of workshop participants was in sharing experiences in conservation tillage (Benites *et al.*, 1998).

In a third step a planning workshop was organised in early 1999, again supported by GTZ, FAO, ARC and the Swedish development co-operation (SIDA). This workshop elaborated the objectives, activities and structure of a regional network and nominated a steering committee. The steering committee was mandated to identify an institution, which could host a permanent secretariat and to raise funds for running the secretariat as well as for financing network activities. It was believed, that at least in the initial period the network would depend on external funding. The network was named «African Conservation Tillage Network» (ACT), even though the founders come from southern and eastern African countries, only. But it was reasoned that the network should expand in future to other parts of Africa. A project proposal was written and submitted to the European Union and the German government. The network became operational with German funds in 2000.

Structure of the network

To achieve these results an appropriate structure of the network is required. The network is directed by a *steering committee* representing some of the main players in the agricultural sector, such as advisory services, farmers' associations, bi- and multilateral development co-operation organisations, NGOs and research organisations) including the private sector (e.g. farm implement manufacturers, chemical companies). The network is managed by a *permanent secretariat led by the network co-ordinator*, which is hosted by the *executing agency*, the Institute of Environmental Studies in Harare. A permanent secretariat was considered as necessary especially for the initial period of the network. Task of the co-ordinator is, to develop.

Table 1: Definition of conservation tillage and examples of methods

All soil and water management practices, which reduce soil tillage (soil disturbance), maintain at least a minimum soil cover from crop residues or green manures/cover crops, and involving rotations. Planting is done through the mulch and whiles chemical may be necessary for weed control in the first few years, this thins out as the system begin to manage a low level of weed infestation.

Principle practices applied in Southern and Eastern Africa are:

- **Ripping**
  ripping only the planting line with help of a tractor or animal drawn rippertine, attached to the ripper can be planters and/or fertiliser hoppers
- **Tied ridges**
  mainly for holding water and facilitating infiltration in low rainfall areas. The tied ridges are also used in high rainfall areas to control and manage excess water and possible runoff. Often done on permanent ridges; the tying ridges are lower than planted ridges.
- **Mulch placing**
  mulch material taken from non-cultivated surfaces; these materials would often last no longer than that particular season
- **Direct planting 1**
  used mainly in commercial farming under irrigation, i.e. where two crops per year are produced and a good ground cover maintained
- **Direct planting 2**
  in hand hoe systems were planting holes are made just before the start of rains and planting done with the rains. This goes with a heavy weeding requirement.

It is also common to find kraal manure applied in the planting hole prior to planting.

- **Pot holing**
  hand-hoe made planting stations designed for harvesting and consentrating water in the planted stations. Fertilizer is also appied in the same station. Planting will always be done in the same stations to allow newer crops to benefits from the fertilizer residue effects.
- **Winter ploughing**
  tillage done soon after harvest. The soil is relatively dry, so the ploughing leaves large clods on the surface. The rough surface is desired to minimise wind erosion and, mostly importantly, collect within the field the first rains and prevent runoff.
- **Surface physical measures**
  Especially in the control and management of erosion, a variety of physical measures have been employed. These include contour ridges, storm drains, grass strips and level bunds. These are normally runoff barriers (controlling runoff velocity and quantity) aligned across the slope.
- **Agroforestry**
  the growing of crops and trees together on the same field in some form of spatial arrangement or temporal sequence and improved overall yields. Common

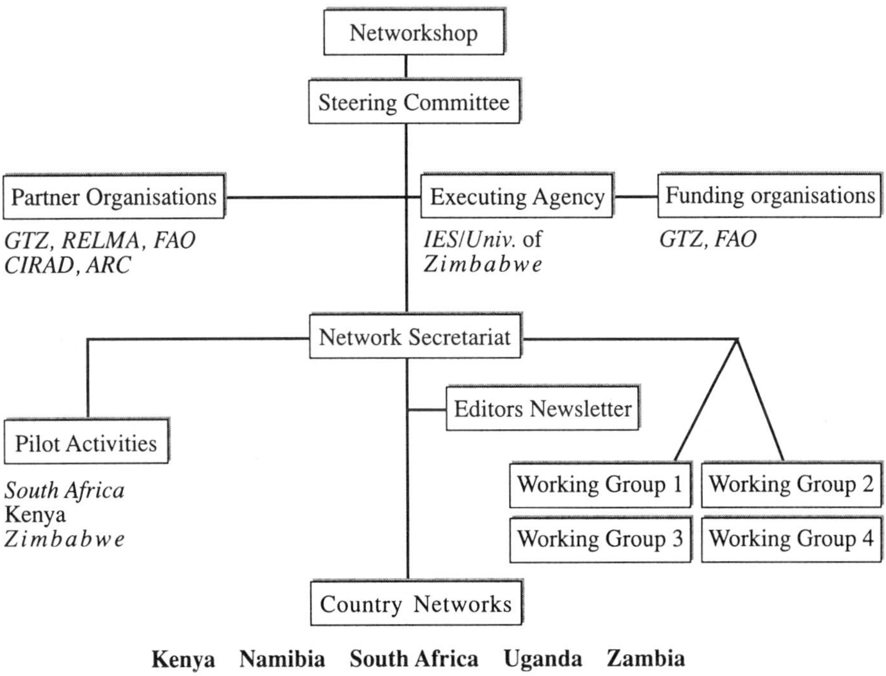

Figure 1. Structure of the network.

practices include the planting of *Sebania sesban*, *Tephrosia vogelii*, in crop fields. This is also sometimes used in what is referred to as improved fallows; tree or shrub plants are planted and left to grow for 2-3 seasons. There are then cut leaving the stamp in the ground. The trees would have also littered the ground surface with leaf material. The plant fixes nitrogen and the roots help break hard pan, while the leaf materials provide soil cover and organic matter

- **Green manure/cover crops:**
  planting of especially leguminous plants, which are later, ploughed under as green manure has been promoted in the region. This includes plants such as sunhemp (*Crotalaria juncea*) and mucuna (*Mucuna pruriens*). However, adoption of such practices is low, as farmers appear to have other priorities other than planting a crop only to be ploughed under. In southern parts of the region with one 4-5 months rain season, possibilities are also considered difficult. In the northern areas one close alternative is they let natural grass grow under fallow and later buried into mounds on which crop is then planted and implement annual work plans based on the planning of the first phase of the network. This is a somehow ambiguous task, as most of the planned activities should be executed by network members, as part of their ordinary work. The regional network is considered as a network of networks, which means, it should be based on *national networks.* These networks exist, however, for the time being in a few countries, only, South Africa, Namibia, Zambia, Uganda and Kenya. *Membership* is open to persons or groups of persons and institutions with interest and desire to actively contribute to the development and promotion of conservation tillage technologies for smallholder farmers in Africa.

Funding of the network

The network does not dispose of own funds. The initial funding is assured by the German development co-operation (GTZ), in the frame of a regional development project. Specific activities, such as participation in workshops, information management or publications are sponsored by network partners, such as FAO, SIDA, or ARC. It is hoped, that in the long run funds will be raised by membership fees, mainly of institutions, and by sales of publications or training materials. For the time being membership is free, except for applicants from overseas without a stake in Africa. It is argued that membership fees can be raised only, when the network can offer real incentives. It is, however, obvious, that the external funding is limited in time, and that the network is not sustainable without own financial sources.

Partners of the network

To build the network, engaged partners are required. For the time being these are more international organisations than national ones. Besides GTZ, SIDA with its regional programme RELMA is a strong support of the network. RELMA covers nearly half of the geographical region of the network, and the objectives of the network are quite similar to RELMAs objectives. FAO, in co-operation with the World Bank, aims as transferring the know-how from Latin America to the African continent. FAO assists in the building and maintenance of data bases, organises training workshops and sponsors the participation in workshops or study tours. The ARC is the only national organisations, which actively supports the network, mainly by allowing one of its researchers to spend a considerable time in network activities. It is hoped, that other national institutions, will follow this example.

NETWORKING FOR CONSERVATION TILLAGE

Core functions

During the last networkshop ACT's core functions were defined as:

1. Stimulate and support formation/strengthening of national networks
3. Dissemination, mobilisation, provision and inventory of information
4. Creation of enthusiasm and momentum among the actors - establish ownership
5. Documentation and dissemination of the evidence of CT benefits
6. Organisation of learning through exposure/seeing
7. Support to lobbying at national level
8. Support to the development of dissemination material

Major network activities started initiated since 2000

In line with the above core functions and the project document submitted to the funding agency, the networks, i.e. the members and partner organisations have started a series of activities.

- ACT publishes monthly electronic *newsletters*, which are widely disseminated, not only to members but to many other interested individuals and institutions.
- In co-operation with FAO a *website* was established. This website provides actual news and gives access to *data bases* which are actually build. The literature database, focusing on Africa, is already accessible, while for the databases on green manures/cover crops and conservation farming implements the structures exist but the data have still to be collected and filled in. This task can be achieved only with joint efforts of all the stakeholders active in the respective fields.
- In view of collecting and compiling information *topical work groups* will be formed. In total seven work groups are planned:
- Green manure/cover crops/mulch
- Weed control in conservation tillage systems
- Impact of conservation tillage on soil quality
- Conservation tillage implements
- Conservation tillage dissemination approaches
- Curricula for schools and tertiary learning institutes
- Links to input/output markets from both manufacturers and farmers perspective

The group on conservation tillage implements has taken up the work during the last months, while the group and green manures/covercrops is still in the process of formation. Both groups are involved in the set up of the databases on conservation tillage implements and green manure/covercrops/mulch.

- *Pilot activities* are conducted together with partner organisations in South Africa, Zimbabwe and Kenya. A major reason for these ground activities is the concern that sole networking might benefit small farmers only in the long run. Purpose of these activities is the development and testing of a «conceptual framework for enhancing the dissemination of conservation tillage». Major objective is to develop and test approaches for the acceleration of the adoption of conservation tillage practices by resource poor farmers in Southern and Eastern Africa. This objective should be achieved through the
  - Development or adaptation of alternative conservation tillage practices for different farming systems and environments
  - Development and application of participatory extension approaches
  - Initiation of farmers groups or strengthening the performance of existing groups
  - Promotion of farmer innovations
  - Awareness creation of teachers, students and local authorities for the need of a change of tillage practices

Figure 2. Conceptual framework for the dissemination of conservation tillage technologies.

Table 2. Country Networks and Initiatives in Southern and Eastern Africa

Country	***Name***	***Region covered***	***Principle Partners***	***Funding***
South Africa	*No-Till Club*	*Kwazulu-Natal*	*Commercial Farmers*	*Members*
Namibia	*SoilNet*	*country-wide*	*Min. of Agric, University and Polytechnic colleges, Dev. Projects*	*Min.of Agric. Donor organisations.*
Zambia	*Conservation Farming Unit (CFU)*	*country-wide*	*Min. of Agric., Zambia National farmers Union, Dev. Projects*	*SIDA/US-aid*
Kenya	*Kenya Conservation Tillage Initiative*	*country- wide*	*Min.of Agric., Univ., Agric. Res., NGOs, Dev. Projects, Private Sector*	*SIDA/RELMA*
Tanzania	*Planning stage*	*Starter workshop in July*		
Uganda	*Sustainable Soil Management*	*Country-wide*	*Ministry of Agric. National Agric. Research Org (NARO)*	*Ministry of Agric*

Primary target group of these pilot activities are women farmers and women groups. This mainly, as the number of women headed farm households in steadily increasing in the region due to AIDS and male out migration to industrial and urban centres. Women headed households are overburdened with work, and therefore technologies which might reduce the labour input should profit in the first rang women.

- Partners in the region are supported in forming and/or strengthening *country networks*. Networks exist already in some Provinces of South Africa, e.g. the Contill Club in Kwazulu- Natal, the SoilNet in Namibia, the Conservation Farming Unit in Zambia and the Kenya Conservation Tillage Initiative in Kenya (Biamah *et al.*, 2000). Efforts are actually undertaken to assist stakeholders in Tanzania and Malawi to form their own country network and to strengthen the newly formed network in Namibia (Republic of Nambia, 2000) and Uganda.
- Hardly any information is available on the surface under conservation tillage and the number of farmers practising conservation tillage in the various African countries. Therefore the network has started a *baseline survey* in some countries, were members volunteered to collect the required data. In addition, little and only scattered information is available on the benefit of conservation tillage, such as reduction of labour input and total production costs as well as increase of yields and income. Especially long-term data are missing. Such data are, however, required for any lobby work for conservation tillage. Political decision makers will be only convinced when confronted with hard facts.

- Baseline survey work is underway in Zimbabwe and Namibia and is expected to start soon in Kenya, Uganda, Malawi, Zambia and Tanzania. The survey should provide information on the status of conservation tillage (surface, no. of farmers) which is actually lacking for all African countries.

CHALLENGES AND CONSTRAINTS

There is a growing interest in conservation tillage in most African countries. While commercial farmers practice conservation tillage already on a relatively large scale (e.g. about 75% of commercial farmers in Zimbabwe are practising some form of conservation tillage) government services and development organisations have only recently discovered that conservation tillage is also an option for small farmers. This is partly the result of efforts of international organisations like World Bank, FAO and Sasakawa 2000, who undertake efforts to transfer the knowledge (technical know-how and lessons learned) from Latin America to Africa. However, most government services are rather reserved with regards to these initiatives. This is, e.g., the case in a country like Zimbabwe, where ploughing with draft animals was successfully introduced into the smallholder sector during the last 20 years (about 80% of the farm service is ploughed), and where especially the extension service is rather reluctant to change the extension messages. As the issue of conservation tillage is not just another new technology, but also a complete reversal in the way farming is perceived, both private/NGO and government agricultural research, extension and training programmes will need to be specially targeted to enable them well back-up farmer innovations in the adoption of conservation tillage practices.

One of the lessons learned from Brazil is, that new technologies spread only fast, when farmers feel the need to change their practices and when they take the lead in the development of alternative technologies. This requires efficient farmers organisations, which are, however, lacking in Africa. Here commercial farmers union exist in countries with a substantial number of commercial farmers, but hardly any small farmers union except of some unions, which are more or less political or formed around the distribution/acquisition of farm inputs. Thus the empowerment of farmers, the initiation of farmers clubs, conservation tillage or friends of the earth clubs, combined with management training is a precondition for the enhanced dissemination of conservation tillage.

And last but not least, the private sector, implement manufacturers and repair services needs to be strengthened. For the time being implements for conservation tillage are produced only in Zambia and Zimbabwe with, however, still very confined distribution. These implements are hardly available in other countries. The development of new implements based on experiences from Latin America is insufficient. While commercial farmers are able to command direct planting machines from United States, New Zealand or Brazil, it is hardly possible to import e.g. animal drawn direct planting machines. This is to a large extent caused by the low purchase power of small farmers, who do not dare to invest in new technologies. . Therefore, collaboration and networking efforts should interest and integrate participation of the private sector. Private sector participation will be crucial in farmer initiatives especially as concerns issues of supply.

By far the most important constraint is, however, the harsh environment the majority of smallholder farmers face. Low soil fertility combined with unreliable rainfall, a high risk of crop failure due to drought and limited access of agricultural markets make farming an unprofitable and risky enterprise. An Africa specific constraint to conservation tillage is the traditional communal land tenure system, which limits the land use right to the growing period. Fields in the savannah regions turn into communal grazing grounds during the dry season, leaving them completely bare at the beginning of the season. Because of this hardly any farmers is ready to grow green manures or covercrops. Direct planting is more or less reserved to commercial farms, which are able to prevent uncontrolled grazing. Possible solutions to this situation are binding local agreements on controlled grazing of fields, and the introduction of feedlots. These feedlots could be planted with leguminous cover crops, which are grazed to a certain extent. The negotiation of such agreements takes, of course, quite a time and requires an external mediator. But it is possible, as the experience of GTZ projects in Senegal shows (GTZ, 2000). The situation becomes, however, complicated in regions, where migrating pastoralists graze their cattle during the dry season on crop residues, such as the Masai East Africa or the Fulani (Pheul) in West Africa.

OUTLOOK

Conservation tillage definitely has a high potential in African smallholder agriculture, despite all constraints mentioned above. It may even be that the environmental constraints: recurrent water scarcity, low fertile soils and limited biomass/low organic matter in soils, actually form the basis for the major argument why conservation tillage is particularly important in sub-Saharan African drylands. Conservation tillage focuses on maximising rainfall infiltration and plant water uptake of soil moisture. This is not the case with former soil and water conservation technologies, which focus on erosion control. There is ample evidence showing that even in African drylands, water scarcity in agriculture is not necessarily caused by absolute lack of water, but due to large losses in the water balance (as runoff, soil evaporation and drainage) and the poor distribution of rainfall over time. Conservation tillage addresses these issues.

The rapid spread of direct planting in Latin America has raised hopes, that these technologies might be also find their place in Africa and contribute to combat land degradation and desertification. The African Conservation Tillage Network, founded by optimistic promoters of conservation tillage practices, undertakes efforts to keep the momentum and pave the way for less soil destruction farming practices. The network addresses itself primarily to government services and development projects active in the field of research and extension. Due to the different ecological and socio-economic conditions on the African continent, the network does not focus on direct planting but promotes all soil management practices, which help to stabilise soils and increase water infiltration and storage.

The activists of the African network and the country networks are aware of the difficult task they are undertaking and that it will be a long way, till conservation tillage is practised on a significant percentage of agricultural lands, however, they are convinced that this engagement is worthwhile and will contribute to a change of farming practices in the next future.

REFERENCES

Benites, J., Chuma, E., Fowler, R., Kienzle, J., Molapong, K., Manu, J., Nyagumbo, I., Steiner, K. and Veenhuizen, R. (eds.). (1998) Conservation tillage for sustainable agriculture. *Proceedings of the international workshop in Harare, Zimbabwe, 1998.* **Part I and II**. GTZ, Eschborn.

Biamah, E.K., Rockström, J. and Okwach, G. E. (2000) Conservation Tillage for dryland farming. Technologiocal options and experiences in Eastern and Southern Africa. RELMA Workshop Report No.**3**, SIDA/RELMA, Nairobi.

Douglas, M.G., Mughogho, S.K., Saka, A.R., Shaxson, T.F. and Evers, G. (1999) Malawi. An investigation into the presence of a cultivation hoe pan under smallholder farming conditions. TCI Occasional Paper Series No. **10**, FAO Investment Centre Division, Rome.

GTZ (2000) Codes locaux pour une gestion durable des ressources naturelles. GTZ, Dept. **45**, Eschborn.

Kaumbutho, P.G. and Simalenga, T.E. (eds.). (1999) Conservation tillage with animal traction. A resource book of Animal Traction Network for Eastern and Southern Africa *(ATNESA)*, Harare, Zimbabwe.

Republic of Namibia (2000) Proceedings of the workshop Soil Fertility Management and Conservation Tillage for Sustainable Land use in Namibia. 16-18 May 2000, Min of. Agric., Water and Rural Dev, Windhoek

Scherr, S. J. (1999) Soil degradation. A threat to developing country food security by 2020? Food, Agriculture, and the Environment Discussion. Paper **27**, IFPRI, Washington

Steiner, K.G. (ed.). (1998) Conserving natural resources and enhancing food security by adopting no-tillage. GTZ, Eschborn.

M. AHMAD GILL

NEED FOR ESTABLISHMENT OF SOUTH ASIAN CONSERVATION AGRICULTURE NETWORK

Director General Agriculture (Water Management), Agriculture Department, Government of the Punjab, Lahore Pakistan

Food security and development of sustainable system for use of land and water resources will remain key concern for the 21st century in many regions of the world including South Asia. Continuous and concerted efforts are needed to alleviate poverty, create employment, and boost agricultural production while preserving the natural resources.

The most viable approach for obtaining higher agricultural production from limited available resources is adoption of Conservation Agriculture technologies which offer not only higher productivity but also efficient and effective utilization natural resources as well as avoid environmental degradation. Asia region requires no emphasis on need of adoption of conservation agriculture as it supports 61 percent of the world population. Although technologies are available for efficient utilization of resources and are being practiced in some parts of the world but practitioners in most of the regions have no access or awareness about the same. There is, therefore, a dire need to establish a Network for providing a platform to channelize information amongst researchers, equipment manufacturers, extensionists, farmers committed to promote and use Conservation Agriculture practices. Such a Network will provide an opportunity to all stakeholders to share their knowledge, information, experiences, and problems which will stimulate promotion of Conservation Agriculture in this part of the world for production of more grains per unit of land, water, energy, and other rapidly depleting resources. It was proposed by the workshop on Conservation Tillage held in Kazakhstan in 1999 to establish a Network of various stakeholders in South Asian region to promote Conservation Agriculture. Subsequent workshop of Conservation Agriculture held in 2001 in Pakistan also endorsed the earlier decision and recommended for establishment of South Asian Conservation Agriculture Network (SACAN). Proposed SACAN will be a turning point in making this region self-sufficient in food through adoption of Conservation Agriculture practices. This paper deals with the need of this Networking, its mission, objectives, management structure and future roadmap to launch it.

BACKGROUND

Asia covers about one fourth of the globe's land mass and accommodates over 60 percent of the world's population. It constitutes one quarter of agricultural area and 36 percent arable land of the world. The continent produces 43 % of the total wheat of the world and shares over 90% of total rice production. Nature has blessed the region with abundant agricultural production and manpower resources. The diversity in na-

L. García-Torres et al. (eds.), Conservation Agriculture, 123–132.

tural resources, climate, people, and socio-economic conditions has given rise to varying farming systems in the region. South Asia's Indo-Gangetic plains constitute one of the most agriculturally productive areas of the continent.

The «Green Revolution Technologies» introduced in this region during mid 60s and 70s led to dramatic increase in the yields and production of wheat and rice, the two most important staple food grains. These crops occupy 54% of the total world's area planted to cereals and provide 41% of the calories consumed by over six billion people around the globe. In the first decade of Green Revolution, grain production grew rapidly mainly due to high yielding varieties (HYV) and increased cropped area. Afterwards, the land devoted to rice and wheat has been stabilized and there is very little scope for further expansion in area for these crops.

POST GREEN REVOLUTION CHALLENGES

The evidence is accumulating that growth in rice and wheat yields has started slowing down in the high potential agricultural areas of South Asia. This may have resulted from degradation of the resource base devoted to rice-wheat systems. The factors which stabled the Green Revolution in South Asia are mining of soil nutrients, declining organic matter, increasing salinity, fluctuating watertables, and buildup of weed infestation, and pest population. Another important factor in stagnation of productivity could be traditional way of cultivation leading to heavy tillage. Contrarily, population of the region is increasing at an annual rate of over two percent. These conflicting realities i.e. declining rate of production and increasing population growth are a serious source of concern for all of us. Luckily, productivity enhancement technologies exit which do not pose threat to natural resources. Effective linkages are, however, required to be established for transmissions of these innovations to the user community. At present, total population of the region comprising of Bangladesh, India, Nepal and Pakistan is 1.30 billion. It is estimated that the same is going to be doubled by the year 2025. Thus posing a serious threat for food, fiber, shelter and environment.

SHRINKING WATER RESOURCES

Freshwater ecosystems occupy less than 1% of the earth's surface but deliver goods and services of enormous value. Irrigated agriculture supplies about 40% of the world's food crops and hydropower provides nearly 20% of the world's electricity. An estimated 12% of all animal species live in freshwater ecosystems and most other depend on some way on freshwaters for their survival. Compared with domestic and industrial water uses, agriculture has a disproportionate impact on water flow, water quality, and alteration of freshwater habitats. About 70% of the world's all water withdrawals are for agriculture but more than half of the water entering irrigation distribution systems does not reach the crops because of leakage and evaporation.

As population grows, dependence on irrigation will become even more for meeting food requirements. This will place extraordinary stress on freshwater systems, particularly in arid and semi-arid regions. Despite their value, freshwater ecosystems are being intensely modified and degraded by human activities. So much water is consumed for domestic, industrial, and agricultural uses that the natural flow of major rivers such as the Colorado, Yellow, Indus, and Amu Darya no longer reach the sea during dry season.

New estimates of water scarcity calculated by the World Resources Institute in collaboration with the University of New Hampshire show that 2.3 billion people live in river basins under water stress with annual per capita water availability below 1,700 m^3. Of these, 1.7 billion people reside in highly stressed river basins where water availability falls below 1,000 m^3 per capita annually and chronic water shortages threaten food production and hinder economic development. Assuming that current consumption patterns will continue, at least 3.5 billion people or 48% of the world's projected population will live in water-stressed river basins by the year 2025.

Water is, therefore, certain to be a major topic of discussion during coming years and it seems likely that climate change will be a perennial topic at global gatherings of resource conservation and environmental policy-makers. As such, it is highly important to develop better understanding for water scarcity and its trend in future. It is also necessary to consider possible strategies for more efficient management of the water resources.

CONSERVATION AGRICULTURE

Proponents of Green Revolution generally argue that developing countries should opt for an agro-industrial model that relies on standardized technologies and ever-increasing use of external inputs to provide additional food supplies for growing populations. A vast majority of farmers and researches, however, propose that developing countries should favour an agor-ecological model such as Conservation Agriculture which emphasizes biodiversity; recycling of nutrients; synergy among crops, animals, soils, and other biological components; and regeneration and preservation of natural resources. For an example, «More you till, more you harvest» will remain no more sustainable. The priorities and options will, therefore, have to be re- fixed.

Conservation Agriculture is an approach for the design and management of sustainable and resource-conserving agricultural system. It seeks to conserve, improve, and make more efficient use of natural resources through integrated management of available soil; water, crop, and other biological resources in combination with selected external inputs. It represents a resource saving and efficient/ effective agriculture which contributes to environmental conservation and at the same time enhances production on sustainable basis. Elements of Conservation Agriculture, interalia, include organic soil cover, minimum tillage, direct seeding through the crop residue, and appropriate crop rotations to avoid disease and pest problems. Conservation Agriculture has several advantages over conventional agro-industrial approaches, a few of which are as follows.

- It relies on indigenous farming knowledge and selected modern technologies to manage diversity, incorporate biological principles, and low-cost locally available resources into farming systems in order to intensify agricultural production without causing environmental damage.

- It offers a practical way to restore agricultural lands that have been degraded by conventional agronomic practices.
- It provides for an environmentally sound and affordable way for small holders to intensify production in marginal areas.

WHY CONSERVATION AGRICULTURE IN SOUTH ASIA?

Total land area of South Asia is 437 million hectares with agricultural area of about 210 million hectares. A little expansion in agricultural area of Nepal and Pakistan took place during last forty years but total area under agriculture is the almost same as some decrease has occurred in agriculture area of other countries. On the other hand, the population of the region become 1.323 billion from 0.55 billion during last 40 years which is further expected to be doubled during next 20 years. Although the water resources have been expanded due to construction of dams, canal etc. during the last four decades and irrigated area has increased but development in the water resources is not consistent with population growth for sustainable agriculture.

This region supports 35 percent of the total population of Asia and produces 35 percent and 31 percent of the total wheat and rice produced in the Asian continent, respectively. The crop productivity is very low in the region as majority of the farmers are still practicing traditional production techniques. Moreover, the cost of production has increased many times due to rising prices of fuel and other agricultural inputs. The existing production technologies do not offer effective and efficient utilization of natural resources, especially water. Extremely low efficiency of input use has led to wastage and depletion of natural resources besides environmental degradation.

The above facts reveal that if an abrupt change is need for adoption of new technologies for crop production by making efficient use of available natural resources. Otherwise, it will become extremely difficult to fulfill food demand of the population of this region. The best option for effective utilization of resources and increasing productivity is adoption of resource conservation technologies.

PAKISTAN'S EXPERIENCE OF CONSERVATION AGRICULTURE

Agriculture in Pakistan has generally performed well and it grew by more than 3 percent a year for the past 40 years. Sources of growth, however, remained changing over the years from introduction of improved seeds, use of chemical fertilizers, and irrigation package during 1960s, to intensification of water and fertilizer use in the 1970s, to improvements in crop management and incentives in the 1980s. Although the past growth has been quite impressive but a careful look suggests that the momentum is running out. There is no chance of a significant increase in total cultivable land and irrigation infrastructure. Stagnation in productivity is mainly due to factors such as resource degradation, failure to adapt technological changes, and poor incentives. At best, a 10 percent expansion in water resources can be expected, that too at huge costs.

The growth in the future will have to be achieved through increased productivity through more efficient use of limited resources. There is a particular need to ensure efficiency and sustainability in management of the most important natural resources i.e. water and arable land. There are good prospects for enhancing productivity of all major crops, particularly that for wheat and rice where there is potential for increasing yield by about30-50 percent.

It is natural that tremendously increasing population competes with agriculture on land and water resources. More land is needed for the people to build their residential and other amenities and more water is required to meet their various demands. The population of Pakistan was only 50 million in 1960 which has increased to 156 billion during last 40 years i.e. almost three times increase. Resultantly, per capita land availability has been declined from 1.59 to 0.50 ha during last 40 years. Similarly, per capita agricultural land availability has been reduced form 0.44 to 0.17 ha. during this period. The aggregate per capita availability of essential food items in physical as well as in terms of calories and proteins has, however, increased during past forty years but these are not up to satisfactory level. Moreover, majority of farmers are still following traditional crop production techniques, which cannot cope with increasing food demand.

The prices of fuel and other agricultural inputs have increased exorbitantly during last 40 years due to which cost of production has become unbearable for the farmers to produce food crops. They are, therefore, compelled to change their crop rotations shifting from food corps to cash crops. In the long run, it will create serious problem for the government to provide food to the growing population. It is, accordingly, imperative to look for alternate crop production technologies that offer less cost of production as well as conserve the natural resources.

The trade policies in the past were conducive for developing countries, including Pakistan allowing liberal disposal of their agricultural products in the world market. It will, however, become difficult for these countries to remain competitive for their exports after implementation of WTO policies because of lower productivity levels viz-a-viz developed nations. Ways and means will, therefore, have to be adopted to lower the cost of production and enhance productivity simultaneously. The only possible option is that of adoption of new resource conservation technologies, which offer both of these opportunities.

Development of Conservation Agriculture technologies in Pakistan dates back in late sixties. Studies on factors hindering agricultural growth and possible solutions for the same started at MONA Soil Reclamation Center, Bhalwal of Water and Power Development Authority (WAPDA) in collaboration with Water Management Synthesis Research Project and Colorado State University of USA. After several years of hard work, a technological package was conceived for effective and efficient use of water resources in conjunction with other inputs. This comprised of Conservation Agricultural practices, interalia, including Watercourse Improvement, Precision Land Leveling, and Improved Irrigation Agronomic Practices for various crops. The package was pilot tested under On Farm Water Management Development Pilot Project implemented in various agro-climatic zones of the country during 1976-80. Starting with improvement of five watercourses in 1976-77, almost 50% (27,000) of watercourses of Pakistan have so far been improved. Likewise, Precision (subsequently LASER) Land Leveling, another resource conservation technology, was introduced on an area of 555 acres in 1976-77 and about 0.5 million acres have been precisely leveled in the country. Adoption of improved water management practices in conjunction with other inputs is an important element for water resource conservation. Starting form 174 demonstration plots in 1981-82, and so far these practices have been demonstrated on about 30,000 sites throughout Pakistan.

Minimum/zero tillage is another innovation that not only offers conservation of water and energy resources but also results in better crop yields. This technology was being practiced since long in many parts of the world and it was introduced in Pakistan during 1980. It could, however, not be popularized among the farming community inspite of a lot of benefits mainly due to lack of vigorous follow up. OFWM took responsibility of its introduction to farmers during 1996-97 and wheat was grown on 50 acres at 12 sites with only five government owned Zero Tillage Drills. The technology became acceptable among the farmers due to its contribution in reducing cost of production, conservation of resources and improving yields. It is an encouraging fact that during 2000-01, wheat was grown with this technique on an area of about 100,000 acres in Pakistan and there are more than 500 Zero Tillage drills owned by farmers.

Bed and Furrow planting technology permits growing of crops on beds with less water requirement. This technique has been tested for various crops and it has proved successful for many crops. Recently, its introduction has been started amongst the farming community. Wheat was grown on 137 acres during 1999-2000 with only four bed planters that have expanded to 411 acres during 2000-2001 and there are now 20 such machines with farmers.

SOUTH ASIAN CONSERVATION AGRICULTURE NETWORK
WHY NETWORKING FOR CONSERVATION AGRICULTURE IN SOUTH ASIA?

Many countries of the world have adopted resource conservation technologies. In USA. Brazil, Australia, Canada, Argentine, New Zealand, and Mexico, the Conservation Agriculture/Zero-Tillage is being adopted on an area of 125 million hectares. The reason for the success is the establishment of Networks in these regions/countries like South Africa Conservation Tillage Network (ACT), European Conservation Agriculture Federation (ECAF) etc. which made it possible to share the information among various stakeholders. This exchange of experiences ultimately led to rapid adoption of the new developments. Although the conservation agriculture is being promoted in South Asia region but its pace of adoption is very slow due to absence of any institution that stimulate dissemination of information to farming community and exchange of experiences among various stakeholders. Zero-Tillage is being practiced on about 3 million acres in this region while LASER Land Leveling has been done on only 10,000 acres. Furrow-Bed technology is yet practiced on very small scale. These achievements have been made during last six years. This pace of adoption is not encouraging as there is a great potential for adoption of these technologies. It is, therefore, need of the day to start immediate step for improving the adoption rate.

RATIONALE

The alternative technologies are available to replace conventional agriculture for enhancing productivity together with conservation of natural resources. Rate of adoption of such innovations is, however, quite slow. The key reason in this regard, among others, is the lack of effective forward & backward linkages amongst various stakeholders i.e. researchers, scientists, experts, institutions, extension workers, and user community. Some countries/ regions have bridged this gap very effectively and successfully by establishing Networks. Pace of adoption of Conservation Agricultural practices has accelerated at a phenomenal rate in domains of these Networks.

It is believed that there is a valuable experience and expertise in the practices and principles of Conservation Agriculture in the South Asia also and that effective collection, synthesis, and sharing of this knowledge will greatly stimulate adaptation and adoption of resource conserving practices and systems throughout the region. It is, therefore, imperative to provide a similar platform in South Asia as well in order to realize benefits of the available technological innovations.

The importance and need for a Network for promotion of Conservation Agriculture practices in the region was realized by the international workshop on «Conservation Tillage: A Viable Option for Sustainable Agriculture is Eurasia» held in Shortandy, Astana, Republic of Kazakhstan during September 19-24, 1999. Subsequently, its exigency was endorsed by «International Workshop on Conservation Agriculture for Food Security and Environment Protection in Rice-Wheat Cropping Systems» held at Lahore, Pakistan, during February 6-9, 2001. One of its recommendations was that:

«A Regional Network for Conservation Agriculture in rice-wheat ecosystem in Asia should be established similar to other networks around the world. Such a network would generate momentum for the adoption of CA in the region relevant to the diverse range of socio-economic and agro-ecological conditions. It would encourage exchange of expertise, experiences in conservation agriculture among stakeholders and member countries. This network should promote an active integration of CA and IPM approaches in order to achieve a more holistic and system response to the sustainability and productivity required of farmers, in particular small farmers. The proposed network could be called the South Asia Conservation Agriculture Network (SACAN). It is proposed that such a network be coordinated by On-Farm Management at Lahore, Pakistan».

Accordingly, a broad structure of SACAN outlining its mission, objectives, and management structure has been proposed as a further step in this direction. A road map for follow up actions is also suggested.

SACAN-MISSION STATEMENT

The mission of the South Asia Conservation Agriculture Network (SACAN) is proposed to alleviate poverty and improve environment through technology transfer and capacity building of different stakeholders by increasing production and productivity as well as nutritional quality of food alongwith preserving and enhancing the natural resources base. SACAN will be committed to the advancement and promotion of Conservation Agriculture through research together with free exchange of experiences, knowledge, information, and data.

PROSPECTIVE PARTNERS

South Asia Conservation Agriculture Network will be an international network of organizations and institutions. Its members will be equal partners interested in promotion of Conservation Agriculture practices in the region committed to its goals. As a reinforced network, SACAN will promote interaction among its members by promoting cross sectoral and multi-stakeholder dialogues at global, regional and national levels. SACAN will pursue its mission in partnership with international, regional, and national agricultural researchers, scientists, experts, extension workers, policymakers, government institutions/organizations, NGOs, machinery manufacturers, farmers and all other stakeholders.

OBJECTIVES

The main objectives proposed for the Network are:

- To act as a forum/ platform/ focal point for sharing/exchange of information and experiences among all stakeholders, interalia, including researchers, scientists, experts, policymakers, extension workers, manufactures, and farmers to increasingly apply methods of soil, water, and energy conservation that increase productivity, ensure environment protection, and are economically viable.
- Arrange workshops, symposia, discussion sessions, field days, and demonstration trails for adaptation of Conservation Agriculture.
- Collect as well as prepare promotional material for various aspects of Conservation Agriculture and disseminate it amongst the farmers to facilitate adoption of its practices.
- Carryout studies, evaluations, and research behalf of institutions and donor agencies
- To support action at local, national, regional, and global level that follows principles of sustainable agriculture resources management.
- To identify gaps and stimulate stakeholders interaction to meet critical needs within available land, water, and energy recourses.
- To encourage formation of national Networks and Associations aiming at promotion of Conservation Agriculture practices.
- Interact with other regional and International Networks for Conservation Agriculture, United Nation Agencies, CGIAR Institute, NARS, CIMMYT, IRRI, IWMI, ICARDA, RWC, Universities, FAO etc.

GEOGRAPHICAL MANDATE

South Asia Conservation Agriculture Network will promote its objectives in Pakistan, Bangladesh, China, India, Iran, Kazakistan, Mianmar, Nepal, Sri Lanka, and Uzbekistan.

ENVISAGED MANAGEMENT STRUCTURE

There will be a steering committee (SC), a Technical Committee (TC), various Technical Assistance Committees (TACs), and a Secretariat for operating the SACAN. SC will comprise of representatives of the partners and will be responsible for making decisions on operational issues while TC will provide professional and scientific advice to the partners and will also safeguard the technical quality of SACAN. TACs will be the centerpieces and of the future partnership. SACAN secretariat will provide all managerial and coordinating services for smooth operation of the Network activities.

ACTIONS REQUIRED BY MADRID CONGRESS

It is proposed that draft mission statement, proposal objectives/ functions, and suggested management structure of the SACAN may be deliberated in a special session of the Madrid Congress to finalize these aspects by making additions, alterations, amendments, and modifications for successful launching of the proposed Network. If found feasible, recommendations may be made to the:

- International Donor community to provide technical and financial support for launching and future functioning of the Network
- The host country/Government of Pakistan: a) to provide international status to the Network for providing travelling facilities (e.g. visa issuance) for experts visiting Pakistan in connection with its affairs as well as for providing other privileges extended to such organizations; and b) to provide support or establishment of Network secretariat.

It is also proposed that an Adhoc Committee may be constituted to take start up actions for establishment of the Network. Donors may be requested to provide seed money required for the purpose.

FUTURE ROAD MAP

A road map has been prepared to take start up actions for launching the proposed South Asia Conservation Agriculture Network. Some of its key elements are as follows.

- Establishment of Network secretariat.
- Development of Network Web Site highlighting its mission, objectives, functions, and start adding relevant information for electronic retrieval of stakeholders.

- Arrange registration of the Network as Non Government & Non-Profit organization.
- Establishment of linkages with prospective sources of related information and similar Networks.
- Draft Network Rues and Byelaws.

A. KURISHBAYEV

A STRATEGY FOR SOIL CONSERVATION FARMING IN NORTHERN KAZAKHSTAN

Kazakh Research Institute of Grain Farming, Shortandy-1, Akmola region, Kazakhstan

The arable land area in Northern Kazakhstan is approximately 19 million hectares including 3.3 million ha with complex of solonetz soils. Only the high quality of Kazakhstan wheat can ensure its marketability. Strong north-kazakhstani wheat can be used for improving the bread making quality of wheat grown in Belarus, Baltic countries, European part of Russia, some countries of Middle Asia and so on.
Northern Kazakhstan can produce not less than 3.0 million tons of strong wheat grain every year. Crop diversification is important to stabilize farm income due to erratic production and marketing conditions. The climate conditions and the agrolandscape of Northern Kazakhstan make it possible to divide the territory in the following zones according to grown crops:

- ordinary chernozems: feed and malting barley, bread wheat, rape, peas, oats, winter rye, buckwheat, soybean, the mixtures of grain legumes and forage crops, combined sowings of perennial legumes and cereal grasses, perennial legumes, perennial grasses, early-ripening hybrids and varieties of maize;
- southern chernozems: strong and valuable bread wheat, durum wheat, barley, peas, chickpea, mustard, oats, buckwheat, millet, sunflower, peavine, perennial grasses, combined sowings of perennial grasses and legumes;
- dark chestnut soils: strong bread wheat, chick pea, mustard, millet, Sudan grass, foxtail millet.

A national or regional agricultural practice policy should address the following points:

- the use of site specific management taking into consideration agro-landscape;
- the use of recommended agricultural practices without a routine approach considering existing climate and marketing conditions;
- the maintenance of a balance in soil fertility;
- to provide crop and production alternatives for farmers according to their input and marketing logistics and degree of diversification.

Key words: strong wheat, crop diversification, agrolandscape, agricultural practice

The sustainability of grain production in Northern Kazakhstan is related to the adoption of soil conservation farming systems which are the basis for ecologically sound, balanced and stable farm management and which satisfy to the greatest extent to the strict requirements of dry farming.

So the problems faced before the grain production branch of the Republic are accounted not only for droughts during last years but most of all for the non-observance of scientifically ground cultural practices.

What are the main problems in agriculture in our region?

L. García-Torres et al. (eds.), Conservation Agriculture, 133–138.

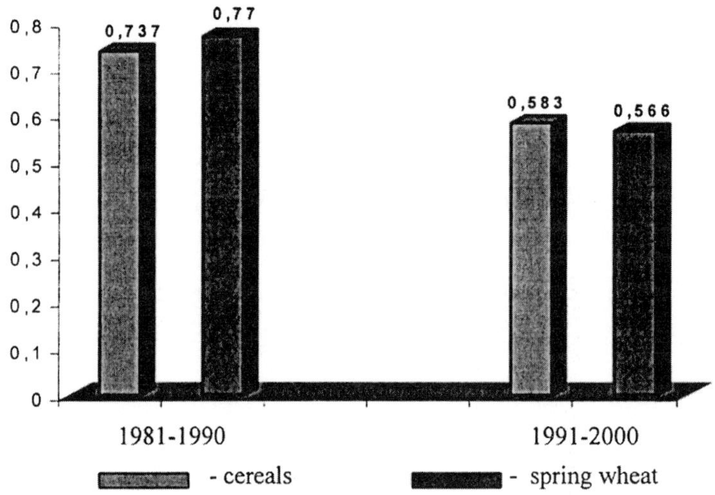

Fig.1. Coefficient of yield sustainability

- The decreased yield potential of grain crops accompanied by reduction of their acreage.
- The stability of yield formation that is the grain production sustainability is falling down year after year (Fig.1).
- The tendencies to grain quality deterioration of Kazakhstan wheat due to the inefficient use of soil-climate resources.
- The reductions in soil fertility that is considered the base for grain production sustainability. Non-observance of soil conservation system resulted in water and wind erosion. The humus content in the soil arable layer decreased by 25-30% due to labile humus substances determining supplies of easily mobile nitrogen.
- The application of mineral fertilizers in Northern Kazakhstan as compared to the eighties is about 2% that is it decreased 13 times including 32 times decrease under wheat. Therefore, 80% of arable land will have low availability of labile phosphorus within the next few years (Fig.2).
- Weed infestation of cereal crops remains severe. About 6.9 m ha are infected with perennial root offset weeds, 2.8 m ha with wild oat, 3.0 m ha with bristle grass and different sorghum-like weeds (Fig.3).
- Such diseases as root rots, loose smut, Septoria blight, brown and stem rusts, Helminthosporium blight are widely spread that reduces grain yield by 20-25% in some years.

The main reason for these problems is the shortage of funds needed to implement the recommended technology of cropping which is the result of price disparity for technique, fuel, chemicals and agricultural production and of limited budget funding of agriculture. Current farming systems should answer the following requirements:

1) to meet consumer and market demands;
2) to meet the biological requirements of agricultural crops;

2) to efficiently utilize the soil-climate potential;
3) to take into consideration the technical logistics of commodity producers.

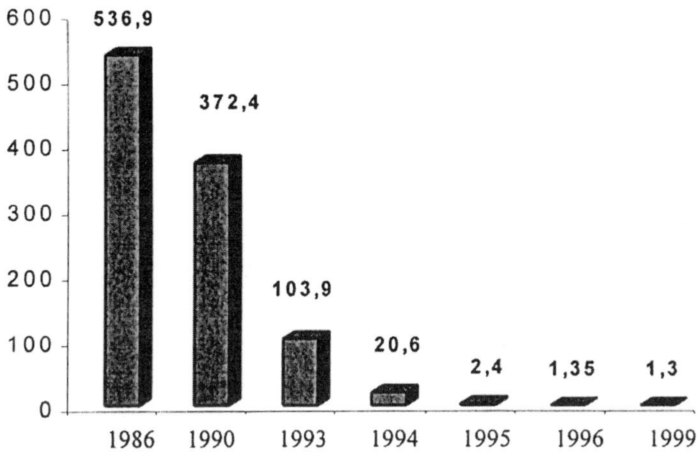

Fig.2. Dynamics of mineral fertilizers application in Northern Kazakhstan in 1986-1999 (thousand tons of active substance)

The main reasons of the existing ecological contradictions in agriculture are related to the consequences of directive land management in Kazakhstan, unfounded expansion of seeding acreage under cereal crops that resulted in a disparity of the used farming system and environment. At the end of 80th about 25 million ha of land were cultivated including arable land and meliorated pastures. 15.5 million ha or 63% were sown with cereal crops, 2 million ha were sown on non- arable lands.

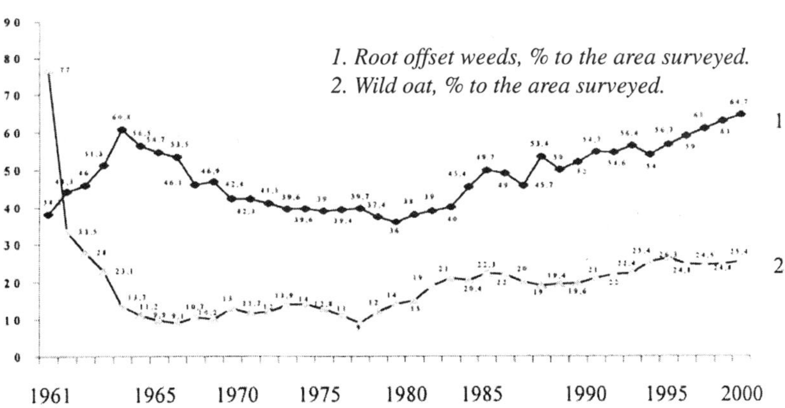

Fig.3. Dynamics of weed infestation and herbicide application in Northern Kazakhstan

Therefore, one of the key objectives of sustainable farming is to optimize the structure of arable land and to grow crops according to the conditions of specific region or agrolandscape.

The arable land area in Northern Kazakhstan after removing alkaline soils and the soils of radical improvement is 19 million hectares including 3.3 million ha with complex of solonetz soils (about 30%). The productivity of alkaline soils is 30-50 % less as compared to the productivity of zonal soils, so the perennial salt-resistant grasses and barley should be grown there.

The area of most fertile soils - ordinary and southern chernozems and also dark chestnut soils is 16 million ha. This arable land should be used for guaranteed production in Northern Kazakhstan: cereal crops, legumes, oilseed crops, forage crops and other high profitable crops can be grown there. The spring wheat is grown on the most fertile soils which make up over 8-9 million ha.

Such arable land conversion will increase the stability of grain production to the greatest extent. Moreover, the results of research and the experience of some foremost farms show that the improvement of technological discipline can give 30-40% increase in grain yield that is to secure the yield from 1.5-1.7 to 2.5-3.0 t per a ha on chernozem soils and from1.2-1.5 to 2.0-2.2 t per a ha on dark chestnut soils.

Only high quality of Kazakhstan wheat can ensure its marketability. Strong north-kazakhstani wheat can be used for improving the bread making quality of wheat grown in Belarus, Baltic countries, European part of Russia, some countries of Middle Asia and so on.

According to our assessments Northern Kazakhstan can readily produce not less than 3.0 million tons of strong wheat grain every year. To achieve this we should grow strong wheat in the zone of southern chernozems and dark chestnut soils where the annual average precipitation is 250-360 mm.

But the farming experience of highly developed countries demonstrates that in market conditions it is not profitable to grow only one crop, for example, wheat. The diversification of grain production branches should be done.

The potential of Northern Kazakhstan in growing oil seed crops and grain legumes can be clearly seen as compared to the structure of seeding area in Canadian province of Saskatchewan which has the similar soil and climate conditions. The planned area under crops in Northern Kazakhstan is given in Fig.4.

The climate conditions and the agrolandscape of Northern Kazakhstan make it possible to divide the territory in the following zones according to grown crops:

- ordinary chernozems: feed and malting barley, bread wheat, rape, peas, oats, winter rye, buckwheat, soybean, the mixtures of grain legumes and forage crops, combined sowings of perennial legumes and cereal grasses, perennial legumes, perennial grasses, early-ripening hybrids and varieties of maize;
- southern chernozems: strong and valuable bread wheat, durum wheat, barley, peas, chickpea, mustard, oats, buckwheat, millet, sunflower, peavine, perennial grasses, combined sowings of perennial grasses and legumes;
- dark chestnut soils: strong bread wheat, chick pea, mustard, millet, Sudan grass, foxtail millet.

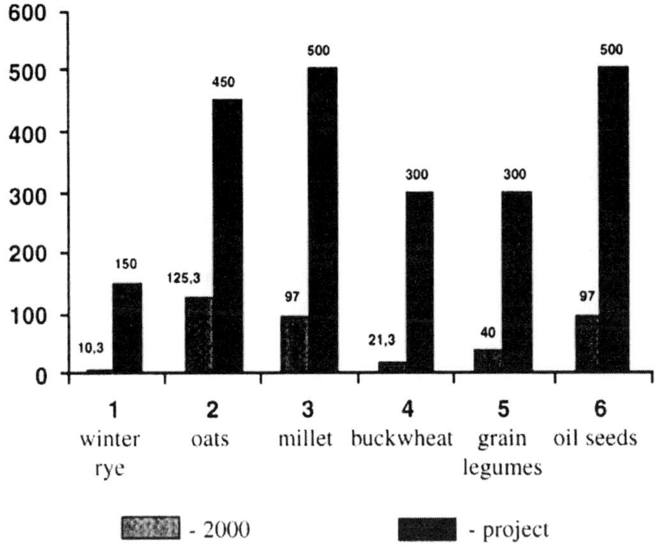

Fig.4. Structure of sowing area in Northern Kazakhstan, current and planned, thousand ha

In addition 4 categories of agrolanscape on chernozem soils can be mapped according to their susceptibility to erosion, soil fertility level and natural water accumulation. According to the institute recommendations, the production of high quality commercial grain, oilseed, legume and other lucrative crops should be concentrated in the first two categories of agrolandscape (where the soils are less erosive and more fertile). The forage grain should be grown in the third category of agrolandscape and the fourth category (where the soils are the most subjected to erosion and the least fertile ones) should be occupied by perennial grasses.

It is apparent that instead of the common tendency to increase fertility of all soils which, as a rule, doesn't answer the actual possibilities, the strategy of land management in Northern Kazakhstan should be directed first of all to the efficient use of the best soils. In agrolanscapes with high fertile soils it is worthwhile to grow the crops and varieties of intensive type that will result in increasing the profit costs of farming and ensuring the yield stability year after year. Moreover, the governmental investments should be put into the intensive use of the most fertile soils that can give the greater economic benefits.

So the main points of a national or regional agricultural practice policy should be as follows:

1) to use site specific management;
2) to use the recommended agricultural practices without a routine approach considering existing climate conditions;
3) to maintain the positive balance of soil nutrients by using the biological factors for restoring soil fertility;

4) to ensure weed, pest and disease control at the level which is lower than the threshold of their harmful effect with minimizing the mechanical soil tillage and chemical ways of control;
5) to provide the alternative for the farmers that is to make it possible for commodity producers to choose the different options of technology according to their logistics and diversified farming.

To provide for the sustainable and ecologically sound development of agriculture the crops breeding and seed production should be given a priority.

The main directions in this field are:

- to develop the new drought resistant, high yielding and high quality varieties of grain, forage, oilseed, groat and legume crops, perennial cereal and legume grasses which have different ripeness, are resistant to main diseases and pests and are responsive to different cultural practices;
- to improve the seed production of these crops providing for the speeded strain changing and renovation first of all by restoring the elite and commercial seed growing;
- to grow varieties according to different zones and agrolandscapes which allows to realize their biological potential.

The development of the ecologically sound, power saving agricultural machinery and technique must consider the following points:

- high adaptive implements for subsoil tillage (without forming clods on the surface) which decrease the power consumption by 15-20% and improve snow trapping by 20-30 %;
- the combined implements for seed bed tillage which allow to decrease power consumption, moisture evaporation and to increase the productivity and quality of work;
- different types of seeders adapted to the specific agrolandscape and climate conditions;
- wide-cut swathers for high cutting with minimal losses of grain harvested;
- power saving high productive combing machines for forming stubble strips;
- high efficient machines for chemical protection of plants ensuring the optimal application rates;
- the complex of machines for cleaning, drying and treating grain and seeds.

Thus the provision for sustainable grain production in the region is related to agrolandscape farming, zonal crop growing, optimization of soil fertility, introduction of high yielding and tolerant varieties, power and labour saving technologies and machinery that will result in the better realization of our bioclimatic resources and increase the efficiency of soil conservation tillage system, the base for grain production in steppe regions of Eurasian continent.

J. R. BENITES & J. E. ASHBURNER

FAO'S ROLE IN PROMOTING CONSERVATION AGRICULTURE [1]

Land and Water Development Division, FAO, Viale delle Terme di Caracalla, 00100 Rome, Italy

FAO Regional Office for Africa, PO Box 1628, Accra, Ghana

Serious problems of land degradation, falling soil fertility, rapidly declining production levels and desertification are occurring in large parts of Latin America, Asia, Eurasia and Africa, a main cause being the plough-based or hoe-based agricultural practices. These inappropriate land management practices cause the soil to become more dense and compacted, the organic matter content to be reduced and water runoff and soil erosion to increase. They also lead to the effects of droughts becoming more severe and the soils becoming less fertile and less responsive to fertiliser.

There is now a wealth of evidence from examples throughout the world of these sustainable production systems and their dramatic effects that can be achieved when the basic principles of good farming practice are applied. The terminology being adopted for such systems by FAO and other organisations is «Conservation Agriculture» (CA) and this implies conformity with all three of the following general principles:

- no mechanical soil disturbance and direct seeding or planting
- permanent soil cover, making particular use of crop residues and cover crops
- judicious choice of crop rotations

Particularly renowned practices of conservation agriculture have been applied on a large scale on both large and small farms in Brazil and Paraguay. Other examples bear out the contention that minimal disturbance of the root zone, once this has developed, substantially contributes towards raising mean yields.

Globally, CA is now being practised on about 60 million ha, mostly in the Americas. Its use is growing exponentially on small and large farms in South America, encouraged by economic and environmental pressures. Over the last decade, FAO has been implementing field projects that place greater emphasis on helping farmers to improve their care of the land through practices of CA rather than through efforts solely to combat erosion. Its CA work initially concentrated in Latin America but FAO is now also focusing on Africa, Asia and Eurasia.

FAO has now established an interdisciplinary Conservation Agriculture Working Group (CAWG) and has made a wealth of information available through its specialist conservation agriculture web-site. It is actively supporting several CA networks such as RELACO, ACT, SACAN and ECAN.

Key Words: Development co-operation, sustainable agriculture, tillage, conservation agriculture, cover crops, participatory planning, network

L. García-Torres et al. (eds.), Conservation Agriculture, 139–154.

INTRODUCTION

Serious problems of land degradation, falling soil fertility, rapidly declining production levels and desertification are occurring in large parts of Latin America, Asia, Eurasia and Africa, a main cause being the plough-based or hoe-based agricultural practices. These inappropriate land management practices cause the soil to become more dense and compacted, the organic matter content to be reduced and water runoff and soil erosion to increase. They also lead to the effects of droughts becoming more severe and the soils becoming less fertile and less responsive to fertiliser.

During the last 50 years there has been a profusion of agricultural development programmes throughout the world trying to arrest erosion, which is perceived as one of the main causes of land degradation. Sadly, though not surprisingly, the physical erosion control methods have failed to remedy the situation. What has often been judged initially as a success during the lifetime of a development programme, has frequently failed to be effective a few years later and after the departure of programme personnel, sustainability has failed to be achieved (Barber, 1999).

In the early seventies, farmers in Paraná, Brazil recognised that continuing erosion and declining crop yields were risking to force them to abandon their land and move into a marginal existence. They discussed what could be done to reverse these trends and at first attempted to rigorously adopt conventional terracing systems. The mixed and often disappointing results led them to later decide to approach the erosion problems at their source, considering the direct impact of rainfall on the bare soil. They abandoned the plough, scarified or subsoiled their compacted soils to break through induced 'plough-pans', actively discouraged the burning of crop residues and developed cutting rollers to flatten and break or cut the residues of the harvested crop. They also developed special seeders that could place the seed – and later, also the fertiliser– in a narrow slot cut through the flattened crop residue. This procedure eliminated rainfall impact on the soil, reduced the speed and amount of runoff and virtually eliminated erosion. It also significantly reduced the cost of land preparation for each new crop (Benites, 2001). The rate of adoption of these techniques has been remarkable. In Brazil alone, it is estimated that since 1995, an additional 2 million hectares of land is managed each year using these practices and today there are over 13 million ha conforming to what is referred to in this paper as conservation agriculture systems.

This gives occasion to comment on this particular terminology, as many other terms are also commonly used but which can have different connotations and which might not necessarily comply with a conservation agriculture system. Conservation tillage, reduced tillage, zero tillage and direct drilling are just a few of the terms frequently used. The terminology «Conservation Agriculture» that is being adopted by FAO and other organisations implies conformity with all three of the following general principles:

- no mechanical soil disturbance and direct seeding or planting of the crop. Soil inversion in particular is to be avoided

- maintenance of a permanent soil cover, making particular use of crop residues and cover crops
- judicious choice of crop rotations

It is thus seen that direct drilling is an essential component of a conservation agriculture system, but alone, it would not necessarily comply unless the practice is combined with permanent soil cover and crop rotations.

There is now a wealth of evidence from examples throughout the world of these sustainable production systems and their dramatic effects that can be achieved when these basic principles are applied. The Brazilian experience (and indeed that in other countries) demonstrates that this change of paradigms is possible. Instead of battling against nature, farmers have now found ways to produce in harmony with it. They have changed their production systems completely, by trying to imitate natural habitats. Instead of using heavy equipment and fossil energy to mechanically cultivate the soils, they leave this task to natural elements such as earthworms, arthropods and other soil organisms (Steiner, 2001). They also have shown that it is preferable to switch from 'arresting erosion' to assisting farmers to achieve a higher, more conservation-effective and stable production, complementing this as necessary with physical works.

The following paragraphs provide a brief overview on past and recent efforts of FAO in supporting conservation agriculture. Lessons learned from successes and failures are drawn and avenues for future support to the promotion of conservation agriculture are projected.

FAO'S WORKING GROUP ON CONSERVATION AGRICULTURE

Globally, conservation agriculture (CA) is being practised on about 60 million ha, mostly in South and North America. The rate of adoption of these practices is growing exponentially on small and large farms in South America, stimulated by both economic and environmental pressures. In Europe, the European Conservation Agricultural Federation (ECAF), a regional lobby group, has been founded. This body unites national CA associations in the UK, France, Germany, Italy, Denmark, Belgium, Switzerland, Portugal, Spain and others. Table 1 summarises the present status of no-tillage and although the area concerned is now very appreciable, it is only 4 % of the total world-wide arable land area.

FAO has been promoting the CA concept for more than 10 years, particularly in Latin America. Now that CA is becoming a success story in that region, FAO is expanding the programme to others such as Africa, Eurasia and Central and South Asia (particularly the Indian plains of the Ganges).

The formation of groups with a common interest concerning the concepts and practices of conservation agriculture have now led to the establishment of an interdisciplinary Conservation Agriculture Working Group (CAWG) within the Agriculture Department of FAO. The main actors are within the Services or Divisions of Land and Water, Agricultural Engineering, Crop and Animal Production. The CAWG has become very active developing field projects, creating awareness, networking and advising on policy. It promotes the dissemination of information through workshops and international meetings, the 2001 Madrid Congress being a current high-profile example. The CAWG has also overseen the development of technical publications, bulletins, guidelines and training manuals. Detailed information and numerous publications relating to conservation agriculture are available on the Conservation Agriculture Web-site now prepared by FAO, which may be accessed at: *http://www.fao.org/ag/ags/AGSE/Main.htm.* The CAWG also has a number of publications in the pipeline concerning the principles of CA and the importance of soil moisture and it is overseeing the preparation of a CD-rom containing conservation agriculture literature.

Table 1. Area under no-tillage in different countries (Derpsch, 2001)

Country		*Area under No-tillage in ha*		*% of total Cultivated Area*
		1990	*1999/2000*	
USA [1]	*(1987)*	*4.050.000*	*21.120.000*	*17.5*
Brazil [2]		*1.000.000*	*13.470.000*	*25*
Argentina [3]		*300.000*	*9.250.000*	*37*
Australia [4]			*8.640.000*	
Canada [5]			*4.080.000*	
Paraguay [6]		*10.000*	*960.000*	*52*
Mexico [7]			*650.000*	
Bolivia [8]			*350.000*	
Venezuela [9]			*150.000*	
Chile [10]			*100.000*	
Colombia [11]			*70.000*	
Uruguay [12]			*50.000*	
Others [13]			*1.000.000*	
Total			*59.890.000*	*Estimated total arable land area = **1.4 billion ha*** [14]

Source: *1) Dan Towery, CTIC, 2001; 2) FEBRAPDP, 2000; 3) AAPRESID, 2000; 4) Bill Crabtree, WANTFA, 2000; 5) Hebblethwaite, CTIC, 1997; 6) MAG – GTZ Soil Conservation Project, 2001; 7) Ramón Claverán, CENAPROS, 1999; 8) Carlito Los, 2000; 9) Carlos Bravo, , 2000; 10) Carlos Crovetto, 1999; 11) Roberto Tisnes, Armenia, 1999; 12) AUSID, 1999; 13) Estimate 14) FAOSTAT 1999*

Although the working group focuses on CA as the essential element for the remodeling of land husbandry, the interdisciplinary composition of the group means that sustainable management approaches can also be considered for other components in watersheds, such as pastures and forests. One of the main issues concerning the integration of livestock within CA systems and the associated competition for utilization of the crop residues, is an issue being addressed separately by Mueller *et al* (2001). FAO is also promoting the introduction of multipurpose and forage shrubs and trees into grass-based pastures as a means to remodel extensive livestock production systems in LA so that they are molded towards silvo-pastoral systems. This is a parallel animal production programme, which complements the other CA activities of the CAWG.

FIELD PROJECTS

FAO has been implementing field projects over the last decade that place greater emphasis on helping farmers to improve their care of the land through practices of CA rather than through efforts solely to combat erosion. This CA approach provides a more effective response to an old problem and specifically recognises the natural desire of farmers to increase both their yields and incomes. They are now achieving these goals as they learn to develop their own abilities to stabilise and eventually reverse the trend to continuously deplete their natural resources. The CA approach also provides opportunities for governments to harmonise certain national objectives, in particular the attainment of better management of the natural resources and development of sustainable agriculture with particular focus on securing the livelihoods of farming families. One of the most important lessons learnt over this period has been that by seeking improvements to the land at the same time as seeking increased production, the approach requires many adjustments to be made to the more classical and traditional thinking regarding the subject (Hinchcliffe et al., 1995).

Africa

FAO is currently assisting a number of countries to analyse and re-appraise the entire approach to food production, which has been inherited the efforts of numerous aid-donors, particularly over the last five decades. The focus now is directed towards trying to move on and completely rethink the conventional 'erosion-control' approach to development work that so far has led to little success. «Improving land husbandry» is a fresh approach that FAO projects in Malawi, Kenya and Lesotho are successfully following. Land husbandry is defined as the active process of implementing and managing preferred systems of land use in such a way that there will be an increase - or, at worse, no loss of productivity, stability, and usefulness for the chosen purpose (Shaxson, 1993). It directly involves the farmers and other land users in the active management, primarily of rainwater, vegetation and crop residues, but also of the terrain slopes, the plant nutrients and soils, including their inherent biota. The scale of the interventions is variable and may be at a level of single fields or the entire landscape. It embraces not only land under crops, pastures and plantations, but also that under native vegetation of every sort.

CA has tremendous potential in Africa because it not only allows the control of erosion and runoff, but it also offers more stable yields and reduces labour requirements, of particular significance in these times of high incidences of HIV/AIDS. There are a number of ongoing initiatives promoting different practices ranging from conservation (or reduced) tillage through to conservation agriculture *per se.* Activities with FAO support are currently being planned in Eritrea, Ethiopia, Kenya, Uganda, Tanzania, Guinea, Ghana and Burkina Faso. The World Bank has also shown interest in incorporating the concept within the Soil Fertility Initiative (SFI) for Africa, a proposal currently being under development.

Asia

Over the period from 1976-1980, agronomic improvements or better land husbandry were introduced to Indore, India and proved to produce superior results to those resulting from traditional soil conservation practices. The agronomic improvements affected soil conditions and soil moisture, amongst other factors, in the ways that we now describe as a conservation agriculture approach (Shaxson, 2001). Although this work was not specifically undertaken by FAO, the section on networking below highlights current support towards SACAN, established during the recent CA Workshop held in Lahore, Pakistan earlier this year.

In the light of the Pakistan Workshop, the next major challenge lies with systems for rice puddling operations, which should be abandoned in favour of true CA and water saving approaches. Cover crops will also necessarily have to be introduced into the crop rotations. There is ample scope in the region for integration of livestock into agriculture through the judicious inclusion of forage crops in the rotations. More adoptive research is required on the non-mechanised options of CA for the marginal land-holders, tenant farmers and areas where mechanisation is rarely available. Greater emphasis needs to be given to addressing the wider uptake of CA by small-holders, especially as this concerns institutional requirements in each country to support farmer-led approaches with appropriate extension services.

Following a request of the Government of Mongolia, FAO is implementing a TCP project on «Improved Cereal Production Technology». The main objective is to introduce CA technologies to the country. This is seen as a means to improve both cereal productivity and the profitability of farming on a sustainable and environmentally friendly basis. Current experience with the introduced CA technologies looks promising and the participating farmers are very keen to test the new production approach (Bachmann, 2001).

FAO, together with CIMMYT and other partners is currently promoting direct drilling of wheat in rotation with paddy rice. Information being fed back from the farmers' fields in the South Asian rice-wheat area has revealed that increases of 50% in net benefits are possible when the wheat is drilled directly into the rice crop or stubble, as opposed to those attained after conventional tillage operations. About half of these benefits arise from production cost reductions and the remaining half from the increased yields due to water saving and improved water-use efficiency (Hobbs, 2001).

Another vast area where the adoption of CA would be extremely beneficial is Central Asia. The practice of conventional agriculture is now virtually impossible in countries of the former USSR owing to serious environmental problems including soil compaction, surface pulverization and the inevitable erosion. There is also a generalised lack of farm machinery, much of which now requires renovation or replacement. Unless CA is adopted, the investment costs of replacement machinery will to be very high.

Latin America

FAO projects in Costa Rica, Honduras and El Salvador have been promoting three simple practices: leaving residues on the surface instead of burning them; direct sowing of grain crops along the contour at higher population densities and the practice of crop rotations. In cases where the residues have been left on the surface, farmers have experienced yield increases that have doubled or tripled. They have also observed better water infiltration, more vigorous plant growth and a better response to fertilisers because these could then be applied to soils with increased moisture. Other elements have also contributed to these successes, including an enhanced awareness by some governments of the particular relevance of a CA approach in hilly areas and the efforts dedicated to farmer training and even the introduction of CA concepts into general school education curricula.

FAO worked to help strengthen the «Zero-Tillage» revolution in Brazil from 1981 to 1988, undertaking field visits, training courses and round-table meetings and generally acting as a catalyst to encourage people to look at long-standing and traditional problems, but in a different way. Major support came from the Ministry for Environment for promotional work with farmers' «Friends of the Land» clubs that had sprung up. This proved to be a key factor in promoting national management of watersheds. It resulted in improved water quality as characterised by less pollution and silting due to erosion and regularisation of stream flows and aquifer recharge due to an improved rainfall infiltration and a consequent reduction in the severity of floods. Further support for the promotion of zero tillage and direct drilling came through collaboration with the Brazilian Association for Higher Education in Agriculture (ABEAS) and the University of Brasilia, which in 1999, conducted the first correspondence course in Brazil at graduate level on zero tillage.

A sub-regional project covering Argentina, Bolivia, Chile and Paraguay promoted an approach designed as a transition from traditional and physical soil conservation practices towards a conservation agriculture approach. The objective was to optimise the soil's potential for plant production, but in a sustainable manner. Particular attention was paid to enhancing rainwater infiltration and only adding, when necessary, more conventional soil conservation structures and techniques as support measures. The project supported the implementation and execution of field activities, workshops, training programmes, publications and the definition by national counterpart institutions of policies and legal frameworks for «Participatory Planning of CA».

Most recently in Tarija, Bolivia, the hand jab planters known in Brazil as «*matracas*» with hoppers for seed and often also for fertiliser, together with simple animal drawn direct planters were introduced. Some modifications to the planter are presently being carried out to make the draft animal equipment more adapted to the rugged conditions that farmers often have to contend with in Bolivia.

In Mexico, FAO assisted in the definition of a strategy and policies designed to improve the present production system, these being incorporated into the National Development Plan. The objective was to promote adoption of up-to-date technologies concerning the use, management and conservation of soil and water, delivering the message through an efficient and operational technical assistance service. The main outputs of the assistance included implementation of a participatory planning programme based on the farmer's needs, opportune use of available technology and improved access to technical personnel and production support programmes.

FAO assisted the Government of Cuba in the formulation of the National Action Programme (NAP) of Cuba, financed by the international co-operation and according to the guidelines within the UN convention to combat desertification and mitigation of drought effects (CCD). A number of appropriate CA technologies for the Cauto river basin area were identified, being then implemented by users in selected pilot areas. In another region and financed by a two-year TCP project, more than 1,000 hectares of sugar cane were established using CA technology. Diffusion of the results was achieved through the activities of five farmers' field schools. There are plans proposing changes to sugar production techniques in areas close to sugar cane mills, where there is justification for introduction of rotations of sugar cane with alternative crops or for changing entirely to an alternative crop.

NETWORKS

The transfer of the concepts, principles and technologies of CA needs mechanisms to develop network interchange between countries, so as to better share known solutions to avoid some of the problems identified during the continual learning process. Such networks can accelerate the rates of advancement of knowledge and techniques being steadily accumulated by both national institutions and farmer-fed community groups in their efforts to reverse land degradation on a global scale during the 21st century. A selection of activities in which FAO has been involved with is described below.

Southern and Eastern Africa

In 1998 a group of concerned organisations including FAO[2], convened a workshop in Zimbabwe on the topic «Conservation Tillage for Sustainable Agriculture». There were some 70 participants from 15 African countries and 12 participants from other continents brought a global perspective to this, perhaps the first event focussed on CA in Africa. The workshop resulted in the initiation of the African Conservation Tillage network (ACT), which started its activities in April 2000. It is supported by GTZ and FAO and aims to identify, disseminate and promote the adaptation and adoption of resource-conserving tillage practices in Africa. The network sees as its primary task the opening up of channels of communication, but also is planning activities to stimulate the establishment of national conservation tillage networks and the identification, adaptation and adoption of conservation tillage techniques[3] in the continent. It plans to concentrate its resources in Eastern and Southern Africa initially (2000-2002) and then later extend activities to Western and Central Africa (2003-2005). Northern Africa is programmed for the period 2006-2008. Despite these chronological plans for expansion, many members from all regions in Africa are already subscribed to ACT and have already participated in a number of its activities.

The ACT Newsletter is circulated regularly (electronically) and an inscription can be registered very simply by accessing the ACT web-site at *http://www.welcome.to/*

ACT-Network. The web-site also provides general information on the network, access to a literature database, an expert database, a news page, an events page and a page with all the newsletters. FAO is also assisting ACT to develop a database of CA equipment and tools and another one for cover crops.

Latin America

The Latin American Network of Conservation Agriculture (RELACO) started during a training course on Tillage Systems organised by The National Institute of Agricultural Technology (INTA) in 1989. The course was held at the Sáenz Peña Agricultural Experimental Station in the Province of Chaco, Argentina and attended by persons from nine other countries. As described in another Madrid Congress paper, RELACO now includes all Latin American countries in its membership list (da Veiga *et al.*, 2001).

Over the last fourteen years, RELACO has provided technical assistance and stimulated the interchange of information between its participating national institutions. A system of feedback and information exchange between participants, collaborators and co-ordinators of the Network has also been established through the publication of regular newsletters. Examples of successful validations and demonstrations have also been published and teaching material used during a training course on «Tillage Systems» has been published and distributed as the FAO Soils Bulletin No. 66 «Manual of Tillage Systems for Latin America», dated November 1992. To date, the efforts to plan, to research and to organise the necessary technology transfer concerning CA in Latin America is giving pleasing results, although there are varying degrees of success. Continued efforts are however, still needed to ensure that the real and potential contribution of CA is effective. If this can be achieved on an enhanced scale compared with the situation as of today, then it may hopefully, contribute very significantly towards greater food security for the growing populations of Latin America (Benites, 1997).

South Asia

An International Workshop on Conservation Agriculture, co-sponsored by FAO, CIMMYT and various national partners, was held at Lahore, Pakistan earlier this year (6-9 February 2001). One important issue pertinent to this section emerged when the eleven participating countries[4] accorded the establishment of the South Asia Conservation Agriculture Network (SACAN), focussing on the rice-wheat ecosystem in the region. The network is already generating momentum for the adoption of CA technologies in the region, selected as relevant to the diverse socio-economic and agro-ecological conditions. It is encouraging to note the exchange of expertise and experiences in CA experiences amongst the stakeholders and member countries. The network is also promoting the active integration of CA and IPM approaches, in order to achieve a more holistic and systematic response to the sustainability and production needs of the farmers concerned, particularly the many classified as small farmers. This network is being co-ordinated by the On-Farm Management Research Institute at Lahore, Pakistan.

Central Asia

The promotion and adoption of conservation agriculture in Central Asia should be pursued as a top priority. One of the first steps taken so far was supporting an international workshop in Kazakhstan in September 1999. The workshop aimed to stimulate discussion and contact between national stakeholders and potential CA experts from similar agro-ecosystems in other parts of the world, in particular from North America, donor countries and international organisations. There were also representatives of agricultural research centres from various Eurasian countries where they are engaged in conservation work. FAO is now supporting the recently created Eurasian Conservation Agriculture Network (ECAN) as proposed by the workshop participants[5].

The first activity of ECAN was to organise a three-day Conference supported by FAO and GTZ in Altai from 17-19 July 2000. Leading researchers were invited from a number of arid territories of Russia[6], together with Mongolia and Kazakhstan. The objectives were to better co-ordinate research concerning means of conserving energy and resources in the arid territories and also to identify possibilities to minimise cereal production costs through optimisation of the technological elements and systems. The meeting contributed towards developing mechanisms for more effective use of soil management research and integration of accumulated knowledge concerning technologies and systems.

TECHNICAL CO-OPERATION AMONGST DEVELOPING COUNTRIES (TCDC)

In order to speed up the adoption of CA principles and practices by farmers in the developing world, the TCDC programme of FAO can assist in facilitating the efficient utilisation of regional and national expertise. The TCDC concept is part of the Partnership Programme and encourages visits by professionals to other developing countries participating in the scheme. The costs involved in effecting the visits are fully supported by FAO. The objective is to encourage the exchange of knowledge and experience from one country to another, so benefiting and complementing that of each country. The programme has proved useful in the FAO efforts to promote CA concepts, principles and technologies and to strengthen the capacity of national organisations in the development of sustainable land use systems.

One example of use of the programme is a training course jointly organized by FAO and other collaborators[7] on the subject «Soil Management and Conservation: Efficient Tillage Methods for Soil Conservation». It was held at IITA, Ibadan, Nigeria in 1999 with the participation of Spanish and Portuguese-speaking technicians from 6 African countries[8]. The objective was to indicate some of the problems of soil and water conservation, to prepare strategies and plans and to organize action programmes that take account of integrated planning for soil management.

Another example of the programme is well illustrated by a mission comprising three FAO staff and two Brazilian experts from the Institute of Agronomic Research of Paraná (IAPAR) to Uganda, Zambia and Tanzania in January 2000. The mission not only demonstrated a range of CA direct planting equipment, but also designed proposals concerning how to introduce CA concepts in a participatory manner to the traditional farming systems.

The interchange of experiences accomplished during this TCDC mission has, it seems, succeeded in the creation of a critical mass which hopefully, and following the Brazilian model, will lead Uganda, Zambia and Tanzania into a new era of environment-friendly prosperity.

CONSERVATION AGRICULTURE LINKAGES WITH INTERNATIONAL INITIATIVES

In countries where there are many donor agencies involved in supporting the agricultural sector, a concerted approach to the CA issue is urgently needed to ensure coherence with other components of the agricultural programme. FAO provides such mechanisms for inter-country networking, as do a range of other research networks. However, care is needed to avoid that activities at an international level take priority over those focused at a country or local level.

The Soil Fertility Initiative (SFI) for sub-Saharan Africa (SSA) was launched during the World Food Summit (WFS) in 1996. It is a collaborative programme of the World Bank, FAO, IFDC and the CGIAR system. There was consensus during the WFS that special effort was needed to identify and apply solutions to arrest the increasing decline in soil fertility, progressive land degradation and all the associated problems leading to increased food insecurity and poverty. Solutions need to include physical, hydrological and biological soil issues as well as aspects of soil chemicals and fertilisers. CA concepts and systems should therefore be incorporated into the SFI programme. A number of socio-economic issues also need to be addressed such as the availability of farm power, labour shortages, lack of investment due to insecurity of land tenure, restricted access to inputs, services and financial resources, to highlight only a few. CA systems approach solutions to many of these issues, a firm foundation being exemplified by the situation in Brazil and other South American countries. It is for these reasons that it is important that there are strong links between the SFI and CA activities.

ACTION NEEDED

There is a need to change mentalities, to overcome the fears and the gaps, and to abandon common and traditional practices that endanger the natural resources, soil, water and plant nutrients. Science offers scope for reviewing, re-interpreting and re-combining past results and observations. There is also considerable scope for introducing the non-conventional approaches that this paper has briefly discussed, to staff involved in training and thence to students, new recruits and in-service staff of agricultural advisory services in the public and private sectors. In this way, the support services to farmers can be strengthened.

Experience has shown that single-pointed and isolated interventions are unlikely to address the biological and socio-economic realities of farmers' situations. This brings into focus the importance of true dialogues and participatory work between partners in deciding on practicable ways of improving particular ecosystems and livelihoods.

Policies to encourage the implementation of CA practices on a wide scale will need financial and advisory support to farmers from the public sector to offset the costs and difficulties of making the needed transitions. The inclusion of positive and negative externalities in performing the economic calculations of real benefits and costs to individuals, to the agricultural sector and to state or nation as a whole, will help to compensate for farmers receiving public benefits. Examples of important externalities justifying such support would normally include reduction to pollution levels, sequestration of carbon, minimisation of downstream damages by flooding and erosion, amongst others. There is also a need to provide research and advisory functions that are essential but which cannot excite the interests of commercial firms to provide them because of a lack of financial attractiveness (Fresco, 2001).

Many challenges still remain and successful techniques to apply CA principles in some situations still remain to be solved, particularly in the very dry areas and by resource-poor farmers.

CONCLUSIONS

Conservation agriculture has demonstrated that high production levels can be combined with enhancing the natural resource base and conserving the environment. It will spread further, but its benefits will spread rapidly and widely only where Government policies, services and infrastructure facilitate the conversion to these systems.

In Brazil, conservation agriculture has been successfully adopted over the last 20 years. The agro-ecological zones of Brazil are quite similar to certain regions in sub-Saharan Africa and in Asia. However, there are important differences in the farming systems. Nevertheless, the basic principles of conservation agriculture are equally valid in Africa and Asia. Indeed, in some countries such as Zimbabwe and Zambia in Africa and in India and Pakistan, they have been proven to work very successfully. Farmers need to work together with research workers to identify appropriate solutions and to adapt the technology to the specific local conditions in each country or region.

A farmer-driven participatory technology development approach that considers the farming system as a whole, needs to be adopted as a tool for studying, utilising and adapting CA systems for the potential farming areas. Initial priority should be given to activities in low risk areas where it may be easier to transfer the technologies.

Interdisciplinary teams of researchers and extension workers can facilitate the implementation process. These teams need to have experience both in CA technologies and techniques of participatory technology development. In most cases, appropriate training is needed for the teams.

Component research is necessary, including aspects such as agronomy, engineering and economics so as to back-up the results of on-farm activities. This is particularly important to appreciate, as the experiences in South America have clearly demonstrated successful use of different approaches in different areas and circumstances but as the

CA system is dynamic, there can be no single recipe that can guarantee success (Ashburner, 2001). Adaptation to local circumstances may require particular research on species of cover crops, soil fertility management, management of pests and diseases, termite control, weed management, adaptation of tools, implements and equipment, together with socio-economic and gender aspects.

Special links need to be established to small enterprises in the private sector that are needed for development and production of the necessary special implements and tools such as knife rollers, direct planters, boom sprayers, etc. There also need to be links established to ongoing development projects and donor activities in the project area to ensure good networking and exchange of experiences and knowledge within the country and the region. Exchange of knowledge between African and South American, particularly Brazilian stakeholders must be promoted. This is particularly important for the implement-manufacturing sector so that joint ventures and licensing agreements for instance, can be established.

Awareness creation campaigns using the radio, pamphlets, newspapers, study tours, field days, workshops within country and to neighbouring participating countries, are just some of the important tools that are useful for sharing information and alerting decision-makers and society in general to the potential benefits of the technologies.

The development of farmers' groups and movements with a common interest in the concepts and practice of CA, such as the «Friends of the Land» clubs in Brazil and the «Land Care» clubs in Australia, have always served to stimulate and support other members to face the change. In this way, groups such as these have become action groups, transmitting the new ideas and technologies from farmer to farmer. In addition, they have also become important local pressure groups, managing to obtain improvements at an institutional and political level so as to receive broader and more significant support for the technologies.

Another important role of farmers' groups is to share resources. The application of CA technologies normally requires the use of special and different equipment, which can mean significant investment. Sharing, contracting or other multi-farm approaches to reduce operational costs are very advantageous and can be promoted by these farmers' groups.

One additional and closing conclusion is to highlight the challenges that the CA approach poses for resource poor farmers. This aspect should receive priority attention in all efforts to improve food security and address the issue of poverty alleviation.

NOTES

[1] The views expressed in this paper are the personal opinions of the authors and do not necessarily comply with the official policy of FAO

[2] FAO, German Technical Development Co-operation (GTZ), FARMESA, Zimbabwe Farmers Union (ZFU) and the South African Agricultural Research Council (ARC)

[3] The terminology «conservation tillage» and «conservation agriculture» has already been raised earlier in this paper. It is seen that the aims of ACT lie within the CA concepts described in this paper.

[4] Australia, Bangladesh, China, India, Indonesia, Mongolia, Nepal, New Zealand, Pakistan, Sudan, Turkey and Uzbekistan

[5] Membership currently includes Armenia, Azerbaijan, Georgia, Belarus, Kazakhstan, Kyrgyzstan, Republic

of Moldova, Russian Federation, Tajikistan, Turkmenistan, Ukraine and Uzbekistan
[6] Irkutsk, Novosibersk, Kemerovo, Omsk, Kurgan, Chelyabinsk, Saratov, Orenburg regions, Krasnoyarsk and Altai territories, Khakasia and Altai republics
[7] The collaborators included experts from the Brazilian Enterprise for Agricultural Research and Rural Extension, Santa Catarina S.A. (EPAGRI), the Soils Department of the Faculty of Agriculture, Federal University of Río Grande do Sul (UFRGS), the Brazilian Enterprise for Agricultural and Livestock Research (EMBRAPA) and the International Institute for Tropical Agriculture (IITA), Ibadan, Nigeria
[8] Angola; Cape Verde; Equatorial Guinea; Guinea-Bissau; Mozambique and Sao Tome and Principe

REFERENCES

Ashburner, J.E. (2001) Some personal experiences of conservation agriculture. Proceedings of the National Workshop on Conservation Farming for Sustainable Agriculture in Tanzania, Arusha, Tanzania, 23-25 June 2001 (forthcoming).

Bachman, T. (2001) Improved Cereal Production Technology: FAO project on Conservation Agriculture in Mongolia. Paper presented at the 1st World Congress for Conservation Farming, Madrid, 1-5 October 2001 (forthcoming)

Barber, R. (1999) Land and crop management in the hilly terrains of Central America: Lessons learned and farmer-to-farmer transfer of technologies. *FAO Soil* Bulletin **76**. Rome, Italy. 76 p.

Benites, J.R. (1997) Conservation tillage and sustainable agriculture in Latin América. A paper presented in the IV Meeting of RELACO, 17-20 November 1997, Morelia, Mexico; 20 p.

Benites, J., Friedrich, T. and Bot, A. (2000) Conservation Tillage: An Entry Point For Sustainable Agriculture in Eurasian Countries. Proceedings of an international workshop, Astana, Kazakhstan, 19-24 September 1999. Almaty, Kazakhstan.

Benites, J.R., Friedrich, T., Bot, A. and Shaxson, F. (1999) From Soil Degradation to Stable Productivity: the importance of better land husbandry. A FAO view, for discussion. AGLS Working Paper. Land and Water Development Division. FAO. Rome.

Benites, J. (2001) Conservation agriculture: a key to food security and poverty alleviation. Draft note for a press release. (forthcoming)

Bot, A. and Benites, J. (2001) Sustaining rural livelihoods through conservation agriculture. FAO, Rome, 77 p. (forthcoming)

da Veiga, M., Wildner, L. do Prado, Benites, J.R. (2001) Latin American Conservation Agriculture Network – RELACO. Paper presented at the 1st World Congress for Conservation Farming, Madrid, 1 – 5 October 2001 (forthcoming)

Derpsch, R. (2001) Conservation Tillage, No-tillage and related technologies. Paper presented at the 1st World Congress for Conservation Farming, Madrid, 1 – 5 October 2001 (forthcoming)

FAO/GTZ/ZFU/Farmesa/ARC (1998) Conservation tillage for sustainable agriculture. *Proceedings of an international workshop*, Harare, Zimbabwe, 22-27.7.1998. Volumes **I** and **II.** GTZ, Eschborn

Fresco, L. (2001) Agriculture and Conservation: Key For Food Security and Poverty Alleviation. Paper presented at the 1st World Congress for Conservation Farming, Madrid, 1 – 5 October 2001 (forthcoming)

Hinchcliffe, Guijt, F. I., Pretty, J.N. and Shah, P. (1995) New Horizons: the ecpnomic, social and environmental impacts of participatory watershed development. Summary Workshop in Bangalore, India, Nov. 28 to Dec 12, 1994. Internatl. Inst. for Envt. and Devt. (IIED). *London Gatekeeper Series SA* **50** 22pp.

Hobbs, P., Giri, G. and Grace, P. (1997) *Reduced and zero tillage options for the establishment of wheat after rice in South Asia,* RCW Paper No. **2**. Rice-Wheat Consortium for the Indo-Gangetic Plains and CIMMYT. Mexico, D.F. 18 p.

Hobbs, P., Gupta, R., Malik, R.K. and Dhillon, S.S. (2001) Conservation Agriculture for the rice-wheat systems of the Indo-Gangetic Plains of South Asia: A case study from India. Paper presented at the 1st World Congress for Conservation Farming, Madrid, 1 – 5 October 2001 (forthcoming)

Mueller, J.P., Pezo, D.A., Benites, J.R. and Schlaepfer, N.P. (2001) Conflicts between conservation agriculture and livestock over the utilisation of crop residues and cover crops. Paper presented at the 1st World Congress for Conservation Farming, Madrid, 1 – 5 October 2001 (forthcoming)

Shaxson, T.F. (1993) *Conservation effectiveness of farmers' actions: a criterion of good land husbandry. In: Topics in Applied Resource management in the Tropics*, eds. E. Baum, P.Wolff, M. Zobich. Vol. **3**: ''Acceptance of Soil and Water Conservation: Strategies and Technologies» Witzenhausem (Germany): Deutsche Inst. Fur. Trop. Und Subtrop. Landwirt./DITSL. Pp. 103-128.

Shaxson, T.F. (2001) Personal communication

Steiner, K. (2001) Direct Planting through Mulch – Gateway to Food Security and Sustainable Rural Development. Proceedings of the National Workshop on Conservation Farming for Sustainable Agriculture in Tanzania, Arusha, Tanzania, 23 – 25 June 2001 (forthcoming).

N. YOUNG

WESTERN AUSTRALIAN NO-TILLAGE FARMERS ASSOCIATION

President, Western Australian No-Tillage Farmers Association (Inc)

Frustration and innovation are at the heart of WANTFA. The frustration of farmers watching their land blow away and wash away, with no sustainable means of preventing it, lead to the formation of our association in 1992. As individuals the farmers had seen first the possible benefits of cropping without preliminary tillage, and then the potential of a cropping technique which required no tillage at all. At the time, in Australia, this was considered fanciful and idealistic, something that might work elsewhere but our soils apparently needed different treatment to the soils elsewhere in the world. After all, the evidence was all around us – 9m ha of crop planted each year in our state with yields still improving. The impact of better agronomic practices made up for the decline in inherent fertility, so that the degradation was being masked. What could possibly be wrong? Why would anyone want to change? Our farmland is only very recently cleared, so the pioneering attitude is still very common. No-Till should be seen as the logical next step in establishing an agricultural system.

I speak with most knowledge about Western Australia, the one third of Australia on the left. The agricultural land is confined to the 20% in the bottom left corner of that one third –beyond that the rainfall is too erratic or the evaporation too high to sustain anything other than pastoral operations, where sheep, cattle or goats graze the natural scrub on very large stations. There are some pockets of irrigated country within this dry area as well, but I will focus on the South West Land division. Our climate is considered classic Mediterranean, meaning winter dominant rainfall when it is cool, with daily temperature range from 5deg to 20 deg, and summer drought with daily maxima up to 40 deg Celsius. Snow is extremely rare– three times in my lifetime.

Our country was settled by the Noongar people for an unrecorded time, perhaps 40,000 years, prior to the arrival of Europeans. The British established their first colony in Australia at what is now Sydney in 1788, and established a separate colony in the West in 1829. They brought European agriculture with them, and proceeded to clear the land of its established forests and bushes so that wheat, barley and oats could be sown. Initially sheep and cattle were grazed on the natural fodder, but the existence

L. García-Torres et al. (eds.), Conservation Agriculture, 155–158.

of plants poisonous to these animals, though not to the kangaroos and other native animals, later led to the widespread clearing and removal of such bush. My farm is amongst the earlier selected places, being first cleared of its trees just over 100 years ago, but other areas were still being cleared only 20 years ago.

Our land is the remnant of ancient mountains, which have been quietly eroding ever since they were formed, so that it is now a gently undulating plain, making it the oldest land form in the world. This means our soil is now very leached and of low natural fertility, very low in phosphorous, potassium, sulphur and many trace elements including copper, zinc and manganese. We are accustomed to a topsoil of 10cm over a tight clay layer which extends to bedrock, perhaps at 20m. Large areas have a leached sand layer up to 5m deep over the clay and bedrock.

We are exposed to strong winds, up to 80km/hr, in the autumn prior to our first rains of the winter. Rain fall can come in a series of storms often delivering up to 25mm in a 24 hour period, and in the extreme 25mm in 1 hour, interspersed with weeks without any rain. Dry periods lasting 2 to 3 months are expected over summer.

So you can see that ground exposed with cultivation in order to plant a crop at the start of winter was highly likely to erode. It frequently did. Farmers were tired of this and some were determined to find a better way of farming, one which left the soil in their own paddock. The stage was set for the introduction of no-till sowing.

A number of things came together at this time to allow this introduction to happen.

A Bettinson triple disc seeder was imported in 1975 by a farmer with very fragile sandy soil who had seen the machine whilst travelling in England. Another thirty followed, and in spite of the imperfect machine and inadequate chemical choices, no-till was underway.

Sprayseed, a mixture of paraquat and diquat, had been used since the late 1960's as a broadacre knockdown prior to sowing with full cut cultivation –a technique we knew as direct drill–. Glyphosate came down in price such that it was a genuine option for broadacre spraying. Glyphosate was more robust and did not require the plants to be disturbed in order to make them die.

Another enterprising machinery dealer imported a Great Plains disc seeder, which he hired out at very low cost. This machine was more robust than the earlier Bettinson, and farmers were pleased with its results. It was still very expensive compared to the commonly used existing tined seeders, but demonstrated that no-till sowing could be used on a wider range of difficult soils.

A researcher with the State Agriculture Department, Kevin Bligh, measured the different amounts of soil lost by water erosion under different crop establishment systems. This showed that with tillage soil was being lost at a much faster rate than it could possibly ever form, with the inevitable result that farming would soon be forced to stop. In 1990 he was awarded a Churchill Fellowship to study no-till techniques in USA, Canada and New Zealand. As a result he recommended the formation of a No-Till Farmers Association, which eventuated in 1992, and that a West Australian agronomist undertake a one year exchange with a Canadian no-till specialist. Bill Crabtree, then with the State Agriculture Department undertook this exchange with Bob Bradley of Agriculture Canada in 1996.

Two inventive farmer brothers had satisfied themselves that crops planted with only a narrow strip being cultivated by a knife point, and most of the soil left undisturbed, were equal to that planted with a full disturbance by a normal cultivation point. These men then adapted some smart metallurgy from the mining industry to their points so that wear rates were much reduced. This knife point could be bolted to any existing seeder, so that no-till sowing became possible for just the cost of a set of points.

First steps were tentative, with the emphasis on stopping either wind or water erosion. Other benefits quickly became evident. Paddock trafficability was improved in wet conditions, and fewer weeds germinated post sowing. The ability to incorporate trifluralan with the knife points, which suppressed a major problem weed of annual ryegrass was a bonus. Best of all crop establishment in dry autumn conditions was actually better than with full cultivation.

The established wise people were still sceptical, and highlighted the reasons why it shouldn't work. This division in attitude lead to the rapid expansion of our farmers association which lobbied for research into the specific problems they were finding, and to share that information amongst themselves for the common good. They were not interested in hearing why it shouldn't work –they knew that it had to if they were going to continue farming–. This association has now grown to 1200 WA members and 100 from the rest of Australia and overseas. There are only 6000 broadacre farmers in the State, so you can see we have a high acceptance. No-till sowing has grown from an estimated 0.1% in 1990 to an astonishing 60% of cropped land in 2001 in WA.

Exchanges have continued to be important. 10 farmers went on a month long no-till visit to North America in 1994, with a further 30 travelling in 1996 and 40 in 1998. Return visits by Manitoba-North Dakota Zero-till Farmers Association members took place in 1996, in addition to numerous visits by individual North American farmers and scientists. WANTFA has hosted no-till specialists from North and South America annually since 1995, including Professor Dwayne Beck of South Dakota State University twice. This year we hosted Dr Rolf Derpsch and Mr Carlos Crovetto for a week for our Annual Conference.

WANTFA has been able to secure funds from farmer's grains research levies to now employ Bill Crabtree as its Scientific Officer. Bill visited no-till farmers and scientists in South Africa and South America in 2000, and was able to then lead a tour group of our farmers to those countries this August. In the last year we have had visits by groups from France and South Africa, plus individuals from South America.

The ability to compare notes with overseas farmers and scientists on the development of no-till has proven very useful, particularly when faced with beaurecratic indifference or obstacles. West Australian farmers were able to be reassured that their experiences were consistent with observations elsewhere, and to learn the important principles in use world-wide.

Many issues still face us as farmers. We have weeds resistant to many of the control methods we would like to use to control them. This year Glyphosate resistant ryegrass has been recorded for the first time. We see the ready expansion of pests as being the consequence of very limited crop choices, as we only grow winter cereals and winter broadleaf crops. The State is faced with a rising saline water table, such that so far 10% of cropland has been lost and another 20% will be lost in the next 30 years to dryland salinity. Warm season crops are being suggested as a partial solution to both these problems, and the search is on now for suitable crop species. We believe cold hardiness will be important so that crops can be well established before the summer drought sets in.

Our Association is totally farmer controlled, managed by a representative committee who provide their time free of charge. We maintain close links with the scientific community. Our members participate in annual and special conferences, local field days, receive a quarterly newsletter edited by our Scientific Officer, and along with the rest of the world can visit our Website at wantfa.com.au . Membership has been kept to $A100 per annum, a price which all farmers can afford.

We have assisted in the establishment of a similar association in South Australia – the South Australian No Tillage Farmers Association, are linked with like minded small groups in the rest of Australia and were actively involved in the establishment last year of the Pacific Northwest Farmers Association in Pendleton, Oregon. WANTFA would be delighted to help any individuals or group with information at our disposal, and extend an invitation to you all to visit us in Australia.

L.W. HARRINGTON[1]

A WORLD OF CONSERVATION AGRICULTURE

Director, Natural Resources, CIMMYT, and Interim Chair of the Voluntary Action Group for the GFAR Global Program on DMC

A NEW CENTURY

As we enter the twenty-first century, we find ourselves surrounded by wonders. Astronauts construct an international space station. Biotechnologists unravel the human genome and talk seriously about cloning humans. Instantaneous and inexpensive digital communications connect the globe. In agriculture, farm equipment linked to satellite positioning systems allow farmers to fine-tune soil management practices for each square meter of their field.

But all is not wonderful progress.

- In the western Indo-Gangetic Plains, farmers pump longer hours from deeper wells to obtain less irrigation water of poorer quality. Near Lahore, the dams are dry for extended periods.
- In Bangladesh and West Bengal, groundwater used for human consumption is becoming contaminated with arsenic, in part due to excessive pumping of groundwater for agriculture.
- In the hillside slopes of Mindanao, three crops of maize per year are grown, one after the other, resulting in a relentless upsurge in problem weeds and pests and diseases.
- In Zimbabwe, vivid satellite images show extreme degradation of communal lands. Many observers feel that depletion of soil fertility in eastern and southern Africa is the biggest barrier to achieving sustainable food security in this part of the world.
- In Guatemala, Mayan farmers shave their bare hillside fields with a machete in preparation for the next meager crop of rainfed maize.
- In Bolivia, farmers watch scarce rainwater run away to the river, along with a good portion of their soil. The see the crusting of their soils that follows from intensive tillage, but they are not sure what to do about it.
- In the Mexican Mixteca, many farm families simply give up on agriculture, and move to Oaxaca, Tehuacán, Puebla, Mexico City or farther north.

L. García-Torres et al. (eds.), Conservation Agriculture, 159–164.

Finally, at a global level, additional challenges are emerging, among them global warming and climate change. These new challenges find their place alongside older and familiar ones – how to eliminate poverty, hunger, and the silent and invisible desperation of starvation.

INNOVATION IN THE NORTH, STRUGGLE AND SUCCESS IN THE SOUTH

Along with poverty and hunger, resource degradation is not new. Neither are many of the practices used to combat it – a large number of them featured in what has come to be known as conservation agriculture. However, some practices found to be useful in conservation agriculture were only developed over the past several decades.

A good example is conservation tillage or zero tillage. While the principles of zero tillage may be centuries old, their practical application to large scale commercial farms in the developed world waited on the development of suitable implements (to establish crops into standing crop residues) and suitable herbicides (to control weeds when plowing and cultivation are not options). The result in the USA and Australia has been a huge area managed with these practices, a «mainstreaming» of what began as a curiosity.

While the developed world moved conservation agriculture and conservation tillage practices into the mainstream (at least in some regions) developing countries have not been inactive. The amount of work done to tailor conservation agriculture practices to the circumstances of the South has been astonishing. And this work has yielded a good return.

Perhaps the best-known example is the spread of direct sowing and mulch systems across millions of hectares of cropped area in the Southern Cone of South America – especially Brazil, Argentina, Chile, Paraguay and lowland Bolivia. This has been the outcome of a combined effort by farmers groups, private enterprise, NGOs, extension workers, and national and international scientists, among them several truly inspired individuals. Yields have improved, costs have been slashed, and soil and water resources have been saved and renewed.

But other examples abound. Note the substantial work on:

- Green manure cover crops (especially *mucuna* and *canavalia*) and conservation tillage in Mesoamerica (southern Mexico and Central America). This features work by farmers, NGOs, NARSs, CIAT, and the PRM (the CIMMYT-led *Programa Regional de Maíz*).
- Rotations and stubble management in West Asia/ North Africa, with substantial input by NARSs and ICARDA.
- Agroforestry in East Africa, especially the enormous potential from biomass generated from *tithonia*. Much of this work has been fostered by ICRAF and the wide-ranging set of partners within the African Highlands Initiative.
- Legume –grain rotations and green manure cover crops in southern Africa, e.g., the work coordinated by the CIMMYT– NARS Soil Fertility Network.
- Conservation tillage and soil water management in eastern and southern Africa, with inputs from GTZ, Silsoe, NARS, farmers, with recent coordination taken up by the African Conservation Tillage Network (ACT).

- Mulch management in hillside crop systems in Vietnam and other areas of Indochina, with CIRAD in a facilitating role.
- The rapidly expanding tillage revolution in the Indo-Gangetic Plains of South Asia, with input from farmer groups, private enterprise, NGOs, NARSs, and international centers – with coordinated facilitated by the Rice Wheat Consortium for the Indo-Gangetic Plains. (See below.)

The list is exciting – and all the more so because it is incomplete. Many other examples could be cited. Clearly, these endeavors share a lot in common. Yet there has been surprisingly little interaction among these them.

How much more might have been achieved if the champions in these endeavors had had the opportunity to share experiences, methods, insights, and technologies in a systematic way with colleagues from other regions?

A TILLAGE REVOLUTION IN SOUTH ASIA

South Asia constitutes one of the world's great concentrations of rural and urban poor, and one of its most serious developmental challenges. And in the Indo-Gangetic Plains, the breadbasket of South Asia, four adverse trends are converging.

- The demand for food is increasing, driven in part by population growth.
- Harvested area is shrinking, as urbanization proceeds and degraded land is abandoned.
- Traditional sources of productivity growth are largely exhausted, at least in those areas that benefited most from previous rounds of technical change.
- Resource degradation continues, reducing productivity. Problems include salinization and sodification in irrigated areas, groundwater depletion, soil fertility loss, reduced agroecosystem diversity, and environmental pollution from the burning of crop residues.

Any one of these trends has the potential to be serious. A convergence of all four of them has the potential for disaster.

Enter conservation agriculture, in the form of zero tillage and the direct sowing of wheat into standing rice stubble. This practice was originally introduced into Pakistan from New Zealand. Its purpose was to help advance sowing dates for wheat in rice-wheat systems, thereby avoiding the yield penalty associated with late establishment. Then followed a fifteen-year process of implement testing, improvement, farmer experimentation, fine-tuning, development of complementary practices, and commercial manufacture and individual farmer purchase of implements. This process involved a huge number of farmer experimenters, private entrepreneurs, university scientists, state level directors of agriculture and water management, extension workers, and international scientists from CIMMYT, IRRI and other centers of excellence.

For the last seven years, this effort has been coordinated by the Rice Wheat Consortium for the Indo-Gangetic Plains, a NARS-led Ecoregional Program of the CGIAR convened by CIMMYT. This brings together the efforts of the national systems from four South Asia countries (Bangladesh, India, Nepal, Pakistan); five international centers (CIMMYT, IRRI, IWMI, CIP and ICRISAT); farmer experimenters; and experts from numerous advanced research institutes, e.g., IACR in the United Kingdom, CABI International, Cornell University, and many others.[2]

In the end, the outcome has been astonishing. For a seemingly simple practice, farmers find that zero tillage has an astounding range of benefits. Among these are:

- Lower costs,
- Higher yields,
- Improved water and nutrient use efficiency (resulting in less pumping of groundwater).
- Reduced labor input,
- Better control of problem weeds, e.g., *phalaris minor,*
- Earlier sowing dates, allowing more time in the rotation for diversification crops (that can help improve soil fertility).
- The potential to use crop residues as mulch, further helping soil fertility.

Farmer response has been dramatic. From an estimated 150 ha in 1998, commercial zero till area grew to 1200 ha in 1999, about 20,000 ha in 2000, and over 100,000 ha in 2001. For the 2002 season, zero till area in the Indo-Gangetic Plains is expected to be in the neighborhood of 400,000 ha, with several thousand zero till seed drills in the hands of farmers.

It is expected that conservation tillage (and related practices)[3] will be the most important source of improvements in productivity and sustainability of agroecosystems in the Indo-Gangetic Plains over the next decade.

Finally, a generalized use by farmers of zero tillage will help slow climate change. Tillage is a main contributor to carbon dioxide emissions to the atmosphere from rice-wheat systems. This comes from diesel fuel use as well as biological decomposition of soil organic matter fostered by tillage. In the 12m ha of rice-wheat systems in the Indo-Gangetic Plains, fuel use alone leads to the emission of 4.3 million tons per year of carbon dioxide. If even half of this area is converted to zero tillage, more than a million tons of carbon dioxide emissions can be avoided.

How many conservation agriculture enthusiasts from around the world are aware of this tillage revolution in South Asia? How much faster might success have come to South Asia if researchers there had benefited from input form South America? What are the benefits of sharing knowledge?

KEY ISSUES

The well-known example of South America –and the newer example of the Indo-Gangetic Plains– are truly exciting. However, despite unmistakable progress, there are many instances where key issues remain unresolved. Some of these issues tend to hinder the swift and widespread use of DMC practices. These include:

- *Farm implements.* Selection, improvement, adaptation and dissemination of farm implements effective in practical implementation of conservation agriculture.
- *Weed management.* Integrated weed management practices that enable farmers to take up reduced tillage and direct sowing into mulch – especially practices that use a minimum of herbicide input, or none at all.
- *Mulch management.* Practices that make the most of scarce crop residues, green manure biomass or other sources of mulch – especially when biomass production is low.
- *Crop/ livestock integration.* The integration of crops and livestock in a systems perspective – especially when grazing pressure on crop residues is high and alternative fodder sources are scarce.
- *Legume selection.* Use of local legume species in green manuring strategies.
- *Germplasm needs.* The development of crop varieties that perform well with DMC practices – especially when varietal requirements are markedly different than those needed for conventional practices.
- *Short- vs. long-term profits.* The design of conservation agriculture practices that are attractive to farmers because they offer immediate and large benefits in the near-term, not just in the long-term.
- *Policy support.* A policy context that favors widespread the use of DMC practices, for example, lower subsidies on purchased external inputs, research and extension priorities that embrace conservation agriculture, legume seed multiplication and distribution arrangements, programs of farmer experimentation with zero till implements, etc.
- *Environmental benefits.* The design of conservation agriculture practices that have economic, ecological and environmental benefits that go beyond the level of the plot or farm, for example, hillside farming practices that reduce erosion-driven downstream problems of water quality.

How much more effectively might these issues be addressed in specific instances if better communication, knowledge management and sharing existed with colleagues and peers from around the world?

A PROGRAM ON DMC

Arguably, conservation agriculture practices such as direct sowing, mulch systems, conservation tillage and green manure cover crops offer the best, bright hope for an equitable, sustainable and profitable smallholder agriculture in the developing world. This hope can be achieved more effectively and efficiently if conservation agriculture proponents around the world collaborate and communicate, systematically sharing experiences, methods, insights, and technologies.

Under the auspices of GFAR (the Global Forum on Agricultural Research), a Global Program on DMC (Direct Sowing, Mulch Systems and Conservation Tillage) has been launched. This is part of GFAR's effort to foster a new global system on agricultural research. The heart of this program lies in «pulling together and transforming decentralized initiatives into global initiatives using a bottom-up approach.»

The goal of this new DMC initiative is to:

Help improve food security and alleviate poverty, while conserving natural resources and encouraging more durable forms of agriculture, by fostering broader use of sound agroecosystem management practices, especially those centered on direct sowing, mulch-based systems and conservation tillage.

From its goal springs its purpose:

Strengthen the capacity of key stakeholders to develop suitable DMC systems and to accelerate their widespread adoption.

And from the purpose of this DMC program emerge five objectives:

1. *Develop a framework for comparing experiences.*
2. *Synthesize and systematize lessons learned.*
3. *Identify gaps and encourage stakeholders to fill them.*
4. *Provide support and feedback to decentralized initiatives.*
5. *Foster the multiplication of successful experiences.*

This DMC program is a framework, not a new institution. It is open to all. A voluntary action team from CIRAD, CIMMYT, ASPTA (a Brazilian NGO), Madagascar and Iran have sought to carry this forward on a voluntary basis. But it is clear that this program will not achieve its potential until it is properly resourced and staffed. It is in this area that CIRAD and the French government are taking a strong and welcome lead.

There are other networks conducting complementary work. Among these are:

- ACT: the African Conservation Tillage network.
- RELACO: a similar network for Latin America.
- ISTRO: the International Soil Tillage Research Organization.
- FAO: through its program on regional workshops on conservation agriculture, and its fostering of knowledge among conservation agriculture enthusiasts.

This conference presents an opportunity to revitalize this important initiative for a GFAR Global Program on DMC, and to forge suitable links with link-minded networks and institutions. The world of conservation agriculture has come together in Madrid. And not before time. By joining hands in global collaboration, we can make contribute to the fight against poverty and hunger by helping create, in this new century, a new world, a world of conservation agriculture.

NOTES

[1] DMC (Direct Sowing, Mulch Systems and Conservation Tillage).

[2] Financial support from the Netherlands, Great Britain, Australia, New Zealand, IFAD, USAID and other donors is greatly appreciated.

[3] Bed and permanent bed planting, surface seeding, minimum tillage with specialized implements attached to two-wheel tractors.

*M. SULEIMENOV, **M. PALA & **F. KARAJEH

ICARDA'S NETWORK ON CONSERVATION AGRICULTURE IN CENTRAL ASIA

* *ICARDA, P.O. Box 4564, 700000 Tashkent, Uzbekistan*

** *ICARDA, P. O. Box 5466, Aleppo, Syria*

Based on the analysis made by NARS and ICARDA, the following research priorities in the field of conservation agriculture were identified: crop diversification, including development of advanced agronomic practices; improvement of crop rotations for better sustainability of cropping systems; improvement of conservation agriculture practices, including soil tillage, strip cropping and shelterbelts; soil fertility management within the concept of conservation agriculture. During 1999-2000, ICARDA in collaboration with NARS established sites for the Conservation Agriculture program in Central Asia countries within the Asian Development Bank (ADB) supported project «Soil and Water Management for Sustainable Agriculture in Central Asia». Crop diversification looks very promising under both rainfed and irrigated conditions of respective Central asian republics. Conservation tillage practice is more advanced in semi-arid steppes of Kazakhstan, and saves substantial amount of water under irrigated conditions. However, further studies on this including zero-tillage are priority in the region.

Key words: Conservation, agriculture, minimum tillage, crop diversification, crop rotation

INTRODUCTION

The Central Asian Republics of Kazakstan, Kyrgyzstan, Tajikistan, Turkmenistan, and Uzbekistan became independent following the breakup of the former Soviet Union and since then have been in transition from a centrally organized economy towards a market driven economy. For a long period, the conservation agriculture cropping system was developed and generally adopted in the northern Kazakhstan in the 1970s. It was based on stubble mulch tillage within the fallow-wheat cropping system (Barayev, 1978). Its adoption saved the soil against wind erosion and stabilized grain yields of spring wheat at the level of 1 t/ha. But the stubble mulch tillage system wasn't adopted in the dryland and irrigated agriculture areas of other Central Asia countries (Lavronov, 1973).

ICARDA has also had a considerable experience in conservation agriculture through its work in Syria and in other parts of West Asia and North Africa (WANA) region

L. García Torres et al. (eds.). Conservation Agriculture, 165–168

with respect to appropriate soil and crop management practices with proper crop rotations as reported elsewhere (Jones, 2000; Harris *et al.* 1991; Ryan & Pala, 1996; Christiansen *et al*, 2000; Pala *et al.*, 2000a and 2000b). Considering the need in Central Asia and its long-standing experience, ICARDA had planned to initiate collaborative research on different aspects of conservation agriculture in Central Asian Republics.

Based on the analysis made by NARS and ICARDA, the following research priorities in the field of soil and crop management were identified: crop diversification, improvement of crop rotations for better sustainability of cropping systems; improvement of conservation agriculture practices, including soil tillage, snow management, strip cropping and integrated pest management; soil fertility management within the concept of conservation agriculture. Therefore, the first ICARDA's activities started in 1999 in the northern Kazakhstan and Turkmenistan and expanded into other countries in the following year.

MATERIALS AND METHODS

During 2000 the Crop Diversification and Conservation Soil Tillage component was established within the ADB-supported «Soil and Water Management for Sustainable Agriculture in Central Asia» project.

In dryland agriculture of semiarid steppe of northern Kazakstan, it includes the following activities: crop diversification through comparative studies of alternative crops; agronomic studies with food legumes; improvement of crop rotations from «fallow- grains» to more diversified cropping; three factorial studies including soil tillage, crop rotations and soil fertility management; comparative studies of rotations of summer fallow or chickpea with small grains under low input and intensive agriculture conditions.

In dryland agriculture of Central Asia (southern Kazakstan, Kyrgyzstan and Uzbekistan), activities include the following: crop diversification through introducing food legumes and oilseed crops into fallow-wheat rotations; conservation soil tillage= reduced and stubble mulch tilllage against moldboard plow. *In dryland areas of Tajikistan*, an activity was initiated on productive and sustainable use of eroded sloping land by terracing for soil conservation and water harvesting.

In irrigated agriculture (Kyrgyzstan, Uzbekistan, Tajikistan and Turkmenistan), activities include: soil tillage (plow, reduced and stubble mulch tillage); crop diversification through introducing alternative forage crops into cotton-wheat rotations, rapid growing varieties of cotton and wheat allowing double cropping, sugar beet, maize, soybean, dry pea, and safflower into wheat-based cropping systems.

RESULTS

In northern Kazakstan, promising data was obtained on food legumes. Chickpea (1.7 t/ha), lentil (1.0 t/ha) and dry pea (1.96 t/ha) under favorable conditions of 1999-2000 gave comparable yields with wheat (average yield of 1.82 t/ha). The yields of oats and chickpea as replacement of fallow showed a very definite importance of advanced technologies to succeed in this idea. Oats have shown much more remarkable returns (from 1.63 to 4.33 t/ha) to advanced technologies than chickpea (from 1.36 to 2.33 t/ ha), although both crops demonstrated spectacular results for non-fallow crop rotations. The reduction of the summer fallow area will have substantial conservation effect because the fallow is a major soil degradation factor in cropping systems of dryland agriculture. Data on minimum tillage indicated that a limiting factor for wider introducing of minimum tillage is a shortage of nitrogen in the soils or more nitrogen is needed when minimum tillage is applied compared with the conventional deep stubble mulch tillage.

In southern Kazakstan weather conditions were extremely dry and most crops in rainfed conditions failed. The average yield of chickpea was 360 kg/ha, whilst continuous barley and wheat were not harvested at all. Safflower succeeded in using some late rainfall occurred in May and June. The yield of safflower (1030 kg/ha) was higher than winter wheat yield after fallow (800 kg/ha). Tillage methods didn't affect the yields significantly, but in previous research, it was concluded that conservation tillage could be used at least for winter wheat.

In Kyrgyzstan, at both sites the year 2000 was used to establish the field experiments in accordance with the approved experimental design. Under rainfed conditions, preliminary data showed that even in very dry conditions there is a possibility of replacement of summer fallow with forages, pulse (1.08 t/ha of dry pea) or oil seed crops (0.98 t/ha of safflower), although fallow insures the highest grain yields (1.74-1.86 t/ha of wheat) compared with the wheat yield of 1.42 t/ha after safflower. Previous research showed that plow can be replaced by stubble mulch tillage. New research is necessary to find out the best possible conservation tillage treatment including no-tillage practice.

In Uzbekistan, an activity with soil tillage under irrigation was established. For the planting of winter wheat in this year a fraiser cultivator, a stubble cultivator-drill and a zero-till drill from New Zealand will be studied. In rainfed conditions, preliminary data showed importance of immediate tillage after winter wheat harvest. Traditional deep plowing is a very energy-consuming practice. Conservation tillage is a must, but equipment should be tested in a very dry and hard soil.

Under irrigated conditions of Tajikistan, in a soil tillage and fertility management experiment major treatments include application of commercial fertilizers against organic farming with two soil tillage treatments: plow and sub-tillage. Various combinations of double cropping were planted during 2000-2001 year: two wheats, wheat and barley, wheat and maize, wheat and soybean, wheat and tobacco under irrigated conditions in several sites. Under rainfed conditions, activity on terracing of eroded slopes was established in Faizabad and control of soil erosion was efficient. On the other slopy site in Obikiik, persimon fruit tree has been studied in the following treatments on run-off plots: natural grass, mulch, alfalfa, wheat and control. Activity on crop diversification was established at three sites with different duration of growing period.

In Turkmenistan an activity on soil tillage under irrigation was established in 2000 with promising results on replacement of deep plowing with disking. This saved significantly irrigation water and energy. Chiselling at 15-16 cm, and disk harrowing at 10-12 cm used irrigation water of 1690 and 1430 m^3, respectively compared with 2452 and 2060 m^3 after plowing at 30-32 cm and 20-22 cm. Thus, shallower the tilage the lower the water applied, which is a great option for water conservation although the wheat yield did not differ from each other ranging from 5.5 to 5.7 t/ha.

DISCUSSION

Crop diversification and conservation tillage activities under rainfed and irrigated conditions of five Central Asian Republics are established with the priorities identified with the National Programs.

Crop diversification and replacement of fallow-wheat systems looks very promising in rainfed conditions of northern Kazakstan. In other four countries there are opportunities to introduce safflower or legumes into the common cropping systems of fallow-continuous cereals. There are options to improve wheat-cotton rotations in irrigated areas. Conservation tillage practice is much more advanced in semiarid steppes of Kazakstan, where further studies on zero-tillage are priority. However, conservation tillage is also important under irrigated conditions as reduced tillage coud save substantial amount of irrigation water.

REFERENCES

Barayev, A.I.. (1978) Main components of the soil conserving system and its influence on development of spring wheat yield. In *Spring wheat,* pp. 158-198 A.I. Barayev, ed. Moscow: Kolos, (Rus).

Christiansen, S., Bounejmate, M., Bahhady, F., Thomson, E., Mawlawi, B. and Singh, M. (2000) On-farm trials with forage legume-barley compared with fallow-barley rotations and continuous barley in north-west Syria. *Experimental Agriculture* **36**(2): 195-204.

Harris, H.C. Cooper, P.J.M. and Pala, M. (eds.) (1991) Soil and crop management for improved water use efficiency in raianfed areas. ICARDA, Aleppo, Syria, 352 pp.

Jones, M.J. (2000) Comparison of conservation tillage systems in barley-based cropping systems in northern Syria. *Experimental Agriculture* **36**: 15-26.

Lavronov, G. (1973) Scientific basis of the dryland agriculture. Proceedings of the Uzbek Research Institute of Dryland Agriculture, Issue 9, pp. 3-19 (Rus).

Pala, M., Armstrong, E. and Johansen, C. (2000a) The role of legumes in sustainable cereal production in rainfed areas. In *Linking Research and Marketing Opportunities for Pulses in the 21st Century: Proceedings of the Third International Food Legumes Research Conference, 22-26 Sept 1997, Adelaide, Australia* pp. 323-334. Ed. R. Knight. Kluwer Academic Publishers, 3300 AA Dordrecht, The Netherlands.

Pala, M., Harris, H.C., Ryan, J., Makboul, R. and Dozom, S. (2000b) Tillage systems and stubble management in a Mediterranean-type environment in relation to crop yield and soil moisture. *Experimental Agriculture* **36**(2): 223-242.

Ryan, J. and Pala, M. (1996) Changes in organic carbon in long-term rotation and tillage trials in northern Syria. Page 99. In *Carbon Sequestration in Soil: an International Symposium* Abstract Vol. 22-26 July 1996, Ohio State University, Columbus, Ohio, USA.

P. R. HOBBS, R. GUPTA, R. K. MALIK & S.S. DHILLON

CONSERVATION AGRICULTURE FOR THE RICE-WHEAT SYSTEMS OF THE INDO- GANGETIC PLAINS OF SOUTH ASIA: A CASE STUDY FROM INDIA[1]

CIMMYT regional representative for South Asia and co-Facilitator of the rice-wheat consortium for the Indo-Gangetic Plains of South Asia

Facilitator of the RWC

Agronomist/weed scientist at Haryana State University, Hisar, India

Agronomist PAU Ludhiana, India

One of the major cropping systems of South Asia is rice-wheat grown on 13.5 million hectares in the Indo-Gangetic Plains (IGP). It is a major system for food security in the region and provides livelihoods and income to millions of farmers and workers.

The rice-wheat consortium (RWC) is a CG eco-regional program that combines natural resource management with production development in the IGP in geographically defined areas and targets different socio-economic groups. Its members include the national programs of Bangladesh, India, Nepal and Bangladesh, International centers (CIMMYT, IRRI, ICRISAT, IWMI and CIP) and various advanced institutions.

The RWC has been promoting conservation agriculture in the IGP for the past 10 years, especially 0-tillage and reduced tillage systems and more recently permanent bed systems. This paper is a case study of the remarkable revolution that has taken place in the region with data from the State of Haryana in India. In this State, farmers have adopted 0-tillage wheat after rice and more recently bed planted wheat with gusto and the acreage has risen from zero in 1997 to 40,000 ha last year. Local artisans at a cosst affordable to farmers manufacture all of the equipment. A participatory technology development paradigm was used to successfully achieve this impact. Benefits from this technology include reduced costs, less wear and tear on equipment, less weeds, improved water and fertilizer efficiency, less diesel use and therefore a reduction in greenhouse gas emissions and even an increase in yield. This improves farmer profit, improves his livelihood and eventually reduces poverty. All economic class of farmer also uses the technology since most farmers have to hire service providers for ploughing and this technology needs just one call to get the job done. Bed planting where the benefits of 0-tillage are combined with a bed and furrow system is also described. The use of conservation agriculture is probably one of the best ways to increase production, increase food security, improve farmer livelihoods and create environmental benefits while undertaking efficient natural resource use. The example of the information from the Sate of Haryana in India amply supports this statement.

L. García Torres et al. (eds.). Conservation Agriculture, 169–174.

BACKGROUND AND INTRODUCTION

Rice and wheat are grown sequentially in a mainly irrigated, double cropping pattern in the Indo-Gangetic Plains of South Asia on about 13.5 million hectares in Bangladesh, India, Nepal and Pakistan (Ladha, 2000). It is a major cropping system for sustaining food security in the region and there are millions of farmers and agricultural workers dependent on this system for employment and livelihoods.

The rice-wheat consortium (RWC) is an eco-regional program of the consultative group on international agricultural research (CGIAR), convened by CIMMYT that combines natural resource management with production development in the Indo-Gangetic plains of South Asia. It is a forum made up of many partners, including staff from national program research and extension, International agricultural research centers (CIMMYT, IRRI, IWMI, CIP and ICRISAT) and various advanced research institutions (Cornell, Rothamsted, CABI-UK, University of Melbourne, IAC, Wageningen). It promotes regional and global linkages for the development and deployment of improved, sustainable, productive rice-wheat cropping systems for the region.

The RWC has been promoting the use of conservation agriculture practices in this eco-regional zone for the past decade. These resource conserving technologies are based on reducing costs of production, improving the efficiency in the use of natural resources, providing environmental benefits, reducing costs while at the same time increasing production.

GEOGRAPHICAL AREA AND CHARACTERISTICS

The case study will use data from the Indian State of Haryana situated in the Northwestern side of the sub-continent in the Districts of Karnal, Kurukshetra and Kaithal (Harrington, 1993). The soils in this area are mainly alluvial in nature and many are recently reclaimed saline/sodic soils. The climate is a semi-arid, sub-tropical one with a distinct wet rainy season from June to September and a dry season from October to May. Rainfall is mostly in the rainy season (85%) and averages about 4-500 mm per year. Irrigation from canals and tubewells is needed to allow cropping. The main cropping system in the three Districts for this case study is rice in the monsoon season followed with wheat in the cooler dry season. Both crops are grown in the one calendar year. The rice crop is usually planted into seedbeds in May/June and the uprooted seedlings transplanted into flooded, puddled soils in June/July. Irrigation is used to supplement the rainfall. Long duration, *Basmati* type rice (40-50%) and shorter duration modern varieties are grown. The short duration varieties are harvested from late September through 3 October. The *Basmati* rice is harvested later in November and can cause late planting of the next wheat crop.

Wheat is planted after the rice crop is harvested starting in late October through November and sometimes into December. Multiple plowings (6-12) are usually done with a disc harrow or 9-tine cultivator before planting the wheat. Often the wheat seed is broadcast by hand before incorporating with the cultivator because of residue problems. Many farmers also burn the residues, especially after combine harvest and create severe air pollution. This multiple plowing not only increases the cost of production but also delays wheat planting especially where *basmati* rice is grown. This reduces yield by 1-1.5% per days delay after November 20 (Ortiz-Monasterio 1994). Maximizing yields requires timely planting and good plant stands. Wheat is harvested in April to early May. *Phalaris minor* is a major grass weed in the wheat crop and recently this weed developed resistance to *Isoproturon,* the most commonly used herbicide in the State.

Rice is the major crop in the monsoon season but brassica oilseeds, potatoes, legume fodders (*berseem* clover) and other crops can be grown instead of wheat. Sugarcane is grown on some coarser textured soils in 3-4 year rotations with rice and wheat.

NEW RESOURCE CONSERVING TECHNOLOGIES

This paper will describe two RCT's being promoted in the IGP's:

- Zero-tillage establishment of wheat after rice – this system uses a modification of the opener for the traditional *rabi* (winter) crop seed drill to allow planting of wheat into fields following rice harvest without plowing the field. The modification is base on an inverted-T opener used on seed drill imported from New Zealand to Pantnagar University in 1988. This system works well after hand harvested rice where there are anchored rice residues but few loose residues. It's utility is less in fields that are combine harvested where large amounts of loose straw creates a raking problem. Farmers traditionally burn the loose straw residue and do the same when planting with the 0-till Pantnagar drill.
- Bed planting of wheat, rice and other crops on top of a ridge and furrow system. This technology was introduced after Indian scientists visited the CIMMYT program in Mexico and learnt about this technique eagerly adopted by farmers in the Yaqui valley of Sonora State. This system is being promoted mainly for the benefits that accrue from water savings, but also in areas where grassy weeds are a problem since this system allows mechanical weeding and a reduction in costly herbicide applications. This system also allows for fertilizer placement, both basal and topdress, and increased efficiency of these inputs.

FARMER PERCEPTIONS ON THE RC TECHNOLOGIES

Zero-tillage

The approach used by the RWC partners in introducing the 0-till technologies to farmers in Haryana was one of enhanced participation of farmers, extension, scientists and local manufacturers. The technology was introduced to the farmers in the first year and then left the farmers to experiment with the technology in the second year as they gained confidence in its benefits. In this way, there was very good feedback of needed improvements to the drill and what worked and didn't work. These changes were incorporated into the machinery and management package for the next year's work. This led to rapid expansion in area for this technology for the benefit of all partners. The farmers are very enthusiastic about this technology and adoption has risen from a few acres in 1998 to more then 100,000 acres in Haryana in 2001. Stories abound about skeptical farmers ridiculing innovative farmers when the technology was first tried. Some farmers conducted the trials at night to avoid being seen experimenting with this technology. However, once the crop emerged farmers became convinced it would work and rushed to borrow or custom-hire the equipment to sow their fields. The main constraint in further accelerating this technology is availability of sufficient drills to satisfy the demand of the farmers. This will be corrected in coming years as more local manufacturers provide drills for farmers. It is now been accepted as a recommended practice by the State extension service.

Data shows that small landholders without tractors benefit from this technology since they only have to contact the service providers once to get their fields planted whereas for normal land preparation this had to be done many times. That frees up time for other employment. It also saves 4 million liters of diesel and 40 billion liters of irrigation water in the State.The carryover of the stemborer larvae has been the one major reason why some scientists and extension agents have not approved this technology. Data from monitoring shows that this is not the case. In fact, 0-tillage promotes beneficial insects by leaving a favorable habitat in the form of un-burnt residues that help control stemborer populations.

Bed planting

Bed planted wheat acreage has also grown the past two year as more drills were made and the number of farmers experimenting with this technology increased. This included planting wheat onto beds used for a previous rice crop and after minimal bed shaping. Farmers are interested in bed planting as a way to reduce irrigation. They would be even more enthusiastic if the beds could be permanent and just require some simple reshaping between crops. The bed planting would then combine the benefits of beds with those of zero-tillage. Bed panting was only introduced to farmers a few years ago and already almost 100 acres are planted in Haryana.

The most astounding finding was the performance of rice on beds. This was unexpected and was only undertaken to see if we could use permanent beds in rice-wheat areas (this would significantly help in reducing the cost of forming beds and make the technology more appealing to farmers). One farmer got 8.3 t/ha of rice from an acre planted to this system, and even more encouraging he did it with 65% less water!! Data shows that costs of bed planted rice (2900-3300 Rs/ac) were less than for normal transplanted rice (4500). Of 11 fields planted, yields for bed planting was 4500 kg/ha while normal transplanted rice was 3680. Average water savings was 50% or 50cm water per acre. Interestingly, we went straight to the farmers with this technology, the same farmers who benefited first from the 0-till introduced to them.

RESULTS

Five sets of farmer fields are being monitored with one acre field using 0-till and one conventional. Farmers have agreed to continue with the same practice for 3 years although they really want to shift the entire farm to 0-till. Average wheat yields for the 5 monitored sites was 5.56 for 0-till and 5.20 t/ha for conventional. In only one site out of 4 was 0-till less than conventional. The highest 0-till yield was 6.82 t/ha while that for normal was 6.0. The extra yield in 0-till was the result of timelier planting and fewer weeds. In fact, the monitoring data shows that 0-till helps reduce weed population over time. Data is also showing that 0-till fields have a higher soil organic carbon content although this needs to be confirmed over time.

Since the yields of wheat are higher and the costs of production are lower, the farmers improve the profitability of their wheat and this in turn improves their livelihoods. Since resource poorer farmers can also avail this technology they also benefit from being adopters. There are also many benefits to the country in that diesel imports can be reduced. If this technology were used on just 5 million hectares of the 13.5 million hectares of rice-wheat, 300 million less liters of diesel would be needed.

There are also benefits to natural resources and the environment. Zero-till saves about one third of the water needed. This is even higher with bed planting (up to 50% on wheat and more in rice). This is equivalent to about one million liters of water per hectare savings or if practiced on 5 million hectares, 5 billion cubic meters of water each wheat season. That would fill a lake 5 km wide, 10 km long and 100 meters deep. Since water is becoming a major natural resource constraint in the region, this benefit has major implications if the field level savings translate into basin level savings. Bed planting has even more water savings and there is a possibility it can be combined with drip irrigation systems to create even more benefits.

Environmentally, these technologies have profound effects on greenhouse gas emissions. The savings in fuel (0.3 billion liters from 5 million hectares of wheat) translates into almost a million tons of carbon dioxide. Further savings are obtained through using less energy to pump water and the energy saved because of more efficient use of fertilizers. Zero-tillage and permanent bed planting also allow residues to be left on the soil surface and provide an alternative to burning. This is the research thrust and 6 challenge for the future – to provide farmers with a suitable system for handling loose residues. If the residues could be left on the surface without burning on 5 million hectares of land, more than 40 million tons of carbon dioxide would not immediately enter the atmosphere. It would also have an immediate effect on improving air quality at the time of harvest.

Data from various on-station experiments also show that the benefits of 0-till can be even higher if wheat is planted after non-puddled rice. Future research will look at ways to grow rice without puddling both on the flat and on beds. Weeds will be a major problem that will need attention.

CONCLUSIONS

0-till wheat is now an established crop management activity for farmers in Haryana rice-wheat areas. It will continue to grow as more machines are made available and will save millions of rupees in costs, fuel and tractor wear and tear. Environmental benefits will increase as the issue of handling loose residue instead of burning becomes feasible. Bed planting will also grow in popularity as more machinery and more farmers experiment with it. Permanent bed planting will cut costs, improve yields and drastically reduce natural resource use, especially water but also fertilizer and other inputs. There is still a lot to do. The technology needs to be adapted to smaller two-wheel and animal drawn systems that would be more feasible for the resource poor farmers of the eastern IGP where plot size is also much smaller. Issues of weed carryover from the rice crop to the wheat crop will also be more important in these warmer eastern areas. Fields need to be monitored over time to establish whether any unforeseen problems arise that will need attention, but also to document the benefits of this technology on soils, biotic and socioeconomic variables. Data is needed on the effects of these two practices in saline/sodic soils. Leveling of lands can also bring about further benefits in water savings and yield and this needs to be promoted in the region. The use of conservation agriculture is probably one of the best ways to increase production, increase food security, improve farmer livelihoods and create environmental benefits while undertaking efficient natural resource use. The example of the information from the Sate of Haryana in India amply supports this statement.

NOTES

[1]Paper presented at the 1st World Congress on Conservation Agriculture Madrid, 1-5 October, 2001

REFERENCES

Harrington, L.W., S. Fujisaka, M.L. Morris, P.R. Hobbs, H.C. Sharma, R.P. Singh, M.K.Chaudhary, and S.D. Dhiman. 1993b. Wheat and Rice in Karnal and Kurukshetra Districts, Haryana, India: Farmers' Practices, Problems and an Agenda for Action. Mexico, D.F.: Haryana Agricultural University (HAU), Indian Council for Agricultural Research (ICAR), CIMMYT, and IRRI.

Hobbs, P.R., Y.Singh, G.S.Giri, J.G. Lauren and J.M. Duxbury. 2000. Direct Seeding and Reduced Tillage Options in the Rice-Wheat Systems of the Indo-Gangetic Plains of South Asia. Proceedings of an IRRI workshop entitled "Direct Seeding in Asian Rice Systems" held in Bangkok Thailand from 25-28 January 2000.

Ladha JK, Fischer KS, Hossain M, Hobbs PR, Hardy B (2000) Improving the productivity and sustainability of rice-wheat systems of the Indo-Gangetic Plains: a synthesis of NARS-IRRI partnership research. Discussion Paper No. 40.

International Rice Research Institute, Makati City, Philippines, 31 p Malik,R.K.; Gill,G.; Hobbs,P.R. 1998. Herbicide resistance - a major issue for sustaining wheat productivity in rice-wheat cropping system in the Indo-Gangetic plains.

Rice Wheat Consortium Research Paper series number 3. 36 pp. New Delhi, India: Rice Wheat Consortium for the Indo-Gangetic Plains.

Ortiz-Monasterio, J.I., Dhillon, S.S. and Fischer, R.A. 1994. Date of sowing effects on grain yield and yield components of irrigated spring wheat cultivars and relationships with radiation and temperature in Ludhiana, India. Field Crops Research. 37: 169-184.

W. G. STURNY & R.SCHWARZ

ZERO TILLAGE IN SWITZERLAND: THE BERNESE INCENTIVE PROGRAM

Bernese Soil Protection Service, Ruetti, CH-3052 Zollikofen

Swiss Soil Conservation Association, Oberdorf 7, CH-2514 Ligerz

Farmland is increasingly being stressed with high axle loads and very intense soil tillage techniques. The structure of the soil is thereby becoming more and more unstable – causing lots of soil degradation problems – whereas, above all, its load capacity has to be markedly in-creased. Zero tillage – defined as a plant production system without any soil tillage from previ-ous harvest to direct seeding – offers a solution to this dilemma: it goes easy on soil and water, at the same time it is labour-saving and cost-cutting.

Zero tillage is a demanding system which requires some rethinking on the part of the farm manager. During the period of transition crop yields may fluctuate somewhat, until about five years later a new "dynamic equilibrium" is reached in the soil – with a high porosity and earthworm population. This was found on zero tilled plots of a long-term demonstration experi-ment as compared to the plow treatment (cf. Figure 1 and 2, respectively). The site is a medium textured soil located at a college farm close to Berne.

An equable crop rotation (alternating grain and foliage plants) is crucial for the success of a zero tillage system. It is also essential to maintain a permanent soil cover with straw, plant residues, and living plants as a basis for high soil biological activity, including efficient weed control – at the same time minimizing all pressure on the soil.

In order to promote and subsidize conservation agriculture, an "incentive program" was adopted by the Bernese Office of Agriculture for regions susceptible to soil compaction, soil erosion, or nitrate leaching. Farmers willing to give up plowing and to apply reduced or zero tillage techniques on their fields, can conclude a five-year contract with the Bernese Office of Agriculture. According to the contract, they will get yearly contributions which depend on the applied cultivation technique and on the crops they grow (cf. Table 1).

L. García Torres et al. (eds.). Conservation Agriculture, 175–178.

The requirements for the five-year contract period are as follows:

- at least two main crops and half of the catch crops have to be direct seeded (= zero tillage)
- less than 50 % of soil disturbance from previous harvest to direct seeding
- no plowing

Afterwards, a connecting contract, of which the conditions are more strict (payments will only be granted for direct seeding) can be concluded for a second five-year period.

The number of farmers participating in the program was quadrupled since its beginning in 1996 (cf. Table 2). Usually when a farmer is successful in direct seeding, this will encourage his neighbours to try out this new zero tillage technique as well («snowball effect»).

To promote conservation agriculture on a national basis, the Swiss Soil Conservation Association (SWISS NO-TILL) was established five years ago. Actually, we count 230 members. However, the total zero tilled surface in Switzerland corresponds to about 3% of the arable land (cf. Figure 3) – with upward tendency!

Table 1: Contributions for the transitional stage (canton of Berne)

crops	*mulch tillage (transition)*	*zero tillage (target)*
	CHF per ha & year	*CHF per ha & year*
1 winter grain	*150*	*300*
2 spring grain	*150*	*300*
3 winter / spring rapeseed	*300*	*500*
4 strip tillage corn	*450*	
5 corn (silage / grain)	*300*	*500*
6 potatoes	*500*	*600*
7 sugar / fodder beets	*350*	*550*
8 peas, soya / field beans	*250*	*400*
9 sown meadow, green fallow	*0*	*200*
10 sunflower	*300*	*500*
11 further crops as agreed with the Department of Environment and Agriculture		

Table 2: The state of affairs on the performance of the decree on the retention of life basics and cultural landscape (canton of Berne)

year	*farms [n]*	*surface [ha]*	*zero tilled main cops [ha]*	*mulch tilled main crops [ha]*	*payments [CHF]*
1996	*91*	*399.4*	*39.2*	*127.7*	*127'979*
1997	*150*	*713.4*	*273.3*	*280.2*	*266'502*
1998	*195*	*880.8*	*494.2*	*329.4*	*276'434*
1999	*267*	*1258.0*	*804.0*	*363.1*	*321'394*
2000	*352*	*1709.0*	*1067.0*	*439.0*	*412'264*

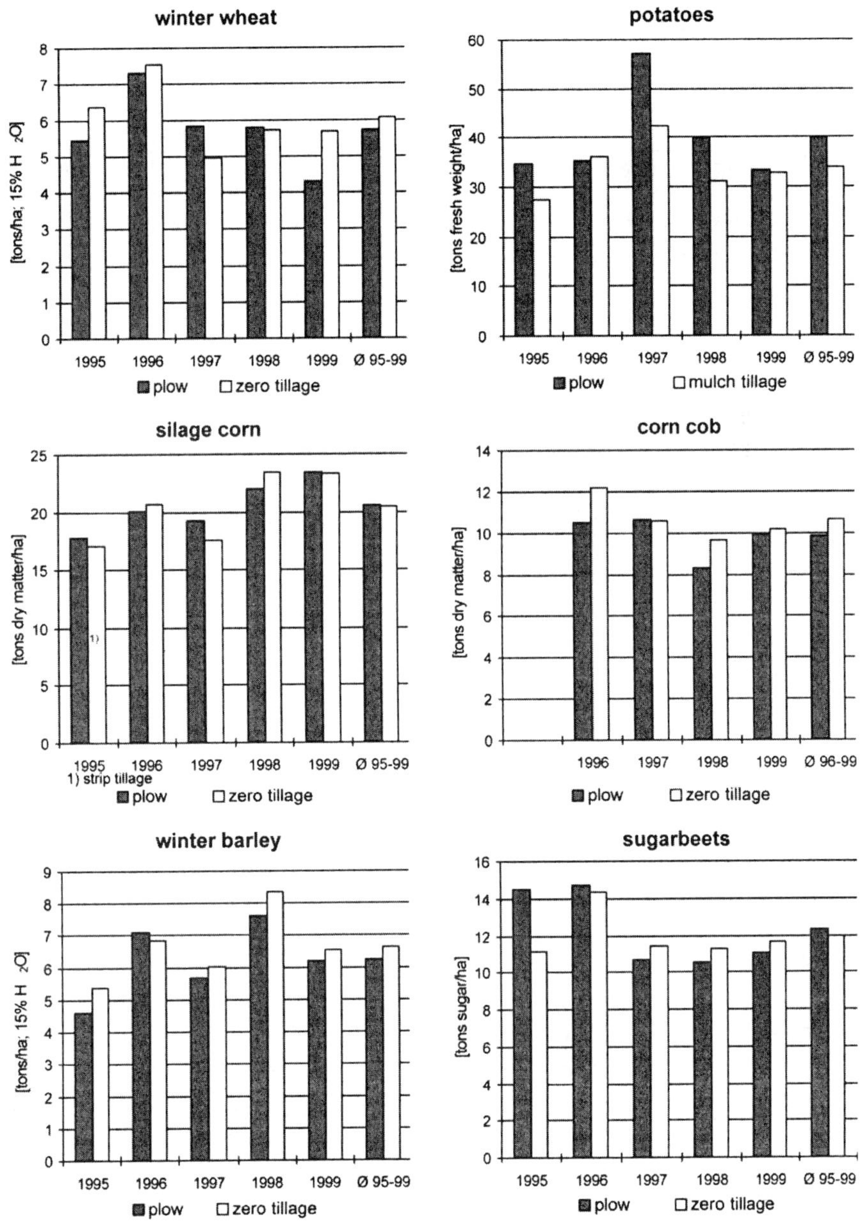

Figure 1: Long-term zero tillage experiment "Oberacker" (CH-Zollikofen): Yields, 1995-1999

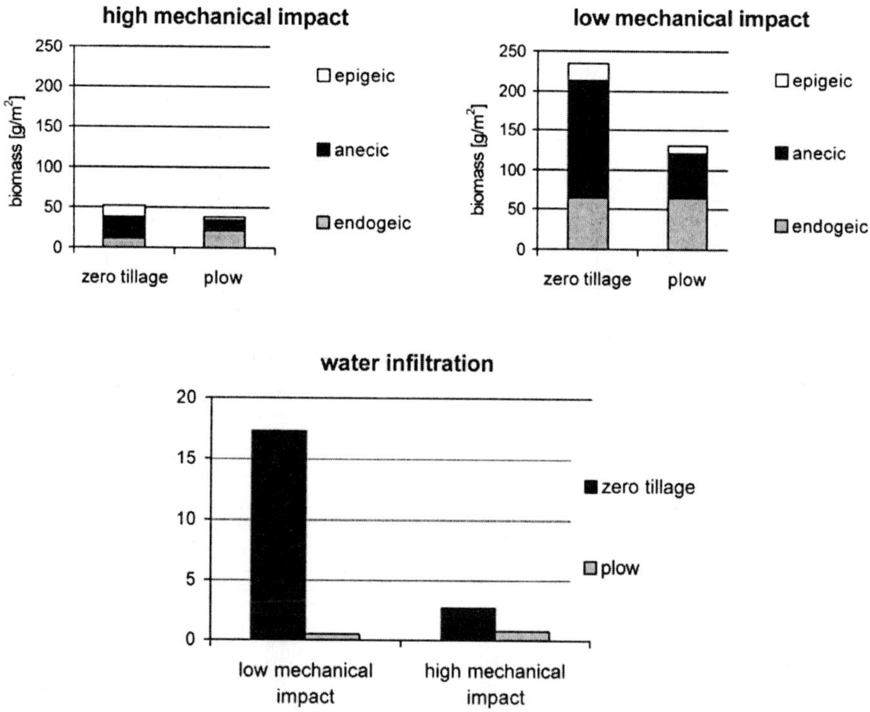

Figure 2: Long-term zero tillage experiment "Oberacker" (CH-Zollikofen): Effects of mechanical impacts on earthworm biomass and water infiltration, 1998

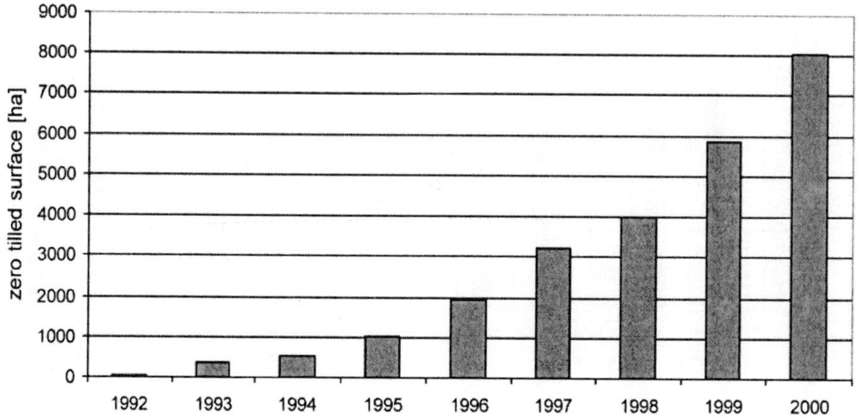

Figure 3: Development of the zero tilled surface in Switzerland

IV. RECENT INNOVATIONS ON CONSERVATION AGRICULTURE

R. DERPSCH

CONSERVATION TILLAGE, NO-TILLAGE AND RELATED TECHNOLOGIES

Deutsche Gesellschaft für Technische Zusammenarbeit (GTZ) GmbH, Eschborn, Germany

INTRODUCTION

There is increasing awareness all over the world of the negative effects of conventional agriculture and the need to change traditional agricultural practices. The key problem of conventional agriculture faces, especially in the tropics, is the steady decline in soil fertility, which is closely correlated to the duration of soil use. This is primarily due to soil erosion and the loss of organic matter associated with conventional tillage practices, which leave the soil bare and unprotected in times of heavy rainfall, wind and heat (Derpsch, 1998). To counter this a new concept of farming is evolving, which has been termed «conservation agriculture». Conservation agriculture maintains a permanent or semi-permanent organic soil cover. This can be a growing crop or a dead mulch. Conservation agriculture aims to conserve, improve and make more efficient use of natural resources through the integrated management of available soil, water and biological resources combined with external inputs. It contributes to environmental conservation as well as to enhanced and sustained agricultural production (FAO, 2001). No-tillage, direct sowing, direct drilling and conservation tillage all aim to achieve conservation agriculture.

Conservation Tillage is a concept that emerged in the United States after the dust bowl of the 1930s. Since this time there has been almost 70 years of developmental history in the USA, involving research, machine development, herbicide development, innovations introduced by farmers, crop and fertiliser management, etc., involving thousands of persons (Morrisson, 2000). Conservation Tillage has been defined as any tillage operation that leaves more than 30% of plant residues at the soil surface after seeding. The most common forms of Conservation Tillage in the USA are no-till, ridge-till and mulch-till. No-till is defined as planting into unprepared soil and not disturbing more than 1/3 of the soil surface, ridge-till is defined as planting on ridges formed during cultivation the previous year without disturbing the inter row

L. García-Torres et al. (eds.), Conservation Agriculture, 181–190.

area, and mulch-till is defined as a full-width tillage that leaves more than 30 percent residues on the soil surface after seeding (CTIC, 2001). Conservation Tillage is being practised on 44.1 million ha (36.6% of total farmland) in the United States.

The second largest area of Conservation Tillage worldwide can be found in Latin America. Here no-tillage is virtually the only form of Conservation Tillage technology being practised. In Latin America no-tillage has been adopted on an estimated area of 27 million ha, mainly in Brazil, Argentina and Paraguay. While in the USA and Europe many people think that Conservation Tillage is a necessary intermediate step before advancing to a no-tillage system, this is not the case in Latin America. Here farmers have found that changing to this very different production system brings clear advantages, and that it is better and cheaper to change completely rather than changing in steps. No-tillage should be understood as a farming system and, when changing from conventional tillage to no-tillage, one has to change the whole system accordingly. It does not help to change system components step by step, as this means that farmers are always using an incomplete system which will take far too long to change. As with rotational tillage, this results in a situation where soils are constantly in the transitional phase, so that farmers do not see the benefits in the soil from continuous no-till.

The World Bank views Conservation Tillage as the gateway to sustainable development. Conservation Tillage is more than a technology; it is a new approach to strengthen farmer organisations and self-help groups. Conservation Tillage is also cost-effective for the taxpayer. It requires fewer tax dollars than traditional soil conservation practices such as contouring or strip-cropping.

Current concerns about the effects of atmospheric greenhouse gases on global warming have broadened the possible importance of the effects of Conservation Tillage on soil organic carbon, since this represents the largest terrestrial carbon pool (Post et al., 1990). Kern and Johnson (1993) have predicted that if Conservation Tillage would be adopted on 76% of the USA's cropland by the year 2010 (compared with 27% in 1990), this would bring a benefit equivalent to approximately 0.7 to 1.1% of the US fossil fuel emissions, due to a combination of sequestering carbon in the soil and the fact that conservation tillage methods require less fossil fuel.

WHAT IS CONSERVATION TILLAGE?

A problem associated with Conservation Tillage is that in different parts of the world there is a different understanding about what this term means. There is confusion in the term conservation as well in the term tillage and also in the term no-tillage. This becomes particularly evident when trying to find figures on the adoption and/or expansion of Conservation Tillage in different countries. Sometimes it appears that each country (or region in a country) has a different understanding of what constitutes Conservation Tillage. It seems that the term is frequently adapted to suit the specific purposes of countries or regions. Often one gets the impression, especially in Europe, that the minute a farmer stops mouldboard ploughing, he or she is already practising Conservation Tillage.

Adding to this confusion, sometimes the term «minimum tillage» is used as a synonym for Conservation Tillage. Minimum tillage has been defined as the minimum soil disturbance needed for crop production. But this is subject to great variations, depending on a range of factors. Therefore this term should be avoided whenever possible. The basic elements of Conservation Agriculture are as follows: little or no soil disturbance, no burning, direct drilling, crop rotation, and permanent soil cover.

Given the impact of globalisation, it would be extremely helpful if tillage terminology could be globalised, because it is very difficult to achieve mutual understanding when everybody speaks a different «language». Even in the USA there are regional and institutional differences, despite its long history in Conservation Tillage (and where the term was indeed first developed). Nevertheless, the most widely-accepted definitions seem to be those of the US Conservation Technology Information Center (CTIC, 2000). CTIC definitions are as follows:

Cnservation tillage is any tillage and planting system that covers 30 percent or more of the soil surface with crop residue, after planting, in order to reduce soil erosion by water. Where soil erosion by wind is the primary concern, conservation tillage is defined as any system that maintains at least 1100 kg/ha of flat, small grain residue equivalent on the surface throughout the critical wind erosion period.

One problem with the term Conservation Tillage is that, in general, too much emphasis is put on tillage. Crop residues on the soil surface seem to be a by-product of tilling less. Research has consistently shown the many benefits of leaving crop residues on the soil surface, such as reducing erosion, improving soil quality, enhancing water quality, etc. Therefore we should be placing more emphasis on crop residue management than on the amount and type of tillage.

Crop Residue Management (CRM) (CTIC, 2000) is a year-round system beginning with the selection of crops that produce sufficient quantities of residue; it may also include the use of cover crops after low residue-producing crops. CRM includes all field operations that affect residue amounts, orientation and distribution throughout the period requiring protection. Site-specific residue cover amounts needed are usually expressed in percentage but may also be in pounds or kg. CRM is an «umbrella» term encompassing several tillage systems including no-till, ridge-till, mulch-till, and reduced-till.

No-till/strip-till (CTIC, 2000): the soil is left undisturbed from harvest to planting except for strips up to 1/3 of the row width (strips may involve only residue disturbance, or may include soil disturbance). Planting or drilling is accomplished using disc openers, coulter(s), row cleaners, in-row chisels or roto-tillers. Weed control is accomplished primarily with crop protection products. Cultivation may be used for emergency weed control. Other common terms used to describe No-till include direct seeding, slot planting, zero-till, row-till, and slot-till.

In the United States, Conservation Tillage represents 36.7% of the total planted area. No-till, ridge-till and mulch-till represent 17.5%, 1.1% and 18.0% respectively of the total planted area in the USA (CTIC, 2001). Reduced tillage is defined as a full width tillage which leaves 15 to 30%, while conventional tillage is a full-width tillage that leaves less than 15% crop residues on the soil surface (less than 550 kg/ha, of small grain residue equivalent).

Concepts in South America: South Americans are not very happy with the CTIC definition of no-tillage. This definition states that «The soil is left undisturbed from harvest to planting [...]». This would allow the soil to be ploughed once a year in a double cropping situation. Alternatively, the soil could be left undisturbed from the harvest of the main crop to the planting of a succeeding cover crop, which in turn would be ploughed in. The CTIC definition also notes that «Cultivation may be used for emergency weed control». In our view this situation simply should not occur if weeds are managed properly. The great variety of herbicides now available generally makes the use of emergency tillage unnecessary. The question how often an emergency is likely to occur is also relevant –once in 20 years, or three times every second year–.

In South America no-tillage is generally defined, according to the initiators of the system in the 1960s, as «planting crops in previously unprepared soil by opening a narrow slot, trench or band only of sufficient width and depth to obtain proper seed coverage. No other soil preparation is done» (Phillips and Young, 1973). Permanent no-tillage is meant, rather than occasionally not tilling the soil. The soil should remain covered by crop residues from previous cash crops or green manure cover crops, and most of these crop residues will remain undisturbed on the soil surface after seeding. As long as this requirement is met, shanks can be used to break compacted soil layers below the seed zone. Therefore the French term «agriculture de couverture du sol» (cover agriculture), seems to be more appropriate than no-tillage, unless we use the latter term in a broader sense.

We have to understand that soil carbon and crop residues are key factors for no-tillage to function. We have concentrated too much and too long on not tilling the soil instead of concentrating on crop residues as main tool for management (Wayne Reeves, personal communication, 1997).

Because no-tillage is the only Conservation Tillage practice that allows maximum amounts of crop residues to be maintained on the soil surface all the year round, this paper is going to give greater emphasis to this procedural approach.

RECENT INNOVATIONS IN NO-TILLAGE FARMING IN LATIN AMERICA

Green Manure Cover Crops (GMCCs): the leading nations in the development of a no-tillage system based on green manure cover crops can be found in Latin America, in particular Brazil, Paraguay and some Central American Countries. Economic analysis has shown that when choosing specific and appropriate cover crops before certain main crops, farmers can achieve higher economic returns compared to traditional systems where no cover crops are used (Sorrenson, 1984, Derpsch, et al., 1991). That is why cover crops are widely used in Latin America..

The functions of green manure cover crops are to: 1. Provide the needed soil cover for the no-tillage system (to reduce water evaporation, increase water infiltration into the soil, reduce soil temperature). 2. Protect the soil against erosion. 3. Reduce weed infestation and the costs of eliminating weeds. 4. Add biomass in order to accumulate organic matter in the soil. 5. Feed soils life. 6. Improve soil structure and avoid compaction. 7. Reduce the incidence of diseases and pests.

When properly chosen the use of cover crops then results in the following benefits. They bring higher economic returns; they reduce the necessity of applying herbicides and pesticides; increase the yield of main crops; avoid erosion; provide nitrogen for the soil (N fixation or N recycling); and, in combination with no-tillage, improve the chemical, physical and biological properties of soil and thus reduce fertiliser requirements. In addition, the inclusion of cover crops helps in the development and use of adequate crop rotations. As a result, the inclusion of green manure cover crops and crop rotations in the no-tillage system finally results in a sustainable farming system.

Research conducted in Paraguay (Kliewer et al., 2000) has shown that crop rotation and short term GMCCs covering the soil for 52 to 57 days can reduce the cost of herbicides drastically, to US$ 43.09/ha in the case of sun hemp, and to US$ 39.27 in the case of sunflowers, compared to costs of US$ 105.10 when only herbicides and double cropping wheat-soybeans were used. Kliewer (personal communication, 2000) has also reported soybean yields after sowing black oats of 2600 kg/ha, without using any herbicides at all. Weed measurements 96 days after seeding soybeans showed 93 kg/ha of dry matter (weeds) after sowing black oats, as opposed to 7390 kg/ha after fallow. In the last scenario, soybeans yielded not more than 780 kg/ha.

Research conducted in Brazil has shown that using black oats as a green manure cover crop before soybeans can increase soybean yield by as much as 63% compared to soybean yields after a wheat crop (Derpsch, et al., 1991).

One of the most recent and fruitful lessons farmers have learned in relation to the no-tillage system in Latin America is that the land should, where possible, never be left fallow. In general, fallow periods of only a few weeks will result in weed proliferation, the seeding of weeds, reduction of soil cover, soil erosion as well as lixiviation of nutrients. Instead of leaving the land fallow, farmers should seed a crop immediately or as soon as possible after harvesting the previous crop. This will reduce weed proliferation, prevent weeds producing viable seeds, increase soil cover, increase the biomass returned to the soil, increase the organic matter content of the soil, avoid soil erosion as well as prevent nutrients from being washed out, and improve the biological conditions of the soil (Derpsch, 1999).

Management of GMCCs is different in Latin America than in most other parts of the world. Farmers there have developed a «knife roller» to roll down cover crops and leave them flat on the soil surface. The knife roller is not designed to cut GMCCs, but rather just to squash or break their stems and in this way kill the plants. In most cases one or sometimes two passes is sufficient to kill plants when the knife roller is used at an appropriate growing stage (full flowering), and during optimal weather conditions (dry). If not, herbicides have to be used to avoid regrowth. The implement is not very expensive and is often made by farmers themselves. The knife roller consists of a hollow steel cylinder about 60 cm in diameter and up to 2 m wide, which allows water in. Knives (7 to 10 cm high, blunt, not sharp) are placed every 19 cm along the cylinder. Empty it weighs about 400 kg, while filled with water it weighs 800 kg. Several of these cylinders may be placed side by side to increase working width.

Black oats (Avena strigosa Schreb) is one of the most widely used cover crops in Latin America. In the States of Paraná and Rio Grande do Sul, Brazil alone, this species is used on approximately 3.2 million ha, while in Paraguay it is used on approximately 300,000 ha.

Liming. Many tropical soils are acidic or contain toxic aluminium. Farmers should apply lime the year before starting on no-tillage because this represents the last opportunity for successful incorporation. Although recent research has shown that farmers can also apply lime without incorporating, as lime moves easily into deeper soil layers in tropical soils, which are generally very permeable with high infiltration rates. In such cases farmers are recommended to apply small rates of lime each year, instead of applying large amounts only ones.

The concepts of liming and fertilisation have greatly changed in Latin America since shifting to the no-tillage system. Experience has shown that most things learnt at university about fertilisation and liming should be reconsidered, and instead the new concepts of fertility management in this system need to be applied. The main principle is that farmers should fertilise their soils rather than their crops.

Soil recuperation of extremely degraded soils. The world is losing at least 10 million ha of land each year because of bad management, soil erosion, soil degradation and salinisation. In developing countries, intensive tillage over decades has led to severe erosion problems and extreme soil degradation because of rapid organic matter depletion. Organic matter content is often lower than 1%, and in some cases as low as 0.2%. Experience with small farmers in Paraguay has shown that extremely degraded soils (which on average produce only 600 kg of maize and 700 kg of cotton per hectare), can recuperate in about 3 to 4 years by using appropriate management measures. These measures include: 1. The use of fertilisers in order to maximise the organic matter production of crops. 2. The use of green manure cover crops in a mixed cropping system aiming at producing 12 t/ha of dry matter per year. 3. Stopping tilling the soil and instead using the no-tillage system. 4. Using appropriate crop rotation. Farmers that practise this technology have been able to harvest more than 5 t/ha of corn and more than 2.5 t/ha of cotton after 3 to 4 years of soil recuperation (Florentín et al., 2001). No special problems have been observed when initiating no-tillage on extremely degraded soils and this is attributed to the fact that the whole system was changed, instead of only abandoning tillage.

Approach to soil compaction. Soil compaction is one of the biggest concerns for newcomers to the no-tillage farming community, as well as sometimes to farmers that have been practising this technology for some time. Tillage traditions are so deeply rooted that many farmers have difficulties in accepting that tillage is not necessary to produce a crop. Of course tillage-induced soil compaction, inherent to conventional tillage tools like plough pans or heavy disc harrow pans (as well as pedogenetic soil compaction), has to be eliminated before no-tillage is implemented. A chisel plough (or, less frequently, a subsoiler) is generally sufficient to solve these problems in most soils in Brazil, Paraguay and Argentina.

Researchers approach the problem of compaction differently to farmers. Since researchers have very sophisticated tools to measure compaction, and can easily demonstrate that soils are more dense under no-tillage than under conventional tillage, researchers often see compaction as a serious problem in the no-tillage system. In general, scientists and researchers in different parts of the world tend to overstate the problem of soil compaction. Unlike researchers, farmers in Latin America measure compaction not in terms of soil density in g/cm^3 or in penetration resistance, but in terms of crop response and yields. Farmers who have been practising continuous no-till for 25 to 30 years in Brazil are not worried about compaction so long as yields are as good or better in no-tillage than in conventional tillage systems (Derpsch, 1999). Farmers also measure compaction in terms of the penetration of seeding equipment into the soil. If soils are too hard to allow the cutting elements of a planter to penetrate well, then the farmer will achieve poor germination. Here of course the question arises whether the problem lies with the soil or with the seeding equipment.

No-tillage is a mental concept: no-tillage is not just a method of land preparation. It is more than simply a change of technology: changing from conventional or reduced tillage to no-tillage means a complete mental change. As Rieck Bieber, a no-till farmer in South Dakota suggests, «No-tillage is not a farming practice – it is a concept of the mind. If you don't believe in it you will fail». (To achieve this mental shift among farmers, technicians, extensionists and researchers, from soil-degrading tillage operations towards sustainable production systems like no-tillage, it is first necessary to change the attitudes of farmers. As long as the mind thinks conventionally it will always be difficult to successfully implement no-tillage in practical farming. If farmers are not able to radically change their approach to farming, they will never make the technology work effectively. This is not only true for farmers but also for technicians, extensionists and scientists as well.

DEVELOPMENT OF CONSERVATION TILLAGE IN SOUTH AMERICA AND OTHER PARTS OF THE WORLD

No-tillage is the fastest-growing Conservation Tillage technology around the world, and for this reason special attention has been given to this technology in this paper. The countries in the world with the biggest areas under no-tillage are the USA, with 21.1 million hectares, followed by Brazil with 14.3 million ha, Argentina with 10.5 million ha, Australia with 8.6 million ha, Canada with about 4.1 million ha, and Paraguay with 1.1 million ha (Table 1). In Australia tine machines are mostly used for no-tillage. Often no-tillage is performed with considerable soil movement, so that the requirements of Conservation Tillage (>30% residues after seeding) are not always met.

Zero tillage (synonymous with no-tillage) is the most important technology adopted in MERCOSUR (the common market consisting of Brazil, Argentina, Paraguay and Uruguay) in the last 50 years. Zero Tillage has reversed soil degradation, allowed an expansion of agriculture into marginal areas, boosted farmers' profitability, and increased the sustainability of agricultural systems. While in the early 1970s the area under Zero Tillage in MERCOSUR was negligible, in 1999 it was being used on about 20 million ha (Ekboir, 2000).

In the MERCOSUR Countries no-tillage technology has experienced a twenty-fold expansion between 1987 and 1997 (from 0.67 to 13.95 million ha) against a 4.6 times increase of the area in the USA (from 4.05 to 18.62 million ha) in the same period. From 1997 to 2001 the area under no-tillage in MERCOSUR Countries enlarged by 86% (from 13.95 to 25.98 million ha), compared with 13% in the USA (from 18.62 to 21.12 million ha) in the same period.

According to CTIC, 44.1 million ha were under conservation tillage in USA in 2000, and of this area some 21.1 million ha were under no-tillage (Dan Towery, personal communication 2001). Conservation Tillage increased in the USA from 26% of the total planted acres in 1990 to 37.2% in 1998, and then declined to 36.6% in 2000. This is because the mulch-till area declined by almost 2.3 million ha. CTIC believes that «much of this decrease may also be due to the new procedure used by USDA/ Natural Resources Conservation Service and other conservation partners in collecting tillage/residue estimates» (CTIC, 2000/2001). Between 1998 and 2001 no-tillage adoption increased from 19.3 million ha to 21.1 million ha. The tendency shows that farmers in the USA also increasingly prefer the no-tillage system over other forms of Conservation Tillage.

In Brazil the area under no-tillage increased from one million ha in 1990 to 14.3 million ha in 2001. In the same period no-tillage in Argentina increased from 300,000 ha to 10.5 million ha, and in Paraguay from about 10,000 ha to 1.1 million ha. Although worldwide the biggest area under no-tillage is to be found in the USA, in the latter no-tillage technology is only applied to 17.5% of the total cultivated area, as opposed to 25% in Brazil, 37% in Argentina and 52% in Paraguay. In relation to the total cultivated area, Paraguay has one of the highest adoption rates of no-tillage in the world.

Crops under no-tillage. To the author's knowledge, all crops can be grown under no-tillage. Experience in Latin America shows that even crops like potatoes, cassava, tobacco, onions and other vegetables can profitably be grown in the no-tillage system.

AREA UNDER NO-TILLAGE AROUND THE WORLD

Few countries in the world (except the United States) have yearly statistics on the different forms of Conservation Tillage. Therefore information on most parts of the world is either very scarce or non-existent: in most countries, statistics on conservation tillage are based on estimates. When narrowing conservation tillage down to no-tillage, however, information is easier to obtain. For this reason, this analysis has particularly focused on this practice. At the same time, it must be noted that information still remains inaccurate and is often unavailable. It has not yet been possible to obtain information about the area under no-tillage in Asia, for example, while little information is available on Africa. According to Jim Findlay, some 100,000 small farmers in Ghana are now practising no-till on 0.3 to 0.75 ha on their farms, after slashing and the application of glyphosate. «Sales of Roundup and similar products in 2001 suggest that there are going to be 140 to 170,000 farmers using some form of no-till this year. The total area under no-tillage in this country is estimated to be around 25,000 to 30,000 ha» (Jim Findlay, personal communication 2001). It has also been reported that in Ghana no-tillage is frequently practised on bare soils, exacerbating the problem of soil crusting, runoff and erosion. Despite the author's best efforts to include more information on the adoption of no-tillage in Africa and Asia, no information beyond what has already been mentioned was received before submitting this paper. On the other hand, the information received about the area under no-tillage in Europe is very confusing (obviously because of definition problems), which is why it has not been included in this presentation.

Table 1: Area under No-tillage and Conservation Tillage in various countries 1999/2000 (hectares)

Country	*No-tillage*	*Conservation Tillage*
USA [1]	*21,120,000*	*44,120,000*
Brazil [2]	*14,330,000*	*14,330,000*
Argentina [3]	*10,500,000*	*9,250,000*
Australia [4]	*8,640,000*	*NA*
Canada [5]	*4,080,000*	*NA*
Paraguay [6]	*1,100,000*	*1,100,000*
Mexico [7]	*650,000*	*NA*
Bolivia [8]	*350,000*	*350,000*
Venezuela [9]	*150,000*	*150,000*
Chile [10]	*100,000*	*100,000*
Colombia [11]	*70,000*	*70,000*
Uruguay [12]	*50,000*	*50,000*
Others [13]	*1,000,000*	*NA*
Total	***62,140,000***	

Source: *1) Dan Towery, CTIC, 2001; 2) FEBRAPDP, 2001; 3) AAPRESID, preliminary information, 2001; 4) Bill Crabtree, WANTFA, 2000; 5) J. Hebblethwaite, CTIC, 1997; 6) MAG – GTZ Soil Conservation Project, 2001; 7) Ramón Claverán, CENAPROS, 1999; 8) Carlito Los, 2001; 9) Carlos Bravo, 2000; 10) Carlos Crovetto, 2001; 11) Roberto Tisnes, Armenia, 1999; 12) AUSID, 1999; 13) Estimates. NA = Not availble.*

Taking into consideration the many gaps in information, it is estimated that no-tillage is practised on about 62 million hectares worldwide. According to the available information, approximately 84% of the technology is practised in the Americas (North and South), about 14% in Australia and only 2% in the rest of the world, including Europe, Africa and Asia. There is considerable potential to bring this soil-conserving technology to Asia and Africa, although limiting climatic and socio-economic factors have to be taken into account. Through the efforts of ECAF, FAO, GTZ, CIMMYT and other organisations, as well as through this «First World Congress on Conservation Agriculture» in Madrid in 2001, there is optimism about the spread of no-tillage technology in Europe and other parts of the world. In order to overcome the information gaps (relating mainly to Africa and Asia, the East European countries, as well as western Europe), the author would welcome any information about the area of no-tillage and conservation tillage being practised in the different regions and countries of the world.

Remarks: Some data on the area under no-tillage in Canada shows 6.7 million ha in that country. These numbers allow for fall tillage with high soil disturbance. When applying the term no-tillage more strictly (i.e. low disturbance and no fall tillage), then the Canadian area is only 4.08 million ha.

Due to the very bad erosion and soil degradation problems in Eastern Europe and parts of Asia, it seems that these regions have the biggest potential for fast growth of no-tillage and Conservation Tillage technology. The main limitations in these regions seem to be the lack of appropriate knowledge, the existing tradition of intensive soil tilling, inadequate equipment to seed in hard soil and through thick layers of mulch, combined sometimes with a dry climate and lack of mulch.

REFERENCES

CTIC (2000) Conservation Technology Information Center, Core 4, Conservation for Agriculture's future. http://www.ctic.purdue.edu/Core4/CT/Definitions.html

CTIC (2000/2001) Conservation Technology Information Center, CTIC Partners Dec 2000 /Jan 2001. 2000 Crop Residue Management Survey, More Farmers Say «Yes» to No-till.

CTIC (2001) Conservation Technology Information Center, West Lafayette, Indiana, CTIC www.ctic.purdue.edu/Cor4/ctsurvey/2000/2000UsSummary.html (01/05/01)

Derpsch, R. (1998) Historical review of no-tillage cultivation of crops, Proceedings, The 1st JIRCAS Seminar on Soybean Research. No-tillage Cultivation and Future Research Needs, March 5-6, 1998, Iguaçu Falls, Brazil, *JIRCAS Working Report* No. **13**, p 1 – 18, 1998.

Derpsch, R. (1998) Implications of No-tillage Versus Soil Preparation on Sustainability of Agricultural Production. Catena Verlag, Advances in GeoEcology 31, 1175-1186,

Derpsch, R. (1999) Frontiers of conservation tillage and advances in conservation practice. Paper presented at the 10th ISCO Conference, 24. – 28. May 1999, West Lafayette Indiana, USA (Proceedings in print).

Derpsch, R., Roth, C.H., Sidiras, N., Köpke, U., (1991) Controle da erosão no Paraná, Brasil: Sistemas de cobertura do solo, plantio direto e preparo conservacionista do solo. *Sonderpublikation der GTZ*, No. **245**, TZ-Verlagsgesellschaft mbH, Rossdorf, 272 pp..

FAO (2001) Conservation Agriculture, Matching Production with Sustainability. What is the Goal of Conservation Agriculture? FAO homepage.

Ekboir, J. (2000) Innovation Systems and Technology Policy: Zero Tillage in MERCOSUR. Paper presented at the XXIV Reunion of the International Association of Agricultural Economy, Berlin, 13-18 August, 2000.

Florentín, M.A., Peñalva, M., Calegari, A. y Derpsch, R., (2001) Abonos verdes y rotación de cultivos en siembra directa. Pequeñas propiedades. In: Proyecto Conservación de Suelos MAG–GTZ, DEAG, San Lorenzo, Paraguay, in preparation.

GTZ (1998) Conserving Natural Resources and Enhancing Food Security by Adopting No-tillage. An Assessment of the Potential for Soil-conserving Production systems in Various Agro-ecological Zones of Africa. GTZ Eschborn, Tropical Ecology Support Program, TÖB publication number: TÖB F-5/e, 53 pp

Kern, J.S. and Johnson, M.G., (1993) Conservation tillage impacts on national soil and atmospheric carbon levels. *Soil Sci. Soc. Am. J.*, **57**: 200-210

Kliewer, I., Casaccia, J., Vallejos, J. and Derpsch, R., (2000) Cost and herbicide reduction in the no-tillage system by using green manure cover crops in Paraguay. Proceedings 15th ISTRO Conference, Fort Worth, Texas, USA, July 2 – 7, 2000

Kuipers, H. (1970) Historical notes on the zero-tillage concept. *Neth. J. Agric. Sci. - Papers on zero-tillage*, Vol.**18** N° 4, 219-224

Morrisson, J.E., (2000) Development and future of conservation tillage in America. Proceedings of China International Conference on Dryland and Water-Saving Farming, 21. - 23. November 2000, Beijing, P.R. China

Phillips, S.H. and Young, H.M., (1973) No-tillage Farming. Reiman Associates, Milwaukee, Wisconsin, 224 pp

Post, W.M., Peng, T.H., Emaniel, W.R., King, A.W., Dale, W.H. and De Angelis, D.L., (1990) The global carbon cycle. Am. Sci. 78: 310-326

Sorrenson, W.J. and Montoya, L.J., (1984) Economic implications of soil erosion and soil conservation practices in Paraná, Brazil. IAPAR, Londrina, GTZ, Eschborn (Unpubl.), 233 pp

SWCS (2001) Conservogram, the Conservationist's newsletter, Soil and Water Conservation Society, Volume 8 N° 1, January/February 2001, p 3

Waydelin, C. W. (1995) Practical experience with reduced tillage farming. In: Proceedings of the EC-Workshop-II-, Silsoe, 15-17 May, 1995, Experience with the applicability of no-tillage crop production in the West-European countries, Wissenschaftlicher Fachverlag, Giessen, 1995, 187-190.

Wittmuss, H.D., Triplett, Jr., G.B. and Greb, B.W. (1973) Concepts of conservation tillage systems using surface mulches. In: Conservation Tillage, the Proceedings of a National Conference, March 28-30, 1973, Des Moines, Iowa

A. CALEGARI

COVER CROP MANAGEMENT

Agronomic Institute of Paraná-Iapar, Rod. Celso G. Cid, Km 375, CEP-86001-970, Fax: 00 55 43 376 2101, Londrina, Paraná, Brazil;

In different parts of the world when traditional management and planting methods are used, ploughing the soils which buries plant residue and leaves, the soil will remain naked and vulnerable to water and wind erosion process. The use of these cultivation methods in heavy sloped areas in the tropics and subtropics, undergoing heavy rainfall, has caused severe erosion and a dramatic loss of fertility in many farming areas. The constant adding organic carbon to the soil through plant cultivation, as well as keeping plant residue, preferably on the soil surface (no-tillage) are important measures to preserve and foster organic matter balance in the soil. Thus, plants used as cover crop given their high capacity to produce vegetal biomass and roots, with an important direct and indirect effects in the soil-water-plant system, play a fundamental role when they are an adequate part of orderly rotation systems with cash and food crops.

INTRODUCTION

The colonization process of Paraná state that started more or less 60 years ago with an absence of adequate occupation planning and use of the territory lead to serious consequences in soil and water use and management. The tropical and subtropical forests were dramatically decreased (from 84% to 24% in 1965 and less than 10% to 1984) with agriculture covering both able as well as marginal areas. According to data, more than 6 million hectares are cultivated with summer species: soybean, maize, beans, cotton, rice, sugarcane, cassava, coffee, fruits, vegetables, etc. Conversely approximately half this area remains uncultivated in the autumn/winter, with serious risks of erosion (soil and nutrients losses) and also improving weeds infestation and consequentely increasing farm production costs.

For many years, water erosion was considered the great environmental problem of the agricultural sector, and the execution of programs having mechanical conservation practices, the main feature of these actions, were insufficient to control the phenomenon. Sorrenson and Montoya, 1984 have reported that in Paraná, average soil losses of 10 to 40t of fertile soil per ha per year when traditional soil tillage systems has been used. These greatly contributed to increase awareness of the problem and to lead growers to organize the search for common alternatives and solutions.

L. García-Torres et al. (eds.), Conservation Agriculture, 191–200.

In Brazil, No-tillage system began more specifically in the state of Paraná at the beginning of the seventies, with a pionner farmer Mr. Herbert Bartz from Rolândia city. The main objective was to control erosion in areas where soybean and wheat were intensively cultivated in southern Brazil. Afterwards, corn also began to be cultivated under this system. After this the researchs system started with experiments and also trying to improve this system according different production systems in South Brazil.

The results obtained throughout the years in Paraná and other parts of Brazil prove that cover crops, as part of the productive systems preferably in no-tillage areas, are very economically feasible as well as ecologically sustainable; proving not only greater crop productivity, but also conservation, maintenance and/or recovery of soil fertility. In addition to this, they promote economy with nitrogenous fertilizers (leguminous plants), greater biologic balance in the soil, decreasing the effects of pests and or disease; in other words they represent a very promising way to manage soils tending to sustainability.

The No-tillage system, entails using different species of cover crops, dry matter crop residues, and crop rotation as fundamentals in the structure of rational and sustainable management, mainly in annual crop areas. The systematization of the areas through work in hydrological micro basins and also in no-tillage today occupies more than 3 million hectares in Paraná and estimates show that the No-Tillage covers more than 14 million hectares in Brazil.

COVER CROPS SPECIES DEVELOPED IN PARANÁ STATE AND OTHER BRAZILIAN STATES

According to different agroecological zones of Paraná state and also other Brazilian regions, with high diversity of farming systems, many species of cover crops are been largely used by farmers. This species besides promote soil properties improvement, and also good response for the next crops, have been used in a multipurpose as like an animal fooder and also some species with potential to be used in a human food. Some of these species are listed below:

- *Crotalaria juncea, Crotalaria mucronata, Crotalaria spectabilis, Crotalaria breviflora, Crotalaria grantiana, Crotalaria paulina Crotalaria ochroleuca, Crotalaria retusa, Crotalaria striata, Crotalaria anagiroydes; Mucuna pruriens (black, grey and dwarf varieties); Canavalia ensiformis, Canavalia brasiliensis; Vigna umbelata, Vigna unguiculata, Vigna radiata; Arachis pintoi; Arachis prostata; Arachis hypogaea, Cajanus cajan, Dolichos lab-lab, Leucaena leucocephala, Indigofera hirsuta,Indigofera endacaphyla, Stylosantes guyanensis, Calopogonio mucunoides, Pueraria phaseoloides, Glycine wiighty, Centrosema pubescens, Macroptilium atropurpureum, Clitoria* ternatea, *Setaria italica* (annual), *Linum usitatissimum, Lotus corniculatus*, Hairy vetch, (*Vicia villosa*), Common vetch, (*Vicia sativa*), *Ornithopus sativus,* Radish (*Raphanus sativus*), White lupin (*Lupinus albus*),

Yellow lupin (*Lupinus luteus*), Blue lupin (*Lupinus angustifolius*), Sweet pea (*Lathyrus sativus*), *Pisum sativum-subespecie arvense* -Iapar-83, *Pisum sativum* –Iapar-74, *Trifolium repens, Trifolium pratense,* Black oat (Iapar-61) (*Avena strigosa*) White oat (*Avena sativa*), *Raygrás (Lollium multiflorum)*, Rye (*Secale cereale*) Sunflower (*Helianthus annuus*), Corn spurrey (*Spergula arvensis*), *Triticosecale, Millet (Pennisetum americanum), Sorghum bicolor*

.

We are working with all of these cover crops species in many regions in different cropping systems. Many cover crops can be planted mixed (2-3 or more species= coktail) to obtain in the same season the all soil positive effects of the different cover crops species, and also take more advantages in the same time. The cover crops are conducted in crop rotation systems with cash crops (maize, wheat, beans, soybean, cotton, cassava, potato, peanuts, sunflower, vegetables and also intercropped with perennial crops= coffee, citrus, fruits tree, grapes, etc.). In the majority of the cases, the seeds production of cover crops species are encouraged to achieve by farmers, which normally have decreased the total production costs. The all cover crops species used normally present different behaviour and effects on the soil according to the edafoclimatic conditions, and also many species can be used to break soil compacted layers and enhance water infiltration, supress some weeds population (alellopathic effects), nematodes, fungi, etc. So, with the variable root system attained different soil profile depth and with the formation of organic acids and other compounds that can contribute to increase soil organisms populations (micro, meso and macro fauna and flora) and consequentely higher biodiversity which promote a better biologic control and great environmental equilibrium. The cover crops can be managed by knife-roller, slash or in some cases with herbicides to promote a mulch formation and facilitate the next crop implantation. Normally residues may be managed by different ways: by removal , feeding to animals, burning, incorporation, or left on the soil surface. The soil productivity are directly influenced by the fate of crop residues and best effects are attained if its are not removed from the field. Residue incorporation effects on soil productivity are difficult to separate from tillage effects because incorporation is achieved through some type of tillage operation. Also, soil water content, soil temperature, and porosity are influenced by the presence of remotion of the residues of surface. Changes that will occur in the microbial community can affect N availability through reduced mineralisation and nitrification, and increased denitrification. Differences in climate, soil type, and residue quality, associated with surface residues will promote changes in the rate and degree of organic matter accumulation.

The different species normally have hability to recycle different nutrients and also nitrogen fixation (legumes) that can be more easily be uptake by the next crop (Table 01 and 02).

Different research studies and farmer experiences have shown the favourable effects of cover crops and no-tillage on soil properties. The different cover crops used in a Southwestern experiment showed an important and significant effects on the organic carbon level for both depths (0-5 and 5-20cm – not presented). The concentration of organic material on surface contributed to increase the organic carbon values (0-5cm) except wheat compared to fallow (Table 3). The means of all winter crops presented greater values than fallow in both depths. This shown clearly the benefitial effect of winter cover crop in increase the organic carbon content in the soil profile.

Table 1. Chemical composition of different cover crops used to cover crop and animal food in the flowering stage.

Cover crops	Nutrients (% of dry matter)						C:N	ppm		
	N	P	K	Ca	Mg	C	Ratio	Cu	Zn	Mn
Mucuna pruriens(grey)	2,50	0,15	1,40	1,20	0,27	52,30	21,12	16	28	183
M. pruriens (black)	2,49	0,13	1,40	1,17	0,27	52,15	21,06	14	29	174
M. pruriens (dwarf)	3,10	0,19	4,49	2,14	0,65	50,83	16,39	9	85	179
C. juncea	2,50	0,19	1,20	2,31	0,47	45,25	18,10	14	44	179
C. mucronata	3,43	0,09	2,30	1,32	0,47	53,70	15,65	13	35	111
C. spectabilis	2,17	0,09	1,59	0,49	0,37	50,83	23,42	8	23	126
C. cajan	2,61	0,14	2,61	1,79	0,45	56,30	21,57	7	22	87
C. ensiformis	3,19	0,15	5,62	1,35	0,63	50,15	15,72	9	62	254
C. brasiliensis	2,49	0,13	1,68	0,20	0,16	51,24	20,57	4	14	17
Vigna radiata	2,09	0,21	4,94	1,48	0,75	52,47	25,10	10	78	127
Vigna unguiculata	2,62	0,20	2,82	0,93	0,28	45,42	17,33	-	-	-
Leucaena leucocephala	4,30	0,22	1,70	0,81	0,50	63,50	14,76	-	45	-
Indigofera sp.	2,17	0,14	1,54	1,20	0,32	40,36	18,60	13	24	53
C. mucunoides	2,16	0,12	1,56	1,40	0,29	46,73	21,63	9	15	172
Pueraria phaseoloides	3,68	0,29	2,14	1,30	0,41	54,10	14,70	11	27	155
Glycine wiighty	2,60	0,23	2,39	0,99	0,35	45,03	17,31	8	32	102
Centrosema pubescens	2,34	0,23	1,19	0,66	0,45	47,60	20,34	10	32	67

Source: Adapted by Gallo et al., 1974, Kluthcouski, 1982; Calegari, 1982 and Chaves, 1989; cyted by Calegari ,1995.

Table 2. Chemical composition of winter cover crops in the flowering stage.

	Dry matter %							C:N	ppm		
Cover crop	N	P	K	Ca	Mg	C	Protein	Ratio	Zn	Cu	Mn
Hairy vetch	3,82	0,30	2,03	0,78	0,27	37,87	23,87	10,05	26	9	61
Common vetch	2,87	0,23	2,88	1,05	0,41	37,1	17,94	12,9	24	9	87
Ornithopus	1,79	0,14	3,55	1,10	0,45	40,14	11,18	22,43	59	13	97
Radish	2,68	0,17	2,80	1,54	0,76	38,58	16,75	14,45	49	8	84
White lupin	3,20	0,09	2,66	0,46	0,38	47,49	20,0	14,84	57	12	330
Yellow lupin	2,94	0,16	2,50	0,59	0,39	42,25	18,37	14,37	66	14	359
Blue lupin	3,19	0,19	2,29	1,20	0,49	37,83	19,93	11,86	24	13	230
Sweet Blue Lupin	2,28	0,10	1,75	0,59	0,42	37,87	14,25	16,61	32	16	147
Field pea Iapar -83	2,09	0,12	1,50	0,70	0,20	39,77	13,06	19,02	8	22	52
Sweet pea	2,23	0,10	2,90	0,39	0,19	41,91	13,93	18,79	22	11	52
Black oat	1,93	0,28	2,15	0,43	0,21	39,69	12,06	20,76	11	7	102
White oat	0,81	0,052	2,40	0,24	0,17	38,52	5,06	47,55	9	6	138
Raygrás	1,34	0,067	2,60	0,41	0,22	59,22	8,37	44,20	23	9	214
Rye	1,22	0,075	1,40	0,18	0,14	44,59	7,62	36,54	15	6	53

Source : Adapted by Calegari et al., 1993.

Table 3. Chemical properties of the sol after eleven years with different cover crops and tillage systems.

	P mg/dm³	C g/dm³	pH cmol/dm³	Al cmol/dm³	Ca cmol/dm³	Mg cmol/dm³	K	Basis Saturation %
No-tillage (0-5 cm Soil depth)								
Hairy Vetch	22,7 A	34,85 B	5,23 A	0,01 B	5,99 A	3,16 A	0,40 AB	61,61 A
Fallow	18,2 A	28,59 A	5,87 B	0,00 A	7,50 C	4,12 C	0,21 A	74,73 B
Wheat	19,4 A	26,84 A	5,63 B	0,00 A	6,19 AB	3,41 AB	0,35 AB	68,15 AB
Radish	33,2 B	29,72 AB	5,77 B	0,00 A	7,40 B C	3,73 ABC	0,49 B	72,89 B
Black oat	27,5 AB	30,70 AB	5,80 B	0,00 A	7,77 C	4,30 C	0,39 AB	74,59 B
Blue lupin	21,2 AB	29,75 AB	5,63 B	0,00 A	6,84 ABC	3,94 BC	0,36 AB	70,71 B
PLOUGHED (0-5 cm Soil depth)								
Hairy Vetch	5,9 A	25,37 AB	4,83 A	0,10 B	4,18 A	2,27 A	0,31 A	48,74 A
Fallow	4,4 A	23,98 A	4,93 AB	0,02 A	4,27 A	2,32 A	0,28 A	51,24 A
Wheat	4,7 A	25,87 AB	5,17 B	0,01 A	4,18 A	2,38 AB	0,39 A	53,87 A
Radish	4,9 A	26,62 AB	5,13 AB	0,03 AB	5,01 A	2,85 B	0,43 A	57,36 A
Black oat	4,2 A	27,12 B	5,10 AB	0,01 A	4,83 A	2,68 AB	0,63 B	56,42 A
Blue lupin	3,8 A	25,73 AB	5,07 AB	0,01 A	4,47 A	2,64 AB	0,31 A	55,54 A

**Values in the same column designated by the same symbol do not differ significantly (LSD 0.15) for each treatment.*
Source: Calegari, 1998.

The experiments result showed that there were difference among different cover crops regime (0-5cm) for calcium, magnesium and potassium mainly in no-till, contributing to higher basis saturation and probably joint with the organic carbon effects complexing the aluminium decreasing its toxic effects, and at the same time may contribute to alter pH values and enhance P release.

An understanding of how crop residue influences nutrient recycling and chemical properties combining strategic integration with residue management in different crop systems is the key to developing good soil fertility management (Schomberg; Ford and Hargrove, 1994).

RESULTS WITH COVER CROPS, CROP ROTATION AND NO-TILLAGE SYSTEM

The value of the legume in the rotation is not only that nitrogen is added to the system but that soil structure is improved by the organisms that decompose that leguminous organic matter. There are many results in different parts of the world with the use of legumes in crop rotations system.

In Northern Nigeria, maize grain yields were found to be greater following a groundnut crop than after crops of cowpea, cotton or sorghum. The yield increase was related to an increased availability of mineral N in the soil after groundnuts (Jones,1974).

In Zimbabwe, the yield of maize was greater after Bambara groundnut (7.6 t/ha^{-1}) than after groundnut (6.2 t/ha^{-1}) , unplanted fallow (4.3 t/ha^{-1}) or maize (3.9 t/ha^{-1}) (Mukurumbira,1985).

Kang,1993 worked a 10-year period with continuous crop on an Entisol in Nigeria where the area was occupied by maize in main season (10 years) sweet potato in the minor season starting from the second year (3 years) ,and followed by cowpea for the remaining 6 years. The results observed shown that the organic C level decline was fastest in the bare uncropped plot, followed by that in the unfertilized plot. With the fertilizer application (N, P, K), a higher level of organic C was maintained than in unfertilized plots. The retention of maize crop residue showed a slightly higher soil organic carbon level compared with crop residue that was had been removed. In addition it was seen that the combination of fertilizer application and crop residue retention lead to higher maize yields than no fertilizer and crop residue remotion.

The use of cover crops like black oat, radish, lupine, vetch, rye, mucuna, pigeon, pea, cowpea, crotalarias, etc. and also crop rotation systems are increasing in different regions of Paraná State (Calegari et al.,1993). These practices are important to complement the no-tillage system that is paramount in lead to sustainability soil management in tropical conditions. The crop rotation is the basis for sustainable agriculture in tropics and also for the no-tillage system, mainly in aspects related to insects and disease control, and also to provide and keep enough amount of crop residues to cover and protect soil surface. These aspects considered are basics for an adequate no-till implantation system and also for the continuity of the system.

According to different regions with the specific edafoclimatic conditions, several crop rotations systems have been developed in Paraná State. For example, soybean can rotate with maize in the summer and in the winter it is possible use black oat or oat + lupin, or oat + vetch, or oat + field pea; maize can rotate with oat-beans/beans-oat + vicia- maize- radish- beans/beans, and onion goes with maize residues/oat- onion-oat+ vicia- onion/black mucuna- mucuna residues- onion/maize.

The crop rotation system used by small farmers in South Brazil, is according to soil, climatic, technical and economical factors. In the majority of the farms basically comprise:

1) In the Fall/Winter season: oats (mainly black oats), vetches (common and hairy), oil seed radish, rye, lupin(white and blue), raygrás; many farmers leave in fallow, a few farmers number use wheat cash crop.
2) In the Spring/Summer season: corn, beans, tabaco, onion, garlic, potatoes, rice, cotton, etc. A very small group grow soybean crop.

Results obtained in a experiment in North of Paraná, Santo Antonio farm (Calegari et al. 1995) showed an increase in soil water infiltration from 20 mm/h in conventional tillage to 45 mm/h in no-tillage (soybean-wheat system) due to several physical factors which were altered by the crop rotation and no-tillage system. Also increased the soil water infiltration rate mainly by effects of organic residues (roots, leaves, etc.) throughout soil micro-organisms activity, including root exudates and fungi hyfas.

Table 4. Average grains yield (kg/ha) in different tillage and management systems.

Santo Antonio Farm - Floresta – PR (Average of 1985/86 until 91/92)

	Crops			
	Soybeans	***Yield %***	***Wheat***	***Yield %***
No-tillage	*2816*	*134,4*	*2121*	*113,7*
Convent. tillage	*2094*	*100,0*	*1864*	*100,0*
**Crop rotations*	*3040*	*119,2*	*2200*	*105,8*
** Monoculture*	*2550*	*100,0*	*2078*	*100,0*

* No-tillage system.
Source: *Calegari et al., 1998a*

The rotations evaluated that included white lupin, contributed with a nitrogen input to the system increasing the maize yield. The areas planted with black oat plus white lupin have been showing a higher accumulation of mulch at the soil surface, when compared with conventional system, which affects positively the crop growth, because of improvements in the physical characteristics of the soil which contributes to greater water availability, less erosion loss, and significant decrease in weed infestation.

Comparing both systems, summer crops yields in no-tillage were 34,4% and 13,7% higher respectively for soybean and wheat in conventional tillage. In addition, the crop rotation system was more efficient, increasing respectively 19,2% and 5,8% the yield for soybean and wheat as compared to monoculture (Table 4).

Therefore no-tillage systems and crop rotation are more sustainable and increased soybean and wheat yields as compared to the conventional tillage.

In natural systems, nutrient release from litter and plant uptake of nutrients generally occurs in synchrony, resulting in efficient use of nutrients. In agroecosystems, the two processes of release and uptake are often separated in time, resulting in low nutrien-use efficiency. This is a particularly acute with N, where excess amounts are lost by leaching, denitrification, and ammonia volatilisation. In agricultural system where cover crops grown and are timely killed and buried or left on the soil surface gradually they will decomposing (mainly graminae crops) and release nitrogen by following crop with less risk of nitrogen lost.

Assessments carried out by Calegari, 1995 on a soil with the following chemical composition: pH = 4,7; Al^{3+} = 0,17 meq; Ca^{2+} = 3,5 meq; Mg^{2+} = 2,04 meq; K^{+} = 0,42 meq; %C = 2,5 and P = 3,0 ppm, in the southwestern region of Paraná (Pato Branco) are showed on Table 5.

The higher maize yields were obtained on legumes residues (vetches, blue lupin, ornithopus and sweet pea), shown the great possibility to economize chemical nitrogen. The no-tillage presented in the majority of the cases higher maize yield than conventional tillage. In general the maize planted after legumes cover crops presented low response to nitrogen application, conversely grasses areas and also fallow presented higher nitrogen response. The favourable tendency shown by the results achieved agrees with the data obtained by Paraná growers who use green manure, crop rotation and No-tillage.

CONCLUSIONS

To guarantee long-term productive, competitive and sustainable agroecosystems, we must undoubtedly seek systems that contemplate, through biodiversity production diversification in time and in space, nutrient recycling and recovery and/or conservation of the physical, chemical and biological properties of the soil. Therefore, integrating duly systematized practices considerably helps improve not only agriculture as a whole, but also the socioeconomical conditions of rural growers. This means that it is crucial to environmentally plan agriculture in places where it has greater chances of success, and where agroecosystems with their self-control mechanics will act decisively to maintain acceptable dynamic balances in production and environmental quality.

Table 5. Maize yield (Ag-513) (kgha^{-1}) after winter crops. Pato Branco-PR. Experimental Station. Average of 3 replications.

Winter cover crop	***No-tillage***		***Conventional tillage***	
	Rate (Kg/N/ha^{-1})		***Rate (Kg/N/ha^{-1})***	
	0	***90***	***0***	***90***
Ornithopus	*6.763*	*7.363*	*5.013*	*5.861*
Hairy vetch	*6.883*	*7.344*	*5.608*	*5.775*
Common vetch	*7.338*	*7.641*	*6.094*	*6.438*
Fallow	*4.441*	*5.991*	*4.827*	*5.938*
Wheat	*5.000*	*5.988*	*4.769*	*5.330*
Raygrás	*4.283*	*6.980*	*5.719*	*6.025*
Sweet pea	*6.425*	*7.558*	*4.736*	*5.341*
Rye	*4.291*	*6.669*	*3.858*	*5.327*
Radish (cv. silletina)	*5.755*	*6.994*	*5.566*	*6.177*
Black oat	*4.586*	*6.836*	*5.436*	*6.127*
Blue lupin- Iapar 24	*6.872*	*6.419*	*5.916*	*6.302*

Source: Calegari, 1995

The no-tillage system including suitable cover crops and crop rotation leads to better work distribution throughout the year, which results in the elimination of soil tilling, harrows and mechanical control over weeds. This condition will provide more time to arrange, plant and manage different activities for better land diversification. With this system there is a significant reduction in soil loss, fertility improves, crops yields increase, there is greater production stability, in addition to the possibility of permanently using the land, thus proving that it also contributes to agricultural system sustainability.

REFERENCES

Calegari, A., Mondardo, A., Bulisani, E.A., Wildner, L. do P., Costa, M.B.B., Alcântara, P.B., Miyasaka, S., Amado, T.J.C. (1993) Adubação verde no sul do Brasil. **2a. ed**.- Rio de Janeiro, AS-PTA.

Calegari, A. (1995) The effects of tillage and cover crops on some chemical properties of an Oxisol in South western Paraná, Brazil. Dissertation Thesis. University of Aberdeen, Department of Plant and Soil Science. Aberdeen, Scotland, UK.

Calegari, A. and Alexander, I. (1998) The effects of tillage and cover crops on some chemical properties of an Oxisol and summer crop yields in Southwestern Paraná, Brazil. *Advances in GeoEcology* **31**: 1239-1246, Reiskirchen. Presented in 9th International Soil Conference. Bonn, Germany, 26-30 August, 1996.

Calegari, A., Darolt, M.R. and Ferro M.. (1998a) Towards sustainable agriculture with a no-tillage system. *Advances in GeoEcology*, **31**: 1205-1209. Reiskirchen , Germany

Jones, M.J. (1974) Effects of previous maize crop on yield and nitrogen response of maize at Samaru, Nigeria. *Experimental Agriculture*, **10**: 273-279.

Mukurumbira, L.M. (1985) Effects of rate of fertiliser nitrogen and previous grain legume crop on maize yields. *Zimbabwe Agricultural Journal*, **82**: 177-179.

Schomberg, H.H., Ford, P.B. and Hargrove, W.L. (1994) Infuence of crop residues on nutrient recycling and soil chemical properties. P. 99-121. I n: Managing Agricultural Residues. Edited by Paul W. Unger. Lewis Publishers, Bushland, Texas USA.

J. D. NALEWAJA

WEEDS AND CONSERVATION AGRICULTURE

Emeritus Professor,
North Dakota State University Fargo, North Dakota, 58105

In crop production agriculture, tillage is used mainly for weed control. The advent of effective herbicides eliminated the need for wide crop rows to accommodate horse or tractor drawn cultivation equipment since herbicides can replace in-crop tillage and allow narrower row widths that maximize crop growth and yield. Until recently, the primary disadvantage of total no-till crop production was an increase in perennial weeds that required primary and seedbed tillage for control. Most herbicides used pre-emergent in crops only controlled germinating weed seeds and not established perennial plants. Most postemergent herbicides lacked selectivity for effective perennial weed control. Post harvest or before planting herbicide treatments for perennial weeds were limited because of possible soil residual or improper weed size for efficacy. The development of glyphosate tolerant crops and the use of pre and post-harvest glyphosate have essentially eliminated the perennial weed disadvantage for no-till agriculture. Glyphosate also controls annual weeds with excellent selectivity in tolerant crops. Glyphosate contributed to an estimated 60% no-till soybean production in Ohio and various herbicides allowed about 20% not-till corn production in Indiana, Illinois and Iowa in 2000. Herbicide usage increases with less tillage; but herbicides require far less energy in manufacture and use than tillage, an important benefit with increasing energy costs. This and other benefits of no-till crop production, such as reduced soil erosion and greater stored carbon, now can be achieved simply, effectively, and economically through herbicides that are safe to the user and the environment.
Key words: No-till, weeds.

INTRODUCTION

Reduced tillage that leaves a crop residual greatly reduces soil loss by wind or water erosion (Bilbro & Fryrear, 1994), sequesters carbon dioxide (Robertson *et al.*, 2000), conserves energy for crop production (Retzlof, 1980), and retains soil moisture (Crutchfield, Wicks & Burnside, 1986). Crop yields are generally equal for conventional and no-till systems provided that weeds are controlled and proper crop stands obtained (Norwood, 1994; Miller & Nalewaja, 1985). Successful conservation agriculture, regardless of definition, is highly dependent upon effective weed control. Growers' respond to conservation practices provided production is maintained and the processes are compatible with individual operations. Today energy inputs in addition to soil conservation are of primary concern to both growers and society. In

L. García-Torres et al. (eds.), Conservation Agriculture, 201–210.

the past, attempts to implement reduced tillage conservation practices have often caused a loss in crop production because the reduced tillage did not adequately control weeds. The advent of effective herbicides provided the opportunity to reduce tillage as well as to plant crops in more narrow row widths for optimum plant growth and crop yield. Before the development of effective herbicides, crops were seeded in row widths that provided space for a horse or tractor drawn cultivator to control weeds. The recent development of highly active post-emergent-applied -herbicides and crops tolerant to broad-spectrum herbicides provide for simple effective weed control for conservation tillage practices. For these reasons, grower acceptance of conservation tillage practices will increase greatly in the near future.

The use of selective herbicides in the crop and non-selective herbicides in place of primary tillage was the impetus for conservation no-till agriculture. In the 1960s attempts were made using dalapon, 2,4-D, atrazine, trifluralin, and cyanazine in various combinations to control weeds in corn and fallow in the Great Plains of the United States. This was mainly to reduce the loss of soil by wind and to help conserve moisture during the fallow year. An ecofallow system using reduced tillage and atrazine was developed and used extensively in western Nebraska. The availability of glyphosate, Roundup^{R} (Monsanto, 358.4 g ae/L) caused a large shift to no-till or reduced tillage fallow. Glyphosate has broadspectrum effectiveness and lacks soil residual to injure the following that made reduced-till fallow simple and hastened adaptation. However, most growers still conduct some primary tillage, usually chisel plowing prior to crop seeding, for no apparent reason other than tradition.

Recently no-till crop production has increased with the development of cultivars tolerant to herbicides or mixtures of herbicides with effectiveness for a wide spectrum of weed species. In 2000, 60% of the soybeans in Ohio were planted without tillage and no-till corn was planted to 24% in Ohio, 21% in Indiana, 18% in Iowa, and 17% in Illinois (Soybean Digest.com, January 2001). This relatively large area of no-till production was possible primarily because of Roundup Ready^{R} soybean varieties and various effective herbicide programs for weed control in corn.

WEEDS AND TILLAGE

Weed seed distribution in the soil is greatly influence by the type of tillage. A survey of two fields, one moldboard plowed and the other chisel plowed for the previous 5 years, indicated that the chisel plow left 60% and plowing left 23% of the wild oat seed in the surface 2.5 cm of soil (Figure 1).

No-tillage obviously would leave all the seed on or near the soil surface. Without tillage, weed seeds may still be protected from depredation by insects, animals, and birds because of self-burial as soils expand and crack with changes in moisture (Somody, Nalewaja & Miller, 1985) thus, allowing weeds to grow from seed without tillage. A portion of the seed of most weed species remain viable when buried in soil and will germinate and infest the field when brought to the surface with subsequent tillage. Most weed seeds can survive in the soil for at least 17 years (Burnside, Wilson, Weisberg & Hubbard, 1996) or even longer (Klingman, 1961). However, some species,

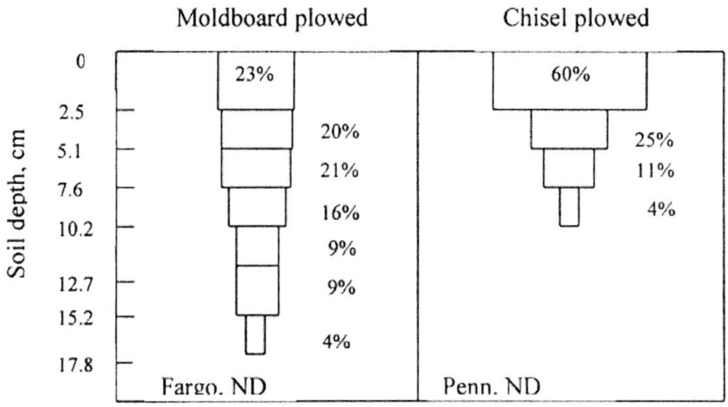

Figure 1. Distribution of wild oat seed in the soil after 5 years of moldboard or chisel plowing.

e.g., downey brome (*Bromus tectorium* L.), kochia (*Kochia scoparia* L.) and Russian thistle (*Salsoa iberica*) rapidly lost viability. Moldboard plowing has been used effectively to control downey brome because the seed viability is lost before being brought to the surface by subsequent tillage. The longevity of wild oat seed viability generally increases as burial depth increases. The seed within the top 4 cm had about half the viability of seed buried 24 to 28cm deep after 7 months (Table 1)(Miller & Nalewaja, 1990). Thus, the long-term seed reserve should be less with no tillage or tillage that keeps the seed near the soil surface. The viable seed emerging from near the soil surface would be soon depleted if in-crop control was complete. This seed viability information indicates that, except for species with a short viability when buried deep in the soil, chisel plowed or moldboard plowed fields would have weed infestations similar to no-till fields in the short term. In the long term, effective weed management that prevents new weed seed and production systems that do not bring old viable seed to the surface would reduce weed infestations.

Table 1. Viable wild oat seed in the soil as *influenced by time and depth of burial (Miller & Nalewaja, 1990).*

	Months			
Depth, cm	*7*	*33*	*60*	*168*
0-4	*15*	*6*	*2*	*0*
12-16	*25*	*15*	*9*	*0*
24-28	*29*	*23*	*15*	*1*

However, conservation agriculture, and no-till, maintain a crop residue on the soil surface that keeps the soil cooler and moister increasing survival of germinating small seeded weeds as compared to conventional tillage.

A chisel plow production system increased total weeds by about five times compared to a moldboard system over five years and three cropping systems (Ball & Miller, 1993). A no- till system was not included, but the three weed control input levels indicated that the increased weeds from reduced tillage would be overcome by increased control inputs. These results indicated that tillage systems that increase weeds need not cause yield-reducing losses, but greater herbicide or other control inputs may be needed. In Western Canada, herbicides eliminated weed changes caused by tillage but did not greatly change the species diversity (Derksen *et al.*, 1995). The herbicides usually were directed at the major species to prevent yield loss. Since herbicides used in conventional cultivars do not control all weeds the uncontrolled minor species may become a major problem and use of another herbicide would be needed for control.

The above research indicated that no-till increased the number of annual weeds but not the number of species. However, other research has indicated a weed species shift, most commonly to more perennial weeds with reduced or no-till systems (Miller & Nalewaja, 1985; Mulugeta, Stoltenberg & Boerboom, 2001; Derksen *et al.*, 1995). The disruption of perennial plant roots by moldboard plowing prevented the establishment of Canada thistle, but chisel plowing was no more effective than no-till (Table 2). Several perennial weed species increased with zero tillage (Derksen *et al.*, 1995). However, another report indicated that perennial weeds were eliminated by both chisel and moldboard plowing, the species were not indicated (Mulugeta *et al.,* 2001). Thus, the effectiveness of the chisel plow in controlling perennial weeds may depend on the prevailing species. No-till, the most conserving system, promotes development of perennial weeds. Perennial weeds are severe competitors with crops and more difficult to control than annual weeds. Control of perennial weeds in crops was limited before the development of Roundup Ready[R] crops.

Weeds respond to environment and no-till greatly reduces root and seed disruption, increases soil moisture, and decreases soil temperature. All of these conditions will influence the number and the type of weed species. Thus, effective weed control for various species is essential for successful long-term conservation agriculture. Herbicides are logical for no-till, as effective biological control methods are not available and tillage is not soil or energy conserving. Further, in-crop control is essential because control by non-selective means when the crop is not growing is not timely for control of many problem weeds.

CROP ROW SPACING

Certain crops have been traditionally planted in rows to facilitate in-between row cultivation for weed control as proposed by Jethro Tull (1731), author of *Horse Hoeing Husbandry*. With the discovery of selective herbicides, weeds could be controlled between crop rows as well as within the rows without cultivation. This reduced the need for wide rows to provide room for a horse or tractor drawn cultivator. Selecting a row width to maximize crop yields and eliminate erosion-causing tillage is now possible. Narrower rows provide more uniform distribution

Table 2. Weeds after 6 years in continuous wheat with various tillage without herbicides and average wheat yield with and without herbicides (Miller and Nalewaja, 1985)

	Weeds[a]						
	AVEFA		KCHSC		CIRA	Wheat yield[a], g/m²	
Tillage	*No.*	*% of C*	*No.*	*% of C*	*No.*	*Not treated*	*Treated*
No-till	1016	75	208	158	88	10.0	21.3
Chisel plow	1887	140	363	277	44	7.5	19.4
Moldboard plow	1350	100	131	100	00	12.5	21.9

[a]*No. = plants/0.4 ha x 1000, % of C = % of conventional, and treatment was diclofop plus 2,4-D. AVEFA = wild oat, KCHSC=kochia, CIRA=Canada thistle.*

of the crop over the area reducing within crop competition and increasing the ability of the crop to compete with weeds (Bilbro & Fryrear, 1994). Yields have generally been increased by the narrower rows when moisture and fertility are adequate (Figure 2) and (Boerma & Ashley, 1982).

The narrow rows alone did not provide adequate weed control as yields without herbicides were only about 45% that of the weed free soybeans regardless of row width (Wiesbrook *et al.*, 2001). Narrow rows increased weed control with herbicides through increased competition with weeds. Thus, the use of herbicides makes possible narrow row crop production and the narrow rows also are beneficial to herbicide performance.

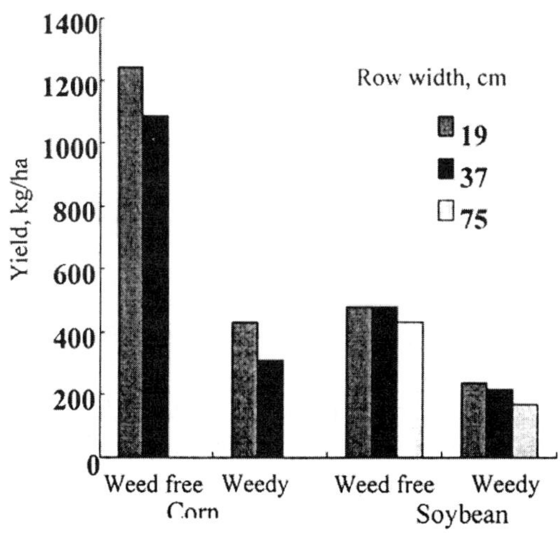

Figure 2. Influenced of row spacing on weed free and weedy corn and soybean. Weed free was Roundup® at 2.25 kg/ha plus 2% ammonium sulfate.

HERBICIDES AND CONSERVATION TILLAGE

The development of selective postemergent herbicides (applied to emerged weeds) greatly increased the potential for conservation no-till agriculture. The earlier developed pre-emergent herbicides (applied to the soil surface) only controlled germinating weed seeds or required a timely rain for efficacy. Some also required incorporation for efficacy and those that did not require incorporation often gave more consistent efficacy when incorporated. Obviously, soil incorporation of a herbicide is not consistent with no-till crop production. The availability of postemergent herbicides made possible application of specific herbicides according to the weeds present. Selective herbicides, by definition, control weeds in crops without injury to the crops. However, variation in herbicide tolerance among cultivars has restricted the use of some herbicides to certain cultivars. A single selective herbicide often controls some grasses or broadleaf weeds and various herbicide combinations are used depending on the weed species. However, no-till requires consistent efficacy for a broadspectrum of weeds because the crop residue prevents supplemental cultivation. The availability of postemergent herbicides did not stimulate extensive no-till crop production because available herbicides or combinations of herbicides did not control enough weeds, especially perennials common in no-till production.

The development of crop cultivars tolerant to herbicides with broad-spectrum efficacy in various environments greatly simplifies and assures effective no-till crop production. Specifically, cultivars with tolerance to glyphosate allowed use of a herbicide that provided excellent crop safety and broadspectrum annual and perennial weed control with one or two in-crop treatments. The large increase in no-till soybean in the USA is because of Roundup Ready[R] cultivars. In corn and wheat, the need for glyphosate tolerance was less essential to no-till production because more effective postemergent herbicides are available for conventional hybrids and cultivars and hybrids are available with resistance to several effective herbicides or herbicide combinations (Table 3). Even with the availability of effective weed control treatments

Herbicides

	Glyphosate	*Glufosinate*	*Imidazolinone*	*Sulfonylurea*	*Sethoxydim*	*Bromoxynil*
Crops						
Soybean	x	x		x		
Corn	x	x	x		x	
Canola	x	x	x			
Cotton	x					x
Flax				x		
Sunflower			x	x		
Wheat	x			x		

[a]With some herbicide groups the crop only had tolerance to specific members in the group and the new cultivar or hybrid allowed for the use of more effective herbicides in the group and in other cases crop tolerance provided for use of certain members of the group (sulfonylurea sunflower and flax).

for corn and wheat, no-till is still not as common as with soybean. No-till corn production will increase as glyphosate tolerant corn becomes more accepted in the market, because of the simplicity of glyphosate for weed control. No-till wheat production will increase as Roundup Ready wheat becomes available and growers become familiar with the system.

Growers are highly interested in producing no-till sunflower, but available herbicides are too limited to control all weeds. Cultivars soon will be available with tolerance to imidazolinone and a specific sulfonylurea herbicide (tribenuron). These will be the first postemergent herbicides for broadleaf weed control in sunflower, in the USA. The imidazolinone will provide a rather broad spectrum of grass and broadleaf control and broadleaf weed control from the sulfonylurea plus a graminicide will provide a wide weed control spectrum. However, weeds are known with resistance to imidazolinones and sulfonylureas. Control of the resistant weeds will require continued use of a pre-emergent herbicide, such as sulfentrazone for control of ALS resistant kochia.

Effective postemergence herbicides are available for use in canola. In North Dakota about 80% of the canola is Roundup Ready[R] and no-till canola production will increase. The firm moist soil with no-till is positive to canola establishment and the canola residue does not interfere with subsequent crop seeding. Sugarbeet may have potential for no-till or strip tillage should the glyphosate or glufosinate resistant hybrids be commercialized. With effective weed control, most crops are adaptable to some form of conservation tillage production with proper crop rotations to reduce diseases and insects.

HERBICIDES AND ENERGY CONSERVATION

No-till crop production also conserves fossil fuel energy in addition to conserving soil. Herbicides are now effective at low rates/ha. A gram of herbicide requires an estimated 25 kcal of petroleum energy for production. One litre of diesel contains 9320 kcal. Tillage operations of one moldboard plowing plus three cultivations requires 307,560 kcal/ha or 33.64 L/ha diesel (Retzlaff, 1980). (Table 4). Thus, the petroleum energy used in tillage would produce 12.3 kilograms of herbicide. The energy to apply the herbicide is 8760 kcal/ha per application. The recently developed herbicides often require less than a total of 0.1 kg/ha in two applications. Glyphosate is normally applied at from 0.3 to 0.8 kg/ha. With two applications the maximum usage would be

Table 4. Litres per hectare and kcal energy for various weed control practices(Nebraska and North Dakota on-farm fuel use surveys and calculations).

Operation	*Litres/hectare*	*Kca/hectare*
Moldboard plow	*16.81*	*256669*
Field cultivator	*5.61*	*52285*
Disc	*6.55*	*61046*
Chisel plow	*8.89*	*82855*
Harrow	*3.27*	*30476*
Sprayer	*0.94*	*8761*

1.6 kg/ha or only 40,000 kcal/ha. The energy requirements for herbicides and applications can be further reduced with proper application techniques. Glyphosate at a half rate gave efficacy similar to a full rate when spray volume was reduced from 90 to 23 L/ha (Ramsdale & Messersmith, 2000). The lower volume not only would allow less herbicide use, but also the smaller sprayer load would also reduce energy for application.

For imidazolinone or sulfonylurea herbicides, the required amount per treatment is from 0.074 to 0.170 kg/ha. However, these herbicides require an adjuvant for efficacy that would also require energy for production. Again since adjuvants are usually used as a percentage of the spray volume the input could be reduced in half if the equally or more effective 23 L/ha were used instead of 45 L/ha or the label requirement of 90 L/ha (Nalewaja & Ahrens, 1998). The potential benefits from no-till crop production are becoming very attractive economically and user friendly with development of new highly active and effective herbicides. This is evident from the extensive no-till soybean production in Ohio as mentioned above. The author has successfully grown Roundup Ready^{R} soybean no-till for the past two years on a field heavily infested with many species of annual, biennial, and perennial weeds. Because of the high population and several species of weeds, three applications of glyphosate were required the first year, but the total rate was about 1 kg/ha. Glyphosate was mixed for 200 g/ha in 30 L/ha spray volume, but speed was reduced for patches of perennial weeds to increase application rate. Weeds were less abundant in the second year because of effective control the first year and only two applications were used. Because of the low glyphosate rate and water known to be high in minerals, an adjuvant that provided 0.25% surfactant plus ammonium sulphate was included in the spray. The low spray volume minimized the amount of adjuvant and allowed use of a small spray tank and a small fuel efficient spray tractor.

POTENTIAL PROBLEMS

In the long-term, weeds may either develop resistance to herbicides or species naturally tolerant to herbicides may increase. This is especially important with no-till as effective herbicides are essential to successful conservation agriculture. The tillage alternative for weed control is contradictory to no-till. In practice, herbicides did not change weed diversity mainly because herbicide use changed to control the major competitive species (Derksen *et al.*, 1995). However, continued usage of glyphosate or other herbicide treatment with broad-spectrum effectiveness could cause new resistant weeds to develop in these fields. In the author's soybean example, hoary alyssum (*Berterda ubcaba* (L.)DC.) appeared tolerant to glyphosate. Plants were sparse and limited to only about two hectares. Hand spraying with 2,4-D was used and this is expected to prevent further infestation. Thus, in order to maintain the use of an effective herbicide, growers will need to make an intensive effort to detect resistant or tolerant weeds and prevent their development. This means that methods such as hand pulling or spot treatments with non-selective herbicides would be essential immediately upon detection of a weed with resistance. The 2,4-D used for hoary alyssum was not selective, but loss of soybean plants was minimal with the hand application. Had more plants been present, a second herbicide selective in soybean would be an option, if effective for the specific weed.

Growing crops with tolerance to a herbicide will prevent that herbicide from controlling volunteers in the subsequent crop. Thus, another herbicide effective on the crop would be needed the following year. Crops tolerant to glufosinate, a non-residual herbicide with broad-spectrum effectiveness and a mode of action different from glyphosate, provide an option to control glyphosate tolerant crops and weeds. The rotation of glyphosate with glufosinate or other herbicide would provide for control of crop volunteers as well as reduce the development of weed resistance.

Soils in no-till production remain cooler in the spring and this may adversely influence warm season crops. Strip tillage in the crop row has been an adaptation to overcome cool soils. Often this has been in conjunction with a cover crop that is seeded with conventional tillage. However, the concept could be applied to no-till. It was suggested that this might have potential in sugarbeets (Dexter, North Dakota University, Fargo). Such systems would have weeds adapted to no-till in-between the rows and those adapted to tillage in the rows. The areas could change from year to year. Never the less, sugarbeet with tolerance to broad spectrum glyphosate or glufosinate would easily make the system possible.

In the example of glyphosate resistant hoary alyssum in Roundup Ready soybean, information on effectiveness of various herbicides for hoary alyssum was not found. No-till agriculture causes development of many uncommon weeds that could soon reduce the effectiveness of successful weed control systems. Weed scientists must be proactive in determining potential weeds and their control for long-term successful no-till crop production. The development of herbicide resistant crops used in conjunction with methods to prevent weed resistance will allow for large long-term increases in conservation agriculture.

REFERENCES

Ball, D.A. and Miller S.D. (1993) Cropping history, tillage and herbicide effects on weed flora composition in irrigated corn. *Agronomy Journal.* **85**: 817-821.

Bilbro, J.D. and Fryrear, D.W. (1994) Wind erosion losses as related to plant silhouette and soil cover. *Agronomy Journal* **86**: 550-553.

Boerma, H.A. and D.A. Ashley. (1982) Irrigation, row spacing and genotype effects on late and ultra-late planted soybeans. *Agronomy Journal.* **14**: 995-999.

Burnside, O.C., Wilson, R.G., Weisberg, S. and Hubbard, K.G. (1996) Seed longevity of 41 weed species buried 17 years in Eastern and Western Nebraska. *Weed Science.* **55**: 75-87.

Crutchfield, D.A., Wicks, G.A. and Burnside, O.C. (1986) Effect of winter wheat (Triticum Aestivum) straw mulch on weed control. *Weed Science* **34**: 110-114.

Derksen, D.A., Thomas, A.G., Lafond, G.P., Loeppky, H.A. and Swanton, C.J. (1995) Impact of post emergence herbicides on weed community diversity within conservation-tillage systems. *Weed Research* **35**: 311-320.

Klingman, G.C. (1961) *Weed Control as a Science.* John Wiley&Sons, Inc., New York. 421pp.

Miller, S.D. and Nalewaja, J.D. (1990) Influence of burial depth on wild oats seed longevity. *Weed Technology* **4**: 514-517.

Miller, S.D. and Nalewaja, J.D. (1985) Weed spectrum change and control in reduced-till wheat. *North Dakota Farm Research* **43**(1): 11-14.

Mulugeta, D., Stoltenberg, D.E. and Boerboom, C.M. (2001) Weed species-area relationships as influenced by tillage. *Weed Science* **49**: 217-223.

Nalewaja, J.D. and Ahrens, W.H. (1998) Adjuvants and spray volume affect herbicide efficacy. In Proceedings *Adjuvants for Agrochemicals challenges and opportunities.* pp. 434-441. Ed. P. McMullan. Fifth International Symposium on adjuvants for agrochemists, Memphis, TN, Aug 17-21.

Nice, G.R.W., Burhring, N.W. and Shaw, D.R. (2001) Sicklepod response to shading, soybean row spacing, and population in three management systems. *Weed Technology.* **15**: 155-162.

Norwood, C. (1994) Profile water distribution and grain yield as affected by cropping system and tillage. *Agronomy. Journal.* **86**: 558-563.

Ramsdale, B.K. and Messersmith, C.G. (2000) Spray volume and adjuvants effect on glyphosate efficacy. *NCWSS Research. Report.* **57**: 449.

Retzlaff, R.E.J. (1980) Winter wheat fallow rotation: an analysis of energy savings. *The Wheat Grower* **3**(1): 10-13.

Somody, C.N., Nalewaja, J.D. and Miller, S.D. (1985) Self-burial of wild oat florets. *Agronomy. Journal* **77**:359-362.

Wiesbrook, M.W., Johnson, W.G., Hart, S.E., Bradley, P.R. and Wax, L.M. (2001) Comparison of weed management systems in narrow-row,glyphosate- and glufosinate-resistant soybean. *Weed Technology.* **15**: 122

A. EL TITI

NON-INVERSION TILLAGE IN INTEGRATED FARMING CONCEPTS: PROSPECTS AND CONSTRAINTS OF CROPPING SYSTEMS IN THE SOUTHWEST OF GERMANY

State Institute For Plant Protection Stuttgart
Reinsburgstrasse 107, 70197 Stuttgart/ Germany

An integrated farming concept (IFS) based on non-inversion tillage was compared with current farming practices (CFS) in two long-term studies. Either paired fields (Lautenbach) or entire farms (Bruchsal) were subject of this comparison. Gross margin of IFS at the first site averaged 5% higher than that of CFS. Despite more weeds IFS achieved 29-90% reduction of pesticides and 20% of N-fertilisers, minimizing NO3-leaching. Both soil physical properties and bioindicators showed significant improvements in IFS. At the other site, 15-IFS-farms were compared with 30 ploughing enterprises, cropping cereals, sugar beets, oilseed rape, sunflower, legumes and maize. The results revealed reductions of No3-leaching, pesticides inputs and soil erosion and improvements of soil structure, diversity of indigenous species and of farm-income. Higher *Fusarium*-incidence occurred only when wheat follows maize in non-ploughing systems. Similar applies for slug attack on oilseed rape after winter barley. Recommendations for sound crop rotation are discussed.
Key words: Non-inversion tillage, Integrated Farming, farm-income, bioindicators, N-leaching.

INTRODUCTION

There are different reasons for adoption of reduced tillage practices in Germany. Erosion control Kainz, et al (1999), more effective use of farm machinery, cutback of production costs, minimising nutrient losses and improving farm-income are some of the common motivations. Apart of this, Integrated Farming Systems (IFS) – conceptually founded on suppressing of noxious and enhancing of beneficial species (El Titi et al 1993)– strongly recommends minimal cultivation strategies that support ecosystem functioning.

However, replacing traditional ploughing e.g. by. Non-inversion tillage is suspected to create serious farming constraints. Root crops in farm rotation are said to aggravate acceptance, whereas regional climatic traits may restrict the time-windows left for field operations.

L. García-Torres et al. (eds.), Conservation Agriculture, 211–220.

In this contribution two long-term case studies discuss feasibility of non-inversion tillage as significant component of integrated farming, conducted under commercial conditions between 1974 and 1996. Impacts of cropping systems on possibilities and limits of minimal tillage will be outlined, highlighting both economic and eco-environmental implications. Conclusions to resolve specific tillage problems are drawn.

MATERIALS AND METHODS

Site. The studies were accomplished at the following two sites:

Lautenbach

This is a commercial arable enterprise of 230 ha with fields-elevation between 185 and 235 m NN. Average temperature is 9.4 ° C, and the annual precipitation 745 mm/ cm2. The dominant soil is a «parabrown soil», being rather shallow at hill-tops (pararendzina) or clay «brown» at bottoms.

Two small farms of 36 ha each were installed in Lautenbach. One was managed due to Integrated Farming principles –IFS -(El Titi et al 1993), whereas the other was farmed due to conventional practices -CFS. Each small farm was subdivided into 6 fields, building six field pairs. Non-inversion tillage, reduced agrochemical inputs, undersowing or green manuring and hedgerows / marginal strips distinguished IFS, whereas annual ploughing, recommended rates of fertilisers and pesticides CFS. Both farming systems had the same crop rotation, which included winter wheat, spring wheat/ barley, oats, sugar beet and legumes (peas, beans).

Bruchsal

This district comprises the regions of Kraichgau and Rhine valley, with elevation between 326 –98 m NN. The average temperature is slightly higher than at Lautenbach, whereas only 650 l/m2 rains falls annually. Winter and spring cereals, sugar beet, oil seed rape, maize, sunflower and forage peas are the main crops in the region. The investigations were accomplished 1988–1997 on fields of 17 motivated farmers with 1200 ha (Arbeitskreis integrierter Landbewirt-schaftung AKIL). Except for potato, all study farms had adopted non-inversion tillage within an IFS-concept (Vereijken 1995). Husbandry measures are considered as adopted, when farm records/book keeping data documented multiannual implementation. For comparison purposes 30 ploughing farms of the surroundings were included. Economic, operational parameters as well as faunal and vegetational surveys were conducted on the course of this study.

RESULTS

Crop yields at Lautenbach documented remarkable differences between IFS and CFS, depending on crop species and year (Fig. 1).

IFS crops produced comparable or even higher yields than those of CFS, except wheat (winter and spring), yielding 0,3-0,5 t/ha/year less (El Titi 1999). The gross

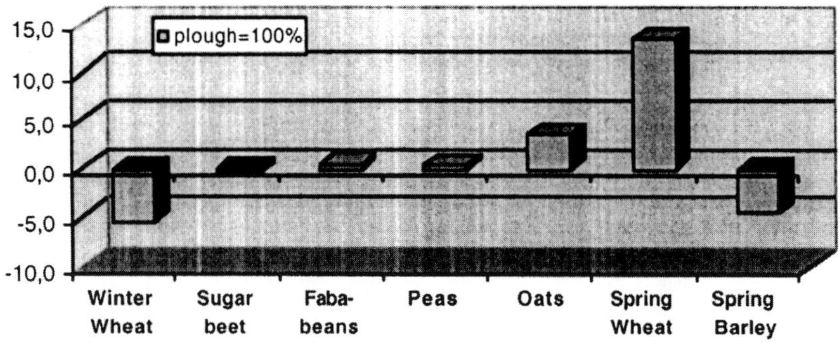

Fig. 1. % deviation of crop yields of IFS (non-inversion tillage) from the counterparts of CFS (annual ploughing) at Lautenbach /Germany in 17-years average.

margins of IFS was slightly higher: IFS € 1186.5/ ha In the 17 years-average/ CFS € 1145.6/ha. Variable machinery costs of IFS ran up to € 120.7/ha in average (CFS: €125.8/ha), whereas those for pesticides in non-ploughing system € 71.7/ha (CFS= €127.3) (Zaddies et al 1986).

Earthworms revealed higher numbers and biomass in IFS-fields (Fig. 2) than on the ploughed CFS-plots. Their abundance on non-inversion tilled fields averaged up to six folds of the ploughed neighbour.

Mesostigmatic mites, Collembola, Dipterous larvae, Enchytraeids, spiders, Carabids, and Staphylinids showed similar responses but not always of the same scale. Soil structure – providing the habitat for soil organisms performed clear

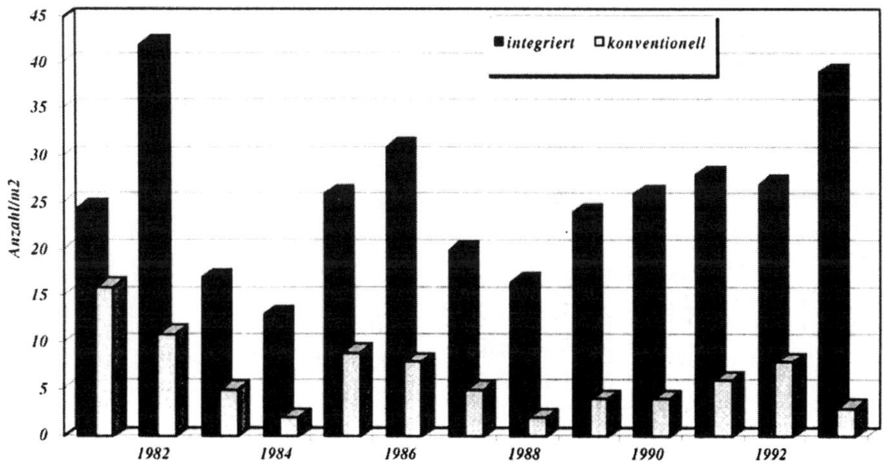

Fig 2. Average number of Earthworm/m2 (Formalin-Extraction) in minimal cultivated «integrated» and in annually ploughed soils «conventional» between 1981 and 1993 at Lautenbach /Germany.

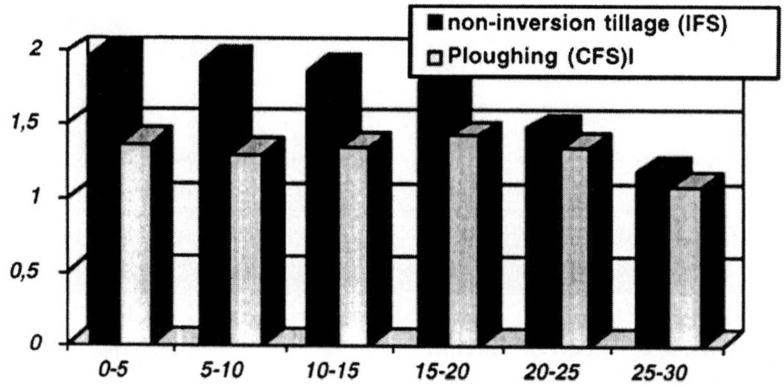

Fig. 3. Distribution of humus contents in topsoil up to 30 cm of both minimal tilled (IFS) and annually ploughed (CFS) soils after 17 years period at Lautenbach /Germany.

improvements. In addition, there was a significant increase of soil organic matter contents in IFS – averaged 26%. The distribution of humus in top soils (Fig.3) shows accumulation in IFS-top layer.

Pre-control monitoring revealed more weeds on the non-inversion tilled fields (El Titi 1994), with slight species shifts in weed communities. Grass weeds and thistles (*Ciricium arvensis*) occurred in higher frequency (8% in average) there. Significant reductions of herbicides (-29%), fungicides and insecticides -80% and -90% respectively were achieved in IFS.

At Bruchsal-site IFS comprised a wider range of crops. Farm-income in this environmentally sensitive region can be very much determined by the legal regulations attributed to water protection & soil conservation, species diversity, etc. Many directives

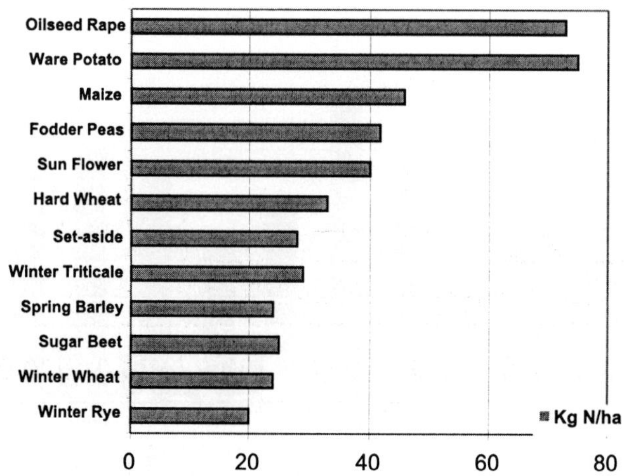

Fig. 4. Three years average of residual nitrogen in profile of minimal cultivated soil –non-inversion tillage- of IFS- AKIL- farms (autumn sampling 1992-1995).

are closely related to tillage regime. Some of them are mandatory, whereas others are voluntary. The SCHALVO-Directive is mandatory for fields in water vulnerable zones (WVZ), enforcing restrictions of pesticides and fertiliser inputs. SCHALVO- Directive is based on a Bonus-Malus-principle. (Anonymous 1992). It defines 45 kg N/ha in autumn as threshold of residual nitrates in soil profile (~90cm).

AKIL-farms adopted the WCZ- Nitrate-guidelines without being obliged to follow. Three years sampling during the crucial period attested residual nitrates maintained below the required threshold, except for potato and oilseed rape.

Soil Protection Directives provide a second option to improve farm income. They aim at sustaining fertility and productivity on farmland. In adopting non-inversion tillage, IFS-farmers met the requirements.

Overproduction & Landscape Management Option (MEKA-Programme (Anonymous (b) 1992) known, as MEKA is a voluntary option offered to encourage environmentally safer farming. Farmers may choose single measures due to point catalogue, accumulating at maximum € 266, -/ha/a. Non-inversion tillage is a part of this programme to which. AKIL-Farms successfully participated. Contrasting gross margins of AKIL- crops, against those of surrounding (ploughing) farms, illustrates the financial profits of adopting non-inversion tillage (Fig.5).

IFS-farms achieved additional financial profits through participating to public subsidy programmes. For those crops with otherwise negative financial results, e.g. winter wheat - the moderate tillage regime mitigated the financial loss.

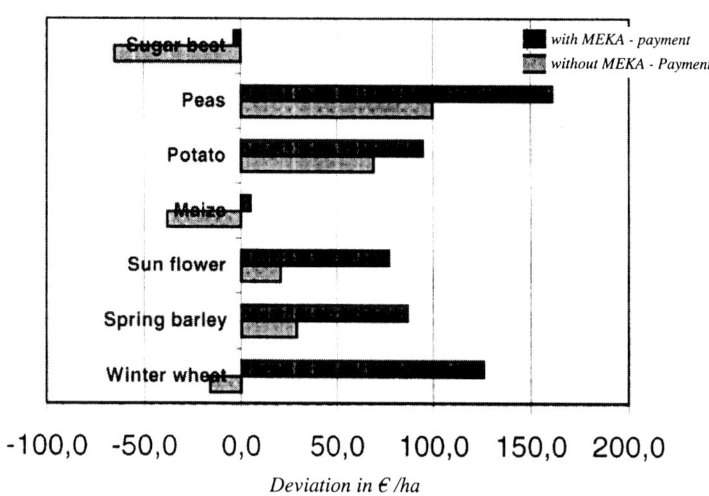

Fig.5. Actual deviations of gross margin (€ /ha) –with and without tillage-based payments– of the main crops of the AKIL farms in four years average from the corresponding margins of ploughing farms.

agrarian uses, altering its composition, structure and natural biodiversity as little as possible and defending it from degradation processes (e.g. soil erosion and compaction).

*Table 1. Relation between residue cover and maize (*Zea mays *L.) yield (Source: M. G. Wagger, North Carolina State University, unpublished data).*

Tillage Method	*Residue Cover (%)†*	*Grain Yield ‡ (kg/ha)*
No-tillage	*93*	*6518*
No-till/Rip	*68*	*6586*
Autumn Chisel	*51*	*5645*
Spring Chisel	*43*	*5107*
Spring Disk	*23*	*4502*
Autumn Chisel/Disk	*25*	*4234*
Spring Chisel/Disk	*17*	*3763*
Autumn Moldboard/Disk	*3*	*3158*
Spring Moldboard/Disk	*5*	*3427*

† Typical covers during the experiment ‡ 8-yr. avg.

Dickey *et al.* (1997) reported that when residues cover 20% of the soil, soil erosion is expected to be 50% less than when no residue is present, and coverage of 90% can reduce erosion by as much as 93%. Besides reducing the threat of erosion, maintaining residue cover can positively impact crop yields. Table 1 demonstrates this potential for maize (*Zea mays* L.) when various types of tillage were used to produce a range of residue cover. In the North Carolina Piedmont, eight-year average maize yields were over 90% higher from no-till compared with moldboard plowing.

Traditionally, crop residues have been used for multiple purposes, i.e. fuel, building materials, mulch, feed and bedding (Smil, 1999), most of which conflict with their use for soil conservation. Among these, livestock related use (feed and bedding) is probably the most widespread in developing countries.

IMPORTANCE OF CROP/LIVESTOCK INTEGRATION

Globally, small-scale, mixed crop-livestock farming systems occupy about 2.5 billion hectares and are the main source of meat and milk (CAST, 1999). Farming systems that successfully integrate crop and livestock enterprises stand to gain synergies that directly impact production and agro-ecological efficiencies. Sanchez (1995) has reviewed the case for the integration of livestock (mainly ruminants) with perennial crops. Some advantages listed include: diversification of income through animal products (milk, meat, fibre, hides, and manure), weed control, soil erosion control, increased yield of main crop, and income during «start-up» period for tree crops.

Several options of crop-livestock systems practised in South East Asia have been described, mostly involving annual crops and ruminants, but also pigs, ducks, and fish (Devendra, 1993; Pezo *et al*, 2000; Devendra *et al*, 2001). Similar reports are available for Latin America (Quiroz *et al*, 1997) and sub-Saharan Africa (Powell and Williams, 1995). Ruminant animals are especially desirable due to their ability to convert forages, browse and crop residues high in cellulose to useful food and fibre products. Animals in smallholder crop-livestock systems contribute to the food security of the household, provide for system diversification, generate cash, spread risk, recycle nutrients, provide draft power and transportation, and are important assets for investment and/or savings (de Haan *et al*, 1997).

A general model for mixed farming systems in which ruminants and other livestock play an important role is shown in Figure 1. In these systems, land can be allocated either to pasture production, crops and/or conservation, but as pressure for land increases, most arable land is devoted to crops; intensifying interactions and conflicts between crops and animals. In Figure 1 the connections between crop residues, soil nutrient availability and animals are highlighted to emphasise a potentially important conflict over the allocation of these residues.

The removal of crop residues for or by livestock, either through grazing or cut and carry, is a common practice in most crop-livestock systems. In many cases residue removal by animals is excessive, leaving insufficient vegetation for soil enhancement and conservation purposes, and compromising the sustainability of the systems.

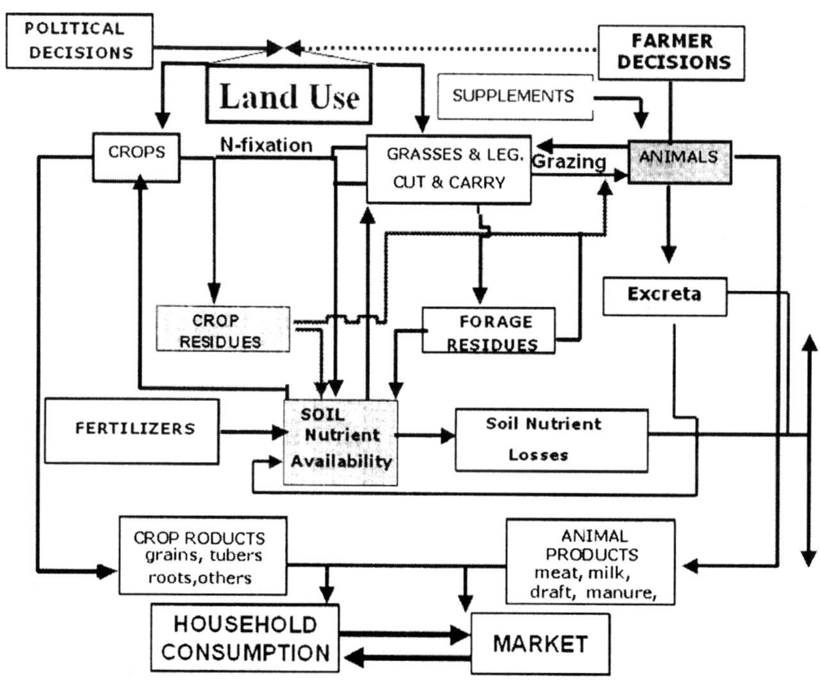

Figure 1. General Model for Mixed Farming, Crop/Animal Systems (Adapted from León Velarde and Quiroz, 1998).

Furthermore, in some systems these residues may have greatest long-run value as soil amendments, because many crop residues have very low feed value, often not meeting animal maintenance requirements. On the other hand, livestock are an important part of production in mixed farming systems and in the absence of alternative feeds, farmers are usually unwilling to forgo this critical feed source.

Although there are many technical options to solve this problem, a final solution may require more than a technological «fix». It is necessary to understand the root cause for such practices, as they usually involve socio-economic or cultural issues. In El Salvador, *conservation agriculture* was successful only when it was presented as an inseparable part of a program including enhanced productivity and economic and institutional incentives (Calderon *et al.,* 1991). In Honduras, an indigenous agroforestry system (Quesungual) devised and practiced by smallholders, was reported to eliminate burning, enhance soil moisture, yield multiple products, and prolong the cropping period (Hellin *et al.*, 1999). In the Northern Zone of Tanzania, West Africa, where maize residues are transported long distances for feeding, a system of baling the leaves and retaining the stalks for conservation purposes has been implemented (Massawe, personal communication). Examples of the successful integration of livestock with cropping systems exist and these case studies can serve as the basis for a concentrated effort to search for possible solutions to this important problem.

CHALLENGES FACING SMALL-SCALE FARMING SYSTEMS

Land Tenure

The legal ownership status of land often provides the onus for management strategies and land-use practices. Sombroek and Sims (1995) have identified numerous forms of land tenure. Many of these forms present formidable obstacles to *conservation agriculture.* Land use by individuals or groups in non-ownership categories such as rented, leased or illegal, pose serious threats to conservation agriculture, particularly when tenancy agreements are short or insecure. Under those conditions, farmers have little motivation to holistically manage crops, livestock and natural resources (Thiesenhusen, 1994), their main concern being day to day survival and maximum extraction. Before attempting to propose or impose technical solutions to crop/livestock conflict, it is necessary to determine the degree to which land tenure issues are driving inadequate agricultural practices.

Farm Size

Smallholdings are farms with no more than a few hectares, in which the household controls most of the capital and farm labour and makes management decisions. According to Devendra, (1993), in smallholding farms the family and the production unit cannot be considered separately due to the complex interplay between crops, animals and people with minimal land, labour and capital. Lal and Pierce (1991) estimated the per capita minimum area required for an adequate diet is 0.5 ha when managed under a modest level of inputs, but could be reduced to 0.15 - 0.25 ha with intensification. Projections by FAO (Alexandratos, 1995) for the per capita availability of arable land in developing countries between 1980 and 2010 show a decrease from 0.65 to 0.4 ha. Access to minimal land area makes the integration of large ruminants into the system difficult, due to their relatively large demand for biomass just to support maintenance. The usual result is over-utilization of crop residues and cover crops for grazing or fodder. Barber (1996) found that 90 % of the farms in a site in El Salvador had 1.4 ha or less, and the modal-sized farm (0.7 ha) was just sufficient to produce the household annual maize needs. Dibissa and Peters (1999) found that in the Ethiopian highlands farm size varied from 0.4 to 3.8 ha (1.9 ha average), which in many cases was insufficient for subsistence. Under sub-humid and humid ecosystems, in farms larger than about 2 ha, there is much more flexibility to allocate some area to pasture and forage crops, to prevent the conflict of excessive crop residue removal, when large ruminants are raised. Small ruminants, particularly goats, fit well in smallholdings due to their small body size and lower biomass requirements (i.e. roughly 8 goats are equivalent to 1 cow); however the decision of which species to raise depends on farmers' beliefs, traditions, and market opportunities.

Landscape

In many areas of the tropics, smallholders have been forced to occupy the steep land (> 12% slope) because it is least desirable for agriculture. More than 20 % of the land in the tropics can be classified as steep and a significant proportion (> 10%) of the world's poor occupy these areas (Shaxson, 1999). Nevertheless, even on land with less than 12 % slope, degradation continues at unacceptable rates. Practices that continually compact and expose the soil surface to raindrop impact result in degradation (Shaxson, et. al, 1989). This is the case with uncontrolled grazing or excessive removal of crop residues for any purpose. In areas with a prolonged dry season, livestock demand for crop residues is often greatest (McDowell, 1988; Sandford, 1989; Quiroz *et al*, 1997), and opportunities for alternative forage production limited, thereby presenting the greatest challenge for the successful integration of crops and livestock within the framework of *conservation agriculture.*

FEEDING SYSTEMS AND NUTRIENT CYCLING

Crop Residue Availability

The ratio between grain and/or tubers and crop residue yields can be expressed either by the harvest index (HI), which is defined as the ratio between grain yield and total above-ground biomass, or as proposed by de Leeuw, 1997, the ratio of crop residues to grain yield or RI. When crop residues are used as feed, the latter is more practical for feed budgeting purposes, since the potential availability of crop residues can be estimated based on the grain yields multiplied by the RI. However, the proportion of those residues potentially consumed by animals should also be considered.

RI's vary with crop species (Devendra, 1976; Devendra and Raghavan, 1978) and genotypes within species. Plant breeding efforts have tended to improve grain yield, sacrificing the accumulation of photosynthetic products in leaves and stems. In general, RIs are higher in Africa than those in Asia (for wheat: 2.0 *vs.* 1.3) or Latin America (for maize: 3.0 *vs.* 2.0) (Kossila, 1988). The development of dwarf genotypes completely changed the «well-accepted» ratios for sorghum but in general, there is a conflict in crop breeding programs between grain and straw yields (McIntire *et al.*, 1988). Small crop-livestock farmers value crop genotypes for both grain and straw yields; and the genotypes they usually grow have higher RIs than hybrids or improved cultivars (de Leeuw, 1997). In response to the demand of those farmers, some plant breeders are beginning to add breeding objectives that include dual-purpose crop varieties.

It should be noted that the RIs do not account for the root biomass. Root mass contributes in numerous ways to enhance soil chemical, physical and biological properties, however, the obvious difficulties involved in its quantification explain the paucity of data. Cereal crop root mass is usually less than above ground biomass (30-50%) with most roots located in the upper 20 cm of the soil (Buyanovsky *et al.*, 1986).

Grazing and Manure Management

Although at least 66% of ingested nutrients are returned via excreta (Table 2), the effective return may be less than 50% due to losses from volatilisation and leaching. These losses can be mitigated to some degree through diet formulation, and management of grazing animals and their excreta (Romney *et al*, 1994). In areas with high population density, where fallow cycles are very short or non-existent, and there are limitations for the use of chemical fertilisers, manure is often the only means to sustain soil fertility and maintain organic matter levels. Under those circumstances manure is ranked as one of the most important outputs from livestock production (Murwira *et al.*, 1995; Tanner *et al.*, 1995). In semiarid conditions of the southern Sahel, it has been estimated that 4-42 ha of rangeland are needed to maintain soil fertility in 1 ha of cropland through manure (Speirs and Olsen, 1992; Bationo *et al.,* 1995; Murwira *et al.,* 1995). However, due to population pressure, crop land is expanding, and those ratios cannot be maintained; therefore, farmers begin to look for substitutes for maintaining soil fertility such as fodder shrubs and trees (Fall et al., 2000).

Most cereal stovers have very wide C:N ratios (> 20-30), and nitrogen concentrations below 1.0%, resulting in a temporary immobilisation of N and very slow release of nutrients from the residues. Feeding crop residues to animals is frequently seen in conflict with their use for soil conservation, but appropriate strategies for nutrient transfer could take advantage of animal manure as a means to enhance nutrient cycling. This could be achieved by synchronising the application of animal manure and crop residues to soils, so that they decompose and release nutrients in a pattern that coincides with crop nutrient demands (Powell and Unger, 1997). Crop residues will be more resistant to the action of microorganisms and mesofauna responsible for mineralisation, if the residues have been subjected to selective

Table 2. The proportion of nutrients returned by lactating cows grazing pasture.

Element	*% Returned in Faeces*	*% Returned in Urine*	*Total Returned*
Nitrogen	*26*	*53*	*89*
Phosphorus	*66*	*0*	*66*
Potassium	*11*	*81*	*92*

From: Hutton, et al. 1967.NZ. J. Ag. Res. 10:367 – 388.

defoliation by animals, leaving only stalks. Under these circumstances the role of animal manure in enhancing nutrient cycling is clear. Nevertheless, grazing is often viewed (with good cause if it is mismanaged) as a destructive and inefficient method of recycling nutrients. Where livestock are kept on common grazing areas outside of the farming unit and their manure is collected and spread on croplands, these rangelands can experience nutrient depletion and degradation (Turner, 1995).

Feed consumed by the cattle may require 30 to 40 hours to pass through the digestive tract and be excreted as faeces (Poore, et al., 1990). When crop residues are used for grazing, animals tend to collect nutrients from relatively large, widely scattered areas and deposit them in concentrated areas as excreta. The degree to which grazing patterns are controlled can have profound impacts on how nutrients are distributed and how efficiently they are recycled. Under controlled grazing management the distribution of excreta can be quite good, comparable to that accomplished by machine or manual spreading and superior to uncontrolled grazing (White *et al*, 2001). In this study it was determined that about 85% of the faeces and urine was distributed in the grazing paddock and the balance (about 15%) was deposited in feeding areas, milking parlour, and lanes. Approximately 10-12% of the paddock area was actually covered by faeces and urine, but as indicated by During and Weeda (1973) as much as five times the area covered by the actual excreta spot is affected. This means that in one year, the excreta could influence 50-60% of the area. The practise of using night paddocks (sometimes called night corralling or kraaling) is a controlled grazing method that concentrates nutrients from manure and urine in a specific area, usually in crop fields or near households (Murwira, et al.1995). Night paddocks can be systematically moved over crop fields to concentrate and distribute manure at once (Arias, 1987) or manure may be accumulated in corrals located near households for distribution later. Boonman (1997) states that where the farmer has the choice, direct grazing is the better for the soil and that it offers rapid recycling of nutrients.

Zero Grazing

Total confinement of animals with feed brought to the confinement facility (zero grazing) is quite rare for ruminants in developing countries. Nevertheless, this system is often used with dairy cattle in the East African highlands, with sheep and goats in Benin and south-eastern Nigeria (Hart and Knipscheer, 1987; Mdoe and Wiggins, 1997), and with sheep, goats and fattening cattle in West Java, Indonesia (Devendra *et al*, 1997, Tanner et al, 1995). Routine animal handling is simplified and the system has many seemingly positive attributes. Animals are completely restricted from cropland and fragile soils thus avoiding the possibility of negative impacts of overgrazing, compaction, excessive run-off and erosion. Manure and wasted or refused feed can be accumulated at the confinement facility for redistribution on cropland or for composting. In Zimbabwe, pen rearing produced 7 tons of collectable manure per livestock unit per year, compared to only two to three for night kraaling (Murwira et al. 1995). Manure stored in piles (from stalls) loses less nitrogen than scattered pats because of the lower surface area to volume ratio, but even so may lose up to 50% of its nitrogen before being spread on cropland. The mineralisation of nitrogen in manure piles makes it more readily available to crops when the manure is spread (Reynolds and de Leeuw, 1995). Another advantage is that the removal of on-farm crop residues and cover crops can be precisely controlled once critical levels needed for soil improvement are known and precise amounts of imported feeds can easily complement on-farm feeds and nutrients.

Disadvantages centre on higher labour demand (forage gathering, manure redistribution) and investment in pens. Even though manure is accumulated at the confinement facility, serious incidental nitrogen loss may take place through urine soaking into underlying soil or running off of the confinement area. Furthermore, almost all of the advantages stated above related to soil conservation and nutrient cycling can be negated if collection and redistribution of excreta and refused feed is mismanaged or ignored.

SEARCHING FOR SOLUTIONS

Despite the presence of world-wide data that indicate the advisability of maintaining soil surface coverage of crop residues, the primary use of these residues in many areas remains as animal fodder to offset dry-season shortages (Sain and Barreto 1996). In most traditional societies there is a direct positive relationship between animal ownership and status in the community. Under those conditions, the lack of communal and national policies to regulate the use of forage resources leads to overgrazing.

Alternative and Complementary Strategies to Feeding Crop Residues

Establishment of alternative and complementary forage sources (legumes, grasses and tree fodder), strategic application of inorganic fertilisers and manure, conservation of surplus forage, supplementation, treatment of crop residues, controlled grazing, zero, or combinations of these, must be flexible enough to adjust to the needs of each farming situation. Equally important is an accurate analysis of all sources of feed inputs, internal and external to the farming system (e.g., rangelands, communal pastures, roadsides or forests). If "free" inputs from external sources result in a gradual nutrient depletion of rangelands or communal grazing, severe land degradation from erosion, or biodiversity losses, these actions must be taken into account in formulating strategies to prevent or reduce livestock-crop-natural resource base conflicts.

Quiroz, *et al.* (1997) emphasised the rational use of adapted germplasm to increase livestock productivity and soil fertility. For example, improved genotypes of grass (*Brachiaria decumbens, B. dictyonuera, B. brizanta, B. humidicola, and Andropogon gayanus*) and legumes (*Arachis pintoi, Stylosanthes guianensis*) have been identified and are commercially available for this purpose. On steep land, the use of living erosion barriers consisting of grasses and/or leguminous trees that can serve as livestock feed is an efficient strategy if the species selected are palatable feeds as well as effective erosion barriers (Barber, 1999). Legumes are particularly important in mixed farming systems because of their role in N cycling and as sources of protein for human and animal nutrition (Devendra, *et al.*, 2001). Nevertheless, when land is limited, farmers are reluctant to take land out of crop production to establish forages. Strategic application of inorganic fertilisers and or manure to eliminate elemental deficits and replace off-take can result in sustainable increases in biomass. Unfortunately because of cost and unavailability, inorganic fertilisers are not an option for many small farmers. Increased emphasis on the efficient use of manure in these systems seems worthy of much effort. The ability to conserve and transfer forage from periods of surplus to periods of deficit appears to be a logical approach to efficient production. Nevertheless, the humid tropics bring special challenges to the practice of forage conservation (hay & silage) due to frequent rains and high humidity during the season(s) when surplus forages are available for conservation. This technology is rarely adopted by small farmers who see it as a costly process requiring machinery and infrastructure that is unavailable to them (Quiroz, *et al.*, 1997). In combination with other strategies, it would seem worthwhile to develop forage conservation methods appropriate for small and mid-sized farmers. How efficiently crop residues and other forages are used by animals depend on the voluntary intake and digestibility of the residues. Efforts to improve both of these factors should receive high priority for research. Cereal residues contain high concentrations of cell wall material and associated lignin. These constituents reduce the value of residues as livestock feeds. Possibilities exist through genetics and treatment to modify these constituents rendering them more nutritious. For example, brown midrib mutants of maize and sorghum contain a single recessive gene that results in reduced lignin concentration of the cell wall and increased digestibility of the forage (Muller *et al.*, 1971). Ammonification of cereal residues with anhydrous ammonia (Klopfenstein and Owen, 1981) or urea treatment (Pezo *et al.*, 2000) can significantly increase their crude protein concentrations, digestibility, intake and animal performance. The adoption of these technologies by small farmers has been limited because some of these materials are somewhat hazardous and require access to specialised equipment, appropriate technical assistance, credit, or consistent supply of affordable inputs (Dolberg and Finlayson, 1995).

Allocation and Efficient Utilisation of Crop Residues

Crop residues contain about one half of the nutrients potentially exportable during harvest. If all residues are returned, nutrient losses can be slowed down, but «nutrient mining» will still occur, because residues alone can replace only a portion of the nutrients exported with the harvest. Therefore, additional nutrient input is needed to sustain the system's productivity (Latham, 1997). Soil degradation as a consequence of «nutrient mining» may not be detected in the short- and mid-term due to soil resilience (Greenland and Szalbocs, 1992), but rehabilitation of degraded soils is a long and costly process, therefore management practice designed to avoid degradation is a sound strategy.

Crop residues left in the field provide a physical barrier to soil erosion, allow soil and organic matter to accumulate, contributing to improved soil structure and macro-porosity that favor root penetration. Return of crop residues also enhances soil chemical properties and conserves water in the root zone by increasing infiltration rates, reducing evaporation and run-off, and moderating soil temperature (Srivastava *et al*, 1993). The use of crop residues as mulch prevents the decline of pH and base saturation and the increase of aluminum saturation in poorly buffered soils. Residue return also increases soil cation exchange capacity, the population of N2-fixing bacteria, and root length and density, leading to increased P uptake by the crops (Powell and Unger, 1997).

A key question that requires an answer for each situation is how much cropland cover or residue is needed to conserve or enhance soil resources? According to Latham (1997) there is still no general understanding of how much (or how little) organic matter is needed to protect the soil structure. It will vary with climatic conditions, soil characteristics such as texture and baseline level of organic matter, and with the crops *per se* and their management. The latter will determine the relative proportion of different fractions left in the field. In general terms, the benefits attributed to the return of crop residues to the soil increase with increasing amounts deposited (Geiger *et al.*, 1992), but even small amounts provide some benefits (Powell and Unger, 1997). In El Salvador Barber (1996) reported that a minimum of 75% residue coverage, equivalent to about 3.5 t/ha, was required to ensure low erosion risk.

Once the acceptable level of vegetative cover is known for a particular soil-crop-environmental situation, the proportion available for livestock feeding can be estimated. Failure to consider the livestock component as an integral part of the agricultural system creates an immediate conflict because system resources must be used to support the animal enterprise. Knowledge of annual seasonal livestock demand for nutrients allows for initiation of appropriate feed budgeting measures. For example, a tropical livestock unit (a 250-kg cow) has a daily dry matter (DM) maintenance intake of roughly 1.5 – 2 % of body weight (BW), which is equivalent to 5.0 kg DM/day or about 1.8 tons annually. This gives an idea of the demand of biomass as feed, but does not address the specific animal nutrient needs or a strategy for utilisation. Energy (total digestible nutrients, TDN) and crude protein (CP) concentration for maintenance should be 56 and 12 %, respectively (NRC, 1989). Unfortunately most crop residues do not meet those requirements. For example, maize and sorghum stovers average between 50-54 % TDN and 5.2-5.9 %CP (NRC, 1989). In this case additional feed inputs will be required to meet animal needs.

Where fodder is being cut and carried to confined livestock, the amount of residual field vegetation can be precisely controlled. If crop stubble is grazed, the herder must maintain control over the animals such that the needed residual is maintained. This can be done by controlling the grazing time permitted in a given area and by allowing animals to express selective defoliation of edible plant fractions. In cut and carry systems, non-edible parts can be returned either as such or as part of compost prepared by mixing with animal excreta. According to Sain and Barreto (1996) farmers in Guaymango, El Salvador who allow grazing of crop residue restrict consumption to about 50 % of the total amount available. As average residue yields of the maize/sorghum system are near 10 t/ha, the amount left (about 5 t/ha) is well above the 3.5 t/ha threshold reported by Barber (1996).

CONCLUSION

Competition for crop residues and cover crops between livestock and *in situ* recycling represents a widespread and serious threat to realising the benefits of *conservation agriculture*. Fundamental aspects of sound crop residue management include conservation tillage and cessation of residue burning. A careful situation analysis involving stakeholders to determine the root cause of conflict and the use of proper combinations of appropriate technologies are required. After assessing the critical residue needs for soil cover/conservation and livestock, some possible technologies include: cultivar selection and/or fertiliser input to increase the amount of residues or the efficiency of their use, establishment of improved forages and living erosion barriers consisting of grasses, legumes and fodder trees, controlled grazing management to maintain critical residue cover, and forage conservation (hay or ensilage) to transfer surplus forage to periods of deficit. Even when all crop residues are allocated for soil improvement and nutrient cycling, they will not completely replace off-take from the harvest. Unless a very low level of output is the goal, inputs will be required to sustain productivity.

According to poet/writer/farmer, Wendell Barry, *'science' means knowing and 'art' means doing, and one is meaningless without the other*. Perhaps with *conservation agriculture* we are long on the knowing and short on the doing.

ACKNOWLEDGEMENTS

We wish to thank FAO field staff and co-operating agencies and projects in Tanzania (Selian Agriculture Research Institute), Costa Rica (MAG), El Salvador (CENTA) and Honduras (Lempira Sur Project) for accommodating field visits and sharing case study information.

REFERENCES

Arias, R. (1987) Identificación y caracterización de los sistemas de producción caprina predominantes en la región del Altiplano Occidental de Guatemala. Thesis Mag. Sc. CATIE (Centro Agronómico Tropical de Investigación y Enseñanza), Turrialba, Costa Rica. 155 p.

Alexandratos (1995) *Agriculture Towards 2010*. FAO, Rome and John Wiley, Chichester.

Bationo, A., Buerkert, A., Sedogo, M.P., Christianson, B.C. and Mokwunye, A.U., (1995) A critical review of crop-residue use as soil amendment in the West African semi-arid tropics. in Powell, J.M., Fernández-Rivera, S., Williams, T.O. and Renard, C. (eds). 1995. *Livestock and sustainable nutrient cycling in mixed farming systems of sub-Saharan Africa.* **Volume II**: Technical Papers. Proceedings of an International Conference held in Addis Ababa, Ethiopia, 22-26 November 1993 (pp305-322). ILCA, Addis Ababa, Ethiopia. 568pp.

Barber, R.G. (1996) Linking the production and use of dry-season fodder to improved soil conservation practices in El Salvador, documento de campo Nº **8**.

Barber, R.G. (1999) Integrated crop and land management in the hilly terrains of Central America: concepts, strategies, and technical options. Vol. **2**. FAO, Rome , Italy. ISBN 92-5-104350-7.

Berry, W. (2000) Life is a Miracle. Counterpoint, Washington D.C. ISBN 1-58243-1412-8

Boonman, J. G. (1997) Farmers' success with tropical grasses: Crop-Pasture rotations in mixed farming in East Africa. Netherlands Development Assistance, The Hague, Netherlands. ISBN 90-53-28-114-2.CAST. 1999. Animal agriculture and global food supply. *Task force report* **135**. Ames, Iowa, USA. ISBN 1-887383-17-4. Pp. 1-92.

Buyanovsky, G.A. and Wagner, G. H.. Post-harvest residue input into cropland.(1986) *Plant Soil.* **93**: 57-61.

Calderón, F., Sosa, H., Mendoza, V., Sain, G. and Barreto, H. (1991) Adopción y difusión de labranza de conservación en Guaymango, El Salvador: Aspectos instrucionales y reflexiones técnicas. In: Agricultura sostenible en las laderas centroamericanas: Oportunidades de colaboración interinstitucional. San José, Costa Rica: Instituto Interamericano de Cooperación para la Agricultura (IICA). pp. 189-210.

CAST. (1999) Animal agriculture and global food supply. *Task Force report* No. **135**, Pp 88. ISBN 1-887383-17-4.

De Leeuw, P.N. (1997) Crop residues in Tropical Africa: Trends in supply, demand and use. *In*: C. Renard (ed.). *Crop residues in sustainable mixed crop/livestock systems.* CABI, Wallingford, U.K. Pp. 41-77.

de Haan, C., Steinfeld, H. and Blackburn, H. (1997) *Livestock and the Environment: Finding the Balance.* European Commission. Directorate-General for Development. Brusells, Belgium. 115 p.

Devendra, C. (1976) The utilization of agro-industrial by-products in Asia and the Far East. *In*: *New Feed Resources. FAO Animal Production and Health Paper* No. **4**. Pp. 111-124.

Devendra, C. and Raghavan, G.V. (1978) *Agricultural by-products in South East Asia: Availability, utilization and potential value. World Review of Animal Production* **14**: 11-28.

Devendra, C. (1993) Sustainable animal production from small farm systems in south-east Asia. *FAO Animal Production and Health Paper* **106**. FAO, Rome. ISBN 92-5-103168-1. 143 pp..

Devendra, C., Thomas, D., Jabbar, M. and Kudo, H. (1997) *Improvement of Livestock Production in Crop-Animal Systems in Rainfed Agro-ecological Zones of South-East Asia.* ILRI (International Livestock Research Institute), Nairobi, Kenya.. ILRI (International Livestock Research Institute), Nairobi, Kenya. 116 p.

Devendra, C., Sevilla, C. and Pezo, D. (2001) Food-feed systems in Asia: Review. *Asian-Aust. J. Anim. Sci.* **14**: 733-745.

Dibissa, N. and Peters, K.J. (1999) Livestock species composition and herd size in relation to household farm resources in mixed farming systems in the Ethiopian highlands. *Outlook on Agriculture*, **28** (1), pp 51-54.

Dickey, E.C., Shelton, D.P. and Jasa, P.J. (1997) Residue management for soil erosion control. NebGuide G81-544-A. Electronic version *pubs@unlvm.unl.edu*

Dolberg, F. and Finlayson, P. (1995) Treated straw for beef production in China. World Animal Review (FAO) 82(1).. [ILRI/FAO. 1999. Farmers, their Animals and the Environment. Multi-document CD-ROM]

ECAF. (1999). Conservation Agriculture in Europe: Environment, Economic and EU Policy Perspectives. Brussels , Belgium pp.1-23.

Fall. S.T., SambaA. N'Diaye and Traoré, E. (2000) Exploitation des arbres à usages multiples dans le systemès d'élevage des zones soudanienne et sahélienne. *In*: Gintzburger, G., M. Bounejmate, C. Agola and K. Mossi (eds.). Production and utilization of multi-purpose fodder shrubs and trees in West Asia, North Africa, and the Sahel. ICARDA, Aleppo, Syria; ILRI, Nairobi, Kenya. Pp. 47-58.

Geiger, S.C., Manu, A. and Bationo, A. (1992) Changes in a sandy Sahelian soil following crop residues and fertilizer additions. *Soil Science Society of America Journal* **56**: 172-176.

Greenland, D.J. and Szalbocs, I. (eds.). (1992) *Soil resilience and sustainable land use*. Proceedings of a Symposium held in Budapest, 28 September–2 October 1992. CABI, Wallingford, U.K. 561 p.

Hart, R.D. and Knipscheer, H.C. (1987) Characteristics and socioeconomic aspects of small ruminant production systems: an analytical framework. in Devendra, C. (ed), 1987. Small ruminant production systems in South and Southeast Asia. Proceedings of a workshop held in Bogor (Indonesia), 6-10 Oct 1986. Ottawa, Ont. IDRC.

Hellin, J, Welchez, L.A. and Cherrett, I. (1999) The Quesungual system: an indigenous agroforestry system from western Honduras. *Agroforestry Systems*. **46**:229-237.

Haynes, R.J. and Williams, P.H. (1993) Nutrient cycling and soil fertility in the grazed pasture ecosystem. *Adv. Agron*. **49**:119-190.

Hutton, J.B., Jury, K.E. and Davies, E.B. (1967) Studies of the nutritive value of New Zealand dairy pastures V. *The intake and utilisation of potassium, sodium, calcium, phosphorus, and nitrogen in pasture herbage by lactating dairy cattle. NZ. J. Ag. Res.* **10**:367-388.

Klopfenstein, T.J., and Owen, F.G. (1981) *The value and potential use of crop residues and by-products in dairy rations. J. Dairy Sci.* **64**: 1250-1268.

Kossila, V. (1988) The availability of crop residues in developing countries in relation to livestock population. *In:* J.D. Reed, B.S. Capper and P.J.H. Neate (eds.). *Plant breeding and the nutritive value of crop residues*. Proceedings of a workshop. ILCA, Addis Ababa (Ethiopia). Pp. 29-39.

Lal R and Pierce, F.J. (1991) The vanishing resource. In: *Soil Management for Sustainability*. Lal, R and F. J. Pierce (eds.) Ankey, USA Soil and Water Conservation Society. ISBN 0-935734-23-6. pp. 1-5.

Latham, M. (1997) Crop residues as a strategic resource in mixed farming systems. *In*: C. Renard (ed.). *Crop residues in sustainable mixed crop/livestock systems*. CABI, Wallingford, U.K. Pp. 181-196.

León-Velarde, C. and Quiroz, R. (1998). Crop-livestock systems research in the Andean region: Ecoregional approach, methods and procedures. *In*: P.K. Thornton and A.N. Odero (eds.). Proceedings, Workshop on Ecoregional Research, held in ILRI, Addis Ababa, 5-8 October 1998. ILRI, Nairobi, Kenya. Pp. 27-41.

Muller, L.D., Lechtenburg, V.L., Bauman, L.F., and Colenbrander, V.F. (1971) Variation in lignin and other structural components of brown midrib mutants of maize. *Crop Sci.* **11**:413.

McDowell, R.E. (1988) Importance of crop residues for feeding livestock in smallholder farming systems. *In:* J.D. Reed, B.S. Capper and P.J.H. Neate (eds.). *Plant breeding and the nutritive value of crop residues*. Proceedings of a workshop. ILCA, Addis Ababa (Ethiopia). Pp. 3-27.

McIntire, J., Reed, J.D., Fedla, A., Jutzi, S. and Kebede, Y. (1988) Evaluating sorghum cultivars for grain and straw yields. *In:* J.D. Reed, B.S. Capper and P.J.H. Neate (eds.). *Plant breeding and the nutritive value of crop residues*. Proceedings of a workshop. ILCA, Addis Ababa (Ethiopia). Pp. 283-304.

Mdoe, N. and Wiggins, S. (1997) *Returns to smallholder dairying in the Kilimanjaro region, Tanzania. Agricultural Economics*.**17**: 75-87.

Murwira, K.H., Swift, M.J. and Frost P.G.H. (1995) Manure as a key resource in sustainable agriculture. in Powell, J.M., Fernández-Rivera, S., Williams, T.O. and Renard, C. (eds). 1995. *Livestock and sustainable nutrient cycling in mixed farming systems of sub-Saharan Africa.* **Volume II**: Technical Papers. Proceedings of an International Conference held in Addis Ababa, Ethiopia, 22-26 November 1993 (pp131-148). ILCA, Addis Ababa, Ethiopia. 568pp.

National Research Council. (1989) Nutrient Requirements of dairy cattle. National Academic Press, Washington, DC.

Pezo, D.A., Lanting, E.F., Wong, C.C. and Kerridge, P.C. (2000) Feed resources for ruminants in smallholder farming systems in South East Asia. *In*: W.W. Stur, P.M. Horne, J.B. Hacker and P.C. Kerridge (eds.). *Working with farmers: The key to adoption of forage technologies. ACIAR Proceedings* No. **95**. ACIAR, Canberra, Australia. Pp. 97-111.

Poore, M. H., Moore, J.A. and Swingle, R. S. (1990) Differential passage rates and digestion of neutral detergent fiber from grain and forages in 30, 60 and 90% concentrate diets fed to steers. J. Anim. Sci. 68:2965.

Powell, M. and Ikpe, F. (1992) Fertiliser factories. *ILEIA newsletter* vol.**8** no.**3**.

Powell, J.M. and Williams, T.O., (1995) An overview of mixed farming systems in sub-Saharan Africa. in Powell, J.M., Fernández-Rivera, S., Williams, T.O. and Renard, C. (eds). (1995) *Livestock and sustainable nutrient cycling in mixed farming systems of sub-Saharan Africa.* **Volume II**: Technical Papers. Proceedings of an International Conference held in Addis Ababa, Ethiopia, 22-26 November 1993 (pp21-36). ILCA, Addis Ababa, Ethiopia. 568pp.

Powell, J.M. and Unger, P.W. (1997) Alternatives to crop residues for soil amendment. *In*: C. Renard (ed.). *Crop residues in sustainable mixed crop/livestock systems.* CABI, Wallingford, U.K. Pp. 215-239.

Quiroz, R.A., Pezo, D.A., Rearte, D.H. and San Martín, F. (1997) Dynamics of feed resources in mixed farming systems of Latin America. *In*: C. Renard (ed.). *Crop residues in sustainable mixed crop/ livestock systems.* CABI, Wallingford, U.K. Pp. 149-180.

Reynolds, L. and de Leeuw, P.N. (1995) Myth and manure in nitrogen cycling: a case study of Kaloleni Division in Coast Province, Kenya. in Powell, J.M., Fernández-Rivera, S., Williams, T.O. and Renard, C. (eds). (1995) *Livestock and sustainable nutrient cycling in mixed farming systems of sub-Saharan Africa.* **Volume II**: Technical Papers. Proceedings of an International Conference held in Addis Ababa, Ethiopia, 22-26 November 1993 (pp509-521). ILCA, Addis Ababa, Ethiopia. 568pp.

Romney, D.L., Thorne, P.J. and Thomas, D. (1994) Some animal-related factors influencing the cycling of nitrogen in mixed farming systems in sub-Saharan Africa. *Agriculture, Ecosystems and Environment* **49**:163-172.

Sain, G.E. and Barreto, H.J. (1996) The adoption of soil conservation technology in El Salvador: *Linking Productivity and conservation. J. Soil and Water Cons.* **51**:313-321.

Sánchez, M. (1995) Integration of livestock with perennial crops. *World Animal Review* **82**: 50-57

Sanford, S.G. (1989) Crop residue/livestock relationships. *In*: *Soil, crop and water management in the Sudano-Sahelian Zone.* Proceedings Int. Workshop. January 11-16, 1987. ICRISAT Sahelian Center, Niamey, Niger. ICRISAT, Patancheru. Pp. 169-182.

Shaxson, T.F. (1999) New concepts and approaches to land management in the tropics with emphasis on steeplands. *FAO Soils Bulletin* **75**. FAO, Rome. ISBN 92-5-104319-1. pp. 1-123.

Shaxson, T.F., Hudson, N.W., Sanders, D.W., Roose, E. and Moldenhauer, W.C. (1989) Land husbandry a framework for soil and water conservation. Soil and Water Conservation Society, Ankey, Iowa, USA. ISBN 0-935734-20-1. pp. 1-64.

Sombroek, W.G. and Sims, D. (1995) Planning for sustainable use of land resources: Towards a new approach. FAO Land and Water Bulletin 2. FAO, Rome. pp. 1 - 60.

Smil, V. (1999) Crop residues: agriculture's largest harvest. *BioScience.* **49**: 299-308.

Speirs, M. and Olsen, O. (1992) Indigenous integrated farming systems in the Sahel. *World Bank technical paper no.* **179** Africa technical department series. World Bank, Washington, D.C.

Srivastava, J.P., Tamboli, P.M., English, J.C., Lal, R., and Stewart, B.A. (1993) *Conserving soil moisture and fertility in the warm seasonally dry tropics. World Bank Technical Paper* No. **221.** World Bank, Washington, DC.

Tanner, J.C., Holden, S.J., E. Owen, M. and Gill. (1993) Feeding livestock for compost production: A strategy for sustainable upland agriculture on Java. *In:* Powell, J.M., Fernández-Rivera, S., Williams, T.O. and Renard, C. (eds). 1995. *Livestock and sustainable nutrient cycling in mixed farming systems of sub-Saharan Africa.* **Volume II:** Technical Papers. Proceedings of an International Conference held in Addis Ababa, Ethiopia, 22-26 November 1993. ILCA, Addis Ababa, Ethiopia. Pp. 115-128.

Thiesenhusen, W.C. (1994) The relation between land tenure and deforestation in Latin America. In: J.E oman and D. Pezo (eds.). Proceedings, Symposium/Workshop «*Livestock and Natural Resources in Central America: Strategies for Sustainability*», held in San Jose (Costa Rica) 7-12 October 1991. CATIE, Turrialba, Costa Rica. Pp. 243-256.

Turner, M. (1995) The sustainability of rangeland to cropland nutrient transfer in semi-arid West Africa: Ecological and social dimensions neglected in the debate. in Powell, J.M., Fernández-Rivera, S., Williams, T.O. and Renard, C. (eds). 1995. *Livestock and sustainable nutrient cycling in mixed farming systems of sub-Saharan Africa.* **Volume II**: Technical Papers. Proceedings of an International Conference held in Addis Ababa, Ethiopia, 22-26 November 1993 (pp435-452). ILCA, Addis Ababa, Ethiopia. 568pp.

White, S. L., Washburn, S. P., King, L. D., Sheffield, R. E. and J. T. Green, Jr. (2001) Spatial and time distribution of dairy cattle manure in an intensive pasture system. In Proc. XIX Int. Grassland Congr. São Pedro, São Paulo, Brazil. Pp.340-341.

T. BACHMAN

«IMPROVED CEREAL PRODUCTION TECHNOLOGY» FAO PROJECT ON CONSERVATION AGRICULTURE IN MONGOLIA

FAO, Viale delle Terme di Caracalla, 00100 Rome, Italy

INTRODUCTION

Spring wheat production in Mongolia is mechanized through the introduction of techniques and farm equipment utilized by the Russians in the 1960s. The former state farms covered large areas of up to 20,000 ha under a centralized management with a paid labour force of some one hundred people. Since 1992, most state farms have been privatized and split into smaller units of 1000 ha and leased, for periods of up to 20 years, to private individual owners or to a group of owners as a shareholder company. Following the abandonment of many marginal state farms, the cultivated wheat area has decreased; consequently, there is adequate farm machinery to cultivate 300,000 ha of wheat. However, mounting production costs, the inability to obtain credit, scarcity and high cost of essential inputs (fuel, herbicides, fertilizers, spare parts etc.), inefficient farm machinery, decrease in prices and crop productivity have resulted in heavy indebtedness of farmers and many are on the brink of bankruptcy.

Following a request by the Government of Mongolia, FAO is implementing the TCP project «Improved Cereal Production Technology». The project was launched during the 2000 cropping season and will terminate in 2002. Its main objective is to introduce Conservation Agriculture technologies in Mongolia, in order to improve cereal productivity and farm profitability, on a sustainable and environmentally friendly basis.

AGRICULTURAL PRODUCTION AND ITS MAJOR CONSTRAINTS IN MONGOLIA

Mongolia is known for its harsh, cold and dry continental climate, which places severe limitations on crop production. The short growing period and extremely low winter

L. García Torres et al. (eds.). Conservation Agriculture, 235–242.

temperatures only allow cultivation of early maturing spring crops, such as spring wheat, potatoes and fodder. One of the most serious environmental problems is created by high winds, particularly during April and May, which promote drying of the upper rooting zone and serious land erosion. Unsuitably prepared fallow and arable land, before and after seeding in early May, are also major constraints.

Wheat continues to be the major cereal, accounting for 98% of the cultivated cereal area. The wheat area has dramatically decreased from 650,000 ha in 1990 to 300,000 ha in 1998. The current cropping system is a spring wheat monocropping system. Average wheat yields of 1.4 t/ha were achieved in the 1980s but have currently fallen to 0.6 t/ha. However, yields of 2 t/ha are achievable in favourable years on better-managed farms. Wheat yields have been dropping owing to poor crop and land management, the use of little or no fertilizer, agro-chemicals, lack of improved seed and spare parts.

Lack of soil moisture, especially at seeding and during crop establishment in May and June, is the primary constraint on wheat yields. Common practice is to fallow once every two years (under strip cropping) where cropped and fallowed fields (strips) alternate in order to prevent wind erosion. Wheat grown for two consecutive years yields only 0.5 t/ha during the second planting year. The fallow is cultivated up to five times, with broad-sweeps and/or disc cultivators, to control weed growth and conserve the accumulated soil moisture for the next year's crop. However, frequent mechanical cultivation accelerates the loss of stored moisture from the rooting profile.

Food security is a major issue for Mongolia. It is a landlocked country and border disputes and shortages in international wheat production could disrupt supplies. The Government is particularly concerned with the continuous decline in crop production. In 1995, a Government resolution proclaimed that self-sufficiency in wheat production is a national priority. Since flour is the single major staple food, and in response to the serious risk of interrupted grain supplies, wheat and flour have been set aside to act as buffer stocks against shortfalls in production.

Demand for wheat is mostly for human consumption. Wheat is sold to flour mills and to distilleries. Total milling wheat requirement is 236,000 t. This is based on a population of 2.3 million people, with a per capita consumption of flour of 100 kg p.a. Using the figures of the 1998 harvest (~ 180,000 t) and the total amount of wheat consumption (236,000 t), Mongolia needed to import 56,000 t in order to satisfy demand. In 1998, the actual wheat import was 38,000 t, leaving a deficit of 18,000 t of wheat.

CONSERVATION AGRICULTURE AND ITS IMPACT ON CROP PRODUCTION

One key to farmers' economic survival, in producing cereal crops in a difficult climate, is efficient crop and land management. The current practice of frequent cultivation of fallow land is unsustainable, both from a technical and financial point of view. Weed control is very poor, soil moisture losses are high and fuel and spares for tractors are scarce and costly. The project's objective is to test and validate resource-saving Conservation Agriculture technologies, which address the major production constraints, and to train farm managers and farmers in using these practices for improved and sustainable cereal production.

Creating a soil habitat is the first step to Conservation Agriculture. This means using soil management practices that reduce soil disturbance, reduce rainfall drop impact on the soil surface (using crop residues or cover crops), managing weeds and disease with crop rotation, mixed cropping, underseeding, and increasing soil fertility and moisture retention.

Conservation Agriculture (CA) has been adopted in very diverse farming systems, from small-scale commercial farming in southern Brazil, or large-scale farming in the Cerrado, to the irrigated rice-wheat system of northern India and Pakistan, to the large-scale farms of semiarid Texas, or the emergent small-scale mixed farms in Zimbabwe.

Conservation Agriculture, usually expressed as residue-based zero tillage, offers numerous benefits that intensive or conventional tillage cannot match.

- direct seeding of grain crops;
- improved crop residue management;
- chemical fallow based on the use of non-selective herbicides.

In Mongolia, Conservation Agriculture is only possible with an alterative weed control method, which replaces the conventional mechanical fallow cultivation. Similar to the situation in Canada and the Northern Great Plains (USA), chemical fallow, based on the use of non-selective herbicides, appears to be an appropriate weed control method. Cover crops for weed control do not seem to fit well into the cropping system, especially the fallow, because rainfall is low and the fallow is needed to accumulate soil moisture for the subsequent wheat crop.

Conservation Agriculture, using minimum or no-tillage and direct planting of the crop, leads to reduced labour requirements, time saving, reduced machinery wear and fuel costs. No-till, for instance, requires as little as one trip for planting, compared to two or more soil tillage operations, plus planting for conventional tillage. (Fewer trips also save on machinery wear and maintenance and fuel cost).

Reduced weight and horsepower requirements with no-tillage can help minimize soil compaction. Additional field traffic, required by conventional tillage, breaks down the soil structure, promoting compaction. No-tillage increases soil particle aggregation (soil clumps), which makes it easier for water to move through the soil and, at the same time, allows plants to use less energy to establish roots.

Crop residues reduce water evaporation from the top few inches of the soil and no-till can make additional water available for growing plants during the season. Crop residues act as tiny barriers to slow water runoff from the field, allowing the water more time to soak into the soil. Channels (macropores) created by soil macrobiota and old plant roots that are left intact also increases infiltration. All this helps to significantly reduce or eliminate field runoff. Crop residues on the soil surface reduce water and wind erosion. Depending on the amount of residues, soil erosion can be reduced by 90% compared to an unprotected, conventionally tilled field.

Resource-saving farming is not new to Mongolia. Minimum tillage and appropriate equipment (V-shaped shallow sweeps) had been introduced by scientists from the Soviet Union about 15-20 years ago. However, farmers because of the serious weed problem have rarely adopted this technology. Farmers tend to control the weed problem by frequent fallow cultivation. The introduction of chemical fallow, based on non-selective herbicides, would solve the weed problem and allow the introduction of no-till technology.

These proposed technologies, which have been shown to work under similar conditions in North America, are expected to have a positive impact on crop production in Mongolia. Since virtually all farm equipment is of Russian origin, the project envisages updating mainly Russian-made machinery. The major advantage of local farm equipment is its immediate availability, its low price and farmers' familiarity with it. Equipment from Western countries is probably more efficient but it is more sophisticated, requires a higher standard of maintenance and it is more expensive.

PROJECT ACTIVITIES AND RESULTS

The project focuses its activities in the Central Cropping Region, which is the main cereal growing area in Mongolia, situated north of Ulaanbaatar, the country's capital. Five large-scale farms were selected for participating in the test and demonstration programme. Each farm allocated 200 ha for the programme of which 100 ha are used for testing Conservation Agriculture technologies and 100 ha for multiplying quality seed. All selected farms are privately owned and use farm machinery suitable for Conservation Agriculture practices.

Due to delays in the project start-up, project activities could not begin in time for the 2000 cropping season. As a consequence, proper technology testing was started during the 2001 season, too early to produce relevant quantitative results.

Technologies to be tested and validated together with farmers are the following:

- improved weed control;
- improved crop residue management;
- no-tillage and direct planting of crops;
- seed production and crop diversification.

Improved weed control

In Mongolia the main grassy weeds are quackgrass (*Agropyron repens),* wild oat (*Avena fatua*) and proso millet (*Panicum miliaceum*). The major broad-leaved weeds are sagebrush (*Artemisia dracunculus*), sagebrush/wormwood (*Artemisia sieversiana*), Canada thistle (*Cirsium arvense*) and Tartary wheat (*Polygonum tataricum*).

Currently, most farmers control weeds by cultivating the soil two to three times during the fallow period, especially against the root spreading weeds (e.g. quackgrass: *Agropyrum repens*), which can reduce yields by 50%. This inefficient and costly cultivation is being replaced by the use of non-selective herbicides (desiccants or burndown herbicides) such as glyphosates (e.g. Roundup). The desiccant is usually applied once during the fallow period with an application rate of 2.5 kg/ha and in conjunction with a surfactant, which kills existing vegetation before planting. If the weeds have not been properly killed, an additional early preplant application of 1.5 kg/ha is necessary. If there are only broad-leaved weeds growing, then the desiccant can be replaced with 2,4-D to save costs.

The existing boom sprayers are old and in a poor state of maintenance since spare parts are scarce. As a result, the sprayers do not assure the precise and even application of desiccants. Spraying is even more arduous than cultivating because, if the desiccants have not been applied properly, the effects of the chemical cannot be seen until it is too late, e.g. when the weeds are too big. Therefore, the project provides up-to-date kits, which comprise a new pump, distribution system, controls and spraying nozzles, which allow farmers to upgrade their boom sprayers. Farmers are also provided with non-selective herbicide (Roundup) and surfactant. For maximum effectiveness, farmers are advised to apply the herbicides when the weeds are about 20 cm high.

The first results with chemical fallow have shown good effectiveness in controlling both broad-leaved weeds and grasses. Two sprayings, as indicated above, are more effective than one. A shift in weed composition, from annual to hard-to-control perennials - when changing from conventional to Conservation Agriculture - has not yet been observed but could become a problem in the long-run.

Improved crop residue management

Efficient residue management is the key to good Conservation Agriculture and it begins at harvest. Spring wheat is harvested in Mongolia with combine harvesters in August. Farmers usually cut and thresh in one pass but wheat is frequently swathed before full maturity and threshed at a later stage. Although farmers have to pass twice over the field with their combines, which is an increased expense, this harvesting system has an important advantage: the wheat crop can be harvested early in the season, and the early frost in late August or early September does not affect crop yield.

Wheat straw remains in the field but is frequently grazed in winter and summer during the fallow period by the nomadic herders' livestock. As a result, about half or even less of the crop residue is often not available for residue management under Conservation Agriculture. Another constraint is the uneven distribution of crop residues after harvest, especially under the swath system: two rows of harvested wheat placed in one line. The accumulation of straw in windrows behind the combine causes a number of problems:

- planting into windrows results in an uneven crop stand because the seeds take longer to germinate and grow, resulting in a significant yield reduction;
- unsatisfactory weed control from herbicide interception;
- poor performance of direct seed drill;
- poor protection of soil from soil moisture evaporation and erosion;

- increased demands on planter equipment;
- poor seed-to-soil contact;
- increased pest infestation;
- increased weed seed concentration;
- poor plant nutrient uptake.

The project locally developed a straw spreader, based on two horizontal rotating disks, in order to achieve a uniform straw distribution on the ground. This straw spreading system works relatively well on fields, which have been harvested with the combine in one pass. However, under the swath harvesting system, straw distribution is uneven and impedes the subsequent direct drilling of the crop. Straw chopping, utilizing the combine harvester, has been considered but does not appear to be a feasible solution because of the high power requirements and the underpowered combines.

Residue-based zero tillage and direct seeding of crops

Since the no-tillage system practised under the project avoids any soil cultivation during the fallow period, the grain crop is seeded directly into the non-cultivated soil. The project works with the conventional seeders, of Russian origin (SZS 2.1), used by farmers. The seeder has been modified in order to adapt it to direct seeding of the grain crop. For this purpose the wide-winged duckfoot opener, which disturbs the soil, was modified by cutting most of the wings and reducing it to a «hoe opener». This reduces soil disturbance and allows, at the same time, for a certain weed control. A comparison of the hoe drill with a disk drill has shown that weed control, which is critical at the time of planting, is better with the hoe drill. Direct drilling with the modified hoe drill has encountered some problems in heavy crop residues when seeders were clogged. Coulters (cutting disks), which cut through crop residue and open the soil, will be mounted. However, in general the modified seed drill works satisfactorily under farmer conditions.

Current conventional seeding rates used by farmers of some 180 kg/ha are high and seeding depth of 7-10 cm is very deep. This practice is the result of the frequent fallow cultivation for weed control, which pulverises and dries out the topsoil before and during planting time. In order to allow the emerging seeds to tap the deeper moist soil layers, farmers are forced to sow the seeds at great depth. As a consequence, it is difficult for the seedlings to emerge and they develop into weak plants. To compensate for the high percentage of weak and/or non-emerging plants, farmers plant with high seeding rates.

The benefit of the no-till/direct seeding system is obvious; it preserves scarce soil moisture, allows shallow seeding and helps to develop a vigorously emerging grain crop. Participating farmers have already reduced their seeding depth down to a range of 4-6 cm. However, they keep the high seeding rate because they fear the dry spell during crop emergence/early plant development, which is typical for Mongolian agroclimatic conditions, can kill a high percentage of young plants. A lower seeding rate, under direct drilling, is expected to increase tillering and to save soil moisture. This presupposes, however, a good control of both grass and broadleaf weeds.

Seed production and crop diversification

During the period of transition, towards a market economy, Mongolia's seed production system has broken down. In the last decade, farmers are only using farmer saved seed of decreasingly low quality, and no new varieties are developed by the seed system. The project provides improved wheat seed to participating farmers and trains them on-farm quality seed multiplication.

Farmers are monocropping spring wheat because it is the only crop, which has a market and which offers farmers a reasonable price. It is also the only crop for which improved seed is sometimes available (imported from the Russian Federation). The permanent monocrop system has contributed to the current low crop production level and the long-term unsustainability of the production system.

The project has identified a need for locally produced malting barley. Currently, all malt needed, for the relatively large local brewing industry, is being imported. Spring barley is one of the most suitable grain crops for Mongolia and has been produced by farmers in the past. It is drought-hardier than wheat and also higher-yielding. The project has provided participating farmers with quality barley seed for the 2001 season with the aim to produce malting barley seed for sale to other interested farmers. Other suitable alternative crops are buckwheat, canola, sunflower and chickpeas which, however, need to be tested to identify appropriate varieties.

PROSPECTS OF CONSERVATION AGRICULTURE IN MONGOLIA

Current experience with Conservation Agriculture technologies looks very promising and participating farmers are very keen to test the new production approach. However, since the project has started (there has only been one first complete season in 2001), it is still too early to present quantitative results on the impact this technology has on performance parameters such as crop productivity, erosion control, weed control, soil moisture retention and farm economics.

A number of other donors have also realized the great potential which Conservation Agriculture has on crop production in Mongolia and have initiated projects. The EU, through TACIS, is financing the on-going project ««Crop Development in Mongolia» which will end in November 2001 but which might be extended. The NGO ACDI/VOCA, financed by USAID, operates a «Farmer to Farmer» project which has an important Conservation Agriculture component. CIDA, in close collaboration with the Canadian and Mongolian private sector (farm equipment, seed dealers) and the Mongolian State University of Agriculture, has started to implement the «Introduction of Minimum Tillage Technologies» project, which aims at helping Mongolian farmers apply new farming methods on a commercial scale. The Asian Development Bank is planning an agricultural development project where conservation farming will be an important activity.

All these project activities indicate that many donors consider Conservation Agriculture a promising production method for Mongolia, suitable to significantly raise crop productivity and farm profitability, through conservation and sustainable use of natural resources, and saving on scarce and costly farm inputs.

P.C. WALL*, L. ZAMBRANA**, P. GAMEZ**,
B. SIMS** & A. CALISSAYA**

DEVELOPMENT OF AN ANIMAL DRAWN ZERO TILLAGE SEEDER FOR SMALL GRAINS

* *CIMMYT, CC 2305, Santa Cruz, Bolivia*

** *PROMETA/CIFEMA, Casilla 831, Cochabamba, Bolivia*

Multiple row seeders adequate for direct (zero tillage) seeding of small grain cereals using animal traction have not been commercially available in South America. Given the probable benefits of the use of zero tillage to small farmers in the inter-Andean valleys of Bolivia, an inter-institutional project was initiated to design, construct and test seeders for these conditions. A three-row seeder has been produced that gives higher plant stands, results in considerable savings in time compared to the traditional system of seeding into tilled land, and has a lower draught requirement than the traditional wooden plough. Although adoption of the seed drill is yet to take place, farmer interest is high, mainly because of the time saving the use of the seeder represents.

INTRODUCTION

The small-holder farmers of the inter-Andean valleys of Bolivia possess, on average, 3-4 ha of land, use animal traction for land preparation and manage mixed crop-livestock systems. Wheat is a major crop, especially in the drier and more degraded areas. Farmer surveys have shown that the major limitations to wheat productivity are moisture stress, soil erosion and degradation (Wall *et al.*, 1995).

Land preparation generally consists of ploughing twice, the second pass at 90E to the first, using two oxen and a wooden (ard type) plough. Where wheat is to be seeded, the seed is broadcast on the surface and incorporated with another pass of the wooden plough. This process is costly and time-consuming, taking in total 12 days per hectare, and forcing the farmer and the oxen to walk 100 km for each hectare prepared and seeded. The intensive land preparation also results in organic matter decomposition and soil structural breakdown, and leaves the land exposed to rain and wind, resulting in widespread land degradation.

Zero tillage is successful in increasing moisture use efficiency, reducing soil erosion and reverting soil degradation in the mechanised agriculture of the plains of the

L. García Torres et al. (eds.). Conservation Agriculture, 243–248.

Southern Cone of South America (Derpsch, 1999). Consequently, the Bolivian national wheat programme (PROTRIGO), together with CIMMYT, has been working on adapting this form of conservation agriculture to the conditions of the small farmers of the inter-Andean valleys. In initial trials, single row animal drawn direct seeders from Brazil were used, but these machines, although they resulted in good crop establishment without tillage, were principally designed for seeding maize and beans, and proved cumbersome and slow for seeding small-grain cereals. Multi-row seeders for seeding small-grains using animal traction are not available in South America.

In 1998, the PROMETA/CIFEMA project, CIMMYT and, PROTRIGO, embarked on a project to develop a 3-row small grain seeder for seeding small-grain cereals without previous tillage using animal traction. The first prototype was developed in the 1998/1999 season and tried in farmers fields in the 1999/2000 crop season.

OBJECTIVES

The objectives of the project were to design, construct and evaluate a small-grain seeder for animal traction that could seed in untilled conditions with up to 4 t/ha of dry crop residues on the surface.

Given that the seeders are designed for use by small farmers with scarce resources, the project aimed to produce a seeder that is:

- As light as possible to facilitate transport and reduce draught requirements.
- Easy for the farmer to manage alone when seeding behind a pair of oxen.
- Moderately priced to enable adoption and rapid diffusion.
- Easy to maintain and repair.

SEEDER DESIGN

The First Model

The seeder was constructed using three separate seed boxes, one for each row. Seed is distributed and metered by vertical rotors with peripheral seed cells, whose depth can be altered for different seed sizes and sowing rates by adjusting an internal screw. The seeding mechanism is mounted on a metal frame supported on two wheels. The frame can be raised or lowered with respect to the wheels to allow two positions: one for transport with the seed-furrow openers above the wheels, and one for seeding with the openers below the wheels. Seeding depth is adjusted by a bolt limiting the height to which the ground wheels can be raised. The seeding mechanism is driven from one of the ground wheels, using a 1:1 ratio. Fore and aft leveling of the seeder is achieved by adjusting the angle of attachment of the draw-pole using two U-bolts.

In order to reduce the weight of the machine, chisel openers are used as opposed to disc openers which require more weight (vertical force) for penetration. Freely rotating metal wheels with exterior spokes (straw wheels) were attached alongside each chisel to hold the straw and reduce it's accumulation on the chisels. Trials using the machine with sharp-crested or slightly convex chisels, with or without the straw-wheels, and with or without press-wheels behind the chisels, showed that the highest plant populations were achieved with the crested chisels, together with both the straw wheels and press wheels.

Row spacing can be adjusted from 20 to 35 cm. Experience has shown that under zero tillage conditions, the 20 cm row spacing is best, even though this makes seeding slower as less area is covered with one pass of the drill.

The draught force required by the seeder is only 450 $\pm$ 6 Newtons (N) compared to 760 $\pm$ 20 N required by a traditional wooden plough under the same conditions. Wheel slippage, measured under zero tillage conditions with straw ground cover was only 1% (Calisaya A, unpublished data).

Initial trials showed increased plant stands compared to the normal farmer seeding method of broadcasting seed on prepared ground and incorporating it with a wooden plough. (Still need to get the data for this and put some values).

Although the seeder was designed originally for direct seeding, it may also be used under normal tilled conditions by removing the straw wheels. With this configuration wings can be attached to the chisels to create ridges between the seeded rows, aiding in water capture. To reduce the slippage of the drive wheel in loose soil, a metal wheel with strakes, similar to the straw wheels, is attached to the outside of the drive wheel, with strakes protruding beyond the circumference of the drive wheel.

Despite the positive results of initial tests, participatory evaluations, with farmers and technicians in farmers' fields, showed several shortcomings of the seeder. The principal problem was the accumulation of straw on the chisels and the inadequacy of the straw press-wheels to control this. The straw press-wheels also tended to make a lot of noise, frightening the draught animals. Other problems encountered were: the difficulty of maintaining wheel contact with the ground in uneven surface conditions resulting in stoppage of the seed metering mechanism and unseeded areas; the difficulty in changing the seed rotors for seeding different crops; and some structural weaknesses that could not withstand the forces encountered when turning the machine at headlands.

Based on this iterative process of participatory trials, the seeder has undergone two generations of modification, resulting in a more robust and versatile machine.

The Present Seeder

The major modifications to the original model have been an increase in the general robustness and mechanical simplicity of the seeder, the removal of the straw wheels and their replacement with discs in front of each chisel opener to cut the straw. The seed-metering mechanism is driven by the centre disc which is fluted for reduced slippage The outer cutting discs are plain. Thus the wheelsare used only for transport and depth control, so overcoming the problem of unseeded areas when stones or harder soil raise the machine slightly.

The new model (Fig. 1) can be produced with two different seed distribution mechanisms: the original vertical seed-rotor system, which is cheaper but more difficult to adjust for different seedrates and for seeding different species; and a horizontally

Figure 1. Conformation of the seeder for direct seeding using the horizontal plate seed metering system.

rotating seed-plate system which is more expensive, but far more versatile with respect to seeding different amounts and types of seed.

The seeder may still be used in tilled soil. For this application the outer (plain) cutting discs are removed and the central fluted disc is replaced with a strake-wheel to gain sufficient traction to drive the seeding mechanism. Wings may still be attached to the chisels to form ridges, but the optimum row and ridge spacing when these are attached is 30cm. In optimum conditions this is not a limitation to crop yield, but in poor soils where crop development is not enough to cover the inter-row area the wider spacing is detrimental to yield. Also, when heavy rains occur soon after seeding, as in the abnormally wet 2000/2001 summer season, the ridges disintegrate and cover the seed to an excessive depth, resulting in reduced plant populations.

Results and Farmer Appreciations

Ten machines with vertical seed rotors were tested with farmers in the 2000/2001 summer cropping season in the inter-Andean valleys of Bolivia. In trials and demonstration plots seeded with the animal drawn seeder, average stands were 246 $\pm$ 37 plants/m^2, whereas in the check plots using the normal farmers' technology, the average stand was 166 $\pm$ 39 plants/ m^2. This latter value is representative of farmer's fields: a monitoring study of 73 farmers' fields in the drier 1999/2000 season showed average plant stand in farmers fields to be 136 $\pm$ 41 plants/m^2. (Calle C. *et al.*, unpublished data)

Farmer evaluation, however, concentrated more on time savings rather than plant stand. Using the seeder with 20 cm between rows without any prior tillage, results in the farmer and oxen walking 16.7 km to seed one hectare, compared to the normal 100 km with the traditional land preparation and seeding method. Furthermore a hectare can be sown in 10 hours (two working days for a team of oxen) compared with 12 days for conventional sowing. This saving in time and effort, and the possibility of using the oxen to seed a greater area in the season, has provoked widespread farmer interest. However, as yet there have been no farmer purchases of the seeder. We postulate that although this is influenced by the price of the machine (an estimated $US 330 for a production model of the vertical seed-rot0r model and US$ 390 for the horizontal seed plate model), the major factor is farmer risk aversion, and that further demonstration over several seasons will result in widespread farmer adoption.

ACKNOWLEDGEMENT

This publication results, in part, from a research project funded by the United Kingdom Department for International Development (DFID) for the benefit of developing countries. The views expressed are not necessarily those of DFID. (PROMETA Project R6970, Livestock Production Programme).

REFERENCES

Derpsch, R. (1999) Frontiers of conservation tillage and advances in conservation practice. Paper presented at the 10th ISCO Conference, 24. – 28. May 1999, West Lafayette, USA. Proceedings in press.

Wall, P.C., Claros, F.O., Zegada, L., Calle, C. y Ortiz-Monasterio, I. (1995) Un Resumen de Sondeos de Productores de Cereales Menores en Cochabamba y Chuquisaca - Abril 1994. In *Memorias de la II Reunión Nacional de Trigo y Cereales Menores*, pp 19-28. IBTA-CIAT-ANAPO-CIMMYT, Santa Cruz, Bolivia.

*S.W. DUIKER, *D.A. MILLER, J.M. HUNTER, *E.J. CIOLKOSZ, & **W.J. WALTMAN

NO-TILL MANAGEMENT INTENSITY ZONES FOR PENNSYLVANIA

** The Pennsylvania State University, 116 ASI Building, University Park, PA 16802-3504, USA*

*** Nebraska Research Initiative, Computer Science & Engineering, University of Nebraska, 113 Nebraska Hall, Lincoln NE 68588-0517, USA*

A classification of management intensity zones for no-till maize production was developed for Pennsylvania using GIS. Zones were based on analysis of Growing Degree Days, drainage characteristics, slope, water holding capacity of the root zone, and rock fragment content. Six zones were distinguished, reflecting the relative management requirements for notill. The map was produced as a tool for farmers and extension agents to recognize the challenges associated with no-till in their area.

INTRODUCTION

The benefits of no-till compared to conventional tillage are numerous. No-till reduces soil erosion and evaporation, increases soil organic matter and earthworm populations, reduces labor requirements, and decreases equipment costs. Under some climatic and soil conditions standard no-till practices result in equal or higher yields than conventional tillage. Under other climate and soil conditions, yields with no-till are often lower than with conventional tillage. Although farmers can obtain good yields with no-till, no-till is more management-intense in these challenging environments. Based on research results and farmers' experience, we developed a management-intensity classification of the soils of Pennsylvania for no-till maize production (*Zea mays, L*) using geo-referenced data.

Research results from the Northeast and Midwest of the USA typically show that yields of no-till continuous maize are higher than conventionally tilled maize on well-drained and moderately well-drained soils when the growing season has more than 2800 Growing Degree Days (GDD) for maize (Hill, 2000). If the growing season is

L. García Torres et al. (eds.). Conservation Agriculture, 249–252.

shorter than 2800 GDD, however, yields of no-till continuous maize are often lower than yields of conventionally tilled maize. On very poorly, poorly and somewhat poorly drained soils, yields of no-till maize are commonly lower than those of conventionally tilled maize no matter the length of the growing season (Blevins, 1984; Griffith *et al.*, 1986; Hill, 2000). Under the challenging conditions of short growing season or poor drainage conditions it is necessary to make adjustments to the no-till system to obtain good no-ill maize yields. Examples are: creating beds to improve drainage (Fausey, 1984), crop rotation (Dick & Van Doren, 1985), delaying planting (Herbek *et al.*, 1986), rotational tillage (Dickey *et al.* 1983), moving crop residue out of the row (Kaspar *et al.*, 1990) or row-zone tillage systems (Morrison, 1999). Factors than make no-till less management intensive than conventional tillage are: soil erosion reduction, infiltration increase, evaporation decrease, and less interference of stones with field operations.

MATERIALS AND METHODS

Three separate data layers (bioclimatic characteristics x landcover x soil properties) were integrated to derive the pre-mapping of tillage management zones. The State Soil Geographic Database (STATSGO), the USGS/EPA Multi-Resolution Land Characterization Cover (30 m resolution and derived from Landsat TM), and USGS 1:250000 Digital Elevation Models (3-arc second) were combined to develop no-till management zones at the multi-county level of resolution. These datasets were analysed with the USDA Natural Resources Conservation Service's version of GRASS 4.13 (Geographic Resource Analysis Support System). The GRASS (rasters) datasets were produced in an Albers Equal Area Projection, with a Clarke (1866) spheroid, and 96.00 W as the standard meridian and standard parallels of 23.00 N and 45.00 N. The classification we developed is based on the length of the growing season, natural soil drainage conditions, water holding capacity of the root zone, rock fragment content, and slope (Table 1).

Table 1. Criteria used to distinguish between high and low management intensity of no-till compared to conventional tillage.

	Decreases management intensity	***Increases management intensity***
Growing degree days	*<2800*	*>2800*
Soil drainage	*Good: Most soils of mapping unit well or moderately well drained*	*Poor: More than 20% of mapping unit very poorly, poorly or somewhat poorly drained*
Water holding capacity of root zone	*<15 cm*	*>15 cm*
Slope	*More than 30% of mapping unit >8% Slope*	*Most of mapping unit <8%*
Rock fragment conten	*High: More than 30% of area loamy and sandy skeletal*	*Low: Mapping unit predominantly not skeletal*

RESULTS AND DISCUSSION

The map of no-till management intensity for Pennsylvania is reproduced in Fig. 1. The characteristics of the zones are listed in Table 2. The zones in the Southeast and Southwest of the state have more than 2800 GDD and well-drained soils (zones 1, 2, 3). These zones have low management intensity for no-till. Zone 1 has well-drained soils, low water holding capacity, steep slopes and high rock fragment content. On these soils no-till pays because of yield benefits, the need to conserve water and soil, and the interference of stones with any tillage operation. Soils of zone 2 are similar to those of zone 1 but do not have very steep slopes and therefore reduced erosion potential. Zone 3 encompasses well-drained soils with high water holding capacity on relatively level land that does not have high rock fragment content. Any tillage system performs on these soils, but no-till is preferred because of its environmental benefits. Soils of zone 4 and 5 have shorter growing seasons and it will therefore be more challenging to obtain competitive fields with no-till. However, in zone 4 moisture conservation is an important benefit of no-till and in zone 5 erosion control. High rock fragment content of these soils further reduces the management intensity of no-till in zones 4 and 5. Soils and climates of zone 6 are most management intensive for no-till because of the poor drainage 3 conditions and/or short growing season, whereas moisture conservation, erosion control and rock content are less important constraints on tillage.

Table 2. Summary of characteristics of no-till management zones.

Zone	*GDD*	*Drainage*	*Water holding capacity*	*Slope*	*Rock fragment content*
1	*>2800*	*Good*	*<15 cm*	*>8%*	*High*
2	*>2800*	*Good*	*<15 cm*	*<8%*	*High*
3	*< and > 2800*	*Good*	*>15 cm*	*<8%*	*Low*
4	*<2800*	*Good*	*<15 cm*	*<8%*	*High*
5	*<2800*	*Good*	*>15 cm*	*>8%*	*High/low*
6	*< and > 2800*	*Poor*	*>15 cm*	*<8%*	*Low*

ACKNOWLEDGEMENTS

Many thanks to Philip J Kolb and Greg Hostetter for their assistance.

REFERENCES

Blevins R L. 1984. Soil adaptability for no-tillage. In *No-tillage agriculture,* pp. 42-65. Eds R E Phillips and S H Phillips. New York: Van Nostrand Reinhold Company.

Dick W A, Van Doren Jr D M. 1985. Continuous tillage and rotation combinations effects on corn, soybean, and oat yields. *Agronomy Journal* **77**:459-465.

Dickey E C, Peterson T R, Gilley J R, Mielke L N. 1983. Yield comparisons between continuous no-till and tillage rotations. *Transactions of the American Society of Agricultural Engineers* **26**:1682-1686.

Farahani H J, Peterson G A, Westfall, D G. 1998. Dryland cropping intensification: A fundamental solution to efficient use of precipitation. *Advances in Agronomy* **64**:197-223.

Fausey N R. 1984. Drainage-tillage interaction on Clermont soil. *Transactions of the American Society of Agricultural Engineers* **27**: 403-406.

Griffith D R, Mannering J V, Box J E. 1986. Soil and moisture management with reduced tillage. In *No-tillage and surface-tillage agriculture,* pp 19-57. Eds M A Sprague and G B Triplett. New York: John Wiley and Sons.

Herbek J H, Murdock L W, Blevins R L. 1986. Tillage system and date of planting effects on yield of corn on soils with restricted drainage. *Agronomy Journal* **78**:824-826.

Hill P. 2000. Crop response to tillage systems. In *Conservation tillage systems and management*, pp. 47-60. Ed R. Reeder. Ames IA: MidWest Plan Service.

Kaspar T C, Erbach D C, Cruse R M. 1990. Corn response to seed-row residue removal. *Soil Science Society of America Journal* **54**:1112-1117.

Morrison J E. 1999. Row-zone alternative to no-till crop production. *American Society of Agriculture Engineers Paper* 99-1089.

Shipitalo M J, Edwards W M. 1998. Runoff and erosion control with conservation tillage and reduced-input practices on cropped watersheds. *Soil & Tillage Research* **46**:1-12.

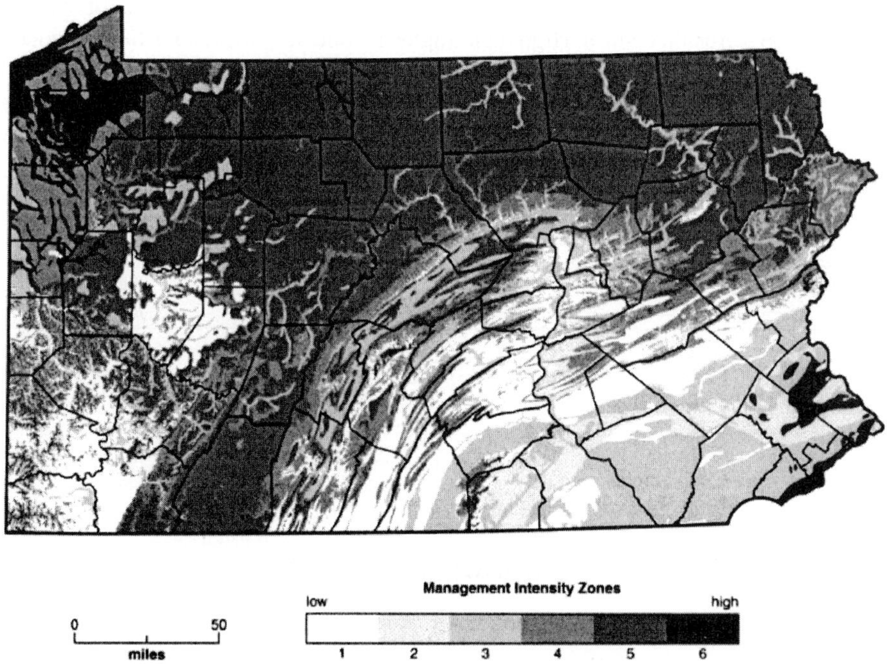

Fig. 1. No-till management intensity zones for Pennsylvania.

V. ADAPTATION OF THE AGRICULTURAL INDUSTRY TO CONSERVATION AGRICULTURE

K.E. SAXTON & J.E. MORRISON

FARM MACHINES FOR LARGE-SCALE CONSERVATION AGRICULTURE

USDA, Agricultural Research Service, Pullman, WA, 99164-6120 USA

USDA, Agricultural Research Service, Temple, Texas, 76502 USA

Conservation agriculture is a new form of farming with emphasis on preserving the soil and water resources of the farmland while maintaining sustainable crop production with increased yields. It is achieved with minimal soil disturbance and managing crop residues to protect the soil from wind and water erosion and to enhance soil quality. With the world's limited farmland resources, there is urgency to perfect and adopt this new farming system.

While conservation agriculture is a significant enhancement to modern farming, it also poses serious challenges to the mechanization of the required processes and functions. Large-scale agriculture is conducted with highly mechanized farming machines and tractors. Many of these machines were designed for optimal performance under traditional clean tilled farm fields. They often do not perform as designed when applied to conservation farming fields where tillage is minimal and crop residues are maintained over the soil surface, thus requiring modifications or replacement.

A wide variety of farming equipment has been developed or adapted for conservation farming in large scale, mechanized agriculture. Major equipment includes that used to: seed, harvest, fertilize, control weeds and pests, and manage soil conditions. The principle reason for many equipment changes has been the omission of significant soil tillage, but this is the most important change needed to conserve the farmland soil and water resources.

While conservation agriculture has required significant changes in large-farm mechanization, more innovation is yet required to achieve the perceived goals. These changes have already increased the precision of the farming processes, reduced the time and energy required per unit of farmed land, and resulted in significant improvement in the preservation of agricultural natural resources.

Key Words: Mechanization, Equipment, Soil, Water, Crop Production, Natural Resources

INTRODUCTION

Farming systems in North America have changed dramatically in scale and methods over the past 100 years developing from early small scale, animal powered units to current large scale, petroleum powered farms. In more recent years, many farmers have endorsed and adapted conservation farming methods to their large-scale agriculture. These changes and trends have come about due to economics, new machine capabilities and concern for the natural resources that sustain agriculture.

L. García Torres et al. (eds.). Conservation Agriculture, 255–262.

Conservation agriculture is a newly emerging form of farming with emphasis on preserving the soil and water resources of the farmland while maintaining sustainable crop production with increased yields. It is achieved with minimal soil disturbance and managing crop residues to protect the soil from wind and water erosion and to enhance soil quality. Numerous studies have shown that residues will significantly reduce wind erosion, (Horning et al., 1998), water erosion (Laflen et al., 1985; Norton et al., 1985;) and soil water evaporation (Pannkuk et al., 1997). With increasing world-wide food demand from diminishing farmland resources, there is an urgency to perfect and adopt this new farming system.

While conservation agriculture is a significant enhancement to modern farming, it also poses serious challenges to the mechanization of the required processes and functions. Large-scale agriculture is conducted with highly mechanized farming machines and tractors. Many of these machines were designed for optimal performance under traditional clean tilled farm fields. They often do not perform as designed when applied to conservation farming fields where tillage is minimal and crop residues are maintained over the soil surface, thus requiring modifications or replacement.

A wide variety of farming equipment has been adapted to conservation farming for large scale, mechanized agriculture. A major reason for many equipment changes has been the omission of significant soil tillage, but this is the most important change needed to conserve the farmland soil and water resources. The principal machines includes those used to perform: 1). seeding, 2). harvesting, 3). fertilizing, and 4). weed control.

The development of effective seeding machines has been one of the most challenging for conservation agriculture. Not preparing the traditional seedbed by tillage leaves the challenge of seeding into largely untilled soils with varying amounts of crop residues over the surface. Many modifications and new designs have been innovated in recent years to improve the performance of the seeding machines and simultaneously incorporate other processes such as fertilizing.

Harvesting equipment, while largely unchanged in basic principles, has had added concerns with soil compaction and residue management. Harvesting root crops with major soil disturbance has yet to be solved in most instances. Fertilizer application is often included as banding with seeding to achieve minimal soil disturbance and limited trafficking. Chemical application equipment for weed and pest control has become more precise and efficient.

HISTORICAL PERSPECTIVE

Cropland agriculture began in the Eastern regions of North America in the 1700's, and progressed westward with new lands being brought into production until the early 1900's (Morrison, 2000). During this relatively short period of time, croplands in the higher rainfall regions were significantly degraded by water erosion. By the 1930's the supply of new lands had been exhausted and the drought years of the infamous «Dust Bowl» turned the wheat land prairies into wind-blown areas of devastation. This brought about the early development of conservation tillage farming systems to maintain wheat stubble for soil surface protection from wind erosion.

Conservation tillage for the middle and eastern regions with more rainfall and mixed-crops started in the 1950's with research experiments and very limited adoption by farmers of methods to seed crops without soil-inversion plowing, or at least to reduce the number of field procedures required to till and seed a crop. In the early 1960's, more weed-control herbicides became available and researchers from universities and farm machinery manufacturers designed «minimum-tillage» and «no-tillage» seeders for both row crops and cereal grains.

The earliest farming in North America was by animal power as was the tradition throughout most of modern countries. This farming style dictated small acreage's per farmer, diversified farming with animals and crops, and significant labor input. A large percentage of the population was employed in agricultural production.

With the development of steam and later petroleum powered machines, a long term evolution began throughout the 20th century towards the large scale farming systems of today employing less than 3% of the population. Each farmer now manages several hundreds to several thousand hectares, specializes in a few crops or animals, and efficiently achieves high productivity with minimal labor inputs. This large scale, specialized agriculture requires high technology in machines, genetics, chemicals, and application skills. Thus, today's North American agriculture is highly mechanized, medium to large scale by world standards, and progressively adopting conservation agriculture for improved economics and resource protection.

CONSERVATION FARMING MACHINES

Conservation tillage farming systems, in which most soil tillage is omitted and crop residues remain on the soil surface, have been shown to be very effective as improved economical methods of crop production while also reducing wind and water erosion (Darby, 1985). To change to a conservation tillage system of farming provides the opportunity to omit many to the conventional farming operations such as plowing, disking and seedbed smoothing. Seed drilling remains as one of the main operations to be perfected because the requirements are more demanding than in conventional bare, tilled soils.

Non-tilled soils are often more firm requiring more machine cutting capability, and surface residues must be accommodated and managed. Conservation tillage is most beneficial when all or most crop residues remain on the surface to protect against wind and water erosion and for soil moisture conservation. However, these surface residues create an impairment to seeding drills which must be overcome by wide spacing of the seed openers, or cutting or brushing the residues aside from the opener track.

Seeding Machines

Precision seed placement into the soil is one of the most critical processes in crop production. Accomplishing this in non-tilled soils with surface residues is a greater challenge than with conventional farming where considerable tillage effort is expended to prepare the soil surface into a «seed bed». A serious constraint on seeder designs is the requirement by large-scale farms for wide machines with both seed and fertilizer bins capacities which are sufficient for several hours of continuous operation. Many of the original conservation tillage seeding machines were modifications of those used with conventional tillage and often were only modestly successful. Consistent soil penetration to seeding depth and avoiding blockages with surface resides were the largest problems.

Seeding machine capability for conservation tillage can be summarized in the following criteria:

1. Precisely meter and place a wide variety of crop seeds at a uniform and intended depth while maintaining the soil condition for effective seed germination and seedling emergence.
2. Seed into a soil surface which has crop residues without burying or removing this residue and leaving it in a configuration for effective soil, water and crop protection.
3. Simultaneous place fertilizer in a parallel band with the seed row for rapid and efficient plant use while minimizing its utilization by weeds.
4. Possess good mechanical durability to maintain precise seeding performance throughout a long economic life with minimal wear and maintenance.
5. Have a viable economical purchase cost relative to the machines capability, durability and life-time usage.

Precision seed placement requires all seeds to be at a uniform soil depth and soil coverage. With this depth control, seeds germinate and emerge uniformly in time, thus younger plants do not compete with older, more mature plants, thus making a very uniform crop stand. Minimal soil disturbance is preferred to conserve soil moisture in the seed zone and reduce soil ridging. The seed coverage must be firm but free of compaction and crusting to allow good seedling emergence and easy root penetration.

The machine component which engages the soil surface to provide the seed furrow and seed placement is the seeder «opener». These openers can be classified into general categories according to their method of operation and type of seed furrow produced. The two most common seeding openers around the world have been the double disk opener and the hoe or chisel openers. Single disk openers with special adaptations are a more recent development.

Double disk openers were originally developed for the tilled surface soil conditions of conventional tillage. The two disks running at a small angle to each other and close together force the soil apart in a V-shape furrow to the seeding depth. A seed drop tube is placed between the disks to allow the seed to drop to the furrow bottom. As the opener passes, the soil is expected to flow back over the seed, and as in the hoe openers, a closing press wheel often follows to firm the soil over the seed. A recent modification to the basic design of double disc openers was the staggered double disc opener, wherein one disc slightly leads the other for improved soil and residue cutting. Fertilizer placement may be with the seed, although this must be minimal to avoid germination and seedling damage. Most often, fertilizer is placed with a separate set of openers. Compaction of the furrow soil can occur in wet soils inhibiting root penetration.

Chisel and hoe drill openers have a furrow opener tool or point mounted on a narrow shank. They produce a U-shaped seed furrow by loosening and lifting a strip of soil from the surface to the seed depth. Most have a seed drop tube on the rear of the shank to deposit seed into the furrow bottom. Some soil falls back over the seed as the opener passes and usually a press wheel follows to firm the soil over the seed. Fertilizer can be simultaneously placed with hoe openers either with separate openers or in combination with the seed opener fitted with separate drop tubes. Increased soil disturbance occurs with these larger shanks, but this can be minimized by the use of covering discs to direct the loosened soil back into the open furrow to cover the seed and to leave adjacent residue uncovered.

Several single disk openers have been introduced for seed drills in recent years. One design utilizes a single disk running at a slight angle to the direction of travel to cut a seed furrow with a U-shape and drop the seed along the rear side of the coulter as it travels forward. Depth is either controlled from a rigid attachment to the drill frame or by a depth gauging wheel attached to the spring mounted disk frame. Since the furrow soil is moved to one side of the furrow as the disk passes, seed coverage requires this soil to be moved back over the open furrow and firmed by wheels following the furrow opener. These openers have been shown to not perform well on steep slopes due to down hill slipping, causing misalignment of seeder components and inadequate seed coverage and packing.

Other single disk openers have been developed which utilize a single large disk running straight ahead to cut a single vertical soil slot. One example has two adjacent side blades to cut horizontal soil slots at seeding depth. This combination of disk and side blades form an extended inverted T-slot, or more correctly a +-slot (Cross-Slot), in the soil (Saxton et al., 1991; Baker et al., 1996). Seed and fertilizer drop tubes place the seed and fertilizer separately, one on each side of the disk. The lifted soil then drops back into its original place as it clears the disk and press-depth wheels firm the soil and residue over the seed furrow.

Each seeding furrow shape has varying capabilities and may be suitable to different soil and residue conditions. Most openers perform best with friable, flowing soils and have more difficulty of seed placement and coverage in wet, adhesive clay soils. Row-crop seeders are typically equipped with either double-disc or staggered double-disc openers. Generally for drill seeders, dry soils can be seeded with U-shaped furrows by hoe and single disk openers, which allows the soil to flow back over the furrow for seed coverage. In these dry, hard soils, single and double disk openers may have more penetration and depth control difficulty.

In wet soil conditions, both hoe and double disk openers often have furrow closing and seed coverage problems. Hoe drill openers gather the wet soil and thus make ragged, uneven furrows. Double disk openers make a compacted furrow which is difficult to close. The single disk drill opener is likely the most effective for wet soils with its minimal soil disturbance and positive furrow coverage. For row-crop seeding in cold, wet soils, the practice of «strip-tilling» is being used to preplant-till very narrow strips or row-zones of soil to loosen, dry and warm it before the use of conventional double-disc row crop seeders (Morrison, 1999).

Surface residues must be managed by the seeding openers to avoid interference with the seeding operation and to leave the residues for soil and crop protection. It is desirable to leave the residue largely uniform over the surface, either standing or flattened, and close to the seeded row for crop protection. This can generally be accomplished by using cutting disks in front of the seeding openers to allow the openers free passage through the residues. For row crops in colder regions, removing the residues from over the seed row results in more rapid warming and germination. Small raking «row-cleaner» spoke-wheels running in front of the opener to move the residue aside from the opener track have been effective.

Conservation tillage drills most often use seed metering to the openers in a similar fashion as conventional tillage drills. Seed boxes with metered dispensers for each opener can be of a variety of styles depending on the seed type and seeding rate. While placing the meter boxes above the openers for gravity flow to the openers has been the most common drill configuration, a more modern method has been to use a single meter into an air stream which moves the seeds through tubes and dividers to the individual openers. These newer air seeders are particularly suited to wide drills used in large, relatively flat fields. While most popular with hoe openers, air delivery has been adapted to almost all of the opener styles. Row-crop seeders are equipped with seed singulation devices for individual rows, but these may be either individual row units or centralized with pneumatic seed delivery to individual row openers.

Harvesting Machines

Harvesting crops with conservation tillage is largely the same as with other forms of crop production with the exception of residue management. Conservation tillage is effective because of careful management and utilization of the crop residues. Leaving crop residues on the soil surface creates a new challenge to subsequent farming operations and other crop hazards such as weeds and diseases.

Reside management begins with harvest of the crop which provides the residues. Complete and uniform residue spreading by the harvester is a major requirement for both the longer straw and smaller particles of chaff. Uniform residue distribution provides uniform soil protection, return of nutrients to the soil, and subsequent machine performances. Most modern harvesting machines can be equipped with chaff and straw spreading devices as it exits the machine. These work satisfactorily except for those with quite wide cutting swaths (> 10 m) which is wider than most spreading devices can achieve leaving more residue near the center of the harvesting swath.

Partial removal or harvesting of residues is often beneficial if they are in large quantity. Complete removal or burning is not encouraged due to loss of conservation protection and nutrients for soil organic matter development. For most large scale farming, all residue remains on the soil surface after harvest, thus uniform spreading by the harvester is essential. Harvesting of root crops which most often require significant soil tillage has yet to be effectively farmed with the same degree of conservation tillage as other crops.

Fertilizer Machines

Applying fertilizers for crop production in conservation agriculture requires machine modifications from those used in conventional agriculture. First, residue covered surfaces pose a difficulty for many machines equipped with tine applicators, due to raking and blockages. Second, many of these machines do significant soil tillage which is generally not desirable for conservation tillage.

The most common fertilizer application adaptation for conservation tillage has been to include this as one function on the seeding machine. This has worked particularly well for grains and close-sown row crops. Parallel banding the fertilizer near the seed row has proven to be very effective for early seedling growth and increased crop production. Wide-row crops with large fertilizer requirements such as maize often have starter fertilizer banded with the seed row followed later by a side-dressing with a low-disturbance tine or disk applicator. This avoids large applications at seeding with the danger of seedling damage and allows a variety of fertilizer forms and composition.

Spray Machines

Applying chemicals for weed and pest control is an especially important function in conservation tillage. Pesticides for conservation agriculture are predominately available as ingredients for mixture and application as aqueous solutions with hydraulic spraying applicators. Weed control is particularly critical because there is no opportunity for tillage weed management. The use of non-selective herbicides for complete plant control before seeding is generally essential to most conservation tillage production systems.

Spray machine technology has advanced significantly to provide more precision of application rates, uniform coverage, and reduced impact by wind. Most sprayers are equipped with electronic controls and monitoring for constant applications regardless of machine speed, and many have rate adjustments possible without stopping. Improved spray nozzles and wind shields have increased spray efficiency and allowed operation with increased wind speeds. Since spraying machines are relatively small weight, except the fluid tank, they often have a wide application swath ranging up to 50 m for larger fields. Compaction by the tank wheels is a concern and generally requires large tires with minimal ground pressures.

CONCLUSIONS

Machines for large-scale conservation agriculture require several adaptations from those used in more conventional tillage agriculture. Harvesting machines require additional straw and chaff spreading devices. Chemical sprayers become an integral component for weed and pest control and as a result require high precision applications and operable over a wide range of soil and environmental conditions for timely applications. Seeding machines have required the greatest adaptations due to the technical problems associated with seeding into residue covered untilled soils. Achieving uniform seeding depth and seed coverage has been difficult to achieve while managing to avoid residue blockages and also the «tucking» of residue into the seed furrow. Seeding machines with both hoe and disk openers have been successful depending on the soil and residue conditions, but further development is still needed to achieve desired capabilities.

While conservation agriculture has required significant changes in large farm mechanization, more innovation is yet required to achieve the perceived goals. These changes have already increased the precision of the farming processes, reduced the time and energy required per unit of farmed land, and significantly improved the preservation of agricultural natural resources.

REFERENCES

Baker, C.J., Saxton, K.E. and Ritchie, W.R. (1996) The Science and Practice of No-Tillage Seeding. CAB International, Wallingford, Oxon, UK. Aug. 1996. 258 pp

Darby, G.M. (1985) Conservation tillage: An important, adaptable tool for soil and water conservation. In *Soil Erosion and Conservation*, S. A. El-Swaify, W.C. Moldenhauer and A. Lo. (Eds.), Soil Conservation Society of America, 7515 Ankeny Rd., Ankeny, IA. pp. 649-653.

Horning, L., Stetler, L. and Saxton, K.E. (1998) Surface residue and soil roughness for wind erosion protection. *Transactions of the American Society of Agricultural Engineering*. Vol. **4**(4): 1061-1065.

Laflen, J.M., Foster, G.R. and Onstad, C.A. (1985) Simulation of individual-storm soil loss for modeling the impact of soil erosion on crop productivity. In *Soil Erosion and Conservation*, S. A. El-Swaify, W. C. Moldenhauer and A. Lo. (Eds.), Soil Conservation Society of America, 7515 Ankeny Rd., Ankeny, IA. pp. 285-295.

Morrison, J.E., Jr. (1999) Row-zone alternative to no-till row crop production. Paper No. 99-1089, *Proceedings of American Society of Agricultural Engineering*, 2950 Niles Road, St. Joseph, MI 49085-9659 USA.

Morrison, J.E., Jr. (2000) Development and future of conservation tillage in America. *Proceedings of China International Conference on Dryland and Water-saving Farming*, November 21-22, 2000, Beijing, P.R. China.

Norton, L.D., Cogo, N.P. and Moldenhauer, W.C. (1985) Effectiveness of mulch in controlling Soil Erosion. In *Soil Erosion and Conservation*, S. A. El-Swaify, W.C. Moldenhauer and A. Lo. (Eds.), Soil Conservation Society of America, 7515 Ankeny Rd., Ankeny, IA. pp. 598-606.

Pannkuk, C.D., Papendick, R.I. and Saxton, K.E. (1997) Fallow management effects on soil water storage and wheat yields in the Pacific Northwest. *Agronomy Journal* **89**: 386-391.

Saxton, K.E., Baker C.J. and Choudhary, M.A. (1991) The cross-slot drill opener for seeding and fertilizing residue covered soils. *International Agric. Mechanization Conference*, Oct. 16-20, 1991, Beijing, PRC, pp. **3**: 206-3:221.

M.F. RIBEIRO

NO-TILLAGE EQUIPMENT FOR SMALL FARMS IN BRAZIL

Instituto Agronômico do Paraná, Caixa Postal 129, 84001-970 Ponta Grossa,PR – Brazil

The development of animal-drawn no-tillage equipment started in South Brazil in 1985 when IAPAR designed prototypes of a planter and knife-roller .With the acceptance of no-tillage by small-scale farmers, the private sector took over the role of designing and providing a wide range of equipment. For cover crops management, knife-rollers are used in order to bend over and crush the plants. For chemical management and weed control, chassis-mounted knapsack sprayers are available. A wide range o options have been developed for planting: from hand-jab to animal-drawn planters. The main functions performed by these are: straw cutting, seed and fertilizer metering, furrow opening, seed and fertilizer placement and furrow closing/seed compaction. Researchers and manufactures, together with farmers and extensionists, have developed and assessed the performance and suitability of equipment for different conditions. In addition to technical aspects, farmer's participatory research has been carried out, so as to establish farmers' criteria and preferences
Keywords: farm mechanisation, conservation agriculture, no-tillage, equipment

INTRODUCTION

Brazil has a well-developed farm machinery sector, with a wide range of equipment for the different operations under No-Tillage (NT) systems for human, animal and mechanical power.

The history of the development of animal-drawn NT equipment started in 1985, when the Agricultural Engineering Area of IAPAR- The Agronomic Institute of Paraná State – designed the first prototype of a NT planter Gralha Azul/IAPAR and a prototype of a knife-roller for residue management. After a series of on-farm trials aimed at the adaptation of NT for the agroecological and socioeconomical conditions of small farms, the private sector took over the role of designing new models of planters, knife-rollers and sprayers. In general, these manufacturers are small family enterprises and are located mainly at the Southern states of Santa Catarina and Rio Grande do Sul.

According to the function they perform in NT systems, equipment can be classified into 3 categories: equipment for crop residues management, planting and spraying.

L. García Torres et al. (eds.). Conservation Agriculture, 263–270.

EQUIPMENT FOR NO-TILLAGE

Crop residues management

The management of crop residues aims to make the soil surface suitable for the establishment of the succeeding crop. With regard to conservation tillage systems, this operation should provide: 1) satisfactory performance of planters; 2) satisfactory germination of seeds and crop establishment; 3) adequate amounts of cover so as to protect soil against erosion and supress weeds as long as possible.

In order to promote the allelopathic/suppressive effects of cover crops on weeds, it is necessary that the allelochemicals be released gradually until the crop attains enough growth to be able to compete with the weeds; one of the contributing factors for the speed of allelochemicals release is the level of decomposition of the biomass. (Almeida, 1991).

In NT systems, the management of crop residues can be done either mechanically or chemically, or with a combination of both.Mechanical management can be done by the use of slashers, knife-rollers or disk harrows. With the exception of slashers, mechanical management is difficult in areas with stumps or stones, or in very hilly areas.

Figure 1 shows the side view of a knife-roller. It is comprised of a roller upon wich knives are mounted transversally and a support, traction and protection structure. When pulled, the knives bend over, crush or chop off the plants. When the management is done for direct seeding the ideal situation is bending the plants off and crushing them. Excessive chopping of the biomass can result in clogging of straw in the components of the planter

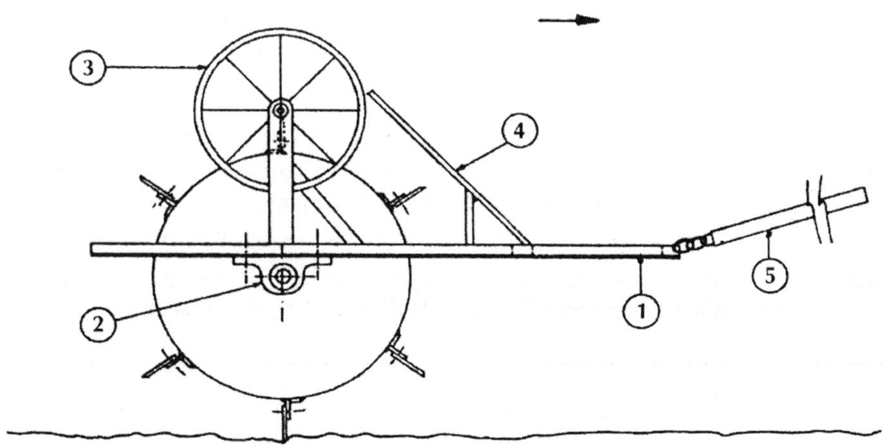

Figure1. Lateral view of the knife-roller Queixada/IAPAR. 1) chassis; 2) ball bearings; 3) transport wheel; 4) protection structure; 5) shaft.

Figure 2 shows the management of black oats with a knife-roller. The effectiveness of management is dependent upon the design characteristics of the implement, the soil and the cover crop. Araújo *et al.* (1993) has given detailed recommendations for the design of efficient knife-rollers.

In the case of bending cover plants off, with the subsequent establishment of summer crops under direct seeding, the cover crop should have a uniform growth and the management should be done in the beginning of the reproductive stage, when the seeds are not matured. It means that leguminous species should be managed at the full flowering stage and the cereals at the milking stage.

*Figure 2. Management of black oats (*Avena strigosa*) with the knife-roller.*

The knife-roller is not appropriate for all situations, but it should be promoted wherever possible, as it can be a substitute for the knockdown herbicides and can promote the suppression effect of cover crops. Table 1 shows the effectiveness of the knife-roller compared to chemical management (Ruedell, 1998). It can be observed that the use of the knife roller for black oats management at the milking stage reduced the amount of weeds for the development of the soybeans crop, compared to chemical management. It happens because of the shade effect of the mulch on the germination of weeds.

Table 1. Effect of mechanical and chemical management of black oats on the establishment of weeds in soybeans crop under NT. FUNDACEP, Cruz Alta, 1995 (Ruedell, 1995)

	Number of weeds/m^2		
	With herbicide	*Without herbicide*	*Mean*
With knife-rolling	6.3	5.0	5.7
Without knife-rolling	165.7	79.0	122.4
Mean	86.0	42.0	

Animal-drawn disk harrows can be also used and are effective for maize residue management and for some cover crops such as *Raphanus sativus* and *Pisum sativum.* Care must be taken during the adjustment of this equipment in order to avoid excessive soil disturbance.

Planters

Animal-drawn NT planters

The basic functions of a NT planter are: handling residues, seed and fertilizer metering and distribution, groove opening and placement of seed and fertilizer, seed coverage and packing. Figure 3 illustrates the main components of a NT planter.

The coulter (1) cuts the straw and drives the seed (2) and fertilizer (3) metering devices. The protection against damage of the coulter during the transport and the animal during the operation is provided by a protection structure. The hitch point mechanism (4) allows the transfer of the weight of the planter onto the coulter. This is an important aspect, since in animal traction the penetration ability by the weight of the planter is limited. The furrow opener opens a groove through the straw so as the fertilizer is placed. Two types of openers are available: the disc (5) and the tine (6), which can be chosen according to the soil and residues conditions. The double disc opener completes the residues cutting done by the coulter, opens a narrow groove and demands less power than the tine. Whereas, its utilization is constrained in heavy soils. The tine-type opens a wider groove, demands more power and has more penetration ability than the disc. Conversely, tine is more susceptible to clogging and itsperformance is constrained in stony areas. The compaction wheel (7) firms the soil around the seeds and enhances the water conductivity from the soil to the seed.

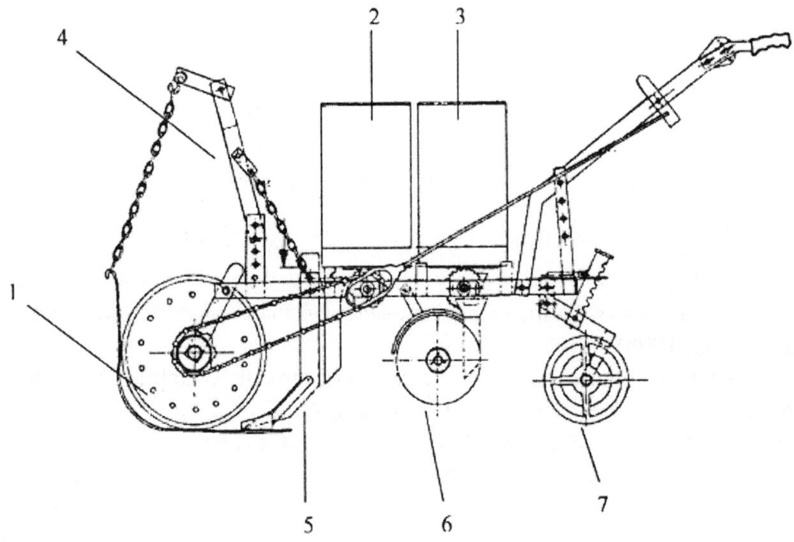

Figure 3. Side view of the prototype Gralha Azul/IAPAR animal-drawn no-tillage planter.

The effectiveness of straw cutting is dependant upon the conditions of the soil, the residues and the planter. Deficient cutting will result in clogging of straw in the seed components, which in turn will result in problems in the seed and fertilizer metering, placement and covering. The soil should provide a steady screen to the action of the coulter; otherwise, it will press the straw down into the soil, resulting in clogging in the furrow openers (Casão Jr. &Yamaoka, 1990). Clogging is also avoided if there is enough space between the coulter and the furrow opener, so as the straw can flow between them.

No-tillage animal-drawn planters should be light but be able to open a groove through undisturbed soil. This can be accomplished by positioning the hitch point – such that there is a weight transfer from the planter onto the coulter - or through the use of tine-type furrow openers. The latter provide more penetration ability the planter and more stability of the equipment during the operation.

There are two basic designs of NT planters: the first evolved from the Gralha Azul/IAPAR prototype and the second evolved from the «fuçador» plow. The first design is represented by the models Mafrense, Iadel and to some extent, Triton. The second by Werner and Fitarelli. The planters that evolved from the «fuçador plow», particularly the Werner model, are lighter, simple in adjustments, less expensive. This model is also preferred by farmers who own oxen and cultivate in stony and hilly areas. However, the options of seed plates are limited, which can result in plant numbers below or above the desired populations. The models evolved from Gralha Azul/IAPAR prototype are preferred by farmers who own pairs of horses. The models with two land wheels are suited only to flat areas. The advantage of those equipment are: more options of seed plates; good penetration ability of the furrow opener (chisel types), particularly the Triton/Ryc and Iadel models. Disadvantages are: high cost; more complex in adjustments, heavier.

Besides the one-row models, there are some manufacturers which produce multiple-rows planters, e.g.Fitarelli .

Sprayers

Sprayers are used for chemical management of cover crops, chemical control of weeds and for applying liquid fertilizers.

The knapsack sprayer is the most common equipment in small farms due to the low cost and ability to work in hilly, stony or stumpy areas. It is suitable for application rates ranging from 250 to 500 l/ha and with a field capacity ranging from 0.6 to 0.9 h/ha (Araújo *et al.*, 1998).

For chemical management of cover crops or post-emergence application, chassis-mounted knapsack sprayers are available (Figure 4). The pump is driven by land wheels and the herbicide is applied through points attached to a height-adjustable horizontal boom. As the land wheel moves, a crankshaft system linked to the wheel drives the piston to create a pressure in the tank and the flow to the spray boom. It consists of a horizontal structure member on which the nozzles are properly mounted and spaced.

Figure 4. Human-drawn chassis-mounted knapsack sprayer

The advantage of this equipment is that it reduces the risk of contamination of the operator, but even in this situation, individual protection equipment is required. Human-drawn models are available with tank capacities between 20 to 50 l and have field capacity between 0.6 and 1 h/ha (Araújo *et al.*1998). Animal-drawn models have tanks with capacities from 130 to 270 l and are more suitable for flat areas.

ROLE OF PUBLIC AND PRIVATE SECTOR IN THE DEVELOPMENT OF NT EQUIPMENT FOR SMALL-SCALE FARMERS

The Brazilian experience has shown that the most effective way for the development of equipment is a partnership between research, the private sector, extensionists and farmers. Research contributes with its technical knowledge and farmers provide their field practice. For instance, when IAPAR conducts planters evaluations, there are some parameters, which are left up to farmers to analyze through participatory evaluations: maneuverability, easiness in adjustments and operation. Quantitative parameters such as regularity of seed distribution, soil disturbance, fertilizer distribution etc. are analyzed through scientific experiments under different soil and straw conditions. This combination of the technical knowledge and scientific knowledge creates a set of criteria under which it is possible to choose – among a wide range of options – the most suitable equipment for a specific condition of soil and source of draft power.

Manufacturers have to play the role of designing and adapting equipment together with farmers, researchers and extensionists. Brazilian manufactures have been developing their equipment in consultation with farmers, and their advantage is that they can make modifications very quickly.

Even with the relatively low cost of animal drawn equipment, these are not easily affordable by small-scale farmers. Most of animal-drawn equipment has been acquired through rural development programmes. The private sector (tobacco industries) has also been financing equipment, and it is important that this support is continued to be provided to groups of farmers.

Wall (1993) discussed the issues of subsidies (or more properly, «social transfer payment») for promoting NT for small-scale farmers under the following perspective: «...the improvement of livelihood of small farmers will not be solved only with technology. The problem should be analyzed also through a social perspective, and should be solved through political and social instruments. It is not possible for small farmers to compete with large-commercial farmers under the same conditions. The second point is who benefits from soil conservation: it is not only the farmer, but whole society, through the reduction of floods, silting of rivers, maintenance of roads, water treatment and the conservation of a national patrimony represented by the land.»

ACKNOWLEDGEMENTS

The author wishes to acknowledge Prof. Ashraf Choudhary for the English review of the paper.

REFERENCES

Almeida, F.S. (1991) Controle de plantas daninhas em plantio direto. *Circular IAPAR* **67**. 34pp.

Araújo, A.G. Casão, Jr. R. and Figueiredo, P.R.A. (1993) Recomendações para dimensionamento e construção do rolo-faca. In: *Encontro Latino-Americano sobre Plantio Direto para a Pequena Propriedade, I Anais.* IAPAR, pp. 271-280.

Casão, Jr. R. and Yamaoka, R.S. (1990) Desenvolvimento de semeadora-adubadora direta a tração animal. In: *Congresso Brasileiro de Engenharia Agrícola, XIX. Anais.* **2**: 766-77.

Ruedell, J. (1995) *Plantio Direto na Região de Cruz Alta.* Convênio FUNDACEP/BASF. FUNDACEP/FECOTRIGO,134 pp.

Wall, P. (1993) Perspectivas de la siembra directa en las pequenas propriedades de América Latina. *Encontro Latinoamericano sobre Plantio Direto na Pequena Propriedade, I, Ponta Grossa, PR. Anais.* pp.13-28.

A.R. LEAKE

INTEGRATED PEST MANAGEMENT FOR CONSERVATION AGRICULTURE

The Allerton Research and Educational Trust, Loddington House Loddington, Leicestershire, LE7 9XE, U.K.

The intensification of agricultural practice has provided opportunities for certain pests and diseases to thrive in a range of crops. Soil cultivation techniques interact with different species in different ways and hence the intensity of tillage adopted (from full inversion to no-till) can result in greater or lesser antagonistic pressure dependant upon the organism concerned. Since in no situation can tillage alone provide the solution, control options rely on integrating alternative methods including cultural, biological, mechanical and appropriate chemical means. Where multiple control mechanisms are employed the interaction between tillage and crop antagonist is unlikely to cause yield loss of economic consequence.
Key words: Tillage intensity, Pest management, Disease management

INTRODUCTION

«The soil must be man's most treasured possession; so he who tends the soil wisely and with care is assuredly the foremost among men» – Sir George Stapleton.

Whatever is done to the soil, whether adding lime, fertiliser or manures, whether ploughing, cultivating or draining, the primary objective has always been to provide all the conditions required to establish and sustain a crop from seedbed to harvest (Batey, 1988). Initially these conditions were met by subsistence farmers who cultivated, manured, weeded and gathered as a manual operation. The development of horse drawn implements during the middle-ages increased the scope of farmers to produce crops as well as increasing the depth and intensity of soil cultivation. The use of manures increased crop yields and chemists such as Humphrey Davy (1778-1829) and Justus von Liebig (1803-1873) calculated the amount of carbon, nitrogen, calcium, phosphorus, sodium and potash removed from the soil in an ordinary crop rotation of hay, wheat, turnips and barley and therefore the theoretical replacements required (Owin and Whetham, 1964). At this time the nutrients used to replace off-take were generally bulky and organic and hence, in the absence of a developed transport system, only moved locally. This meant that nutrient availability, particularly nitrogen, potassium and phosphate, limited the cultivation

L. García Torres et al. (eds.). Conservation Agriculture, 271–280.

intensity and crops grown through the rotation. Hence rotations were both short and diverse and as a consequence pest, diseases and weeds were limited. The advent of inorganic phosphatic fertilisers pioneered by Lawes and Gilbert at Rothamsted (Lawes and Gilbert, 1881) and subsequently developments in the manufacture of nitrogenous fertilisers by Haber and Bosch (Laegreid, Bøckman and Kaarstad, 1999) meant that nutrient supply no longer limited output and the restorative periods of the rotation were no longer required. The same period saw unprecedented increases in mechanisation and increases in engine horse-power enabling larger areas of land to be ploughed to increasing depths. While the combination of these technologies dramatically increased output other problems began to manifest. Annual weeds were given the opportunity to grow and seed every year, increasing the seedbank and level of competition with the crop. Soil and trash borne diseases were able to infect crop plants and build up innoculum while certain pests found continuity of host. Farmers devised a range of strategies to limit the damage, initially cultural, biological and mechanically based solutions supplemented with highly toxic inorganic compounds such as copper, lead, mercury and arsenic. The development of the organochlorine insecticides in the 1930's and selective herbicides in the 1950's heralded a new era in pest management, enabling still greater intensification. However the widespread, and often indiscriminate use of crop protection products has led to environmental problems as well as resistance to the toxin by the target organism. It has become clear, in recent years, that pest control does not simply mean the complete eradication with a general poison but entails careful management of the crop to keep pests below the level of economic damage. This involves integrated control, which is the selective use of both biological and appropriate chemical methods.

Soil cultivation has long been used to control antagonists. The method, depth, timing and frequency of cultivation can provide a way of manipulating the soil environment to disadvantage pests, diseases and weeds. However conflicting demands may occur, for example open and cloddy seedbeds produce fewer weed seedlings but such conditions provide highly suitable habitat for molluscs who predate the crop seed, and can inflict total crop failure. Over a number of centuries the most suitable implement to mitigate against pests, diseases and weeds in the arable phases of the rotation was the plough, which achieves complete soil inversion, followed up by lighter implements such as tines and rollers to create a seedbed. Such an approach buried growing weeds, freshly shed weed seeds, crop residues, pathogen innoculum, eggs, larvae and adult pests, mineralised organic matter in the soil releasing nutrients to aid crop establishment, as well as providing surface conditions conductive to crop germination and growth. On heavy soils ploughing was fundamental to crop establishment while on sandy soil it was required to create structure.

The development of chemical control agents has reduced the importance of tillage to crop establishment and growth. The importance now is to understand the interactions that can occur between tillage and the antagonists which may threaten the crop so that integrated control solutions can be devised. Such solutions will rely on combinations of cultural, biological, mechanical and chemical options.

THE EFFECTS OF TILLAGE UPON PESTS

Many pest insects live or pupate in the soil, selecting a suitable depth in terms of temperature and humidity. Other pests live or shelter in crop debris or weeds before a new crop is planted. When ploughing is carried out, insects in both categories may be buried to a depth from which they cannot emerge. Others die of the temporary drought created in the upper soil layers or are exposed on the surface where they are desiccated or consumed by predators, including birds. There is also some pest mortality from simple abrasion of the soil clumps in motion, for instance where powerharrow or compacting rollers are used (van Emden and Peakall, 1996). Some cultivations perform dual roles. The heavy rotovation of land prior to potato crops has been shown to greatly reduce slug numbers, while the tilth created is finer and prevents surviving slugs from attacking the developing tubers (Leake, 1999). Hence reduced tillage has generally reduced the level of pest control achieved through mechanical means, this being particularly so for general soil pests such as cutworms, wireworms and slugs. Even specialist species such as frit-fly can cause serious problems in zero tillage situations. Where wheat was drilled into herbicide treated pasture, moving from ley to arable phase of the rotation, the emerging seedlings became so seriously attacked by mature frit-fly larvae transferring from the dying sward that the practice had to be abandoned (van Emden, 1989). Under NIT (non inversion tillage) there is often a need to reduce weed pressure by the creation of stale seedbeds which can deplete the seedbank of up to 60% of the seedlings that would otherwise appear in the crop. This technique is most effective where the inter-crop period is relatively long and at times conducive to weed germination, for instance after oilseed rape and set-aside. However the presence of bare soil during July and August is attractive to wheat bulb fly, who lay their eggs in the soil surface and whose larvae subsequently attack the emerging crop.

In temperate climates attacks by molluscs are regarded as a feature of non-inversion tillage, and direct drilling in particular. Ploughing is considered useful in burying both trash and slugs out of the crop germination zone. On heavy soils slot closure behind direct drills may be incomplete and this can act as conduit for slugs. Consolidation can be effective in reducing losses but not in all cases. Glen *et al* (1989) reported that more damage occurred on consolidated seedbeds of fine or medium tilth and least on loose seedbeds of the same texture, although this was deemed to be associated with drilling depth. In a separate trial comparing winter wheat drilled at 20mm and 40mm, 26% of seeds were lost at the shallower depth, compared to only 9% at 40mm (Glen *et al*, 1990). However, while this protects the seed, the emerging crop maybe less vigorous since more assimilate is used before the first leaf emerges and photosynthesis can begin. This can be particularly pronounced in min-till seedbeds since nitrogen mineralised from SOM (soil organic matter) is less. Hence molluscs are able to inflict severe damage through grazing the young plants and in this situation application of a molluscide post crop emergence can be highly beneficial.

Surface trash is implicated in increased mollusc numbers but may also provide protection from attack by other species. In a replicated trial Kendall *et al* (1995) found significant reductions in Barley Yellow Dwarf Virus (BYDV) infections on winter wheat in non-inversion and direct drilled plots compared to ploughing. Where straw was chopped and spread rather than baled and carted off the field, infection was less. There is evidence to suggest (A'Brook, 1968) that in high trash situations the apterous aphid vectors are unable to identify their host and consequently colonisation is reduced. There is also evidence to indicate higher numbers of beneficial predators in reduced tillage systems (Kendall *et al* 1995). Which may play a key role in depleting numbers of aphids entering the crop.

THE ROLE OF BIOLOGICAL CONTROL IN IPM IN REDUCED TILLAGE SYSTEMS

Reduced tillage often results in ecosystem diversification, sometimes including the increased activity of natural enemies. The combined biotic impact of parasitoids and predators, coupled with the reduction in populations of phytophagous insects that may accompany vegetational diversification (Root, 1973, Horn, 1981, 1988) often contributes to reducing field crop pests below the economic injury level (Pimental, 1997).

Little is known about the effects of ploughing on beneficial insects (Van Emden, 1996) although Carl, (1979) observed that while adult cereal leaf beetles (*Oulema melanopla)* disperse from the field to overwinter, their larval and pupal parasites remain in the soil and many are destroyed by ploughing in the spring.

Data from pitfall traps revealed larger numbers of polyphagous predators (*Carabidae, staphylinidae* and *linnyphiid* spiders) present in autumn in crops established using conservation tillage rather than plough based systems (Jordan, Leake and Ogilvy, 2000). However Holland *et al* (1996) observed that carabid activity differed more between farm site and between individual fields than through tillage practices. Carabid beetles may also be useful in depleting weed seed numbers. Tooley *et al* (1999) observed that in the absence of other predators carabids were able to inflict considerable levels of seed loss. Observations of other pests such as grain aphid (*Sitobion avenae*) and orange wheat blossom midge (*Sitodiplosis mosellana*) showed no influence of tillage. Other interactions can occur in the soil with micro-organisms (bacteria and fungi) and soil fauna (protozoa, nomatodes, collembolla, mites, enchytraeids, earthworms and ground beetle larvae). The abundance and function of these organisms will influence soil organic matter content, nutrient status, macro-pore structure, aggregate size, distribution and stability (Zwart and Brussard, 1990). There is evidence that no-till cereal soils have around 40% more aerobic micro-organisms and twice as many facultatively anaerobic and denitrifying bacteria as well as larger soil enzymatic activities than ploughed soils (Doran, 1980). Improved soil structure will enhance crop growth and this will create plants more able to resist pest attack, although lower mineral N levels, especially at crop emergence, maybe detrimental.

One organism which shows consistently higher populations under NIT is earthworms. Sheer biomass and numbers give earthworms an influencing as well as responsive role in soil dynamics (Brown, 1998). They are increasingly recognised as indicators of the health of agricultural soils and as important factors in soil structure development (McCredie and Parker, 1992) and maintaining and increasing the efficiency of nutrient cycling (Bugg, 1994). Earthworm populations measured at the LIFE project (Hutcheon and Iles, 1996) showed earthworms biomass to be 36% higher in conservation-tilled soils than in ploughed soil while Brown (1999) showed that greater biomass under reduced tillage compared to conventional and organic systems.

The effects of tillage upon crop diseases

Changes in tillage practice, especially fundamental changes like a move from ploughing to non-inversion tillage, might be expected to have profound effects upon crop diseases. Precisely how the effect is manifested depends on a number of factors particularly the life cycle of the disease. Factors such as how the pathogen behaves between crops, how it reaches and spreads within the crop and how it affects the crop are all important. Tillage practice will also affect the sources and prevalence of innoculum and through changes in soil nutrient levels, affect the susceptibility of the host. Straw incorporation also effects the physical chemistry and biology of the soil and crop management strategies for straw and stubble during the inter-crop period are likely to have an effect, since diseases most influenced by cultivation practices are those whose pathogens are soil or trash borne. Non-inversion tillage has been shown to have differential effects on a range of arable crop diseases. The effects are largely dependant upon innoculum source and transmission capabilities.

TILLAGE INTERACTIONS WITH SPECIFIC DISEASES

1. Trash-borne pathogens of wheat

a) Eyespot (Pseudocerosporella herpotrichoides).

Eyespot survives on stubble and will survive longer where stubble breakdown is delayed. This can occur in plough based systems where straw is buried deep in oxygen starved zones in the soil then ploughed back to the surface in the subsequent crop. Hence crop rotation can be important in helping to reduce infection by increasing the time between susceptible crops. High levels of infection in the crop trash is likely to give higher levels of infection potential but infestation levels as low as one infected culm per $10m^2$ can initiate infestation levels of economic consequence. Since sporulation begins in October and continues through to March sufficient innoculum will be available to cause an epidemic from even very low levels of infected trash, given the right conditions. Encouraging straw breakdown by early, shallow incorporation can be very effective in removing the host, as breakdown is likely to be more rapid at the surface. A number of studies, which have compared disease incidence in crops established through ploughing and non-inversion tillage, found no significant difference between establishment systems (Shultz *et al*, 1990).

b) Wheat leaf stripe (Cephalosporium graminearum)

Formerly a minor disease in the UK the increase in straw chopping over the past decade has increased the prevalence of this disease. The pathogen is capable of rapid multiplication and sporulation on straw residues on the surface and will survive in the soil on dead and decaying plant material. The hyphae enter through damaged roots, affecting the vascular system giving rise to the characteristic striping effect and eventually «whiteheads». Damaging levels of disease have been recorded following all types of tillage, from ploughing through to direct drilling. However severity is greater in direct drill and chopped straw situations and generally lesser in deep ploughed situations. Eradication of grass weeds in break crops and reducing the number of cereal crops in the rotation can all assist in reducing this disease.

c) Wheat Leaf Blotch/Glume Blotch (Septoria tritici/S. nodorum)
This is the major trash borne pathogen effecting wheat crops in the UK. While primary infection is by airborne spores arriving from previously infected trash, disease incidence is greater where chopped straw is mixed with the soil surface in the autumn compared to plough based systems. However plough based systems are not capable of burying all residues and so the scarcity of infective innoculum is short lived, and by spring time there are no significant differences between cultivation techniques.

2. Soil borne pathogens

a) Take-all (Gaumannomyces graminis)
The infectious agent enters roots directly causing dark lesions which eventually effect the entire rooting system and stem base causing death of the plant, giving rise to characteristic «white head» symptoms. The infectious agent survives as mycelium and the level of infection is dependent not only upon the level of innoculum but prevailing weather conditions, particularly warm and wet conditions at the time of sowing. Since the mycelium do not survive well in the soil, increased intercrop periods may reduce infection while ploughing infected root material out of the germination zone of the new crop can reduce initial infection over minimally cultivated seed-beds. However a series of trials in the UK comparing direct drilling with minimal tillage and ploughing showed that this differential was short lived and by mid-February these differences were less evident. Analysis of the infectivity of crop debris in the top 10cm soil declined more rapidly in non-ploughed soil indicating greater levels of biological activity facilitating breakdown.

b) Sharp Eyespot (Rhizoctonia cerealis)
It is rare for sharp eyespot to cause plant death although lesions may weaken the stem and cause lodging which exacerbates harvest and reduces quality.

The pathogen is soil borne and has a wide host range. It favours dry conditions and lighter soils and does not fare well in wet conditions. This maybe the reason why the disease is more prevalent in ploughed crops since the inversion of soil exposes a greater surface area to drying resulting in lower moisture levels in the seedbed.

c) Foot Rot/Ear Blight (Fusarium spp.)
Fusarium avenaceum, F. culmorum, F. graminearum and *F. nivale* are all implicated in the infection of winter wheat in the UK. They have various infection routes including

from seed, soil or trash. Because of this it is difficult to identify tillage effects although a long term comparative trial did indicate the advantages of non-inversion tillage over ploughing (Hutcheon and Jordan, 1996).

3. Pathogens of Oilseed rape (OSR)

a) Dark Leaf Spot (Alternaria brassicae)
This can be a severe disease of oilseed rape giving rise to significant yield loss where warm humid weather encourages the disease. Infection can occur at any growth stage of the crop. The infectious agent being both seed and trash borne. Limited comparative data of tillage indicates that methods adopted are unlikely to contribute to the disease.

b) Light Leaf Spot (Cylindrosporium concentricum)
Of similar importance to dark leaf spot this disease can also cause yield losses. Although the pathogen survives in trash there is no evidence that tillage methods have any effect on subsequent infection levels.

c) Downy Mildew (Peronospora parasitica)
Infection can cause loss of photosynthetic area and leaf drop where severe, effecting yield. Where pods become infected premature senescence may occur reducing yield. The pathogen exists in crop debris or can be carried over on OSR volunteers. As with a) or b) there is no present link to tillage practices.

d) Stem Canker (Phona lingam)
Stem Canker can effect leaves, pods and stems, although it is severe infection of the latter which causes death. Autumn infection can occur from pycnospores or ascospores released from infected plant debris. There is no evidence of tillage effects.

e) Stem rot (Sclerotinia sclerotiorum)
This can be a major cause of yield loss in OSR. The disease has a wide host range including weeds commonly found in arable crops. The infectious agent survives as sclerotia in stem cavities and these germinate to produce ascospores which are liberated by wind. These infect fallen petals which provide a nutrient source to subsequent infection, appearing as lesions on the main and lateral stems. Infected plants ripen prematurely and are visable as patches of «whiteheads» in fields.

DISCUSSION

The effect of tillage upon disease survival and infection of subsequent crops is unclear. The extent to which infectious agents are able to survive and infect depends on a host of factors, not least the specific disease, but also prevailing weather conditions. It is clear though that higher levels of trash borne disease is likely to be present with higher levels of trash but the extent of subsequent infection may not be correlated.

Tillage also effects the levels of antagonists and nutrients. Intensive tillage will mineralise nitrogen especially under warm and wet conditions. Such conditions favour rapid crop growth and higher levels of disease infection. Microbial activity also tends to be higher where more surface residues and organic matter are present and this can give rise to higher levels of organisms antagonistic to disease.

CONCLUSION

The intensity of tillage effects different pests and diseases in different ways. Farmers adopting low-till techniques may have to adapt other practices to minimise problems. This may include manipulation of sowing date or rotational changes. Disease management strategies, such as the use of disease resistant cultivars and careful attention to nutrient requirements, particularly nitrogen, can be important given the interactions with cultivation intensity. Pest management will require good knowledge of crop susceptibility within the rotation and increased monitoring of pest presence and life cycles. Beyond this the use of inputs targeted to shift the balance in-field towards the crop can be very effective. One thing is clear; it is not sufficient to continue all other practices in the same way as is carried out for ploughing. Where farmers have experienced problems with no-till and direct drill techniques it is usually associated with the failure to adopt new practices to accompany the changes.

REFERENCES

A'Brook, J. (1968) The effects of plant spacing on the numbers of aphids trapped over the groundnut crop. *Annals of Applied Biology* **61**: 289-94.

Batey, T. (1988) Soil husbandry. Aberdeen University Press.

Brown, R.L. (1999) Grass margins and earthworm activity in organic and integrated systems. *Aspects of Applied Biology* **54**, Field Margins and Buffer Zones, Management and Policy.

Brown, R.W. (1998) Earthworms. In Boarded Barns Farm Study, year 8. Rhône-Polenc Agriculture Annual Report pp 69-74.

Bugg, R.L. (1994) Earthwork update. Sustainable Agriculture newsletter Vol 6.3 (Sustainable Agriculture Research and Education Programme, University of California.)

Carl, K.P. (1913) The importance of cultural measures for the biological control of the cereal leaf beetle *Oulema melanopla. Mitt. Schwciz. Entomol.* Ges. **52**: 443.

Doran, J.W. (1980) Soil microdial and biochemical changes associated with reduced tillage. *Soil Science Society of America Journal* **44**: 765-771.

Glen, D.M., Milsom, N.F. and Wiltshire, C.W. (1989) Effects of seedbed conditions on slug numbers and damage to winter wheat in a clay soil. *Annals of Applied Biology* **115**, 177-190.

Glen, D.M., Milsom, N.F. and Wiltshire, C.W. (1990) Effects of seed depth on slug damage to winter wheat.

Holland, J.M., Drysdale, A., Hewitt, M.V. and Turley, D. (1996) The LINK IFS project – The effect of crop rotations and cropping systems on *Carabidae. Aspects of Applied Biology* **47**, Rotations and Cropping Systems.

Horn, D.J. (1981) Effect of weedy backgrounds on colonization of collards by green peach aphid (*Myzus versicae)* and its major predators. *Environmental Entomology* **10**: 285-289.

Horn, D.J. (1988) Parasitism of cabbage aphid (*Brevicoryne brassicae*) and green peach aphid (*Myzus persicae)* on collards in relation to weed management. *Environmental Entomology* **17**: 354-358.

Hutcheon, J.A. and Jordan V.W.L. (1996) Influence of cultural practice and rotation on the incidence of stem-base disease in systems comparisons 1989-1994. *Aspects of Applied Biology*, **47**: 343-350.

Hutcheon, J.A. and Iles, D.R. (1996) Influence of crop management systems and rotation on earthworm populations 1990-1994. *Aspects of Applied Biology*, **47**: 471-474.

Laegried, M, Bøckman O.C. and Kaarstad. O. (1999) Agriculture, Fertilisers and the Environment. CABI Publishing.

Jordan, V.W.L., Leake, A.R. and Ogilvy, S. (2000) Agronomic and environmental implications of soil management practices in Integrated Farming Systems. *Aspects of Applied Biology* **62**, Farming Systems for the new Millennium.

Kendall, D.A., Chin, N.E., Glen, D.M., Wiltshire, C.W., Winstone, L. and Tidboald, C. (1995) Effects of soil management on cereal pests and their natural enemies. Ecology and Integrated Farming Systems. Wiley.

Lawes, J.B. and Gilbert, J.H. (1881) On Agricultural Chemistry, j.RASE Vol 12 (i)

Leake, A.R. (1999) The results of CWS Agriculture's organic farming experiments 1989-1996. j RASE vol. 160.

Owin, C.S. and Whetham, E.H. (1964) This History of British Agriculture 1846 – 1914. David and Charles, Newton Abbot.

McCrelie, T. and Parker, L. (1992) The role of earthworms in Western Australia Agriculture. *Journal of Agriculture* **30**: 160-165.

Pimental, D. (1997) Techniques for reducing pesticide use – economic and environmental benefits. Wiley.

Root R B. (1973) Organization of a plant-arthropod association in simple and diverse habitats: The fauna of collards. *Ecological Monographs* **4**: 95-124.

Shultz, H., Boduer, L., Jorgensen, L.N. and Kristensen, K. (1990) Influence of different cultural practices on distribution and incidence of eyespot (*Pseudocerosporella herpotrichoides)* in winter wheat and rye. Tidsskr. Planteual **94**: 211-221.

Tooley, J.A., Fround-Williams, R.J., Boatman, N.D. and Holland, J.M. (1999) Weed seed predation in arable field margins by carabid beetles (*Carabidae: coleoptera*). *Aspects of Applied Biology* **54**, 1999, Field Margins and Buffer Zones: Ecology, Management and Policy.

Van Emden, H.F. and Peakall, D.B. (1996) Beyond silent spring. Chapman and Hall

Van Emden H F. (1989) Pest Control, Cambridge University Press, Cambridge.

Zwart, K.B. and Brussaard, L. (1990) Soil fauna and cereal crops in: The Ecology of Temperate Cereal Fields Proc. 32nd symposium of the British Ecological Society. Ed. Firbank L G, Carter N, Darbyshire J F, Potts G R. Blackwell Scientific Publications.

B. BASSO

PERSPECTIVES OF PRECISION AGRICULTURE IN CONSERVATION AGRICULTURE

Department of Plant and Soil Sciences
University of Basilicata, Potenza, ITALY

Conservation practices have clearly shown advantages in improving soil, water and air quality as well as reducing costs of operations. Conservation of natural resources is the base for a long-term sustainability of agricultural and natural ecosystems. Conservation Agriculture is not just conservation tillage, but a series of land management practices that include crop residues retention, cover crops, appropriate cropping system rotation, integrated pest management able to minimize land degradation. Because of the complexity associated with natural and agricultural ecosystems, a land management practice found to be sustainable at one site might not be equally sustainable at another site. Agricultural production systems are inherently variable due to spatial variation in soil properties, topography, and climate. To achieve the ultimate goal of sustainable cropping systems, variability must be considered both in space and time because the factors influencing crop yield have different spatial and temporal behavior. Advances in technologies such as Global Positioning Systems (GPS), Geographic Information Systems (GIS), remote sensing and simulation modeling have created the possibility to assess the spatial and temporal variability present in the field and manage it with appropriate site-specific practices. Such approach is commonly called Precision Agriculture or site-specific crop and soil management. The objective of this paper is to evaluate the potential use of Precision Agriculture principles and technology for Conservation Agriculture. Examples of Precision Agriculture application through the integration of various techniques are presented to show the potential benefits of site-specific management of natural resources. Further perspectives are also discussed to link Precision Agriculture to Conservation Agriculture for a mosaic agriculture.
Key words: Natural resources conservation, yield variability, simulation models, mosaic agriculture.

BACKGROUND AND RATIONALE

Most demographers agree that almost one billion people of the world's population today are chronically malnourished. In the developing nations poverty presents increasing challenges; it occurs in more advanced countries as well. As the global population increases, the demand for food expands, with a growing diversity in diets. As a result of population pressure, the world's finite resources are taxed to the limits by the same people whose existence depends on those. This degradation has resulted mainly from soil loss by erosion and from chemical and physical deterioration caused by overgrazing, deforestation, and inappropriate agricultural practices. Many of the same activities that have degraded soil resources have diminished the

L. García Torres et al. (eds.). Conservation Agriculture, 281–288.

quality and availability of freshwater resources as well. Clearly, the long-term productivity of soil and water resources is in jeopardy. According to Rosensweigh and Hillel (1998) almost five billion hectares of land globally have been degraded in the past half-century. The erosion of the natural resource base must be halted and then reversed. If the environmental degradation is to be curtailed, and if the food demands of a growing human population are to be met, agricultural land management practices that sustain and enhance the long-term productivity of the natural resource base must be implemented.

The goal of agricultural research is to improve agricultural productivity and sustainability by increasing crop yields and by using inputs (water, fertilizer, labor, and farm machinery) more efficiently and at less cost to the farmer and the environment. Successful implementation of these practices, referred to as sustainable land management or Conservation Agriculture, will require quantitative evaluation of the factors that determine whether an agricultural system is sustainable or unsustainable. Only by identifying and measuring these factors will it become possible to evaluate the long-term performance of a given management practice. This, however, is not an easy task. The issue of what constitutes sustainable land management is complex and transcends concerns of a physical-chemical-biological nature to include socio-economic, cultural, and political concerns.

Conservation Agriculture refers to a series of land management practices that minimize land degradation and increase soil, water and air conservation. Direct drilling (no tillage), crop residues retention, cover crops, appropriate cropping system rotation, integrated pest management are examples of Conservation Agriculture practices. Such practices and techniques allow for soil protection, improvement of soil structure and soil fertility through soil organic matter increase and aggregate stability, increase in soil water infiltration and reduction of runoff, and carbon sequestration with reduction of green houses gases emission. Because of the complexity associated with natural and agricultural ecosystems, a land management practice found to be sustainable at one site might not be equally sustainable at another site. Agricultural production systems are inherently variable due to spatial variation in soil properties, topography, and climate. To achieve the ultimate goal of sustainable cropping systems, variability must be considered both in space and time because the factors influencing crop yield have different spatial and temporal behavior (Pierce and Nowak, 1999). Advances in technologies such as Differential Global Positioning Systems (DGPS), Geographic Information Systems (GIS) and remote sensing have created the possibility to assess the spatial variability present in the field and manage it with appropriate site-specific practices. Such approach is commonly called Precision Agriculture or site-specific crop and soil management.

The objective of this paper is to evaluate the potential use of Precision Agriculture principles and technology for Conservation Agriculture.

ASPECTS OF PRECISION AGRICULTURE

Precision Agriculture can be defined as the application of technologies and principles to manage spatial and temporal variability associated with all aspects of agricultural production for the purpose of improving crop performance and environmental quality. (Pierce and Nowak, 1999). The term means different things to different people. To some, it means using satellite technology, sensors, and colorful maps to do the job that their fathers did without any technology. To others, it is a glimpse of the future of agriculture where inputs are applied on a site-specific basis to increase profits and environmental quality. The concept of treating small areas of a field as separate management units is not new. Ancient agriculture was based on site-specific management practices by hand applications, but with today's large-scale application, obviously, this is not possible. Current agricultural technology has reached a level that allows a farmer to measure, analyze and deal within-field variability that was known to exist previously but was not manageable.

Precision Agriculture gives farmers the opportunity to increase productivity and reduce the undesirable effect of pollution caused by agriculture. Several are the factors affecting the spatial varibility of production, like topography, soil available water, drainage, depth of the root system, nutrient availability, soil texture, soil organic matter, pH, plant population, pest infestation, weeds, that can act indipendently or as most of the cases simultaneously. Because these properties are spatially variable, the assumption often made in agricultural research of uniform conditions across the field is not a reality. If fields were uniform, there would be no need for Precision Agriculture. Without variability, the concept of Precision Agriculture has little meaning and would never have evolved. Fields usually contain a complex arrangement of soils and landscapes thus spatial variability in soil properties and crop productivity is the norm rather than the exception (Mulla and Schepers, 1997). A successful management system must identify each of these potential limiting processes, and determine which key properties to measure for proper characterization of each limiting process.

Currently most of the management practices ignore variability applying inputs or practices uniformily across the field. Such approach will obviously cause overapplication in some areas and underapplications in others. The site specificity of Precision Agriculture is intuitively appealing and represents a means of improving the economic and environmental performance of cropping system.

Precision Agriculture encompasses a broad array of topics ranging from variability of the soil resource base, weather, plant genetics, crop diversity, machinery performance, most physical, chemical, biological inputs used in crop production, and socio-economic aspects. A successful Precision Agriculture system depends on how well it can be applied to manage spatial and temporal variability in crop production and what benefits could bring. Before this, though, it is fundamental to have a clear understanding of the variability, which is not an easy task. The key factors for a successful implementation of Precision Agriculture techniques are: (i) the extent to which conditions within a field are known and manageable; (ii) the adequacy of inputs recommendations, (iii) the degree of application control (Pierce and Nowak, 1999).

Precision Agriculture consists of georeferenced data collection; data analysis; decision making and variable rate treatments. Precision Agriculture relies on the collection of geo-referenced environmental data to provide relevant information for use in management planning. Direct field measurement and remote sensing are the two most common forms of data collection. Although direct measurement techniques are very accurate, their cost and intense labor requirements may restrict the number of sampling. In an effort to reduce the cost of obtaining spatially referenced data and at the same time increase data resolution, farmers are turning to continuous measurement techniques and remote sensing for more accurate within-field variability information. The development of continuous yield-sensor and DGPS has been perhaps the most important and influential development in Precision Agriculture data collection. Yield rates vary spatially and maps produced by the yield monitors systems are evidence of the degree of within-field variability. The magnitude of this variability is a good indication of the suitability of implementing a spatially variable management plan. Applying variable rates fertlizers to improve crop performance and environmental quality is but only one example of a Precision Agriculture approach. Indeed, Precision Agriculture involves decisions on tillage, species and cultivars selection, irrigation, drainage.

Moran et al. (1997) classified the information required for site specific management in information on seasonally stable conditions, information on seasonally variable conditions, and information to find the causes for yield spatial variability and to develop a management strategy. The first class of information includes conditions that do not vary during the season (soil texuture, pH) and only need to be determined at the beginning of the season. Seasonally variable conditions, instead, are those that are dynamic within the season (soil moisture, weeds or insect infestation, crop diseases) and thus need to be monitored throughout the entire season for proper management. The third category is comprehensive of the previous two to determine the causes of the variability. Site-specific management (SSM) strategies may be able to optimize production, but their potential benefits are highly dependent on the accuracy of the assessment of such variability (Pierce and Nowak, 1999).

Climatic variability is not less important, and may often be even more important than spatial variability. Yield maps may differ significantly from year to year, and yield patterns within a field also change annually. Accounting for yield variability is almost impossible without indication of when, where and why stress developed in the field. Traditional analytical techniques, such as regression of static measurements against yield, have failed to explain the reasons for yield variability because the dynamic, thus temporal, interactions of stresses on crop growth and development cannot be accounted for (Jones and Ritchie, 1991; Sudduth et al., 1996). Process oriented crop simulation models, such as the CERES and CROPGRO models (Ritchie et al., 1985; Boote et al., 1998), integrate the effects of temporal and multiple stress interactions on crop growth processes under different environmental and management conditions. Crop models can be used for understanding yield variability in both space and time, helping farmers in making better decisions on their management through long-term simulations analysis (Sadler and Russell, 1997; Batchelor et al., 1998; Cora et al., 1999; Basso et al., 2001).

Recent advances on the resolution and availability of remote sensing imagery, coupled with a decrease in its associated costs, have allowed the collection of timely information on soil and crop variability by examining spatial and temporal patterns of vegetation indices (Blackmer and White, 1996). Such information can be used to derive inputs for crop models in a GIS environment (Moran et al., 1997; Basso et al., 2001). Basso et al. (2001) carried out a study showing the utility of a reclassified NDVI map to identify spatial patterns, to select uniform zones across the field for target sampling and to execute the models on those areas to assess causes of yield variability (Figures 1 and 2).

Figure 1. Reclassified NDVI image showing three different zones (black, gray and white).

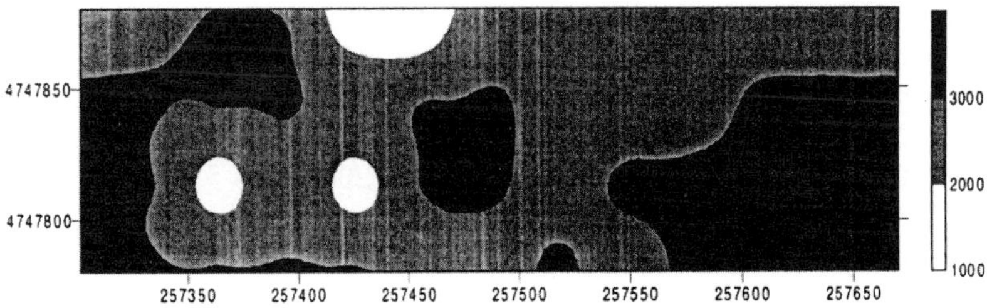

Figure 2. Kriged map of simulated yield for each NDVI-Class.

LINKING PRECISION AGRICULTURE WITH CONSERVATION AGRICULTURE

Although Precision Agriculture is highly dependent on technology in developed countries, its theoretical principles are independent of location and scale. The limitation of a Precision Agriculture approach in developing countries is the difficulty in obtaining spatial information at low cost and without technology. The first step to be able to link Precision Agriculture principles with Conservation Agriculture is the assessment of yield variability and environmental risks across landscapes. Crop simulation models and digital terrain modeling can play a major role in identifying best management practices for natural resources conservation. Crop simulation models require data on soil, weather, genetics and management. Several of these inputs (solar radiation, soil water limits, genetic coefficients) are not available in developing countries but scientific progress has been made to develop functions, equations and models able to generate most of the information required to execute the models.

Long-term simulation with stochastic dominance analysis allows the identification of the most appropriate management practice for natural resource conservation. (Braga et al., 1999).

Conservation Agriculture practices such as reduced tillage and crop residue management can be varied over space only if conditions within the field are known. Varying tillage depth can alleviate soil compaction where it is indicated. Residues level can be varied based on soil characteristics to optimize soil temperature and water content for crop growth in conservation tillage. Different cultivars and seeding rate can also be varied based on soil characteristics planting more seeds on soils that can sustain the higher plant populations. Long term simulation analysis is the key for making these kind of decisions in areas where the availability of sensors on-the-go is not available.

Spatial variation in soil moisture is often the cause of crop yield spatial variability due to its influence on the uniformity of the plant stand at emergence and for in-season water stress. Soil physical properties are highly influenced by terrain characteristics and landscape position. The likelihood of soils becoming saturated increases at the base of slopes and in depressions where there is a convergence of both surface and subsurface flow. The automation of terrain analysis and the use of Digital Elevation Models (DEMs) has made it possible to easily quantify the topographic attributes of the landscape and to use topography as one of the major driving variables for many hydrological models. Figures 3 and 4 depict how topography influences spatial variability of soil water balance and the net surface water flow (runon-runoff). The results were obtained by executing Salus-Terrae, a digital terrain model with a spatial soil water balance (Basso et al., 2000) on a 5 ha field.

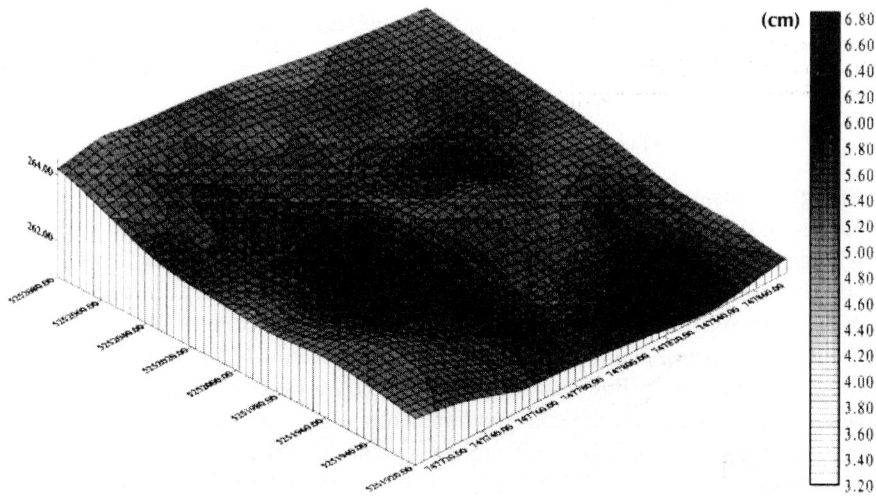

Figure 3. Soil water content map.

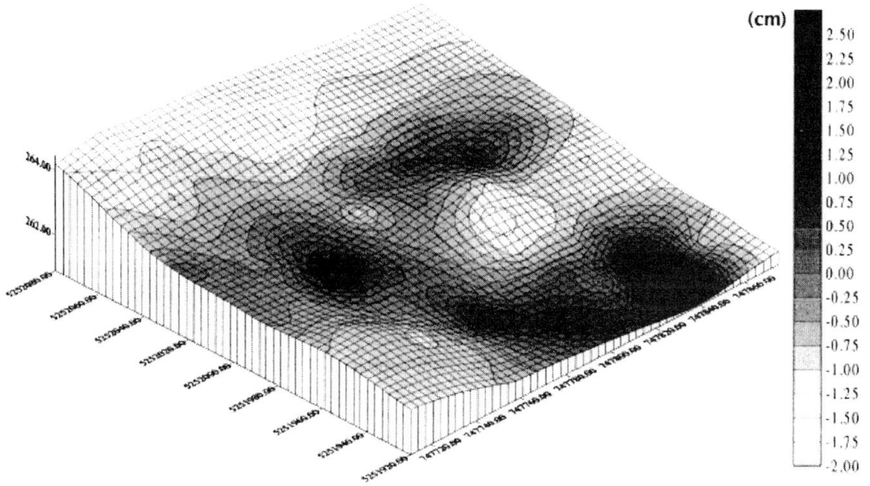

Figure 4. Net surface water flow (runon-runoff) from Salus-Terrae.

The zones of water accumulation can be managed differently (i.e. grass waterways, buffer zones, trees, etc). Adopting the best land use for each of the areas we can assist to site-specific Conservation Agriculture or «*mosaic agriculture*» for larger scale. In agricultural terms, the mosaic is obtained by «paving» landscapes with different crops or natural ecosystems according to the most appropriate land management practices for that specific piece of land or area of the field. Upscaling simulation results obtained from model runs on small areas with higher availability and quality model inputs to areas that have similar characteristics (i.e. landscape positions, drianage characteristics, soil fertility) can be the best approach for selecting the right piece of mosaic.

CONCLUSIONS AND FUTURE PERSPECTIVES

The goals of Precision Agriculture and Conservation Agriculture are basically the same: conserving natural resources, maintaining productivity and decreasing costs. Precision Agriculture is a technology driven Conservation Agriculture but the contest and scale of application are different. The principles of Precision Agriculture of managing each parcel of agricultural land for maximum crop production while protecting the environment can logically and theoretically be applied also in developing countries where technologies are limited. Previous knowledge of farmers, coupled with hand labor and local georeferencing can be used as a technology surrogate to apply the principles of site specific management.

Greater help and support should come from International development agencies that often have single disciplines projects with limited utility to the solution of realistic problems. Interdisciplinary approach integrated with new satellite technology available to the agencies, and simulation models for crops and landscape analysis are able to account for all facets of the management systems assuring much higher chances to succeed in achieving the goal of a conserving natural resources while maintaining profitability.

REFERENCES

Basso, B., J.T. Ritchie, F.J. Pierce, R.P. Braga, J.W. Jones. 2001. Spatial validation of crop models for Precision Agriculture. In: *Agricultural System* ***68*** (2001) 97-112.

Basso, B., Ritchie, J.T. and Gallant, J.C. (2000) Modeling Surface and Subsurface Water Flow in a Spatially Variable Terrain. In: *Precision Agriculture*. Proceeding of 5th International Conference on Precision Agriculture, Bloominghton, MN.USA.

Batchelor, W.D., and Paz (1998) Process-Oriented Crop Growth Models as a Tool to Evaluate Spatial Yield Variability. In: Proceedings of the First International Conference on Geospatial Information in Agriculture and Forestry, Vol. 1. pp. I198-205.

Boote, K.J., Jones, J.W., Hoogenboom, G., and Pickering, N.B. (1998) CROPGRO model for grain legumes. In: G.Y. Tsuji, G. Hoogenboom, and P.K. Thornton (eds.), Understanding Options for Agricultural Production, pp. 99-128. Kluwer Academic Publishers, Dordrecht, Netherlands.

Braga, R.P., Jones, J.W., and Basso, B. (1999) Weather Variability in Site-Specific Management Profitability: A Case Study. In: P.C. Robert, R.H. Rust, and W.E. Larson (eds.), Precision Agriculture, pp. 1853-1863. ASA-CSSA-SSSA, Madison, Wisconsin, USA.

Cora, J.E., Pierce, F.J., Basso, B., Ritchie, J.T. (1999) Simulation of within-field variability of corn yield with Ceres-Maize model. In: P.C. Robert, R.H. Rust, and W.E. Larson (eds.), Precision Agriculture, pp. 1309-1319. ASA-CSSA-SSSA, Madison, Wisconsin, USA.

Jones, J.W., and Ritchie, J.T. (1991) Crop growth models. In: G.J. Hoffman, T.A. Howell, and K.H. Soloman (eds.), Management of farm irrigation systems. ASAE, St. Joseph, MI.

Moran, M.S., Inoue, Y., and Barnes, E.M. (1997) Opportunities and Limitations for Image-Based Remote Sensing in Precision Crop Management. *Remote Sensing of the Environment* Vol. **61**: 319-346.

Mulla, D.J., and Schepers, J.S. (1997) Key Processess and Properties for Site-Specific Soil and Crop Management. In: F.J. Pierce and E.J. Sadler. (eds.), The State of Site Specific Management for Agriculture. ASA Misc. Publ., ASA, CSSA, and SSSA, Madison, WI.

Pierce, F.J., and Nowak, P. (1999) Aspects of Precision Agriculture. *Advances in Agronomy*, **67**: 1-86.

Ritchie, J.T. (1985) A user-oriented model of the soil water balance in wheat. In: W. Day and R.K. Atkin (eds.), Wheat growth and modeling, pp. 293-305. Series A: *Life Sciences* Vol. **86**. Plenum Press, NY.

Rosensweigh, C., and Hillel, D. (1998) Climate Change and the Global Harvest: Potential Impacts of the Greenhouse Effect on Agriculture. (Cary, N.C.: Oxford University Press).

Sadler, E.J., and Russell G. (1997) Modeling Crop Yield for Site Specific Management. p. 69-80. In: F.J. Pierce and E.J. Sadler. (eds.), The State of Site Specific Management for Agriculture. ASA Misc. Publ., ASA, CSSA, and SSSA, Madison, WI.

Sudduth, K.A., Drummond, S.T., Birrell, S.J., and Kitchen, N.R. (1996) Analysis of spatial factors influencing crop yield. In: P.C. Robert, R.H. Rust, and W.E. Larson (eds.), Proceedings of the Third International Conference on Precision Agriculture, pp. 129-140. ASA, CSSA, SSSA. Madison, WI.

VI. INFLUENCE OF CONSERVATION AGRICULTURE ON ENVIRONMENT

D.C. REICOSKY

TILLAGE-INDUCED CO_2 EMISSIONS AND CARBON SEQUESTRATION: EFFECT OF SECONDARY TILLAGE AND COMPACTION

USDA-Agricultural Research Service, North Central Soil Conservation Research Lab,
803 Iowa Ave. Morris, MN 56267

Long-term data shows the effect of intensive tillage on soil carbon loss and suggests the need for alternative management strategies. Conservation agriculture with improved tillage methods can aid in carbon sequestration. This work demonstrated that secondary tillage methods decreased the CO_2 loss immediately following the moldboard plow. Both the disk harrow and the field cultivator reduced the CO_2 flux immediately following the moldboard plow operation by 40 to 50 %. Major soil reconsolidation with one pass of a road packer caused an abrupt decrease in the CO_2 flux after primary tillage by moldboard plow, chisel plow, subsoiler, and paraplow. Further small decreases in CO_2 fluxes were noted with four passes of the packer. The abrupt decreases in gas exchange were related to increases in soil bulk density following the compaction. These results demonstrate the importance of soil physical properties before and after tillage controlling gas fluxes and soil carbon loss.
Key Words: plow tillage, gas exchange, bulk density, carbon loss

INTRODUCTION

Tillage or soil preparation has been an integral part of traditional agricultural production. Tillage is a principle agent resulting in soil perturbation and subsequent modification of the soil structure primarily with soil degradation. Agricultural soils, especially tillage-induced CO_2 losses, play an important role in C sequestration or storage and thus can help mitigate global warming (Lal, Kimball, Follett & Cole, 1998). The moldboard plow has been the symbol of U.S. agriculture over the last 150 years. Intensive tillage in the U.S. has mineralized or oxidized between 30 and 50 % of the native soil C or soil organic matter since the pioneers brought the soils into cultivation. Intensive tillage causes soil carbon (C) loss and tillage-induced greenhouse gas emissions that impact environmental quality. Soil C dynamics can have an indirect affect on climate change through net absorption or release of CO_2 from soil to the atmosphere in the natural C cycle. Minimizing agriculture's impact on the global

L. García Torres et al. (eds.). Conservation Agriculture, 291–300.

increase of CO_2 requires that we modify tillage methods to sequester and maintain high soil C levels.

Recent studies involving a dynamic chamber, various tillage methods and associated incorporation of residue in the field indicated major C losses immediately following tillage (Reicosky & Lindstrom, 1993) with substantial variation across the landscape (Reicosky, 1995). The short-term impact of moldboard plow and various tillage methods on CO_2 loss from the soil was measured using a portable dynamic chamber designed to measure canopy photosynthesis and mounted on the front end of a 4-wheel drive forklift for portability. Reicosky & Lindstrom (1993) found that the moldboard plow had the roughest soil surface, the highest initial CO_2 flux and maintained the highest flux throughout the 19-day study. High initial CO_2 fluxes were more related to the depth of soil disturbance that resulted in a rougher surface and larger voids than to residue incorporation. Lower CO_2 fluxes were caused by tillage associated with low soil disturbance and small voids with no-till having the least amount of CO_2 loss during 19 days. Ellert & Janzen (1999) and Rochette & Angers (1999) found similar results for different soils and less intensive tillage methods. They concluded that short-term influence on tillage and soil C loss was small under semi-arid conditions in agreement with Roberts & Chan (1990) and Franzluebers, Hons & Zueberer (1995a,b). On the other hand, Reicosky & Lindstrom (1993) concluded that intensive tillage methods, especially moldboard plowing to 0.25 m deep, affected this initial soil flux differently and suggest improved soil management techniques can minimize agricultural impact on global CO_2 increase. This interaction of soil and residue mixing enhances aerobic microbial decomposition of the incorporated residue to decrease soil organic C (Reicosky et al., 1995).

Reicosky (1997) reported that average short-term C loss from four conservation tillage tools was 31 % of the CO_2 from the moldboard plow. The moldboard plow lost 13.8 times more CO_2 as the soil not tilled while conservation tillage tools averaged about 4.3 times more CO_2 loss. The smaller CO_2 loss from conservation tillage tools was significant and suggests progress in equipment development for enhanced soil C management. The soil physical properties were not quantified, but suggest conservation tillage tools reduce the extent, frequency and magnitude of mechanical disturbance caused by the moldboard plow and reduces the large air-filled soil pores to slow the rate of gas exchange and C oxidation.

Reicosky (2000) evaluated short- and long-term effects of tillage-induced, soil property changes on CO_2 loss. Soil dynamics changing physical properties has a direct impact on soil gas dynamics in tilled systems. Soil tillage changes soil porosity, bulk density, air permeability, void ratio, and the pore size distribution. The exact mechanisms for impact of these parameters to CO_2 loss from the soil are not completely understood. Thus, the general objective of this work is to better understand the soil dynamics and gas exchange through tillage and re-compaction as it effects C loss. One objective was to evaluate the tillage-induced CO_2 flux where the moldboard plow was followed by secondary tillage using either a disk harrow or field cultivator. The second objective was to evaluate the role of severe re-compaction on the tillage-induced CO_2 flux of soil loosened in primary tillage the moldboard plow, chisel plow, paraplow and subsoil unit.

MATERIALS AND METHODS

The experiments were conducted at the USDA-Agricultural Research Service, Swan Lake Research Farm in West Central Minnesota, USA (45° 41' 14" N Lat. and 95° 47' 57"W Long.) on soils high in soil organic C (Reicosky & Lindstrom, 1993, 1995; and Reicosky, 1997, 1998). Similar soils were used in both studies and will all be described once. The soil selected was a relatively uniform Barnes loam (fine, loamy, mixed, Udic Haploborolls) formed on a glacial till under tall grass prairie vegetation. The surface horizon is generally very dark with relatively high organic C (typically 2 to 3 g C kg^{-1}) and developed over subsoil high in calcium carbonate. The preceding crop was spring wheat (*Triticum aestivum* L. cv. Marshall) in all these studies. Both plowed and adjacent not-tilled plots had surface residue from the previously harvested wheat crop. To minimize weed and volunteer wheat effects on the CO_2 exchange rate, the entire field was sprayed with RoundUp Ultra (glyphosate) herbicide at the rate of 0.8 kg ai ha^{-1} as needed.

The CO_2 flux from the tilled surfaces in these studies was measured using a large, portable chamber described by Reicosky & Lindstrom (1993) and Reicosky (1997 and 1998). Measurements of CO_2 flux were generally initiated within one minute after the tillage pass and continued for various times. The CO_2 flux from the soil surface was measured using this large, portable chamber described by Reicosky & Lindstrom (1993). Briefly, the chamber with the mixing fans running was placed over the tilled surface or the no-tilled surface, the chamber lowered and data collected for one-second intervals for a total of 60 sec to determine the rate of CO_2 and water vapor increase inside the chamber. The chamber was then raised, calculations completed and the results stored on computer diskette. After the appropriate lag and mixing times, data for a 30-sec calculation window was selected to convert the volume concentrations of water vapor and CO_2 to a mass basis, then regressed as a function of time using linear and quadratic equations to estimate the gas fluxes. These fluxes represent the rate of CO_2 and water vapor increase within the chamber from a unit horizontal land area as differentiated from soil surface basis caused by differences in soil roughness. Only treatment differences with respect to tillage methods, tillage type or experimental objectives will be described with the results.

Secondary tillage effects

This secondary tillage implement experiment was conducted in August 1997 following wheat harvest to evaluate effects of secondary tillage on CO_2 loss. Standard operation was to moldboard plow (MP) using a standard six-bottom, 46-cm plow to 25 cm deep (a 2.7 m wide strip for a gas exchange measurements). Within 1 min after the pass with the moldboard plow, portable chamber measurements were initiated. This required between 4 and 5 min for four replicated measurements on the same spot. Within 1 min after the last gas exchange measurements, a secondary tillage implement (disc harrow (DH) or field cultivator (FC)) was pulled over the same area and the CO_2 flux measurements repeated four times. A tandem disk harrow was used as a secondary tillage implement. A tandem disc harrow has two opposing gangs that throw the soil outward from the center of the implement followed by two gangs that throw soil back toward the center. Thus the disk harrow tills the soil twice leaving the soil nearly level. The depth of disk harrow operation in the plowed soil ranged from 10 to 12 cm in this work. The field cultivator is similar to a chisel plow, but is lighter in construction and designed for less severe conditions. The field cultivator used in this work had four ranks of equally spaced flexible shanks. The shanks were spaced 60 cm apart providing an effective spacing of 15 cm. The points were 5-cm wide. The field cultivator was operated at a depth that ranged from 7 to 10 cm. This process of primary tillage, measurement of gas exchange, followed by secondary tillage and measurement of gas exchange was repeated for the different treatments within experimental design at approximate hourly intervals out to six h after the initial tillage.

Compaction study

Loosening of the soil enables gas exchange with the atmosphere. This work evaluated the role of re-compaction immediately after tillage on short-term tillage-induced CO_2 loss. The moldboard plow (MP) was a standard six-bottom 46-cm John Deere‡ model 2800 plow to 25 cm deep was pulled at about 8 km h^{-1}. One pass provided sufficient width for chamber measurements. The moldboard plow lifts and fractures the soil with complete inversion and incorporation of crop residue that left the soil in a loose, friable condition. The chisel plow (CP) is a primary tillage implement with curved shanks that breaks or shatters the soil, leaving the soil surface rough and with crop residue on or near the surface. The chisel plow using this study had three gangs of soil engaging tools. The effective operating depth in this study was 15 cm with the effective shank spacing of 30 cm staggered on each gang. The soil engaging components consisted of five cm wide reversible point spikes. The Howard paraplow (PP) was a model 410B with four shanks on 53-cm centers also to 25 cm deep. The paraplow uses slant legged shanks that lift and fractures the soil similar to the moldboard plow, only without soil inversion. Virtually all the soil between the shanks was loosened and the soil surface elevated 10 to 15 cm as a result of the tillage operation. The subsoil operation was done with four straight shanks of a modified twin-beam Yetter subsoil (SS) unit with shanks on 76-cm row spacing to 35 cm deep. The soil was disturbed on the row about 10 cm high at the peak with 20 to 30 cm-wide depression area between each row. The results would suggest that the degree of soil shattering with the narrow subsoil shank was not as extensive as with the moldboard plow or the paraplow.

Soil compaction after tillage was done with a Bross road packer (inset fig 4) that filled with rocks and gravel weighed 8.2 tonnes and pulled at about 6 km h^{-1}. The weight was distributed on two axles where the smooth rubber wheels were adjacent across the full width of the packer. The compaction resulted in a smooth surface with only a few loose soil particles between closely spaced tires. The bulk density in the surface layer was measured using 5 cm cores following compaction of the moldboard plow, chisel plow and no-till plots only. Bulk density measurements were not taken on the freshly tilled soil, but were estimated to be about 0.9 Mg m^{-3} .

The sequence of events was to do primary tillage (moldboard plow, paraplow, or subsoil unit) and immediately make gas exchange measurements. Upon completion of the initial gas exchange measurements, the road packer was pulled over the tilled area either one or four times based on the specific treatment. Within 30 sec of the last compaction pass, another set of gas exchange measurements was completed on the compacted area. A selected group of plots were compacted 1 h after the initial primary tillage with the paraplow and subsoil unit to investigate the dynamics of the compaction process. These gas exchange measurements were then compared with a no till area with and without the same amount of compaction.

RESULTS

Secondary tillage effects

The effect of secondary tillage method on the CO_2 flux as a function of time after tillage is summarized in fig 1. In each case, the secondary tillage operations caused a decrease in the CO_2 flux. While there is some scatter in the data, the decrease in the CO_2 flux over that of the moldboard plow shows that secondary tillage can be beneficial in reducing the CO_2 loss immediately following a primary tillage. Noteworthy is the very low flux of CO_2 from the no till plot.

The short-term response is better seen in the fluxes immediately following the secondary tillage operations shown in fig 2. The moldboard plow without secondary tillage had the largest flux of 46 g CO_2 m^{-2} h^{-1}. The disk harrow decreased the CO_2 flux the most and had an initial flux of 20 g CO_2 m^{-2} h^{-1}. The field cultivator had a similar value. The combination secondary operations with the disk harrow followed by the field cultivator resulted in slightly more CO_2 loss than either the field cultivator or the disk harrow alone. All of the CO_2 fluxes with the secondary tillage were still substantially higher than that from the no till surface. The cumulative losses for 24 h

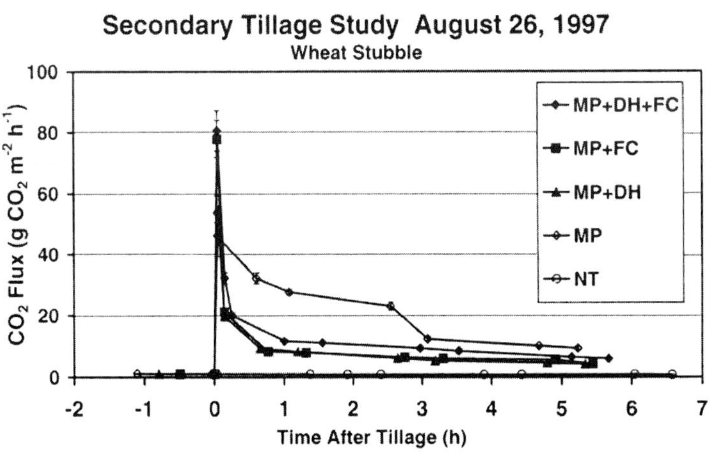

Figure 1. Carbon dioxide loss after plowing followed by secondary tillage.

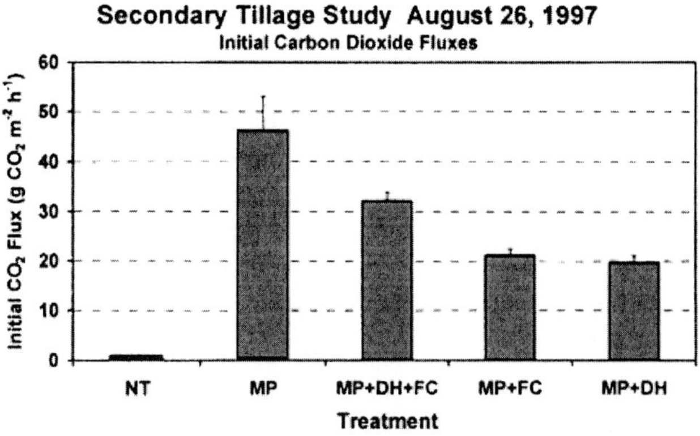

Figure 2. Initial carbon dioxide flux after secondary tillage.

after tillage in fig 3 show the same relative distribution in CO_2 loss. The moldboard plow alone lost most CO_2 and secondary tillage reduced the CO_2 loss as a result of reconsolidation and leveling of the soil. The cumulative loss of CO_2 after secondary tillage ranged from 40 to 60 % of the moldboard plow alone. The corresponding loss for the no till treatment was only 7 % of that from the moldboard plow. These results suggest that secondary tillage can reduce the short-term CO_2 loss following the moldboard plow minimizing soil C loss.

Compaction study

The change in soil physical properties as a result of compaction had a tremendous effect on soil bulk density and air permeability. The results of the secondary tillage

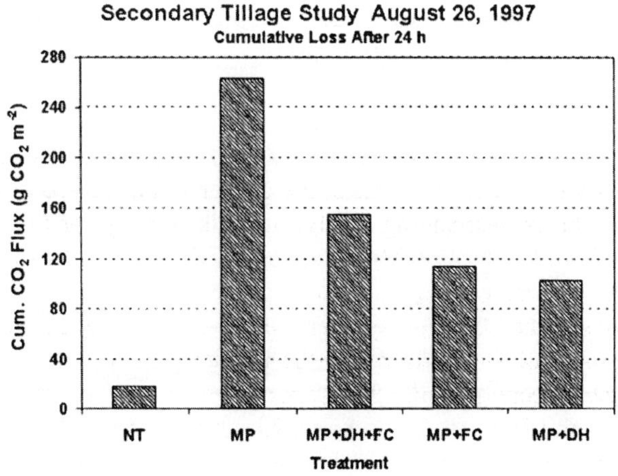

Figure 3. The 24-h cumulative loss of CO_2 after secondary tillage.

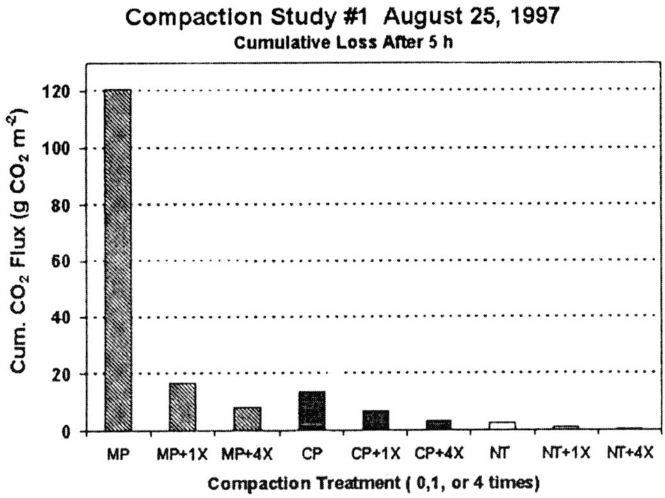

Figure 4. The 5-h cumulative CO_2 loss after tillage and compaction.

study suggest a change in soil properties that reduced the CO_2 loss from the soil. The effect of more severe compaction on the primary tillage methods of the moldboard plow, chisel plow and an area not tilled is shown in fig 4. The data shows that cumulative CO_2 loss for 5 h after tillage after no compaction or after one pass or after four passes of the road packer on all three primary tillage methods. Again, the moldboard plow without compaction lost the most CO_2. The loss following the chisel plow alone was substantially lower than the moldboard plow as result of the soil tillage tool interaction and the shallower depth of tillage. Noteworthy, is the cumulative effect of the severe compaction reducing

the flux of CO_2 from all primary tillage methods. The first pass with the road packer caused the largest decrease in the CO_2 loss. The succeeding three passes of the packer caused a smaller further reduction in the CO_2 flux as a result of the severe compaction.

The associated bulk densities under no till plots ranged from 1.18 to1.31 Mg m^{-3} after four passes with the Packer. The surface bulk density on the moldboard plow treatment was 1.26 Mg m^{-3} after one pass of the packer and increased to 1.49 Mg m^{-3} after four passes. The corresponding change in bulk density for the chisel plow treatments ranged from 1.34 to 1.36 Mg m^{-3}. The bulk density results show that most of the compaction occurred in the first pass of the packer; however, there were substantial increases in the bulk density after four passes with the packer.

While the magnitude of the CO_2 flux from the no tilled plots was very small initially, there was a noticeable decrease in the flux from one pass to four passes with the road packer. This degree of soil compaction is unrealistic and not practical or representative of field compaction from agricultural implements. It does, however, show the rapid dramatic change in soil properties that effects CO_2 loss from and oxygen entry into

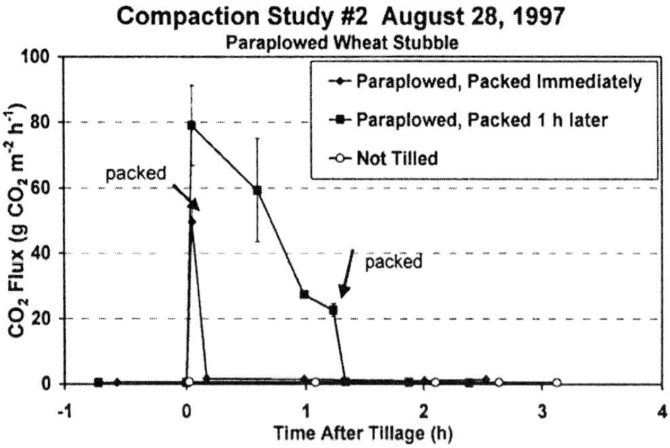

Figure 5. Carbon dioxide loss after Paraplow and compaction.

the soil. The extra compaction on all of the treatment was sufficient to seal the soil surface and minimize further gas exchange.

The temporal trends of gas exchange after primary tillage generally show a gradual decline with time as a soil dries out and the available energy is used by the microbes in response to the large amount of oxygen that enters the loose open soil after the tillage event (Reicosky & Lindstrom, 1993 and Reicosky 1998). The effect of severe compaction on the paraplow and subsoiler 1 h after the primary tillage is summarized in fig 5 & 6. Both paraplow and the subsoiler showed a gradual decline CO_2 flux during the first hour after tillage. The paraplow decreased from 79 g CO_2 m^{-2} h^{-1} immediately after tillage to 23 g CO_2 m^{-2} h^{-1} 1 h later. Immediately following one

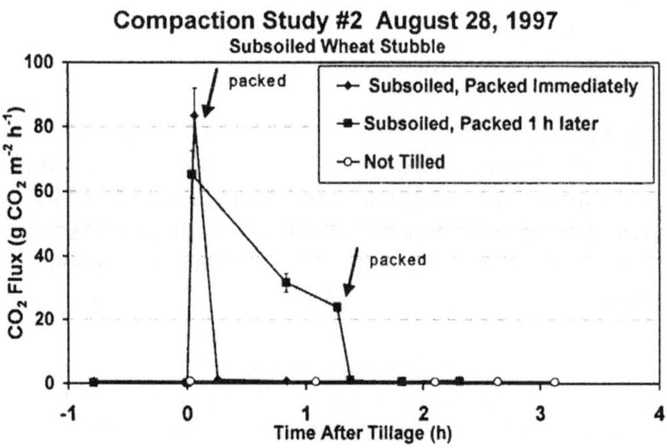

Figure 6. Carbon dioxide loss after subsoll and compaction.

pass of the road packer, the CO_2 flux dropped to 1.0 g CO_2 m^{-2} h^{-1}. Similarly, the subsoiler had an initial CO2 flux of 65 g CO_2 m^{-2} h^{-1} immediately after tillage that decreased to 24 g CO_2 m^{-2} h^{-1} 1 h later. One pass of the road packer further reduced the flux to less than 1.0 g CO_2 m^{-2} h^{-1}. Severe compaction reduced the CO_2 flux to that of the no till treatment without compaction for both the paraplow and subsoiler. The dramatic effect of the rapid change in soil properties on the tillage-induced gas exchange was demonstrated for the paraplow and subsoiler primary tillage tools and further demonstrates the importance of soil physical properties in controlling soil gas exchange and C loss.

DISCUSSION

In conventional tillage systems with the fall moldboard plow as the primary tillage, various secondary tillage operations are performed the following spring prior to planting to further breakdown the large clods and level the soil for a fine, clean-tilled seedbed. The secondary tillage varies widely in the type and the number of operations. If herbicides or fertilizers are incorporated with tillage, the fall-tilled soil is often leveled in spring with a disk harrow or field cultivator. This work demonstrated that secondary tillage immediately following the primary tillage reduced the short-term CO_2 flux. While this timing may not be practical for the farmer, the leveling and sealing of the soil after primary tillage suggests tool additions or modifications that can reduce C loss.

Abrupt change in soil properties by tillage caused ambient CO_2 concentration downwind of a plowed treatment to increase, which was more dramatic than not tilled treatment (Reicosky, 2000). This suggests plowing loosened the soil and created higher soil air permeability resulting in higher gas exchange. Tillage-induced change in soil properties led to higher short-term losses than undisturbed soil. For both studies, the tillage-induced change in soil properties with primary tillage resulted in substantial CO_2 loss that was partially reduced by secondary tillage and virtually eliminated with complete re-consolidation.

Changing the surface soil properties by primary tillage combined with the aerodynamic pressure forces associated with natural wind movement over the soil can result in substantial CO_2 loss. Further work is needed to identify improved secondary tillage methods and to quantify the interactions of changes in soil physical properties and the aerodynamic forces involved in soil gas exchange, especially CO_2, when surface soil properties are changed by primary tillage. Large CO_2 loss differences between plowed and not-tilled treatments reflect need for improved soil management and policies favoring conservation tillage to enable carbon sequestration in agricultural production systems.

ACKNOWLEDGMENTS

The author would like to acknowledge the technical support of Chris Wente, Steve Wagner, Ryan Bright and Alan Wilts in the data collection and analysis and to Chuck Hennen and Laurel Maanum for their help in the field operations.

NOTES

‡ Names are necessary to report factually on available data; however, USDA neither guarantees nor warrants the standard of the product, and the use of the name by USDA implies no approval of the product to the exclusion of others that may also be suitable.

REFERENCES

Ellert, B.H. and Janzen, H.H. (1999) Short-term influence of tillage on CO_2 fluxes from a semi-arid soil on the Canadian Prairies. *Soil Tillage Research* **50**: 21-32.

Franzluebbers, A.J., Hons, F.M. and Zuberer, D.A. (1995a) Tillage and crop effects on seasonal dynamics of soil CO_2 evolution, water content, temperature and bulk density. *Applied Soil Ecology* **2**: 95-109.

Franzluebbers, A.J., Hons, F.M. and Zuberer, D.A. (1995b) Tillage-induced seasonal changes in soil physical properties affecting soil CO_2 evolution under intensive cropping. *Soil Tillage Research* **34**:41-60.

Lal, R., Kimball, J., Follett, R.F. and Cole, C.V. (1998) *The potential of U.S. cropland to sequester carbon and mitigate the greenhouse effect,* Sleeping Bear Press: Ann Arbor, MI, 128 pp.

Reicosky, D.C. (1995) Soil variability and carbon dioxide loss after moldboard plowing. In *Site Specific Management for Agricultural Systems,* pp. 847-865. Eds P.C. Robert, R.H. Rust and W.E. Larson. Proceedings 2nd International Conference, Minneapolis, MN.

Reicosky, D.C. (1998) Strip tillage methods: Impact on soil and air quality. *Proceedings Australian Society of Soil Science. Inc. National Soils Conference,* Brisbane, Australia, Australian Society of Soil Science Inc.: Brisbane, Australia, pp. 56-60.

Reicosky, D.C. (1997) Tillage-induced CO_2 emissions from soil. *Nutrient Cycling in Agroesystems.* **49**:273-285.

Reicosky, D.C. (2000) Tillage-induced soil properties and gas exchange. *Proceedings 15th International Conference of the International Soil Tillage Research Organization, "ISTRO 2000",* 2-7 July, 2000. Ft.Worth, TX, USA. Topic 8 (CD-ROM computer file).

Reicosky, D.C., Kemper, W.D., Langdale, G.W., Douglas, Jr. C.L. and Rasmussen P.E. (1995) Soil organic matter changes resulting from tillage and biomass production. *Journal of Soil and Water Conserveration* **50**(3): 253-261.

Reicosky. D.C. and Lindstrom, M.J. (1993) Fall tillage methods: Effect on short-term carbon dioxide flux from soil. *Agronomy. Journal* **85**(6): 1237-1243.

Reicosky, D.C. and Lindstrom, M.J. (1995) Impact of fall tillage and short-term carbon dioxide flux. In *Soil and Global Change*, pp. 177-187. Eds R. Lal, J. Kimble, E. Levine and B.A. Stewart. Lewis Publishers, Chelsea, MI.

Roberts, W.P. and Chan, K.Y. (1990) Tillage-induced increases in carbon dioxide loss from soil. *Soil Tillage Research* **17**: 143-151.

Rochette, P. and Angers, D.A. (1999) Soil surface carbon dioxide fluxes induced by spring, summer, and fall moldboard plowing in a sandy loam. *Soil Science Society of America Journal* **63:** 621-628.

*D. DE SOUZA ANDRADE, *A. COLOZZI-FILHO,
*E.L. BALOTA & **M. HUNGRIA

LONG-TERM EFFECTS OF AGRICULTURAL PRACTICES ON MICROBIAL COMMUNITY

** Instituto Agronômico do Paraná (IAPAR), Cx. Postal 481, 86001-970, Londrina, PR, Brazil*

*** Empresa Brasileira de Pesquisa Agropecuária (Embrapa Soja), Cx. Postal 231, 86001-970, Londrina, PR, Brazil*

In Brazil, efforts to increase soil organic matter and to reduce soil nutrient losses through erosion have been made including the adoption of the no-tillage system and cover/rotation/intercropping practices with legumes. The better results obtained within these systems could be attributed to erosion control, improved plant residue management, higher rates of biological N_2 fixation, increased contribution of mycorrhizal fungi, nutrient cycling and decreased organic matter mineralisation, as well as lower nutrient leaching and runoff losses. Benefits of the long-term effects of legume crops and no-tillage system on microbial community have been reported, including beneficial effects on mycorrhizal fungi colonization, on common bean rhizobia and soybean bradyrhizobia diversity, nodulation and N_2 fixation rates. The inoculation of grain legume crops every crop season has resulted in increases in nodulation; N_2 fixation rates and grain yield of important legume crops, such as soybean (*Glycine max*) and common bean (*Phaseolus vulgaris*).
Key words: Microbial biomass, microbial diversity, nitrogen fixation, rhizobia, mycorrhizal fungi, tillage systems.

INTRODUCTION

Sustainable agriculture aims the production of suitable yields with economic returns and maintaining soil quality. The effects of land management practices in tropical soils on the microbial activity and diversity are not well known. It has been suggested that agricultural practices create highly selective and homogenous environments, with a reduction in biological activity. The key to the long-term sustainability and productivity of agricultural soils is organic C and N maintenance. Alleviation of environmental stresses or improvement of agricultural systems, including liming,

L. García Torres et al. (eds.). Conservation Agriculture, 301–306.

legume cover/rotations and cropping/livestock integration are approaches that favour soil microbe community, increase soil organic matter and may improve crop production in infertile soils.

SOIL ACIDITY

Soil acidity, often with toxicity of aluminum and low levels of calcium restricts crop production in many areas under tropical conditions. Liming, a common practice used in acidic soils, increased microbial biomass-C, microbial-C/organic-C ratio, number of ammonifiers and nitrifiers microorganisms, ammonification rate and nitrification potential in *Coffea arabica* crop (Andrade et al., 1995). Likewise, PCR-mediated RFLP analysis of the 16S-23S rRNA intergenic spacer (IGS) and of the 16S rRNA gene, indicated that the rhizobial populations from common bean nodules cultivated in an unlimed acidic oxisol were less diverse than those from the limed soil. The richness index (number of IGS groups) increased from 2.2 to 5.7 along the soil liming gradient and, based on species, showed an increase from 0.5 to 1.4. The Shannon index (species diversity) ranged from 0.9 in unlimed soil to 1.4 in limed soil, and based on the number of IGS groups, the diversity increased from 1.8 to 2.8 (Andrade, 1999; Andrade et al., 2000).

SOIL/CROP MANAGEMENT PRACTICES

Different tillage and cropping systems (rotation/intercropping) affect soil sustainability. In tropical conditions conventional tillage, with the traditional practices of ploughing and disking to prepare land reduces soil organic matter with consequences to the biological activity (Colozzi-Filho et al., 2001). An alternative is the no-tillage management, sowing directly through the mulch, which protects the soil life against erosion, high temperatures, besides improving soil structure, stability and moisture content and with time also increases also soil organic matter content.

Potential benefits from legume crop

Cover crops play important roles in no-tillage systems and other systems as well. They provide cover and protect the soil from erosion. Cover crops also serve as sinks for plant nutrients that might otherwise be leached. For example, they act as source of supplemental N (legumes) and slow-release of nutrients. As an example, a considerably greater quantity of net mineralisation in a legume litter (*Lupinus* sp.) was detected when compared with the non-legume crop (*Triticum aestivum*) (Andrade et al., 1993) and benefits of intercropping common bean with maize (Zea mays) were also reported, when compared to the monoculture of the non-legume (Hungria et al., 1997).

No-tillage systems

How important are the tillage systems and crop rotations to the microbial community? In the north region of the State of Paraná, Brazil, soil microbial biomass under no-tillage system showed a significant increase in relation to the conventional tillage. Those results highlight the importance of microorganisms in mediating the release of N from crop residues in different planting systems (Balota et al., 1998; Colozzi-Filho et al., 1999; Colozzi-Filho et al., 2001). Long-term field experiments have also shown that in soils under different tillage and crop rotations systems the ratio microbial-C:organic-C increased in the no-tillage system and in the presence of a legume (Balota et al., 1998).

Some components of microbial community are of interest of agriculture because they also play important roles in nutrient cycling, as for N and P. It has been show that no-tillage system also results in higher population size of specific groups of microorganisms, for instance *Azospirilum*, ammonium and nitrite oxidizers. By contrast, microorganisms involved in phosphate solubilisation did not change over a year of evaluation. As expected, the population size of phosphate solubilizers microorganisms increased after ploughing in the soil under conventional tillage (Colozzi-Filho et al, 2001).

The mycorrhizal fungi have been studied mainly in agroecosystems as a contribution to nutrient transfer in the plant-soil system. The traditional practices of ploughing stress microbial population and promote greater sporulation of mycorrhizal fungi because spores are the resistant structure, protecting them against unsuitable conditions. Different plant species also release different organic compounds, resulting in qualitative and quantitative differences in microbial population. A consequence would be that long-term crop rotation practices could alter the composition of the microbial population and of specific groups of microorganisms. For example, Colozzi-Filho (1999) reported that crop rotation with legumes increase root colonization by mycorrhizal fungi as well as the diversity of those microorganisms.

BIOLOGICAL NITROGEN FIXATION

Emphasis is also placed on the contribution of symbiosis between rhizobia and legume crops recycling nitrogen for use by crops growing in rotation or intercropping. Soybean (*Glycine max*) and common bean (*Phaseolus vulgaris*) are the main legume crops grown in Brazil and in some South America countries. Biological nitrogen fixation (BNF) plays a key role in those crops and a right combination can bring benefits in N status of both legumes and non-legumes grown in intercrop or crop rotation.

Rhizobial soil population and diversity

Soybean is an exotic plant in Brazil and has been cropped intensively after the 1960s, so that today most soils where this legume is grown show a very high population of soybean bradyrhizobia, estimated in 10^3 to 10^6 cells g^{-1} of soil. That population includes strains used in the inoculants since then, and dispersion of bacteria from other cropped fields has been recorded (Hungria et al., 2000b). Some fast-growing rhizobia have been isolated from field-grown soybean nodules, and sequencing of the 16S rRNA genes has detected similarity with *Rhizobium tropici* and *Rhizobium* genomic species

Q (unpublished data). Furthermore, several strains resembling agrobacteria that effectively nodulate soybeans were also isolated from soybean nodules in both Brazil and Paraguay (Chen et al., 2000). However, although many Brazilian soybean cultivars are effectively nodulated by fast-growing strains, and those bacteria are usually found in high number in soils, they compete poorly with *B. japonicum* and *B. elkanii* (Hungria et al., 2001).

The situation is different for the common bean crop, since almost all soils, even when they have never been cropped before with this legume, show a very high population of indigenous rhizobia, estimated in 10^3 to 10^6 cells g^{-1} soil depending on crop and soil management practices (Andrade, 1999; unpublished data). Furthermore, a high level of rhizobial diversity has been reported in studies performed with soils from several Brazilian regions (Grange, 2001).

Selection of efficient and competitive strains and seed inoculation

The approaches of the strain selection programs for the soybean and common bean crops are different, due to the differences found in rhizobial diversity. For the soybean crop, a continuous selection program is mandatory, since bacteria have to meet the increased N demand of more productive cultivars. The main approach consists of re-isolating adapted strains from areas, which have been previously inoculated, looking for variant genotypes with higher competitiveness and BNF capacity (Hungria & Vargas, 2000). Nowadays reinoculation guarantees a mean increase of 4,5% in grain yield, but can reach values as high as 25% (Hungria et al., 2000b), therefore soybean inoculants represent 99% of the Brazilian market and reinoculation is practiced by 60% of the farmers.

A different approach is used for the common bean crop and the strain selection is based on the search for efficient and competitive strains within the diverse indigenous population. Recently, the selection program has identified one strain, *R. tropici* PRF 81 (=SEMIA 4080), isolated from a soil of the State of Paraná that consistently increased yield by up to 900 kg ha^{-1}, reaching yields of 3,000 to 4,000 kg ha^{-1} (Hungria et al., 2000a).

Response to inoculation also depends on soil/crop management practices. As discussed before, the no-tillage management system brings many benefits to the soil microbes, including rhizobia, and consistent increases in the cell number, rhizobial diversity, nodule number and dry weight, BNF rates and grain yield have been reported for both soybean and common bean crops (Hungria & Stacey, 1997; Ferreira et al., 2000; Hungria & Vargas, 2000).

Main limiting factors to BNF

Although inoculation with efficient and competitive strains can increase grain yield of soybean, common bean and other legumes in Brazil some limiting factors to the biological process are often reported under field conditions such as high temperature and low soil moisture. Practices as the use of cover crops and no-tillage management decrease soil temperature and increase soil moisture, benefiting BNF and other microbial processes, as discussed earlier. Other factors can affect BNF, as the use of pesticides, which can drastically reduce nodulation and BNF (Campo & Hungria, 1999). Nutrients such as phosphorus and molybdenum also often limit BNF, mainly under acidic conditions. For example, nodulation of common bean was reduced by low P under acid conditions (unpublished). In another experiment performed in the State of Paraná, Brazil, a supply of molybdenum increased soybean yield from 3,100 to 3,400 kg ha^{-1} (Campo et al., 1999).

CONCLUSIONS

Soil production capacity can be reduced by soil degradation (erosion, organic matter mineralisation, acidification) or improved by soil management practices that promote nutrient cycling, and benefit soil microorganisms as mycorrhizal fungi and nitrogen-fixing bacteria. There appears to be several beneficial effects of the no-tillage system and crop cover/rotation with legumes in the maintenance of microbial community size and diversity. Biological nitrogen fixation is a key process in tropical N-depleted soils and inoculation with efficient and competitive rhizobial strains has resulted in increases in N_2 fixation rates and grain yield in several legumes, as soybean and common bean. Further benefits can be obtained by the control of stressing conditions usually found in tropics, such as soil acidity, depletion of phosphorus and molybdenum, high soil temperature and low moisture.

ACKNOWLEDGEMENTS

To CNPq (PRONEX 41.96.0884.00; 520396/96-0; APQ-463971/00-0-NV) and INCO-EC (ERBIC18CT980321) for financial support.

REFERENCES

Andrade, D.S. (1999) Diversity of rhizobia nodulating *Phaseolus vulgaris* L. in soils from Brazil Wye College, University of London, Wye, Kent, UK. (Ph.D. Thesis).

Andrade, D.S., Murphy, P.J. and Giller, K.E. (2000) Diversity and abundance of populations of bean-nodulating rhizobia as a function of liming and cropping history in acidic Brazilian soils, F.O., Hungria, M., Yates, M.G., Newton, W.E., eds. *Nitrogen fixation: From molecules to crop productivity*. Kluwer Academic Press, Dordrecht, p.546.

Andrade D.S., Colozzi-Filho, A., Medeiros, G.B. (1993) Decomposição da palha de culturas de inverno em solos sob plantio direto e convencional. In: *Encontro Latino Americano Sobre Plantio Direto na Pequena Propriedade*, **1**, 1993, Ponta Grossa. Resumos, Ponta Grossa, p.36.

Andrade, D.S., Colozzi-Filho, A., Pavan, M.A., Balota, E.L., Chaves, J.C.D. (1995) Atividade microbiana em função da calagem em um solo cultivado com cafeeiro. *Revista Brasileira de Ciência do Solo,* **19:** 191-196.

Balota, E.L., Colozzi-Filho, A., Andrade, D.S. and Hungria, M. (1998) Biomassa microbiana e sua atividade em solos sob diferentes sistemas de preparo e sucessão de culturas. *Revista Brasileira de Ciência do Solo*, **22:** 641-649.

Campo, R.J., Hungria, M. (1999). Efeito do tratamento de sementes de soja com fungicidas na nodulação e fixação simbiótica do N_2. Embrapa Soja, Londrina, 7p. (Pesquisa em Andamento, 21).

Campo, R.J., Albino, U.B., Hungria, M. (1999) Métodos de aplicação de micronutrientes na nodulação e na fixação biológica do N_2 em soja. Embrapa Soja, Londrina, 7p. (Pesquisa em Andamento, 19).

Chen, L.S., Figueredo, A., Pedrosa, F.O. and Hungria, M. (2000) Genetic characterization of soybean rhizobia in Paraguay. *Applied and Environmental Microbiology*, **66:**5099-5103.

Colozzi-Filho, A. (1999) Dinâmica populacional de fungos micorrízicos arbusculares no agrossistema cafeeiro e adubação verde com leguminosas. Piracicaba,. 106p. (Ph.D. Thesis); Escola Superior de Agricultura «Luiz de Queiroz».

Colozzi-Filho, A., Balota, E.L., Andrade, D.S. (1999) Microrganismos e processos biológicos no sistema plantio direto. In: Siqueira, J. O.; Moreira, F. M. S.; Lopes; A. S.; Guilherme, L.R.G.; Faquim, V.; Furtini Neto, A. E.; Carvalho, J.G. (Ed). *Soil fertility, soil biology, and plant nutrition interrelationships.* Lavras, Sociedade Brasileira de Ciência do Solo/ Universidade Federal de Lavras,1999. p.487-508.

Colozzi-Filho, A., Andrade, D.S., Balota, E.L. (2001) Atividade microbiana em solos cultivados em sistema plantio direto. *Informe Agropecuário*, **22:** 84-91.

Ferreira, M.C., Andrade, D.S., Chueire, L.M.O., Takemura, S.M. and Hungria, M. (2000) Effects of tillage method and crop rotation on the population sizes and diversity of bradyrhizobia nodulating soybean. *Soil Biology and Biochemistry*, **32**:627-637.

Grange, L. (2001) Diversidade de rizóbio capaz de nodular o feijoeiro (*Phaseolus vulgaris*) isolado de solos da Região Nordeste e da Região Sul do Brasil. Universidade Estadual de Londrina, Londrina. (M.Sc. Thesis).

Hungria, M. and Stacey, G. (1997) Molecular signals exchanged between host plants and rhizobia: Basic aspects and potential application in agriculture. *Soil Biology Biochemistry*, **29**: 819-830.

Hungria, M. and Vargas, M.A.T. (2000) Environmental factors impacting N_2 fixation in legumes grown in the tropics, with an emphasis on Brazil. *Field Crops Research*, **65**: 151-164.

Hungria, M., Andrade, D. S., Chueire, L.M.O., Probanza, A., Guttierrez-Mañero, F. J. and Megías, M. (2000a) Isolation and characterization of new efficient and competitive bean (*Phaseolus vulgaris* L.) rhizobia from Brazil. *Soil Biology Biochemistry*, **32:** 1515-1528.

Hungria, M., Andrade, D.S., Colozzi-Filho, A., Balota, E.L. (1997) Interação entre microrganismos do solo, feijoeiro e milho em monocultura ou consórcio. *Pesquisa Agropecuária Brasileira,* **32**: 807-818.

Hungria, M., Campo, R.J., Chueire, L.M.O., Grange, L. and Megías, M. (2001) Symbiotic effectiveness of fast-growing rhizobial strains isolated from soybean nodules in Brazil. *Biology and Fertility of Soils* (in press).

Hungria, M., Vargas, M.A.T., Araujo, R.S., Kurihara, C., Maeda, S., Sá, E.S., Campo, R.J.,Cattelan, A.J., Mendes, I.C. and Oliveira, M.C.N. (2000b) Brazilian trials to evaluate the effects of soybean reinoculation. In: Pedrosa, F.O.; Hungria, M.; Yates, M.G.; Newton, W.E., eds. *Nitrogen fixation: From molecules to crop productivity.* Kluwer Academic Press, Dordrecht, p.549.

D. McGARRY

TILLAGE AND SOIL COMPACTION

Natural Resource Sciences, Meiers Road, Indooroopilly, Queensland 4068, Australia

Soil structure degradation (SSD), often called soil compaction, is ubiquitous to cropping and grazing lands, worldwide. With soil erosion, compaction is regarded as the most serious environmental problem caused by conventional agriculture. Examples span all soils, small to large farms, developed and developing nations, subsistence to corporate farming enterprises.

Addressed first will be the processes, nature, extent, location, rate and costs of SSD. Next, consideration will be given to the four «best management practices» to first repair and then control SSD - soil inspection, zero till, controlled traffic and cover crops; all central to conservation agriculture. Synergy of the four gives maximum benefit. The aim is to forever rid farmers of SSD and ensure their soils attain optimum physical condition for future generations.

Key words: Tillage, compaction, controlled traffic, zero tillage, cover crops

INTRODUCTION

Soil compaction, commonly described as soil structure degradation (SSD), is often ranked as the most widespread and insidious form of land degradation in agriculture. SSD is the antithesis of conservation agriculture in that it represents a «lose-lose» farming system. Not only does it cost money to form SSD but also its presence creates a high-risk scenario for crop failure, before further finance is required to mechanically remove the problem-commonly unsuccessfully. Exacerbating the problem, SSD is the «Scarlet Pimpernel» of land degradation, because unlike erosion and salinity, SSD is commonly hidden from view. The composite effect makes SSD a major risk in the global «food security challenge».

Fortunately, SSD is a reversible form of land degradation (Scherr, 1999) though repair of subsoils can be challenging (Auerswald & Kutilek, 1998). Once SSD is recognised and the cause(s) of its formation resolved, farming systems can be devised to initially repair, then ensure future prevention of the problem. An important bonus is that the elements used to prevent the problem of SSD are the same elements at the core of practicable, farmer-driven, sustainable land management systems for conservation agriculture. So, the means of investigating, rationalising and solving

L. García Torres et al. (eds.). Conservation Agriculture, 307–316.

the problem of SSD are not only important in their own right but they are also at the core of resolving several land degradation issues towards sustainable, reduced cost and lessened-inputs land use.

In the current script, consideration will first be given to the processes leading to SSD, previous work on the subject, then the nature, extent, location, rate and costs of the problem. These topics represent the negative view of SSD; ie the «rearguard action» of forever combating the problem. The more positive strategies are those of repair and prevention of SSD. Application of these practices demonstrate that tackling SSD leads to more sustainable, lower cost cropping systems, as well as increased awareness and care by farmers for the soil resource.

PROCESSES FORMING SSD AND PREVIOUS WORKS

The emphasis in this script will be on human-induced or -exacerbated SSD. SSD is caused by the energy input from tillage implements, tyres and animal hoofs, particularly when traversing moist to wet soil. There is another type of degradation of soil structure, termed hardsetting, as reviewed by Mullins *et al.* (1990). However, though the effects and the resulting field morphology have similarities, hardsetting is principally a naturally occurring phenomenon with different causes, energy inputs and processes from SSD (McGarry, 1993). Furthermore, hardsetting is most commonly an immediate soil surface phenomenon whereas SSD more commonly forms a subsurface layer (McGarry, 1993; Douglas *et al.,* 1999).

SSD is a «change for the worse» in soil structure condition. Soil structure may be defined as: «The physical constitution of soil material as expressed by the size, shape and arrangement of soil aggregates and pores» (Burke *et al.*, 1986). This emphasise the soil air spaces as the principal effect of SSD on crop performance is through one of, or combinations of, the loss in pore spaces, loss in the continuity of pores and a decrease in the size of pores. These result in reductions in each of the storage capacity, infiltration and extraction of soil water, as well as reduced rooting depth and volume, and increased soil strength.

Reviews of SSD and journal articles on specific case studies are many and have emanated from several different countries, cropping systems and soil types (Soane & van Ouwerkerk, 1994). Additionally, Ouwerkerk & Soane (1995) and Arshad (1999) introduce special issues of Soil and Tillage Research. Lal (1994), reviewing minimum tillage systems, specifically addresses the alleviation and prevention of SSD.

Two processes lead to SSD - compression and shear. The first causes a volume decrease or densification, through the expulsion of soil air and the second causes deformation through the rearrangement of soil particles or microaggregates (Koolen, 1994). There are many, interacting soil factors that play roles in the SSD process and pre-dispose soils to SSD (McGarry, 1993). The principal determinant of the severity and extent of SSD is the soil water content at the time of traversing or cultivating a soil (Kirby & Blunden, 1992). The water acts as a lubricant, permitting soil aggregates and individual particles to move, causing the loss of air spaces and the closer packing of particles and aggregates. The soil water content at key times in a given cropping system is particularly important, for example during primary cultivation and at harvest when tillage tools and laden harvesters are in the field.

THE NATURE, EXTENT, LOCATION AND RATE OF SSD

The nature of SSD can be considered both in terms of soil morphology (soil profile description) as well as soil measurements. Massiveness is the most common visible sign of SSD (Robertson & Erickson, 1980). Soil aggregates are compressed into large and dense blocks that equate to reduced air space and increased soil strength. The second most common, visible form is «platiness» where the soil forms plate-like structure, horizontal to the soil surface; commonly 0.5 to 2 cm thick (McKenzie, 1998). Platy structure is a strong barrier to plant root proliferation into subsoils; young plants commonly failing to penetrate such layers and mature plants being forced to extend horizontally until they locate a zone of weakness to bypass the platy zone. There is no one measure of SSD; any one measure providing insight into a complex phenomenon. In no way should this «difficulty» deter experimentalists from seeking best-discriminator and critical measures of SSD, as the ability to quantify sustainability is critical to making the concept operational (Harrington, 1993). Examples of measures found to be useful discriminators of SSD and also of repaired soil structure are presented in the conferences, books and review papers presented above, and in Coughlan *et al.* (1991), McGarry (1993; 1995) and Topp et al., (1997).

The extent and location of SSD are commonly considered challenging issues, principally as SSD is a subsurface phenomenon, requiring soil excavation to view or sample. Additional «complications» are that SSD has both horizontal and vertical (soil depth) axes, crop performance is a poor surrogate for soil examination (principally as during-season weather can nullify or exaggerate SSD effects), and the magnitude of SSD must be considered. The large variation in the type and intensity of studies into the extent of SSD reflect that different users require answers at different scales. A farmer or scientist requires detail on individual fields or parts of fields such as the quantification of SSD in individual wheel tracks (eg Alakukku, 1996) or from various cultivation tools (eg Rahman & Chen, 2001). A policy maker or regional planner requires composite information on regions or countries, from information in soil pits (eg McGarry, *et al.* 1999; McGarry, 2000). Global statements also exist, eg Oldeman *et al.*, 1991, stated that 68 million ha of the world's soils have SSD, sealing and crusting.

In terms of extent and location, there are beliefs that SSD occurs only on mechanised, large farms and that draft animals cause less problems than tractors. There are, however, published reports of SSD in small-holder, almost subsistence farming enterprises in both Malawi (Douglas *et al.,* 1999) and Bangladesh (Brammer, 2000). Also, Soane and van Owerkerk (1994) state ground contact pressures of up to 150 kPa and 250 kPa for the hoofs of horse and oxen, respectively, and Proffitt *et al.* (1995) demonstrated the almost total loss of soil porosity in the soil surface due to SSD from sheep feet.

Long periods of time are not necessary to cause soil SSD. McGarry (1993) presented examples of SSD that had occurred after each of 70 years of cropping, seven days of cultivations, and «instantaneously», ie following one pass of a tractor or implement.

THE COST OF SSD

The cost of SSD is difficult to assess (Lal, 1994). Central to the problem, primary producers worldwide tend to solely consider the impact of SSD on crop productivity, ie crop yield. If SSD does not affect yield, they consider that there is no problem. For at least two reasons, this is an unrealistic appreciation of the problem.

First, SSD may only increase the potential for yield reduction as crops can grow well in structurally degraded soil, if there is frequent irrigation or rainfall. Anecdotal evidence is often quoted where a rainfed crop, such as wheat, will yield well despite being sown into a severely compacted soil. Periodic rain, up to senescence, permits the plant to survive and yield well - living «hydroponically» in the top 0.1-0.2 m of soil. Similarly, a 70% reduction in cotton yield was not directly caused by SSD; more that the SSD caused poor early crop growth and the farmer withdrew irrigation water from that crop (McGarry, 1990). At a more general level, yield losses in Australian cotton of 20 - 50% (McGarry, 1995) and in Russian grain and fodder production of 8-50% (Lal, 1994) have been reported.

Second, there is strong potential for high costs, both forming SSD and then initiating repair and control strategies. Many farmers pay threefold - the cost of the initial cultivation and traffic that gave SSD, the cost of negative responses to SSD (yield loss, increased irrigations, poor seedbeds, etc) that then require further cultivation to repair, and again run the risk of producing more SSD if conducted in moist soil. This is a typical «downward spiral» associated with SSD and is at the basis of the recognised «significant cost» of reversing SSD (Scherr, 1999). In one agricultural area of Australia, the Murray - Darling basin, SSD alone has cost $Aus144 million damage (Fray, 1991). Lal (1994) states a figure of one billion dollars, annually, due to yield losses caused by SSD in the USA, alone.

REPAIR AND PREVENTION OF SSD-TOWARDS THE «OPTIMISATION» OF SOIL STRUCTURE

Once SSD is recognised and rationalised there is a logical progression from considering and initiating repair programs, then ensuring new management practices prevent future occurrence of the problem (McKenzie, 1998). The long-term aim is to go far beyond simply removing the problem of SSD, towards achieving soil structure optimisation for future generations. As such, this concept is integral to the definition of sustainability where we leave future generations as many if not more opportunities than we had (Serageldin, 1995)

The repair of SSD is commonly based on growing «break crops» in fallows or between summer cash crops. In both cracking (Vertisols) and non-cracking soils the immediate gain is soil drying to depth, to effect subsequent, primary cultivation in dry soil - to shatter the SSD. In cracking soils, the soil drying itself produces cracks (porosity) and initiates the break-up of structurally degraded layers (McKenzie, 1995). Repeated cycles of wetting and drying in cracking soils are an effective soil repair mechanism (Pillai-McGarry & Collis-George, 1990), where the strong potential of repairing SSD with selected «break crops» has been demonstrated (Pillai-McGarry & McGarry, 1999).

The prevention of SSD is achieved through several «best management practices». Each is important in itself but, following the theme of this paper, each one leads to far greater levels of understanding towards integrated, sustainable land management for conservation agriculture. Four can be stated.

The link between soil water content and SSD risk-soil inspection

A vital requirement to achieving sustainable land management for conservation agriculture is farmer-education on the field assessment of soil water content and its relation to trafficability. This is part of an overall, improved interaction with farmers towards identifying land management problems and conceptualising the approach(es) to address them (Lal, 2000) as well as harnessing the power of farming communities for self-help. In terms of SSD, there is a need to improve farmer awareness that soil water content at the time of traversing or cultivating a soil is the principal determinant of the severity and extent of SSD (Kirby & Blunden, 1992). However, this leads to many more, related subject areas that lead to increased awareness and care by farmers for the soil resource - a recognised key component of conservation agriculture (Landers *et al.*, 2001). These subject areas include: empowering farmers to excavate holes in their fields to examine soil structure, root growth and water penetration; and education on the role of soil chemistry and organic matter in affecting soil structure condition, and the potential to influence these properties through gypsum/lime (calcium) applications and green manures/rotation crops, respectively. A vital input is the use of «farmer training and empowerment» manuals and associated in-field workshops (eg FARM Programme, 1998; McKenzie, 1998; Adepetu *et al.*, 2000) that aim to educate farmers on sustainable land management and give them the tools to achieve it. At a higher level, field assessments of soil structure, supported with farmer interpretation, permit statements to be made on the risk-levels of SSD associated with major cropping industries (McGarry *et al.*, 1999). Lastly, the use of soil profile descriptions in the field answers the common demand for «robust and cheap» methods of assessing land degradation (Biot *et al.,* 1995) to facilitate initial, broadscale, rapid but repeatable assessment.

Zero till (ZT)

Commonly advocated and practised for erosion control (Crovetto, 1996; Coughenour & Chamala, 2001) ZT is also a vital tool in SSD prevention as it stops almost all metal - soil contact. Similar to the inspection of soil for current soil water status, ZT leads to many more, related subject areas that in turn lead to more sustainable farming systems. Most commonly cited by farmers, particularly rain-fed growers in semi-arid and arid zones, is the greatly improved water infiltration and water storage in ZT soils - leading to more guaranteed crop growth and harvests. Radford *et al.* (1995) measured a 28% increase in plant-available soil water at sowing with ZT, linked to both a four-fold increase in earthworm numbers and, as measured by McGarry *et al.* (2000), significantly greater soil macropores in the 1.5 - 3mm size, associated with the presence of earthworm burrows and caste material, evident in soil images. Pagliai *et al.* (1994) present similar improvements in soil physical condition after 10 years of ZT compared with conventional tillage. On another front, ZT farmers commonly state that tractor size, use of tractors, and wear and tear on tractors and equipment are all dramatically reduced. Figures of 50% reduction in fuel use and 30% reduction in tractor hours with minimal till compared to conventional practices in sugarcane have been recorded (Braunack *et al.*, 1999). Landers *et al.* (2001) quote fuel reductions of up to 75% with ZT in South America and USA, compared to conventional farming. Furthermore, all such benefits lead to reductions in the use of fossil fuels, greenhouse gas emissions (Robert, 2001), as well as lessened iron ore mining and steel production as with ZT tractors tend to be smaller and have longer life-spans.

At a more general, but important level, ZT farmers have been recognised as having greater awareness of the need to protect and sustain the soil resource (Landers *et al.*, 2001). They are keen observers and advocates of the environmental/biological indicators related to their conservation farming practices, eg darker and more aggregated topsoils from increased soil organic matter from both increased residues and lessened tillage (Balesdent *et al.* 2000).

Controlled traffic (CT)

The aim with CT is to keep the wheels of all in-field equipment always in the same paths, year after year, and to keep these traffic zones separate from the cropping zones, and manage each zone separately (Reeder, 1992; Yule, 1998). Similar to ZT, CT is a vital tool in SSD prevention, principally by removing the random trafficking of fields by a wide array of farm equipment. Again, similar to ZT the practice of CT provides much more than SSD prevention, towards sustainable, integrated land management. Strongly supporting and driving the acceptance and use of CT are strong, on-farm, practical issues. «Efficiency» is the keyword used by farmers currently using the system. This is at many levels. Water use efficiency is a major gain. Soil structure between the permanent wheel tracks becomes so well aggregated that more soil water enters the soil and is held there, so is available to growing crops. Again, this is vital for rain-fed cropping in semi arid to arid zones. Also with CT, soil erosion on sloping land has greatly reduced as rain more easily penetrates the better structured soils, leaving less water to run-off. All in-field operations like sowing, spraying, harvesting, etc are far more efficient with CT as there is greatly reduced overlap (Blackwell, 1998). CT can reduce the energy requirements of in-field operations by 50%, by eliminating the need to cultivate-out wheel tracks and by improving tractive efficiency from unnecessary soil deformation (Tullberg, 1998). One study, over a 3 year period on a commercial, irrigated cotton farm, showed an 11% financial advantage when cotton was grown with CT (Hulme *et al.*, 1996).

Walsh & Jensen (1998) state that Australia leads the world in the adoption of CT farming. In the Central Queensland region, alone, the area under CT farming increased

90% in the 3.5 year period from March 1995 to July 1998 (2,500 ha to 32,500 ha) with the success firmly based on factors such as immediate and significant cost savings, and minimal requirements for machinery modification (Chapman, 1998).

CT is an important, perhaps vital, adjunct to ZT. There are concerns regarding development of adverse soil physical conditions with ZT, alone as on the one hand the soil is no longer routinely loosened by tillage (Ferreras *et al.*, 2000) but the soil continues to be randomly trafficked by a whole range of potentially very heavy equipment (Unger, 1996) in moist to wet soil conditions. Relying solely on the porosity-forming potential of earthworms and other soil macrofauna, though they occur in far greater numbers under ZT, is regarded as a high-risk scenario as soil fauna numbers are strongly affected by weather, on-farm chemicals and cropping cycles (McGarry, 2000). A more sustainable system would use the synergy of both ZT and CT - restricting traffic compaction with CT, hence gaining full benefit of the improved soil structure condition and improved organic matter status, including earthworm activity, from the ZT.

Organic matter enrichment/increased soil cover/rotation crops/cover crops

Consideration of these subject areas is commonly restricted to reducing soil erosion by both maintaining ground cover as well as improving soil organic matter content. However, enrichment of the soil with organic matter is both a direct and indirect way of preventing SSD and more generally improving soil quality (Karlen *et al.*, 1994). Directly, aggregate stabilisation through organic bonds improves soil aggregate robustness to external forces of compression and shear; most important in soils with low, inherent aggregate stability, like sands, loams and non-cracking clays. Indirectly, increasing plant residues on the soil surface, particularly aiming to maintain year-long cover, provides a continuous food source for soil biota, particularly earthworms (Radford *et al.,* 1995). Not only do earthworms increase the chemical and physical fertility of soils through their well aggregated and organic-rich cast material but their interconnected, bio-macropores aid water infiltration and root penetration into the soil. An important role of rotation crops, apart from limiting soil organic matter loss (Pieri, 1995), is the provision of alternate root types and patterns in the soil that, similar to earthworm macropores, break-up SSD layers and enhance water infiltration. So, organic matter enrichment and rotation crops have an important role not only in minimising erosion risk but also SSD risk.

DISCUSSION

Soil structure degradation is a negative state of land condition. It has strong potential to cause yield decline and even crop failure. It is an expensive and unnecessary form of land degradation. It is reversible.

The integration of the four «best management practices» of soil inspection, ZT, CT and organic matter improvements achieve both SSD prevention and practicable, sustainable land management systems towards successful conservation agriculture. The synergy of the four elements leads to both a more «naturally self-sustaining» farm management system than conventional cropping practices as well as «increased total farm productivity» - a preferred choice of agricultural development for developing countries (Pretty and Hine, 2001) In terms of the long-term prevention of SSD, ZT and CT ensure that SSD is forever removed from the crop growth zone; green manures/ rotation crops ensure ground cover for increased soil stability and continuous bio-macropores (from increased earthworm numbers) to facilitate root and water entry; and soil structure inspection/monitoring provide a low cost, repeatable mechanism to check the «soil health» within the integrated system.

There is no one unique formula of the four elements. All four should be present but their detailed nature and weighting will vary between farms, regions, soil types, cropping systems, countries and economic realities. However, whatever situation is being addressed, the integration of the 4 components ensure positive, proactive and farm-practicable skills that aim to *optimise* soil structure for plant growth - freeing farmers from the constant «rearguard» action of battling SSD.

REFERENCES

Adepetu, J.A., Nabhan, H. and Osinubi, A. (2000) Simple soil, water and plant testing techniques for soil resource management. Proceedings of a training course, Ibadan, Nigeria, 16-27 September 1996. Rome: FAO.

Alakukku, L. (1996) Persistence of soil compaction due to high axle load. 1. Short-term effects on the properties of clay and organic soils. *Soil and Tillage Research* **37**: 211-222.

Arshad, M.A. (1999) Tillage and soil quality: tillage practices for sustainable agriculture and environmental quality in different agroecosystems. *Soil and Tillage Research* **53**: 1-2.

Auerswald, K. and Kutilek, M. (1998) A European view to the protection of the soil resource. *Soil and Tillage Research* **46**:ix-xi.

Balasdent J., Chenu, C. and Balabane, M. (2000) Relationship of soil organic matter dynamics to physical protection and tillage. *Soil and Tillage Research* **53**: 215-230.

Biot, B., Blaikie, P.M., Jackson, C., and Palmer-Jones, R. (1995) World bank discussion papers no. 289 - Rethinking research on land degradation in developing countries. Washington: The World Bank. 139 pp.

Blackwell, P. (1998) Customised controlled traffic farming systems, instead of standard recommendations or «tramlines ain't tramlines». In *Second national controlled farming conference*, pp.23-26. Eds J.N. Tullberg and D.F. Yule. Gatton College: University of Queensland.

Brammer, H. (2000) Ploughpans and tillage problems. In *Agroecological aspects of agricultural research in Bangladesh.* pp.151-158. Dhaka: UPL.

Braunack, MV., McGarry. D., Crees, L.R. and Halpin, N. (1999) Strategic tillage for planting sugarcane. *Proceedings of the Australian Society of Sugar Cane Technologists, 1999* pp. 101-107.

Burke, W., Gabriels, D. and Bouma, J. (1986) *Soil Structure Assessment.* Rotterdam: A.A. Balkema.

Chapman, W. (1998) Achieving change: «It's the package that counts». In *Second national controlled farming conference*, pp.146-148. Eds J.N. Tullberg and D.F. Yule. Gatton College: University of Queensland.

Coughenour, C.M., and Chamala, S. (2001) *Conservation tillage and cropping innovation - constructing the new culture of agriculture.* Ames, Iowa: Iowa State University Press. 360 pp.

Coughlan, K.J., McGarry, D., Loch, R.J., Bridge, B. and Smith, G.D. (1991) The measurement of soil structure - some practical initiatives. *Australian Journal of Soil Research* **29**: 869-890.

Crovetto, C. (1996) Stubble Over the Soil - the vital role of plant residue in soil management to improve soil quality. Madison, WI, USA: American Society of Agronomy.

Douglas, M.G., Mughogho, S.K., Saka, A.R., Shaxson, T.F. and Evers, G. (1999) Report on an investigation into the presence of a cultivation hoe pan under smallholder farming conditions in Malawi. Investment centre division FAO/World Bank cooperative programme. Washington, D.C.: World Bank.

FARM Programme (1998) Farmer Field School on Integrated Soil Management, Facilitator's Manual. The FARM programme, FAO-RAP. Rome: FAO. 218 pp.

Ferreras, L.A., Costa, J.L., Garcia, F.O. and Pecorari, C. (2000) Effect of no-tillage on some soil physical properties of a structural degraded Petrocalcic Paleudoll of the southern «Pampa» of Argentina. *Soil and Tillage Research* **54**: 31-39.

Fray, P. (1991) On fertile ground? The sustainable agriculture debate. *Habitat Australia* **19**: 4-8.

Harrington,. L. (1993) Interpreting and measuring sustainability: issues and options. In *Methods of measuring sustainability through farmer monitoring: application to the rice-wheat cropping pattern in south Asia.* Proceedings of the workshop, 6-9 May, 1992, Kathmandu, Nepal, pp. 3-19. Eds L.W. Harrington, P.R Hobbs, K.A. Cassidy. Mexico, D.F.: CIMMYT, IRRI and NARC.

Hulme, P.J., McKenzie, D.C., MacLeod, D.A. and Anthony, D.T.W. (1996) An evaluation of controlled traffic with reduced tillage for irrigated cotton on a Vertisol. *Soil and Tillage Research* **38**: 217-237.

Karlen, D., Wollenhaupt, D.C., Erbach, E.C., Berry, E.C., Swan, J.B., Eash, N.S. and Jordahl, J.L. (1994) Crop residue effects on soil quality following 10-years of no-till corn. *Soil and Tillage Research* **31**: 149-167.

Kirby, J.M. and Blunden, B.G. (1992) Traffic loadings and soil compaction. *Proceedings of the Sixth Australian Cotton Conference, Broadbeach, Australia.* pp. 31-41.

Koolen, A.J. (1994) Mechanics of Soil Compaction. In *Soil compaction in crop production*, pp. 23-43. Eds B.D. Soane and C van Ouwerkerk. Netherlands: Elsevier Science.

Lal, R. (1994). Minimum tillage systems. In *Subsoil management techniques, Advances in Soil Science*, pp.1-33. Eds N.S. Jayawardane and B.A. Stewart. London: Lewis Publishers.

Lal, R. (2000) Soil management in the developing countries. *Soil Science* **165**: 57-72.

Landers, J.N., de Freitas P.L., Guimaraes, V. and Trecenti, R. (2001) The social dimesions of sustainable farming with zero tillage. *The third International Conference on Land degradation, Rio de Janeiro, 17-21 September, 2001.*

McGarry, D. (1987) The effect of soil water content during land preparation on aspects of soil physical condition and cotton growth. *Soil and Tillage Research* **9:** 287-302.

McGarry, D. (1990) Soil compaction and cotton growth on a Vertisol. *Australian Journal of Soil Research* **28**: 869-877.

Mc Garry, D. (1993) Degradation of soil structure. In *Degradation processes in Australia*, pp. 271-305. Eds G. McTanish and W. Boughton. Australia: Longman Cheshire.

McGarry, D. (1995) The optimisation of soil structure for cotton production. In *Challenging the future: Proceedings of the World Cotton Conference-1, Brisbane, Australia, February 1994.* pp. 169-176. Eds. G A Constable and N W Forrester. Melbourne, Australia: CSIRO.

McGarry, D. (2000) Land management for cropping in Argentina and Brazil - an assessment of soil structure under zero tillage and flotation tyres. Brisbane: Queensland Government. DNRQ00090 39 pp.

McGarry, D., Bridge, B.J., Radford, B.J. (2000) Contrasting soil physical properties after zero and traditional tillage of an alluvial soil in the semi-arid tropics. *Soil and Tillage Research* **53**: 105-115.

McGarry, D. and Chan, K.Y. (1984) Preliminary investigation of clay soils' behaviour under furrow-irrigated cotton. *Australian Journal of Soil Research* **22:** 99-108.

McGarry, D., Sharp, G., Bray, S.G. (1999) The Current Status of Soil Degradation in Queensland Cropping Soils. Brisbane: Queensland Government. DNRQ990092. 20 pp.

McKenzie, D.C. (1995) Soil management options for cotton-based farming systems in swelling soils and non-swelling soils. In *Challenging the future: Proceedings of the World Cotton Conference-1, Brisbane, Australia, February 1994.* pp. 177-181. Eds. G.A. Constable and N.W. Forrester. Melbourne, Australia: CSIRO.

McKenzie, D.C. (1998) SOILpak for Cotton Growers (3rd ed.). Orange, Australia: NSW Agriculture.

Mullins, C.E., MacLeod, D.A., Northcote, K.H., Tisdall, J.M. and Young, I.M. (1990) Hardsetting soils: behaviour, occurrence and management. In *Soil Degradation, Advances in Soil Science*, 12, pp. 37-108. Eds R. Lal and B.A. Stewart. London: Lewis Publishers.

Oldeman, L.R., Hakkeling, R.T.A. and Sombroek, W.G. (1991) World map of the status of human-induced soil degradation: an explanatory note. Wageningen, Netherlands: ISRIC.

Pagliai, M., Raglione, M., Panini, T., Maletta, M. and La Marca, M. (1994) The structure of two alluvial soils in Italy after 10 years of conventional and minimum tillage. *Soil and Tillage Research* **34**: 209-223.

Pillai-McGarry, U. and Collis-George, N. (1990) Laboratory simulation of the surface morphology of self-mulching and non self-mulching Vertisols. II. Quantification of visual features. *Australian Journal of Soil Research* **28**: 141-152.

Pillai-McGarry, U. and McGarry, D. (1999) Structure repair of a compacted Vertisol with wet/dry cycles and rotation crops. *Soil Science Society of America Journal* **63**: 201-10.

Pieri, C. (1995) Long-term soil management experiments in semiarid Fracophone Africa. In *Soil management - experimental basis for sustainability and environmental quality, Advances in Soil Science*, pp.225-266. Eds R. Lal and B.A. Stewart. London: Lewis Publishers.

Pretty, J. and Hinc, R. (2001) Reducing food poverty with sustainable agriculture: a summary of new evidence. Final report from the «SAFE-world» research project. Essex, England: University of Essex. 100 pp.

Proffitt, A.P.B., Bendotti, S. and McGarry, D. (1995) A comparison between continuous and controlled grazing on a red duplex soil. I. Effects on soil physical characteristics. *Soil and Tillage Research* **35**: 199-210.

Radford, B.J., Key, A.J., Robertson, L.N. and Thomas, G.A. (1995) Conservation tillage increases soil water storage, soil animal populations, grain yield and response to fertiliser in the semi-arid subtropics. *Australian Journal of Experimental Agriculture* **35**: 223-232.

Rahman, S. and Chen, Y. (2001) Laboratory investigation of cutting forces and soil disturbance resulting from different manure incorporation tools in a loamy sand soil. *Soil and Tillage Research* **58**: 19-29.

Reeder, R. (1992) Controlled traffic. In *Conservation tillage Systems and Managenment - crop residue management with no-till, ridge-till, mulch-till,* pp. 46-47. Iowa: Mid-West Plan Service, Iowa State University, Ames, USA.

Robert, M. (2001) Carbon sequestrtaion in soils - proposals for land management. Rome: FAO. 67 pp.

Robertson, L.S. and Erickson, A.E. (1980) Compact soil - visual symptoms. Extension Bulletin E-1460. Michigan: Michigan State University.

Scherr, S.J. (1999) Soil degradation - a threat to developing-country food security by 2020? Food agriculture and the environment discussion paper 27. Washington: International Food Policy Research Institute.

Serageldin, I. (1995) Sustainability and the wealth of nations: first steps in an ongoing journey. Washington, D.C.: Third annual World Bank conference on environmentally sustainable development. pp 35.

Soane, B.D. and van Ouwerkerk, C. (1994) Soil compaction problems in world agriculture. In *Soil compaction in crop production*, pp. 1-21. Eds B.D. Soane and C. van Ouwerkerk. Netherlands: Elsevier Science.

Topp, G.C., Reynolds, W.D, Cook, F.J, Kirby, J.M. and Carter, M.R. (1997) Physical attributes of soil quality. In *Soil quality for crop production and ecosystem health,* pp. 21-58. Eds E.G. Gregorich and M.R. Carter. Netherlands: Elsevier.

Tullberg, J. (1998) Controlled traffic: progress and potential. In *Second national controlled farming conference*, pp. 1-5. Eds J.N. Tullberg and D.F. Yule. Gatton College: University of Queensland.

Unger, P. (1996) Soil bulk density, penetration resistance and hydraulic conductivity under controlled traffic conditions. *Soil and Tillage Research* **37**: 67-75.

van Ouwerkerk, C. and Soane, B.D. (1995) ISTRO workshop on «The effects of soil compaction on Physical, chemical and biological factors in the environment». *Soil and Tillage Research* **35**: 1-4.

Walsh, P. and Jensen, T. (1998) Towards a standard for controlled traffic machinery. In *Second national controlled farming conference*, pp. 65-69. Eds J.N. Tullberg and D.F. Yule. Gatton College: University of Queensland.

Yule, D.F. (1998) Controlled traffic farming - the future. In *Second national controlled farming conference*, pp. 6-12. Eds J.N. Tullberg and D.F. Yule. Gatton College: University of Queensland.

T.F. SHAXSON

SOIL MOISTURE CONSERVATION

Greensbridge, Winterborne Kingston, Dorset DT11 9BJ, England

Resource-poor farm families in the seasonally-dry tropics suffer increasing insecurity of food and water supplies. Loss of soil porosity, caused by rain impact and/or damaging tillage, limits rates and amounts of water infiltration and percolation, provokes loss by runoff and thus diminishes possibilities for groundwater recharge. Water stress in plants has quicker, more direct and more severe effects on yields than soil erosion. Stress provokes closure of leaf stomata, reducing rates of transpiration and photosynthesis. A better understanding of the dynamic inter-relations, at microscopic level, within the soil/water/plant/atmosphere system enables development of improved soil-porosity management methods, of which organic materials and processes are essential parts. Large-scale in-field experiences in East Africa and South America, show many significant benefits, including more benign river-flow and increased and cheaper crop production, following application of these concepts.
Key words: Soil porosity; roots; plant growth; groundwater; organic processes.

INTRODUCTION

Expanding populations of resource-poor rural people have growing requirements throughout each year for clean water –derived mostly from groundwater– to satisfy domestic needs, and for irrigation if possible. At the same time, for year-round food security they require their crops and pastures to produce enough for daily use, for storage and later consumption during dry periods of the year, and/or for income-generation by the sale of plant and animal products. Achieving this depends on plants being able to access sufficient soil moisture to satisfy transpirational needs over a long enough time for them to complete their life-cycles to harvest. The source of both groundwater and soil moisture is the yearly recurrence of rainfall as a free good.

However, at the same time that the need for greater security of supplies of both water and of food is rising, ongoing degradation of land is diminishing the prospects for achieving it. In many areas with distinctly seasonal rainfall groundwater supplies are becoming less reliable in the dry season and floods are becoming more frequent in the wet season. At the same time underlying soil productivity is falling, which is commonly attributed to soil erosion. 'SWC' ('soil and water conservation') practices which have been applied to halt degradation seem to have been insufficient –in either

L. García Torres et al. (eds.). Conservation Agriculture, 317–326.

quantity or effect– to reverse these declines. In the seasonally dry tropics and subtropics the combination of increasing runoff and declining biomass production is leading to greater aridity of landscapes. How can this land degradation be reversed?

WATER MOVEMENTS INTO AND THROUGH SOIL

Rainfall can have damaging effects when it reaches any unprotected soil surface, through splashing effects of raindrop impact and their detrimental effects on soil porosity by compaction and interstitial sealing of the uppermost surface layer. This problem is more acute in the tropics and sub-tropics , where intensities, drop sizes and kinetic energy of rainfall in individual storms, as well as averaged over the year, are higher than in temperate regions. As a generalisation, in regions of temperate climate about 5% of rainfall occurs at intensities greater than 25 mm hr^{-1}, whereas about 40% of rainfall occurs at this potentially-damaging intensity within the Inter-Tropical Convergence Zone .

Intense high-energy rainfall on an unprotected soil surface breaks down the aggregates of the soil architecture (structure) at this atmosphere/soil interface. This diminishes the proportion of larger voids in the uppermost layer –only a few millimetres thick– and thereby increases the proportion of very fine pores. When these micro-pores become saturated and infiltration rate becomes much reduced, the proportion of rainfall which can enter the soil and become soil moisture is restricted because the water remaining on the surface is removed as runoff on sloping land. Similarly, beneath the surface, damage to soil architecture and porosity by compaction due to tillage or trampling increases the proportion of pores which hold water more tightly and provoke temporary saturation more frequently.

MOVEMENT TO GROUNDWATER

The less water that enters the soil, the less will remain (after transpiration has taken its toll) which can still pass down beyond the root-zone and possibly into the groundwater from which the flows of streams and rivers are derived. Fig. 1.

WATER IN SOIL PRODUCTIVITY

Water is the key ingredient for development processes in both the soil and the plant. The functioning and growth of plants should not be interrupted or curtailed by avoidable drought in the root zone. This does not imply continuous saturation of the soil but the maintenance of optimum proportions of water and air in the soil for roots to function. This requires a flux of water to satisfy transpiration as well as movements of sufficient oxygen to the roots and of carbon dioxide away from the roots. These necessary gas exchanges are achieved not only by slow diffusion through the soil voids but also by the intermittent downward passage of saturated 'fronts' of rainwater and / or of irrigation water . These fronts displace the gases by water as they pass, and then entrain the free-draining proportion of the emptying pore- spaces with atmospheric

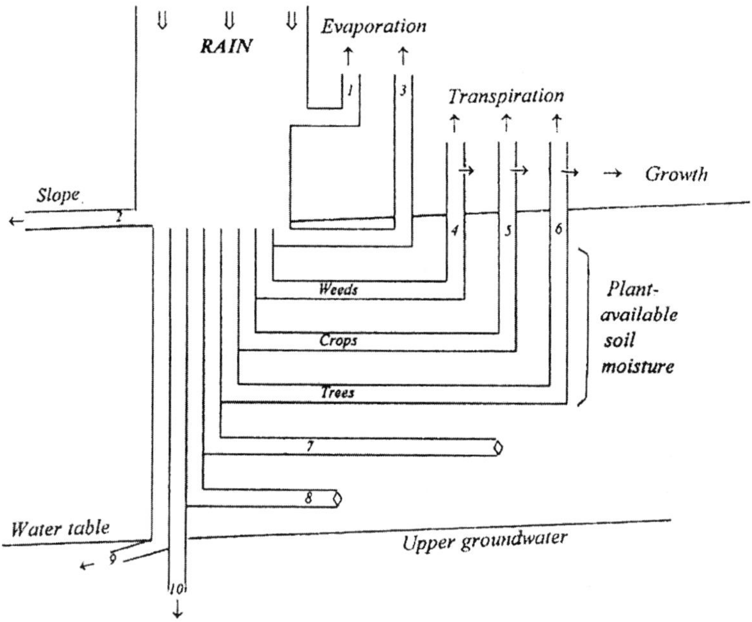

SEQUENCE

1. Direct evaporation from wetted leaf surfaces.
2. Surface runoff / stormflow.
3. Direct evaporation from the soil surface.
4,5,6. Plant-available soil moisture within the root-range of existing weeds, crops, trees.
7. Soil moisture within root-range of existing weeds, crops, trees, but held at tensions unavailable to plants.
8. Soil moisture held at all tensions, but below root-depth of existing weeds, crops, trees.
9. Water not captured by roots and small pores, moving to groundwater and streamflow.
10. Leakage to deep groundwater beneath catchment-floor.

Fig. 1. Destinations of rainwater (expanded from FAO,1995)

gases from above. Thus not only the quantities of rainwater but also the timing and dynamics of their arrival and passage through the soil are both important in considerations of soil productivity.

The genetic potentials of annual and perennial plants are, in rainfed conditions at least, moderated by the combinations of physical, chemical and biological conditions of the soil as a rooting environment. Within that environment, plant functioning and growth depends on the interrelations at microscopic level among root-hairs and roots, the size-distribution of soil pores, the quality and quantity of soil water and soil air and the proportions of each, at various places in the three-dimensional soil profile. Growth functions and outcomes are moderated also by water potentials/tensions and their gradients within the plant (at root/soil interfaces, up the xylem and at the leaf/air interface) as affected by evaporative demand at the leaf surfaces. The closure of leaves' stomatal guard cells in response to rising water potentials in the leaves regulates the transpirational loss of water from the plant when the rate of water movement from root to leaf via the xylem fails to match the rate of evaporative demand. Although growth processes in the plant incorporate only about 5% of the water passing through in the transpiration stream, the closure of guard cells at the same time restricts the rate of income of carbon dioxide from the atmosphere into the leaf and thus hinders the potential rate of photosynthesis and carbon accumulation.

SOIL MOISTURE AND PLANT ROOTS

Once water has infiltrated through the soil surface it enters the realm of plants' roots. The moisture now becomes susceptible to removal back to the atmosphere via the plant's transpiration stream. Since a plant's roots generally permeate the soil most densely near the surface, freely available soil water will be depleted earliest and most rapidly from the uppermost layers. The deeper the roots the less dense is their distribution, so the proportion of the water which has entered the soil and which is potentially available for extraction by plants is usually greater than that which is actually extractable.

Not all the water held in the soil is equally-available to plants. Soil pores fill with water by gravity and capillarity, but those of mean diameter less than 0.0002 mm hold water so strongly that it cannot drain out under the effect of gravity, nor can plant roots extract it. In the size-range 0.0002 mm to 0.05 mm water can be extracted by roots, but water still cannot drain down by gravity. It is only from pores larger than 0.05 mm that water can move both by gravity and via plant roots.

At the soil/root interface, the rate of water-transfer into the plant is governed by how strongly the water in is held by surface-tension forces in the various soil pores. When the soil has drained of free water down to Field Capacity (FC) water can move readily across the root's semi-permeable membrane into the plant because the water-potential in the soil is lower than that in the root. On the other hand, when the soil has become dried to Permanent Wilting Point (PWP) the water potential in the root is equalled or exceeded by that in the soil, and therefore the soil water does not move into the plant.

Between FC and PWP there is some flexibility in guard-cell response to increasing water stress, such that, for e.g. a cotton crop, the evapotranspiration rate may remain unchanged for a few days before water stress in the leaves is sufficient to trigger the guard-cell response which progressively reduces the evapotranspiration rate. The higher the evaporative demand, the more quickly will soil-moisture be depleted to the point where increasing water potential in the plant becomes sufficient to trigger progressive closure of the guard cells.

The huge extent of ramification of root systems in the soil is indicated by measurement of a sample of the root system of a single plant of Rye (*Secale cereale).* Table 1

Root hairs within a soil matrix are almost invisible to the naked eye whereas most visible roots have diameters larger than 0.25 mm.

Table 1. Data on roots of Rye (Secale cereale) in a soil sample 7.5 cm diam. x 15 cm deep

Roots	: length	64 m	*(210 ft.)*
	: surface area	503 cm^2	*(78 in^2)*
Root hairs	: number	12,500,000	
	: length	6,805 m	*(55100 ft)*
	: surface area	7,676 cm^2	*(1190 in^2)*

(Source: Dittmer,1938, *quoted in Russell,1981)*

Water enters a root most rapidly, and roots elongate most rapidly, within a few centimetres of the root tip and among the associated fine root hairs, when these have entered and are depleting zones of readily-available water.

In hot dry conditions where water in the soil is freely-available to roots of a crop with a full canopy of green leaves, evapotranspiration may deplete the water at a rate of 15 mm or more per day, which represents 150 m^3 of water per hectare per day. In such conditions the rate of evapotranspiration is in the region of 85% of the rate of evaporation from a free water surface, as determined by climatic conditions at the time.

A given rate of transpirational loss of water from a deep soil at FC will take longer to dry it out to PWP than a shallow soil of the same type. The frequency, duration and effects of induced drought in the root zone will be greater on the shallow soil than on the deep soil, as is commonly observed.

ASPECTS OF SOIL DEGRADATION AND YIELD DECLINE

Degradation of the soil's architecture within the root zone not only increases the proportion of smaller soil pores but also reduces the proportion of water held at plant-available tensions. Compaction also diminishes the soil's compressibility (which may hinder roots' expansion in length and thickness).

Soil erosion has often been cited as the chief cause of decline in crop yields through its effects of removing soil particles, associated nutrients and organic matter from cropped areas. However, a direct quantified relationship has yet to be convincingly demonstrated. Yield falls after erosion can be better correlated with the poorer condition of a newly-exposed subsoil layer that remains behind than with the quantity of soil eroded The primary problem is the loss of the voids in the soil architecture before erosion starts, rather than the loss of soil physical particles in a consequent erosion process. Physical degradation of soil can have negative effects on plant growth even in the absence of runoff and erosion.

Insufficiency of soil moisture is more significant, and has much quicker negative effects on plant growth and yield, than erosional loss of soil.

Runoff, which may also be transporting eroded soil, represents a loss of a volume of water which could otherwise have contributed to soil moisture, plant growth and possibly to groundwater. This may lead to unnecessary lengthening of in-soil drought and hindrance of plant growth within a growing-season. The more frequent and severe these water shortages, the worse will be the effects on plant functioning, nutrient uptake and final yield.

ROLE OF ORGANIC MATTER AND PROCESSES

The phrase 'organic matter' should be taken to include not just that whose level is determined in a laboratory as a percentage of the topsoil of a cultivated field. Because of its many potential effects the phrase should also include organic materials above the soil surface (the leaves of a canopy), and on the surface (crop residues, leaf litter, manures and composts) in addition to roots and their residues within the soil profile.

Crop and tree canopies both intercept rainfall and contribute to the litter and residues on the surface which, if well-spread across the soil surface, protect it from the force of rainfall and effects of direct insolation. But in addition to the protective effect such organic materials serve also as a substrate for the soil-churning, feeding, burrowing and transformation activities of meso- and micro-organisms –plants, animals, fungi, bacteria et al.– which live in this interface zone between atmosphere and soil. Not only do the smaller organisms break down complex materials into simpler units and build-back other more complex combinations again into their own bodies, but they also generate gums and other excretions which may build into complex humus molecules having positive effects on nutrient capture and the formation of soil aggregates Concerning water-absorption by soils, the floor-conditions of undisturbed and well-managed forests as well as of prairie grasslands in good native condition show a high capacity for enabling entry of intense rainfall. The surface is protected by the on-surface litter of shed leaves, stems and branches. The soil particles joined together by humic gums, inorganic ligands, and rootlets form the aggregates whose enclosed and interconnecting voids provide the soil porosity. Wormholes and other animal burrows into the subsurface layers provide channels for free movement of water and air into the root zone. Tubes left behind after the death and decomposition of roots contribute to the wide range of sizes of spaces into and within the soil profile.

Of the greatest significance to repairing and maintaining the porous physical condition of the soil is the ongoing capacity of active soil organisms (including fibrous plant roots) to re-formulate soil-aggregates and associated voids after damage.

VALIDATING THE APPROACH TO IMPROVING SECURITY OF FOOD AND WATER

More-appropriate attention to getting more water into the soil and thereby minimising runoff offers an integrated approach to solving problems of insecurity of supplies of plant products and of groundwater simultaneously.

BENEFITS FROM IMPROVING SOIL CONDITIONS

In Brazil, Argentina and Paraguay crop production using conventional tillage methods had resulted in serious land damage that reduced soil productivity and increased excessive soil erosion, downstream flooding and sedimentation, and diminished stability and profitability of farming. Since about 1990 these inadequate production methods have been rapidly been supplanted by zero-tillage (ZT) systems. The key technical characteristics of the different site-specific ZT systems are the use of direct-planting methods which (i) cause minimal soil disturbance to soil porosity conditions (which had been optimised as a first action in the sequence); (ii) maintain a dense cover of crop residues on the soil surface as protection against high-impact rainfall and as substrate for activity of soil organisms; (iii) use rotations of crops, including species for green manuring and as cover-crops for weed suppression, and for enhancing organic matter content and processes in the soil These features increase the resilience of the production system to unpredictable changes of weather and of markets and raise efficiency of use of plant nutrients both recycled and applied. Farmers have seen economic advantages following from adoption of these conservation-effective systems which have been great enough to encourage others, on farms both large and small, widely to adopt such practices. Because of this, the area under ZT has risen, during the last 15 years, from less than 2 million hectares to more than 20 million hectares across these three countries alone.

A number of quotations from recent reports indicate the technical, social, economic and environmental benefits which have arisen as a result of adoption of conservation-effective practices such as ZT based on crop residue management:

[In the State of Santa Catarina in southern Brazil] «*Improved land management practices, cqpable collectively of increasing water retention on crop land, improving soil structure, raising fertility and reducing erosion, were adopted on some 400,000 ha ... Spontaneous adoption of improved practices [*including minimum and zero tillage, on farms of all sizes*] occurred on a further 480,000 ha in non-project microcatchments Although* **ex post** *evaluations were complicated by drought, productivity of the main crops – maize, wheat and soyabeans – is estimated to exceed without-project productivity by 20 to 35%. Soil loss was reduced by between 10 and 50%. Runoff water in streams contained less suspended solids, coliform bacteria and pesticide residues, thus lowering silting and water treatment costs in downstream areas, and reducing the incidence of waterborne diseases and pesticide poisoning. Maintenance costs for rural roads were reduced by up to 80%: better all-weather access stimulated both commercial and social activities;*

«*The economic rate of return of the project was estimated at appraisal at 15%, based on analysis of six farm models ...* [at project conclusion] *the revised overall rate of return was estimated ... as 20%.»* (World Bank,1999).

[In a study of the economics of no-till in Paraguay:] «*Soil fertility and organic matter levels rise rapidly when no-till and green manure crops are introduced, immediately raising significantly farm incomes. Under conventional cultivation, in most cases small farmers do not use any fertiliser, very little if any manure and generally no soil conservation measures are taken. ... In the opinion of the no-till adopter farmers, crop yields immediately improve under no-till. Crop yield data are provided in the report for each of the case study farms. These clearly indicate trends of declining yields under conventional cultivation and immediately increasing yields after the adoption of green manure crops and no-till. ... The results for Paraguari indicate that investment in fertiliser and high green-mass producing green manure crops, and the introduction of no-till/green manure crops, on small farms in the extremely degraded zones of Central Paraguari would be highly economic to the nation and to the small farmers»*

«It is concluded that no-till and crop rotations constitute a technological revolution for small farmers. Never before has the senior author analysed such an impressive technology for small farmers in more than twenty years of extensive experience analysing small farm systems in South America, Africa and Asia.. No other farming techniques have been shown to have such a high impact on farmers' incomes, reduce their production costs and risks, and at the same time be environmentally sustainable and generate very considerable net social gains to society»(Sorrenson et al.,1999).

DEFINITION OF THE GROUNDWATER-RECHARGE PROCESS

In 1956 little was known for certain about what factors affected differences in streamflow from season to season, year to year or place to place in East Africa. Because only 5 percent of the region receives annually as much or more rainfall than vegetation might use, safeguarding the headwater regions of rivers whose flow provided water to the rapidly-expanding city of Nairobi was a major concern. However it was not yet possible to predict with any certainty what would be effects of specified land-use changes, provoked by the plant-production needs of a rapidly-expanding human population, on the volume and regularity of river-flow. A series of experiments was set up in headwater catchment areas to measure the effects on the streamflow regimes by four of the main land-use patterns: plantation forestry, plantation agriculture, peasant cultivations and tribal practices of cattle-herding. (Pereira,1962:3). From 1956-1975 detailed, frequent and comprehensive measurements were made in four sets of catchments –two in Kenya, one in Tanzania and one in Uganda–. Conditions of climate, plants, soils and streamflow and their interactions in affecting the hydrological cycle were studied to allow water balances to be predicted and verified.

Assessments derived from the analysis of the 130 catchment-years of data included the following:

«The emphasis was placed on assessing the effects of the land-use changes in terms of catchment water-use ... rather than in terms of streamflow because water use is actively controlled by meteorological conditions and vegetation type, whereas flow can, in a sense, be regarded as a 'residual'. ...

«Recharge of catchment aquifers is affected not only by changes in the balance between surface runoff and infiltration, but also by changes in the processes governing water use by the vegetation. Replacement of one vegetation type by another may lead to: (i) higher dry season water use, with the result that during the following wet

season a greater proportion of rainfall is required to replenish the depleted soil moisture store before recharge of rainwater can begin; (ii) changes in [rainfall-] *interception characteristics which may in turn affect the amount of water available for recharge.*
...

«...It is apparent that soil structure, and the possible effects of land-use change on it, must be considered carefully in any attempt to extrapolate the results. If the change reduces the infiltration rates, wet season water yields will increase, whilst groundwater recharge, and hence dry season water yields, will be reduced. Examples of such adverse effects are all too common in East Africa and elsewhere, particularly in the change from indigenous forest to cultivation on steep slopes.

«The results ... form a basis for the prediction of the hydrological effects of the land-use changes described, within environmental conditions and on soil types similar to those encountered in the experimental catchments. ...The key processes have been shown to be interception, infiltration rates and the control on transpiration exercised by soil moisture deficits». (Blackie & Edwards,1979: 273-277).

CONCLUSIONS

The paper draws together some existing knowledge from different disciplines concerning the nature and functioning of the soil/ water/ plant/ atmosphere system. A number of conclusions follow from this synthesis:

- The status of each component of this system at a given place and time determines the dynamics of the interacting feedback loops between them and the resulting outputs of biomass and groundwater.
- Improving a soil's damaged pore-space and its size-distribution has the effect of prolonging the usefulness of both soil and rainwater, indicated by raised and more stable productivity and greater resilience in the face of adverse weather conditions, and extended periods over which water can be removed via transpiration before PWP is reached and growth and production is halted.
- Much of the water transpired from the soil enters the plant via very large numbers and lengths of root hairs almost too small to see. Providing optimum conditions for their growth and functioning at microscopic scale should be a prime goal of soil management. When addressing problems of soil degradation, think like a root.
- Because soil moisture is not evenly-distributed throughout a soil profile, and this changes over time, it can be used productively by plants only if their roots are close enough to tap it. Degraded conditions of soil porosity can limit not only the plant-availability of the water but also the proximity of highly-absorptive root-hairs and root tips to these three-dimensional patches of moist soil.
- Policies, strategies and training need now to focus the attention - of both non-farm agriculturists and the farmers they assist - on good management of soil porosity. This entails the emphasising of organic materials and processes as integral components of soil fertility, which any necessary external inputs can complement.

- A criterion of success in making better use of rainwater is that the flow of streams and rivers becomes more benign and reliable at the same time as there are greater yields of biomass –from fields, pastures and forests– in the catchments that feed them.
- Contrary to the meaning of the paper's title, we do not need long-term 'soil moisture conservation'. Rather, we need to enable the easier entry of water into the soil followed by its unrestricted movement both upwards through a closed canopy of plant leaves, downwards –past the grasp of roots– to groundwater, but, if we can avoid it, never sideways in runoff.
- Through these perceptions we may be persuaded to manage soil details with increasing respect and care for their well-being, on which the land's continuing usefulness and productivity, and the regular flow of rivers, ultimately depends.

REFERENCES

Blackie, J.R. and Edwards, K.A. (1979) General conclusions from the land use experiments in East Africa. In *East African Agricultural and Forestry Journal* **43** (Special Issue): 273-277. (Blackie, Edwards & Clarke – compilers) pp.273-277.

FAO. (1995) Agricultural investment to promote improved capture and use of rainfall in dryland farming. *Investment Centre Technical Paper* **10**. p.4. Rome: FAO. 46pp. ISBN 92-5-103630-6.

Pereira, H.C. (1962) Introduction to: Hydrological effects of changes in land use in some East African catchments. In *East African Agricultural and Forestry Journal* **XXVII** (Special Issue): 3. (H.C,Pereira, compiler).

Russell, E.W. (1961) *Soil conditions and plant growth* (9th edn.). p.448. London: Longmans. 688pp.

Sorrenson, W.J., Duarte, C. and Lopez Portillo, J. (2001) *Aspectos económicos del sistema de siembra directa en pequenas finca.. Implicancias en la política y la inversion'. = 'Economics of no-till compared to traditional cultivation systems on small farms in Paraguay'.* (quotes taken from author's summary, in English, p.xv). Spanish version pubd. by 'Proyecto Conservación de Suelos MAG-GTZ'. San Lorenzo (Paraguay): DIA/DEAG. pp.84.

World Bank. (2000) *Implementation Completion Report on a Loan in the Amount of US$ million 33.0 to the Federative Republic of Brazil for Land Management II – Santa Catarina Project. WB report no* **20482**, *June 7th, 2000.* Washington: World Bank. 39pp.

F. TEBRÜGGE

NO-TILLAGE VISIONS- PROTECTION OF SOIL, WATER AND CLIMATE AND INFLUENCE ON MANAGEMENT AND FARM INCOME

Institute of Agricultural Engineering, Justus-Liebig-University, Braugasse 7, 35390 Giessen, DE

INTRODUCTION

With the background of political concerns, land utilisation is confronted with the serious need for the protection of soil fauna, landscape, climate and water quality more strongly now than ever before. In public discussions, agriculture is increasingly exposed to reproach, destroying «natural soil fertility», because of today's sensitivity to ecological damage. In addition, the intensity of soil tillage is looked at more critically. Besides the visible signs of soil fertility loss, e.g. soil erosion by water and wind, soil sealing and crusting of the soil surface, there are other problems, including deep soil compaction, reduced water infiltration and deep root penetration, reduced organic matter content of the soil, high emission rates of CO_2 through fast biological oxidation of organic matter and through high fuel consumption as well as drinking water pollution by leaching and run-off of nitrate, water soluble phosphate and other xenobiotica that contribute to overall environmental quality.

In the last decade, the EU Common Agriculture Policy (CAP) has promoted the modernisation of agriculture in Europe. It is recognised that this modernisation has been accompanied by damaging effects to the environment. The obvious interdependence between agriculture (in the EU 50.5% of the total territory is agricultural and 27.9% wooded land) and environment can't be ignored. Agricultural management controls much of the environment.

From the agricultural point of view, decreasing farm profits force farmers to rationalise input costs and time-intensive soil tillage. The question about the necessity of intensive soil tillage, relates to both ecological and environmental concerns of the general public and economic interests of the farmers. These questions have been the

L. García Torres et al. (eds.). Conservation Agriculture, 327–340.

subject of an interdisciplinary research group of the Justus-Liebig-University for the last 20 years.

RESULTS

This paper reviews this work with emphasis on the interactions of different soil tillage intensities on ecosystem soil properties that have demonstrated ecological, economic and environmental benefits of less tillage (Tebrügge & Eichhorn 1992; Tebrügge & Düring 1999). The investigations were conducted on five field sites characterised by different soils, crop rotations and climate with the goal to test and compare the long-term application of the no-tillage (NT) system and conventional mouldboard plough-tillage (CV) system (Tebrügge 1994). No-tillage means the abandonment of turning and loosening of the soil, where plant residues remain on the soil surface and working of the soil is limited only to the drillplanter for direct seed placement. The following comparison (Tab.1) gives an overview of quantitative and qualitative indicators of no-till in contrast to conventional tillage (100%).

Conventional tillage (CV) is characterised by a deep operation of the mouldboard plough in the topsoil, which completely inverts the soil (inversion tillage), destroys the soil structure by overloosening (total pore volume > 60%) and leaves the soil very unstable (Tebrügge & Wagner 1995, Gruber & Tebrügge 1990). Conventional tillage is also characterised by burning of straw, burying plant residues near the base of the tilled layer by inversion tillage, destroying aggregates and by continuous mono-crop based on economic return. The system of soil inversion tillage is still a common practice in Europe and continues to show obvious negative effects on the quality of soil, water and air, the climate, biodiversity and landscape ecosystem characteristics.

Soil degradation in Europe is a serious environmental problem caused by erosion and compaction from conventional-inversion tillage. About 157 million ha (16% of Europe, nearly triple the area of France) are affected by soil erosion and compaction (Oldemann et al 1991). The annual soil erosion rate in Europe averages 17 t ha^{-1} yr^{-1} and greatly exceeds the rate of the soil formation of about 1 t ha^{-1} yr^{-1} (Troeh & Thompson 1993). Most EU countries are affected by soil erosion, though in various amounts depending on soil type and climate extremes. With an average soil loss by erosion of about 2.5 Gtonnes yr^{-1} in the EU, this means that a 25 cm arable layer of soil is lost annually from 700,000 ha (80% of the arable layer of the Netherlands).

Intensification of agriculture is mainly characterised by increasing mechanisation in the soil tillage operations with consequent disturbance of soil structure and aggregate stability produced by microbial activity. As a result, the trafficability of the soil is seriously reduced. This process is critical for soil loading at the time of primary tillage and secondary tillage and seeding, and can cause deep compaction and destroy the continuity of macro pores (Gruber & Tebrügge 1990). As a consequence, infiltration of surplus water after intensive precipitation is limited and causes excess run-off. It is increasingly evident that increased intensity of soil tillage in the last decades has contributed to increased soil degradation (erosion, soil compaction, decreasing organic matter content).

Table 1. Quantitative and qualitative indicators comparing conventional* and no-till**

Quantitative and qualitative indicators for NT **Conventional tillage with mouldboard plough 100%**	**%** **lower**	**%** **higher**	**Literary reference**
Investments for machines (Euro)	38.7		Tebrügge & Böhrnsen 1997a
Power requirement (kW m^{-1})	75		Tebrügge & Böhrnsen 1997a
Working time (h ha^{-1})	79.7		Tebrügge & Böhrnsen 1997a
Fuel consumption (L ha-1)	84.4		Tebrügge & Böhrnsen 1997a
Variable costs (Euro ha^{-1}): Wages	79.4		Tebrügge & Böhrnsen 1997a
Fuel & Lubricants	85		Tebrügge & Böhrnsen 1997a
Repair costs	65.2		Tebrügge & Böhrnsen 1997a
Fixed costs (Euro ha^{-1}): Tractor	85.8		Tebrügge & Böhrnsen1997a
Stubble cultivation	100		Tebrügge & Böhrnsen 1997a
Soil tillage & sowing	27.3		Tebrügge & Böhrnsen 1997a
Performance (ha h^{-1}) incl. Glyphosate application for NT		390	Tebrügge & Böhrnsen 1997a
Driving distance (km ha^{-1})	67.2		Tebrügge & Wagner 1995
Tracks (% ha^{-1})	54.5		Tebrügge & Wagner 1995
Load index (tkm $ha^{-1)}$	69.2		Tebrügge & Wagner 1995
Soil load capacity (soil pressure resistance-triaxial test)		200	Gruber & Tebrügge 1990
Track depth (cm) after traffic load (34.3 kN)	87		Gruber & Tebrügge 1990
Soil density (Mg m^{-3}) in 0-25 cm depth		20	Tebrügge & Wagner 1995
Soil density (Mg m^{-3}) in 25-35 cm depth	7		Tebrügge & Wagner 1995
Earthworms (individuals m^{-2}) up to		700	Tebrügge & Wagner 1995
Earthworm-tubes (number m^{-2}) ø in 25 and 45 cm soil depth		570	Tebrügge & Wagner 1995
Microbial activity in 0-25 cm depth		34.2	Böhm et al. 1991
Aggregate stability (single rainfall simulator)		+++	Groß 1995, 1996
Hydraulic conductivity (cm day^{-1}) during vegetation time		+++	Groß 1996
Sealing index during vegetation time	62.5		Groß 1996
Infiltration (cm day^{-1})		+++	Ehlers 1996, Tebrügge & Abelsova 1999
Sediment run-off (t ha^{-1})	90		Towery 1998, Fischer et al. 1995
Water run-off (mm)	69		Brown et al 1996, Fawcett 1995
Lateral displacement of: herbicide (g ha^{-1})	70		Brown et al 1996, Fawcett 1995, Fischer et al 1995, Tebrügge & Düring 1999
Nitrate (kg ha^{-1})	85		Jordan & Hutcheon 1977, Bonari et al 1995
Soluble phosphate (kg ha^{-1})	65		Jordan & Hutcheon 1997
Leaching of nitrate (NO_3 -N mg L^{-1})	+ +		Kohl & Harrach1991, Beisecker 1994
CO2-emission from fuel consumption (kg ha^{-1})	84.9		Ökoenergie1999, Röver et al. 2000
Organic matter in 0-30 cm (t ha^{-1})		8.4	Tebrügge et al 1997
Organic matter (g 100g $soil^{-1}$)		22	Ball 1995, Ball et al 1998

Quantitative and qualitative indicators for NT Conventional tillage with mouldboard plough 100%	**% lower**	**% higher**	**Literary reference**
Organic carbon Corg accumulation (kg ha^{-1} year^{-1})		+ + +	Ehlers 1996, Mc. Conkey et al 2000
CO2-emission from carbon oxidation (g CO_2 m^{-2})	+ + +		Reicosky & Lindström 1993, Ball et al 1998
Cation exchange capacity (cmol kg^{-1})		+++	Carvalho & Basch 1995; Crovetto 1996
Herbicide costs (Euro) in a rotation of 75% winter cereals		15	Tebrügge & Böhrnsen 1997
Monocotyledons (number m^{-2}) in monoculture of winter cereals		+ + +	Bräutigam & Tebrügge 1996, Ball & Davies 1996
Eyespot infestation (% of plants) (3 years on 3 locations	53		Bräutigam & Tebrügge 1996;
Number of antagonists against pathogens of eyespot		218	Bräutigam & Tebrügge 1996
Number of antagonists against take-all		72	Bräutigam & Tebrügge 1996
W.-wheat infestation of fusarium graminearum (% of plants)		+ + +	Krebs et al 2000
Snails and mice population (number m-2)		+ +	Jordan et al 1996
Economical advantages (Euro ha-1) on 5 soils with different crop rotations , average of 17 years		7-20	Tebrügge & Böhrnsen 1997a
Crop yields (t ha^{-1}) average since 1979 on 5 soils, previous crop residues are chopped and remain on field: winter wheat	0	0	
winter rye		4	
spring barley	6		
winter barley	0	0	
Oats	8		
winter rape		9	
sugar beat		2	
silage and sweet maize	9		

**conventional tillage (CV): 1 labour unit, 100 kW tractor, 4m disc harrow, 4-bottom mouldboard plough,*
3m rotary harrow in combination with seed drill, 12 m sprayer, tilled area 100ha
*** no-till (NT): 1 labour unit, 52 kW tractor, 3m no-till drill, 12 m sprayer, 100 ha*
+ evident, ++ more evident, +++ most evident difference

Soil erosion not only shows considerable economic impacts on arable land by reducing soil fertility, but also causes considerable off-site infrastructure damage to roadways and sewers, basements, drainage, foundations, pavement, earthen dams, harbours and drainage channels etc. It is estimated that production costs for arable land affected by soil erosion increases about 25% per annum (53 Euro ha^{-1}). So including the off site damages (32.5 Euro ha^{-1}), which are unknownaly paid by the public as taxes, the total costs of soil loss and erosion add up to 85.5 Euro ha^{-1} yr^{-1} (Pimentel et al 1995). As a result, the 157 million ha in the EU affected by soil erosion, cause total annual costs of approximately 13.5 billion Euro (30% of the expenses in the EU-15 for agriculture in 1999). This is a significant social and environmental cost that is unnecessary.

Using no-tillage systems leads to a significant reduction in soil erosion (>90% for no-tillage, >60% for conservation tillage, Towery 1998). Assuming that of the 72.5 million ha arable land (AL) for crop production in the EU, 11.6 million ha (16%) show a high risk for erosion, only 4.6 million ha (40%) are suitable for no-tillage (Tebrügge & Böhrnsen 1997b). The loss of valuable topsoil compared to inversion tillage (185.6 Mt) could be reduced by 36%. Referring to a 25-cm thickness of the tilled layer, more than 19,000 ha yr^{-1} are protected from soil loss by no-tillage (fig.1).

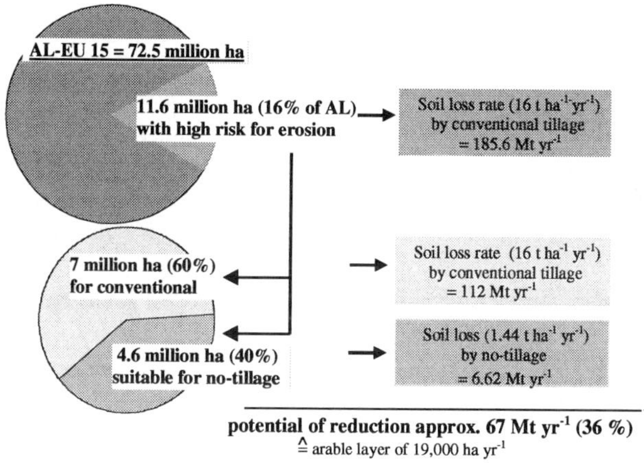

Fig. 1. Reduction of soil erosion by no-tillage in the EU-15 countries.

The total cost savings by the reduction of on and off-site costs (90% of 85.5 Euro ha^{-1}) are about 354 million Euro using no-tillage on only 40% of arable land with high risk of soil erosion (fig.2).

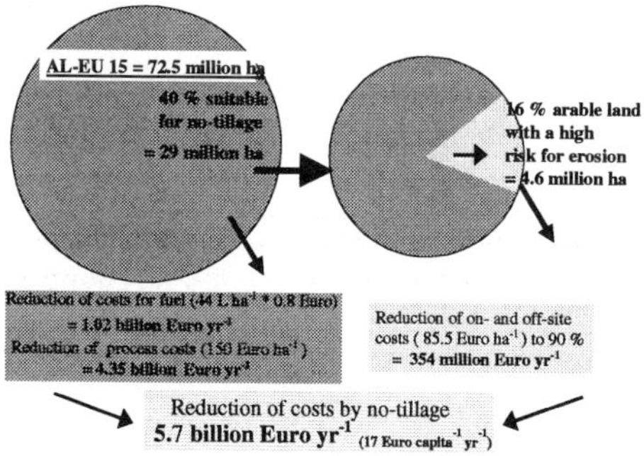

Fig. 2. Reduction of costs for crop production and environment in the EU by no-tillage in comparison to conventional tillage.

According to a recent survey taken in 9 member states of the EU , experts concluded that long-term use of no-tillage on about 40 % of the EU arable land (29 million ha) is possible without major problems (Tebrügge & Böhrnsen, 1997b). The cost saving for fuel (1.02 billion Euro) and process costs (4.35 billion Euro) would be about 5.37 billion Euro yr^{-1}. This adds up to 5,7 billion Euro per yr^{-1}.

According to calculations of the US Department of Agriculture (Glanz 1995), the total cost of wind and water erosion is figured at 7 billion US dollars. Although the application of no till started on a relatively small scale, research in the 1970s showed significant positive effects of no-tillage that showed reduction of wind and water erosion, run-off etc. and increased soil organic matter. Thus, no-tillage has a positive effect on environmental quality.

The key areas of concerns regarding agriculture and water quality are related to nitrate pollution in surface and groundwater, phosphorus levels in surface water and the contamination by pesticides. Recent estimates indicate that, agriculture accounts for more than 40% of all sources of nitrogen emissions and over 30% of phosphorus emissions into surface water (OECD 2001). Conservation tillage systems like no-tillage, cause 69% less run-off in comparison to conventional tillage. Thus, the displacement of herbicide is reduced to 70%, nitrate fertilisers to 85%, soluble phosphate to 65% and sedimentation is reduced to 93% (Blum 1990, Brown et al 1996, Fawcett 1995) and thus positively affects water quality.

In retrospect, it is evident that within the last three decades intensive soil tillage has frequently led to 50% losses of carbon (humus) (Harrod 1994). The application of no-tillage not only counteracts the loss of organic matter but increases its accumulation in the upper layer (Arrue 1997; Ball et al 1998, Tebrügge et al 1991, McConkey et al 2000, Lal 1997). This has a positive effect on the soil fauna activity (Grocholl 1991), cation exchange capacity (Crovetto 1996), soil aggregate and surface stability which inhibits surface sealing, the main cause of soil erosion (Groß, 1995, 1996). Long-term field trials in Canada (Gregorich et al 1995), in Germany, Italy, Portugal (Tebrügge et al 1997) and Spain (Gonzales-Fernandez 1997) comparing organic matter content in the 0-30 cm soil layer of not-tillage (NT) to conventional tillage (CV) verify an accumulation of organic matter of 0.8-1.5 t ha^{-1} yr^{-1} in no-tillage (fig.3).

It is widely believed that the increased atmospheric concentration of greenhouse gases (GHGs) is contributing to the process of climate change and global warming. On a global scale, only about 5% of all CO_2-emissions originate from agriculture (Cole 1996, OECD 2001). Intensive soil inversion tillage causes 6.6-times higher consumption of fuel compared to no-tillage, and also hastens decomposition of organic matter and produces additional carbon dioxide, which is emitted into the atmosphere and contributes to global warming (Kyoto Protocol 1998).

Trials in the USA verify, that 19 days after the soil tillage, the cumulative tillage-induced CO_2-emissions by microbial oxidation of the organic matter with inversion soil tillage was about 5 times higher than a soil not tilled (Reicosky & Lindstrom 1993). Repeated mouldboard ploughing reduces the supply of soil organic matter and corresponding amounts of CO_2 (approx. 1000 mg CO_2 100 g $soil^{-1}$) are emitted into the atmosphere. The loss of organic matter by microbial oxidation after ploughing is

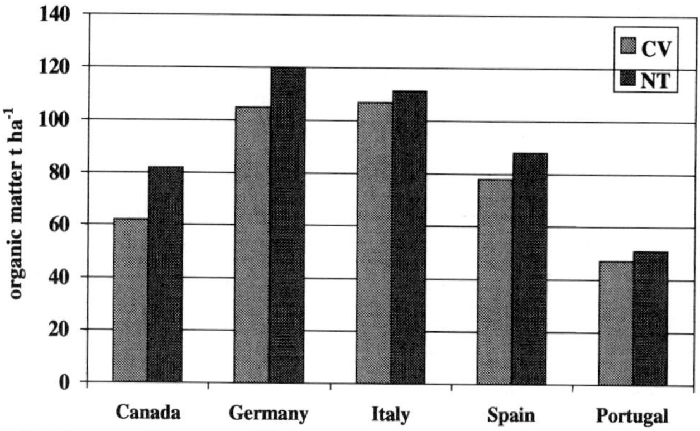

	Canada	Germany	Italy	Spain	Portugal
years of trial:	18	10	5	12	4
accumulation t ha^{-1}yr^{-1}:	1.1	1.5	0.8	0.8	1.0

Fig. 3. Effect of different soil tillage systems on organic matter.

assumed to be 10-times higher than the loss caused by soil erosion (Reicosky 1995). For every ton of carbon lost by decomposition of organic matter, approximately 3.7 t CO_2 are emitted in the atmosphere. In the case of conservation tillage, the soil carbon content increases on average over 0.77 t ha^{-1} yr^{-1} (Mc.Conkey et al 2000) thereby decreasing CO_2-emissions by over 2.8 t ha^{-1} yr^{-1}.

A calculation for the EU-15 countries shows that less microbiological oxidation and decomposition of carbon using NT on 40% of the arable land (29 million ha) leads to a reduction of CO_2-emmision of 82.7 Mt yr^{-1} (fig.4). That is nearly 10% of the total CO_2- emission from energy consumption 1998 in Germany or 3% in the EU-15 respectively (Globus 2001).

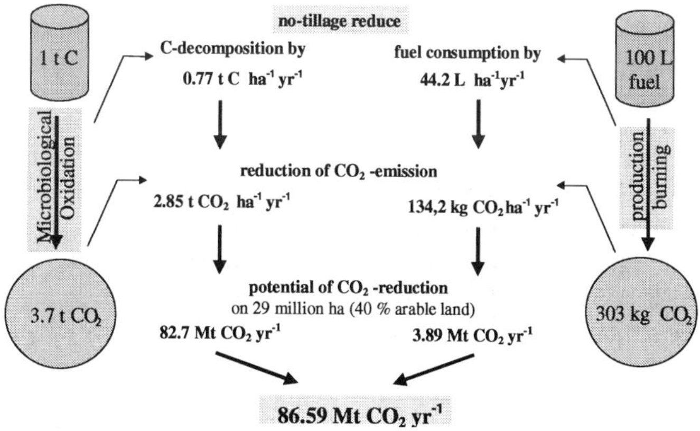

1990 EU-15: 4.33 Gt CO_2 -Emission; Kyoto-agreement 8 % Reduction = 346 Mt CO_2 till 2012 (KOM, 2000)

Fig. 4. Potential of reduction for CO_2-emissions in the EU by no-tillage

Franklin D. Roosevelt, former president of the USA, pointed out in a speech at the signing of the soil protection law in 1936, that the nation's prosperity is characterised by its care for the soil. Roosevelt said, " A nation that destroys its soil, destroys itself". So it's not a big surprise, that conservation and no-tillage in the USA are politically supported by an engaged administration encouraging increased use in agricultural practice. This is partially illustrated by the decline in numbers of mouldboard ploughs sold in North America from 60,653 in 1973 to 1,324 in 1996 (Equipment Manufactures Institute, Chicago/USA) as well as by the answers of experts in Nebraska USA which verify that 75% of the arable land (8 million ha) in Nebraska are suitable for long-term use of no-tillage (Tebrügge & Böhrnsen 2000).

Pollution concerns must not ignore the fact that production of crude oil and burning of 100 L of diesel fuel produces 303 kg CO_2 greenhouse gases (Röver et al 2000). This means that, in case of inversion tillage, seedbed preparation, seeding and stubble cleaning (51 L fuel ha^{-1}) on 72.5 million ha arable land in the EU releases circa 13 million tons of greenhouse gases, mainly as CO_2. No-tillage (6.8 L fuel ha^{-1}), even if applied on only 40% of the EU-arable land, leads to CO_2-emission reduction of 3.4 million tons by less fuel consumption.

Comprising less CO_2-emission by reduced carbon decomposition and by reduced fuel consumption the potential of reduction for CO_2-emissions from arable land in the EU by no-tillage can be calculated about 86.6 Mt yr^{-1}. That is about 25% of the Kyoto-agreement for reduction of CO_2-emissions in the EU-15 till year 2012. The calculation of the potential for CO_2 reduction by application of no-tillage on only 40% of the arable land in the EU-15 suggest, that the EU can fulfil the Kyoto-agreement approximately 4 years, if EU politicians, like those in USA and Canada, are willing to accept that agriculture is able to reduce CO_2 –emission by a changeover from conventional tillage to no-tillage. There is no doubt that conservation tillage sequesters carbon in the soil and protects the soil quality and the climate by less CO_2-emission.

Present global estimates show 48 million ha are cultivated using no-tillage systems (corresponding to approximately 70% of arable land in the EU). This means a global decrease for CO_2-emissions, solely from reduced consumption of fossil fuels of 6.4 million t yr^{-1}. To be added also are 137 million t yr^{-1}, not emitted as CO_2 because of the accumulation of soil organic carbon (0.77 t ha^{-1} yr^{-1}) in no-tillage systems (Arrue 1997).

Inversion tillage affects soil biodiversity and has a substantial influence on the activity of soil microbes, soil fertility, aggregate- and surface stability. In inversion tillage, biodiversity effects are fewer in number, because the lack of plant residues on or near to the soil surface does not provide protection against the energy of splashing rain drops (splash effect) and there is insufficient nutrition and habitat for the soil fauna. Unlike inversion tillage, the no-tillage system, with plant residues on the soil surface, has proved beneficial, supporting various forms of the macro-, meso- and microfauna and their activity in the soil (Böhm et al 1991, Tebrügge et al 1991). Charles Darwin, as early as 1881, recognized the importance of earthworms for soil formation, mixing of the soil and decomposition of plant residues in the upper layer. He mentioned that soil inhabitants unfortunately don't get the same attention as protection of whales or the habitat protection of the Galapagos archipelago. Earthworm excrements decisively take part in the process of mineralization and formation of fertile soil and deserve the same high level of attention.

On the 20-year old sites cultivated with different tillage systems, a few years after changing from inversion mouldboard ploughing, with 20 to 30 earthworms m^{-2}, to no-tillage system, there was an increase to 200 earthworms m^{-2} (Tebrügge & Abelsova 1999). These worms do not only incorporate plant residues into the soil, but mix the different soil and plant materials and create a porous surface and a deep continuous biopore system, which serves to enhance soil aeration and speedy infiltration of surplus water and allows the roots to penetrate into deep layers (Tebrügge & Abelsova 1999).

The arable layer of the no-tillage treatment shows a higher soil bulk density (ca. 1.4 – 1.5 Mg m-3) because of a smaller volume of macropores in comparison to inversion tillage (1.2 – 1.35 Mg m-3). The soil of a no-tillage treatment is able to store a larger quantity of water, because of a change in the pore structure in favour of the middle sized pores and macropore continuity (Tebrügge & Wagner 1995). Under extreme drought situations, this water is available for plant roots and absorption of nutrients (Berengena 1997). Similarly, the better conditions for root growth and yield on a no-till soil (Richter 1995), especially in the case of a water deficit, enable water uptake from deeper soil layers. Application of no-tillage provides the conditions for various chemical transformation processes in the soil. The biological incorporation of various surface plant residues in no-tillage compared to inversion tillage with mechanical incorporation close to the mouldboard plough pan, leads to a drastic change of the supply of energy and nutrients, which is reflected in the completely different distribution of available carbon and nitrogen in the arable layer (Tebrügge et al 1997).

Soil structure develops from the interplay of other abiotic living conditions of the microbes and their food supply. The multiple interactions of soil temperature, aeration and available water strongly varies the period and intensity of the microbial activity. These results suggest, that there is no need for more nitrogen fertilization in comparison to a conventionally tilled soil, except for possibly higher potential yields. Otherwise only an adaptation to the changing nitrogen availability must follow (Richter 1995).

In some EU-countries recently, conservation tillage or no-tillage is regarded as proper farming for the protection of the groundwater. The reason is no loosening of the soil in autumn reduces the danger of excess mineralization, and so in the winter months of high precipitation, a smaller amount of nitrogen leaches into the ground water. These nitrogen loss results are confirmed by the investigations of Kohl and Harrach (1991). With a high level of nitrogen in the upper layer at the beginning of the vegetation period, nitrogen is displaced after inversion tillage to 60 cm after 10-days with 63 mm precipitation. The danger of displacement and the quantity of nitrogen leaching to groundwater can be much higher on a conventional tilled soil without vegetation than on no-till soils, significantly decreasing the initial nitrogen content. The no-tillage system stands out by the fact that surplus water is quickly led away in the continuous biogenic macro pore system, without moving nitrate from the soil matrix because of bypass flow. In addition, organic nitrogen is first bounded to the organic matter in the form of N_{org} and therefore escapes the displacement in the period without vegetation. Mineralized nitrogen, $N_{min,}$ examinations (Hütsch 1991) during the entire vegetation period clearly show that no-till soils have lower N_{min} contents and lower displacement. Worst case scenarios are often simulated in the laboratory or in field by rain simulation. Complete water saturation of the soil and an agro-chemical application followed by massive precipitation would lead to a rapid displacement through the continuous pore system in deeper layers on no-tillage soils (Beisecker 1994). But no farmer would apply fertilisers or herbicides to a saturated soil with the expectation of heavy rainfall. Under normal conditions in no-tilled soils, vertical and lateral movement of plant pesticides are reduced by enhanced adsorption phenomena (Tebrügge & Düring 1999), increased microbial activity (Böhm et al 1991, Grocholl 1991) along with accelerated degradation of herbicides (Levanon et al 1994; Borin et al 1997) and resistance against inhibitory effects on the microflora (Fernau 1996).

The idea that winter wheat is often infested with soil-borne pathogens (e.g. eyespot and take-all) when straw residues are not incorporated, could not be proven by visual examination or laboratory tests (Bräutigam 1996). The values of infestation decreased significantly even with 85% of cereals in the crop rotation due to resistance and antagonists of the soil borne pathogens. The present assumption that permanent conservation or no-tillage is linked to an increasing infestation of ear dry rot (fusarium graminearum) in winter wheat after maize (Krebs et al 2000), could not be confirmed in the crop rotations on the 5 farms investigated.

The soil structural and ecological aspects that contribute to environmental benefits for a farmer in the EU-countries are presently not sufficient to encourage a decision to apply the no-tillage system. First, the yield potential and sustainable yield stand in the foreground of farmers considerations. It is generally ignored that yields and crop selling-price are confounded with process costs. That means, with continuously decreasing selling-prices in the same direction as the world market prices, the yield plus the input costs per product together decide the economic success of a farm.

The economic benefits of no tillage systems are now becoming clearer. In conventional tillage, the sum of all soil tillage operations requires higher investments for machines (by a factor of 1.63), higher costs for maintenance (by factor of 4), for fossil fuel (by a factor of 6.5), longer working time requirement (by a factor of 5) along with smaller performance in the available working period (by a factor of 4) in comparison to no-tillage (Tebrügge & Böhrnsen, 1997a). These economic differences do not include "non market value costs" for environmental damages, like off-site costs by soil erosion, water pollution by agrochemicals or CO_2-emissions, that result from conventional inversion tillage. With these large cost and environmental differences, it is difficult to understand farmer reluctance to accept no till.

The total operating costs for a 100-ha conventional tilled farm are 210 Euro ha^{-1} (Tebrügge & Böhrnsen 1997a). Considering the same size farm, the operating costs with the no-tillage system could be reduced to 50 Euro ha^{-1}. The costs are further reduced by the high performance (15 ha h^{-1}) of the no-till planter. The combined performance shows costs are a 4.2-times higher in conventional tillage compared to no tillage systems.

Comparing total operating costs for crop establishment on all 5 farms with different soil texture and different crop rotations in this long-term study, the yields and selling-prices of the crops summed over 17 years shows economic advantages of 6.9% on the sandy soil (Eutric Cambisol) up to 20.3% on the loamy soil (Eutric Luvisol) for no till compared to inversion tillage by mouldboard plough (Tebrügge & Böhrnsen 1997a) and does not include the reduction of costs by less soil degredation.

In the course of long-term trials, it is important to realize, that the closer the European selling-prices approach the world market prices (e.g. sale price for 1 ton of winter wheat in 1981: 25 Euro, in 1999: 12 Euro, for 1 ton of canola in 1989: 43 Euro, in 1999: 20 Euro), the more significant is the gain from no-tillage systems. The role of today's global markets is very important in determining the final economic return from farmers' management decisions.

SUMMARY

On the basis of 20 years of research on 5 on-farm trials in this study, and extending the results to an international level, it can be concluded that non-inversion no-tillage systems give comparable yields and improve and maintain numerous other soil functions and cross-linkages in the ecosystem. The no-tillage system, therefore, offers the possibility of soil, water and climate protection, and takes into account the economic interests of the farmer. This conclusion is more valid as the selling prices tend in the direction of the world market prices and as agricultural land becomes more highly valued for sustainable production and future environmental quality. Agricultural production systems and environmental quality are very closely linked. The ECAF-EU Life project aims to contribute in this mission. Agriculture must admit that it can afford to help save the environment. In its own interest, agriculture must conserve the soil and at the same time optimise farm profitability to be sustainable.

We (farmers, advisers and researchers) are afraid to give up the familiar, traditional ploughing, not because it is difficult, but because we can't foresee the uncertain future. This long-term study has demonstrated no till as a forward-looking sustainable system that will benefit all mankind by producing food and fibre and maintaining environmental quality. The sustainable effects on the soil-ecosystem with its beneficial effects on the environment are very complex and difficult to understand. However, the hope remains, that this new technology will get increasing public acceptance and the support of government administrations.

ACKNOWLEDGEMENTS

The author gratefully acknowledge critically reviewing the manuscript to Donald C. Reicosky, USDA-ARS , North Central Soil Cons. Res. Lab, Morris, Minesota 56267 USA. Also, appreciation is extended to financial support provided by BMFT/KFJA-033916A (1987-1993), EU-AIR 3-CT 93-1464 (1993-1997), EU-EV5V-CT93-0244 (1993-1995), EU-Life 99-E-308 (2000-2002) and DFG/SFB 299 (2000-2002) respectively.

REFERENCES

Arrue, J.L. (1997) Effect of conservation tillage in the CO_2 sink effect of the soil, p. 189-200. In: Garcia-Torres, L. and P. Gonzalez-Fernandez (eds.), Conservation agriculture: agronomic, environmental and economic bases (in Spanish), Spanish Association for Conservation Agriculture (AEAC/ SV), Cordoba, Spain, pp. 372.

Ball, B.C., Davies, D.H.K. (1996) Weed and pest control in various tillage systems, p. 9-16. In: Terügge, F. & Böhrnsen, A. (eds): Experience with the application of no-tillage crop production in the West-European countries. Proc. EC-Workshop III, Evora-Portugal, Wiss. Fachverlag Dr. Fleck, Giessen, pp. 154

Ball, B.C., Tebrügge, F., Satori, L., Giráldez, J.V. and Gonzalez, P. (1998) Influence of no tillage an physical, chemical and biological soil properties, p. 7-27. In: Tebrügge, F. & A. Böhrnsen (eds): Experience with the application of no-tillage crop production in the West-European countries. Final EU-Report of Concerted Action No Air3-CT93- 1464, pp.89, Fachverlag Köhler, Giessen, pp. 89.

Ball B.C (1995) Soil responses to tillage and their environmental implications in Scotland, p. 7-23. In: Terügge, F. & A. Böhrnsen (eds): Experience with the application of no-tillage crop production in the West-European countries. Proc. EC-Workshop II, Silsoe-Great Britain, Wiss. Fachverlag Dr. Fleck, Giessen, pp. 201

Beisecker, R. (1994) Einfluß langjährig unterschiedlicher Bodenbearbeitungssysteme auf das Bodengefüge, die Wasserinfiltration und die Stoffverlagerung eines Löß- und eines Sandbodens. Ph.D. Thesis Univ. of Giessen. Bodenökologie und Bodengenese, 12, pp.195.

Berengena, J. (1997) Effect of tillage system in soil water content, pp. 53-73. In: Garcia Torres, L. and P. Gonzales-Fernandez (eds.), Conservation Agriculture: agronomic, environmental and economic bases (in Spanish), Spanish Association for Conservation Agriculture (AEAC/ SV), Cordoba, Spain, pp. 372.

Blum, WE.H. (1990) The challenge of soil protection in Europe. *Environmental Conservation*, **17**, 72-74.

Böhm, H.J., Grocholl, H., Ahrens, E. (1991) Mikrobiologische Beurteilung von Bodenbearbeitungssystemen am Beispiel dreier Bodentypen. Z.f. Kulturtechnik u. Landent-wicklung **32**, 114-120.

Bonari, E., Mazzoncini, M., Peruzzi, A., Ginanni, M. (1995) Soil erosion and nitrogen loss as effected by different tillage systems used for durum wheat cultivated in a hilly clayey soil, p 81-91. In: Tebrügge, F. & A. Böhrnsen (eds.), Experience with the applicability of no-tillage crop production in the West-European countries. Proceeding EC-Workshop II, Silsoe-Great Britain, Wiss. Fachverlag Dr. Fleck, Giessen, pp. 201.

Borin, M., Satori, L., Giupponi, C., Mazzoncini, M., Düring, R.A., Basch, G. (1997) Effects of Tillage Systems on Herbicide Dissipation-An Experimental Approach at Field Scale. Unipress, Padova, Italy, pp.247.

Bräutigam, V. (1996) Influence of long-termed no-tillage on soil borne plant pathogens and on weeds, p. 17-29. In: Tebrügge, F. & A. Böhrnsen (eds.), Experience with the applicability of no-tillage crop production in the West-European countries. Proceeding EC Workshop III, Evora-Portugal, Wiss. Fachverlag Dr. Fleck, Giessen, pp. 154.

Brown, L., Donalson, G.V., Jordan, V.W. L., Thornes, J.B. (1996) Effects and interactions of rotation, cultivation and agrochemical input levels an soil erosion and nutrient emissions. Aspect of Applied Biology 47, Rotations and Cropping Systems, 409-412.

Carvalho, M., Basch, G. (1995) Long term effects of two di9fferent tillage treatments on a vertisol in Alentejo region of Portugal, p. 17-23. In: TERÜGGE, F. & A. BÖHRNSEN (eds): Experience with the application of no-tillage crop production in the West-European countries. Proc. EC-Workshop II, Silsoe-Great Britain, Wiss. Fachverlag Dr. Fleck, Giessen sen, pp. 201.

Cole, C.V. (1996) Intergovernmental Panel an Climate Change. Agricultural options for mitigation of greenhouse gas emissions IPCC Workgroup 11. Chapter 23, Washington, D.C.

Crovetto, C. (1996) Stubble over the soil. The vital role of plant residue in soil management to improve soil quality. American Society of Agronomy, 239 pp.

Ehlers, W. (1996) Optimizing the components of soil water balance by reduced and no-tillage, 107- 118. In: Tebrügge, F. & A. Böhrnsen (eds.), Experience with the applicability of no-tillage crop production in the West-European countries. Proceeding EC-Workshop III, Evora-Spain, Wiss. Fachverlag Dr. Fleck, Giessen, pp. 154.

Fernau, G. (1996) Einfluß der Herbizide Isoproturon, Terbuthylazin/Metachlor, Metamitron und Clopyralid auf mikrobiologische Parameter bei drei verschiedenen Boden-bearbeitungssytemen. Ph.D. Thesis, Univ. of Giessen 159 pp.

Fischer, P., Bach, M., Gäth, S., Mollenhauer, K., Frede, H.G. (1995). Geringere Herbizideinträge in Öberflächengewässer durch reduzierte Bodenbearbeitung. Mitt. Dtsch. Bodenk. Ges. 76, 253-256.

Glanz, J. (1995) Saving our soil: solution for sustaining earth's vital resource. ISBN 1-55566-136-X, p. 121.

GLOBUS Infografik (2001) Dc-6874, 56. Hamburg.

González-Fernández, P. (1997) Effect of soil tillage on organic matter and chemical properties, p. 43-49. In: Conservation agriculture: agronomic, environmental and economic bases (in Spanish), Spanish Association for Conservation Agriculture(AEAC/SV), Cordoba, Spain, pp.372

Gregorich, E.G., Angers, D.A. Campell, C.A., Carter, M.R, et al. (1995) Changes in soil organic matter. Ch. 5. In: D.F. ACTON and L.J. GREGORICH (eds.) The health of our soils. Toward sustainable agriculture in Canada, CLBRR Research Branch, Publication 1906/ Agriculture and Agri-Food Canada.

Grocholl, J. (1991) Der Einfluß verschiedener Bodenbearbeitungssysteme auf den mikrobiellen Status von Böden verschiedener Standorte unter besonderer Berücksichtigung der C-Umsetzungen. Ph.D. Thesis, Univ. of Giessen, 169 pp.

Groß, U. (1995) No-tillage as a tool to protect soil surface structure, p. 93-99. In: Tebrügge, F. & A. Böhrnsen (eds.), Experience with the applicability of no-tillage crop production in the West-European countries. Proceeding EC-Workshop II, Silsoe-Great Britain, Wiss. Fachverlag Dr. Fleck, Giessen, pp. 201.

Groß, U. (1996) Einfluß unterschiedlicher Bodenbearbeitungssysteme auf Verschläm-mungsneigung und Aggregatstabilität verschiedener Boden. Ph.D. Thesis Univ. of Giessen. Forschungsbericht Agrartechnik (VDI-MEG) 281, pp.143.

Gruber, W., Tebrügge, F. (1990) Influence of different tillage systems on trafficability and soil compaction. Paper No. 91-1090, ASAE St. Joseph, MI, pp.15.

Harrod, T. R. (1994) Runoff, soil erosion and pesticide pollution in Cornwall. In: Rickson, R.J. (edit.) 1994. Conserving soil resources, CABI, Oxford, U.K.

Hütsch, B., Mengel, K. (1991) Messung von Denitfrikationsverlusten bei Plugarbeiten und Direktsaat. Z. f. Kulturtechnik u. Landesentwicklung. **38**, p. 70-79.

Jordan, V.W.L., Hutcheon, J.A., Kendell, D.A. (1996) Influence of cultivation practices on arable crop pests, diseases and weeds and their control requirements, p. 43-50. In:Tebrügge, F. & A. Böhrnsen (eds.): Experience with the applicability of no-tillage crop production in the West-European countries. Proceeding EC-Workshop III, Evora-Portugal, Wiss. Fachverlag Dr. Fleck, Giessen, pp. 154.

Jordan, V.W.L., Hutcheon, J.A. (1997) Alternative farming method (arable): a study of the integrated arable farming system on level of herbicide and nutients reaching controlled waters. Journal Environment Agency, Bristol, UK, Technical Report No. P17, pp.65.

Kohl, R. and Harrach, T. (1991) Zeitliche und räumliche Variabilität der Nitratkonzentration in der Bodenlösung in einem langjährigen Bodenbearbeitungsversuch. Z. f. Kulturtechnik und Landententwicklung., **32**, p. 80-86.

Krebs, H., Dubois, D. Külling, C., Forrer, H.R., Strei, B. Rieger, S. and Richter, W. (2000) Fusarien- und Toxinbildung des Weizens bei Direktsaat. *Agrarforschung* **7** (6) p. 264-268.

KYOTO PROTOCOL/UNITED NATIONS (1998) Framework Conventions on Climate

Lal, R. (1997) Residue management, conservation tillage and soil restoration for mitigating greenhose effect by CO_2-enrichment. Soil & Tillage Research. 43 (1-2): 81-107. Change, http.//www.un/ccc.de/

Levanon, D., Meisinger, J.J., Coding, E.E., Starr, J.L. (1994) Impact of soil tillage on microbial activity and the fate of pesticedes in the upper soil. *Water, Air and Soil Pollution* **72**, p.179-189.

McConkey, B., Chang Liang, B., Padbury, G. and Lindwall, W. (2000) Carbon sequestration and direct seeding. In Proceeding: Direct seeding "Sustainable Farming in the New Millenium"12th Annual Meeting, Conference and Trade Show of the Saskatchewan Soil Conservation Association, February 9 & 10, 2000.

McIntyre, D.S. (1958) Soil splash and the formation of surface crust by raindrop impact. *Soil Sci.* **85**, 261-266.

OECD (2001) Environmental Indicators for Agriculture, Volume 3: Methods and Results. http://www.oecd.org/agr/env/themes.htm

Ökoenergie (2000) Magazin zur Förderung erneuerbarer Energie und Energieeffizienz. Nr. 30, Ed.: Universität für Bodenkultur Wien, Ökozoziales Forum Österreich.

Oldeman L.R. , Hakkeling, R.T.A. and Sombroek, W.G. (1991) World Map of the status of human-induced soil degradation and explanatory note (2nd. Edition), Global Assessment of Soil Degradation (GLASOD), ISRIC, Wageningen, The Netherlands.

Pimentel, D., Harvey, C., Resosudarmo, P., Sinclair, K. Kurz, D. McNair, M., Crist, S., Shpritz, L., Fitton, L., Saffouri, R., Blair, R. (1995) Environmental and economic cost of soil erosion and conservation benefits. Science, 267, p.1117-1123.

Reicosky, D.C. and Lindstrom, M.J. (1993) Fall tillage method: Effect on short-term carbon dioxide flux from soil. *Agronomy Journal.* **85**, p.1237-1243.

Reicosky, D.C. (1995) Impact of tillage an soil as a carbon sink p.50-53. In: Farming for a better Environment. A White Paper, Soil, Water Conservation Soc., Ankeny, Iowa, pp.67.

Richter, U. (1995) Einfluß langjährig differenzierter Bodenbearbeitunssysteme auf das Bodengefüge und den Stickstoffhaushalt. Diss. Gießen. Boden und Landschaft, 4, pp. 163.

Röver, M., et al.(2000) Bewertung von Verfahren der ökologischen und konventionellen landwirtschaftlichen Produktion im Hinblick auf den Energieeinsatz und bestimmte Schadgasemissionen. Landbauforschung Völkenrode, Sonderheft 211.

Tebrügge, F., Gruber, W., Kohl, R., Böhm, H. (1991) Long-term cultural practices effects on the ecologic system. Paper No. 91-1009, ASAE St.-Joseph, MI, pp.15.

Tebrügge, F., Eichhorn, H. (1992) Die ökologischen und ökonomischen Aspekte von Bodenbearbeitungssystemen, p.7-20. In. Friebe, B. (ed.), Wechselwirkungen von Boden bearbeitungssystemen auf das Ökosystem Boden. Wiss. Fachverlag Dr. Fleck, Niederkleen.

Tebrügge, F. (1994) Beurteilung von Bodenbearbeitungssystemen unter den Aspekten von Bodenschutz und Ökonomie, p. 5-16. In: Tebrügge, F. & M. Dreier (eds.), Beurteilung von Bodenbearbeitungssystemen hinsichtlich ihrer Arbeitseffekte und deren langfristige Auswirkungen auf den Boden, p. 252. Wiss. Fachv. Dr. Fleck, Langgöns, pp.252.

Tebrügge, F., Wagner, A. (1995) Soil structure and trafficability after long-term application of no-tillage, p. 49-57. In: Tebrügge, F. & A. Böhrnsen (eds.): Experience with the applicability of no-tillage crop production in the West European countries. Proc. EC-Workshop II, Silsoe-Great Britain, Wiss. Fachverlag Dr. Fleck, Giessen, pp. 201.

Tebrügge, F., Borin, M. Mazzoncini, M. and Basch, G. (1997) Effects of tillage systems on physical, chemical and biological soil characteristics, p. 41-74. In: Borin, M. et al. (eds): Effects of tillage systems on herbicide Dissipation, Unipress Padova, pp. 247.

Tebrügge, F., Böhrnsen (1997a) Crop yields and economic aspects of no-tillage compared to plough tillage, p. 25-43.

Tebrügge, F., Böhrnsen, A. (1997b) Survey with no-tillage crop production in the West-European countries, p. 55-153. In: Tebrügge, F. & A. Böhrnsen (eds.): Experience with the applicability of no-tillage crop production in the West-European Countries. Proc. EC-Workshop IV, Boigneville-France, Wiss. Fachverlag Dr.Fleck, Giessen, pp.192.

Tebrügge, F., Abelsova, J. (1999) Bioporen fördern die Versickerung – Auswirkung der Pflug-Bodenbearbeitung und Direktsaat auf biogene Durchporung und ungesättigte Infiltrationsleistung des Bodens. *Landtechnik* **54.**,1, p.13-15

Tebrügge, F., Düring, R.A. (1999) Reducing tillage intensity – a review of results from long-term study in Germany. *Soil & Tillage Research* **53**, p.15-28.

Tebrügge, F., Böhrnsen, A. (2000) Direktsaat -Beurteilung durch Landwirte und Experten in der EU und Nebraska. *Landtechnik* **55.**, 1, p.17-19

Towery, D. (1998) No-till's impact on water quality, p. 17-26. In: 6th Argentine National Congress of Direct Drilling (in Spanish AAPRESID), Mar de Plata, Argentina, p. 17-26.

Troeh, F.R. and Thompson, L.M. (1993) Soils and soil fertility, Oxford University Press, New York.

J.N. LANDERS, G. SANT'ANNA DE C BARROS, M. THEOTO ROCHA, W.A. MANFRINATO & J. WEISS

ENVIRONMENTAL IMPACTS OF ZERO TILLAGE IN BRAZIL – A FIRST APPROXIMATION

Brazilian National Science Council Scholar and Executive Secretary, Association of Zero Tillage in the Cerrado, MUDB Conjunto 9 Lote 5 cep 71680-090, Lago Sul, Brasília, DF, Brazil

Professor, Socio-Economics Department, Escola Superior de Agricultura Luiz de Queiroz, Piracicaba, SP, Brazil.

Graduate student, Socio-Economics Department, Escola Superior de Agricultura Luiz de Queiroz, Piracicaba, SP, Brazil.

Graduate student, Biogeochemistry, Centro de Energia Nuclear na Agricultura, Piracicaba, SP, Brazil.

Professor, Sustainable Development Center, University of Brasilia.

The economic values of the principal impacts of Zero Tillage have been estimated for the principal agricultural states of Brazil where ZT is concentrated. Only values of direct use (corresponding approximately to the on-farm values) and indirect use (off-farm values) are estimated. Impacts not yet quantifiable were identified, but option and existence values were not determined. On-farm benefits were estimated as equivalent to the value of soil depreciation; potential yield increases were not computed. The principal indirect benefits estimated lie in the huge potential for mitigating deforestation through intensification of crop/livestock production on degraded pastures and the dramatic reduction in the off-farm effects of soil erosion (road maintenance, water treatment, silting), lower emissions of greenhouse gases (GHG's), aquifer recharge and stream-flow perennization, due to increased rainfall infiltration. The implications for government policy to stimulate Zero Tillage adoption *per se* and through rewards to farmers for responsible land stewardship are discussed.

Key words: Zero Tillage, economic impacts, co-responsibility and deforestation.

INTRODUCTION

The degradation of world soil resources and its implications for world food security have been well documented (FAO, 1999). Brazil's new Zero Tillage (ZT) technology

L. García Torres et al. (eds.). Conservation Agriculture, 341–350.

may have the key to reversing this degradation spiral (Landers 1999, in press). In the crop year 1999/2000, there were 14,33 million hectares of crops under Zero Tillage including winter grains (FEBRAPDP, unpublished data, 2001), or 35 per cent of annual crops plus sugar cane. Over 90% of this happened since ECO ´92. This sustainable system represents a reversal of the natural resource degradation process engendered by Conventional Tillage (CT).

To ensure continued adoption of Zero Tillage, significant government investment is required for this practice to expand from a current base of 35 % of annual crops plus sugar cane (1999/2000) to a proposed target of 80% of this area over a period of 10 years. New technology for integrating crop x livestock enterprises allows normal soybean and maize yields with ZT in degraded pasture (Broch et al.,1997 & Mello, 1995). Degraded carrying capacity of 0.5 U.A./ha may be trebled over a three-year pasture ley following three annual crops (Broch et al., 1997). This could accommodate all demand-led expansion of Brazil's agricultural production within the existing agricultural frontier for the next 20 years, or more, preserving native vegetation, historically sacrificed to this end.

The authors found no published scientific data for Brazil which confirmed negative impacts of ZT on the soil or environment and specifically, no increase of the danger of soil pollution through ZT substituting Conventional Tillage was encountered.

MATERIAL AND METHODS

Existing information only was used for this first attempt at placing economic values on the impacts of Zero Tillage in Brazil. The admitted limitations of the data base are taken into account by conservative assumptions, where needed. Pointing up such limitations is considered as a contribution to further, and necessary, refinement of these calculations.

The economic values of the principal impacts of Zero Tillage were estimated for the states of RS, SC, PR, SP (South region), MG, MS, MT, GO, DF, TO, BA, RO, MA, (Central Region) responsible for over 98% of 1999/2000 Zero Tillage area (15.4 million ha). This excludes most of the drier Northeast and the humid Amazon states, manioc, permanent crops and forestry were not included, representing a conservative factor in the benefits estimated. The estimates of Zero Tillage area include some areas of intermittent minimum cultivation alternating with Zero Tillage.

Benefits were calculated for Zero Tillage adoption on the present 35 % and a future 80% of the present annual crop area under study. The area for overhead irrigation, 830,327 ha in 1998, (Secretariat of Water Resources/Ministry of the Environment, Water Resources and Legal Amazônia - unpublished data) was used to compute economy in the consumptive use of water, since drip irrigation is a relatively small area and already shows considerable water economy *per se*. A 50% adoption of ZT was assumed for this area. The area under furrow irrigation methods, not known to include ZT, was not included in this calculation.

Only values of direct use (corresponding approximately to the on-farm values) and indirect use (off-farm, or downstream, values) were estimated. Values of indirect use occur both as economies in public spending and in conservation of natural resources. Impacts not yet quantifiable were identified, but option and existence values were not determined. Additionally, the current trend to renovate degraded pasture, with an annual crop break under ZT, could permit an expansion of 1 million ha/yr of annual crops for 25 years and also mitigate deforestation, by accommodating both crop and herd expansion through doubling or trebling stocking rates (Broch et al., 1997).

Besides these direct benefits to farmers, the principal indirect use benefits constitute reductions in public spending. These derive from the dramatic reduction in the off-farm effects of preservation of native vegetation, reduced soil erosion (silting, road maintenance and water treatment), lower emissions of GHG's, aquifer recharge and stream-flow perennization, due to increased rainfall infiltration.

No attempt was made to correct for historical dollar inflation in dollar values utilised and values in the Brazilian currency, the Real were converted at the exchange rate of the period of validity of the data. There follow details of the assumptions utilised and estimates of the individual and consolidated impacts. Other positive effects on the conservation of natural and man-made resources, not quantifiable at present, were identified. In the U.S.A., Colaccico (1989) quantified off-farm benefits of soil conservation to recreation, navigation, reduced flooding, etc.

CALCULATIONS FOR THE PRINCIPAL ON-FARM BENEFITS

Reduced soil depreciation or incremental net margins

Most of the direct use benefits could be estimated either in terms related to reduced soil loss or to improved land prices.. On a latossol (oxysol) at the Instituto Agronômico de Campinas, similar to cerrado soil conditions, Tengberg et al. (1997) reported the impact of seven years of erosion induced by four levels of artificial cover, soil loss and runoff. Without cover, crop yield decline was equivalent to a loss of 4 kg/ha of maize per ton of total soil loss, which would translate into about US$5/ha/yr in terms of maize yield loss, at an average reduction of erosion under ZT of 17.7ton/ha/yr (De Maria, 1999). Bassi (1999) valued the reduction in nutrient losses in the river water of a catchment, which received soil conservation practices and where ZT was adopted on 48% of the arable area, at US$8,40/ha/yr. According to Stobehn (1986), the average present value of the loss of a ton of soil in the U.S.A. was US$0.49 at 1982 prices, of which US$ 0.29 in permanent impacts and only US$ 0.20 due to nutrient loss. This implies that lower nutrient losses may not be the major factor in the enhanced yields found for ZT (Hernani et al.1996, Muzilli et al., 1994, Torres et al., 2000) probably related more to lower drought stress, earlier planting and greater nutrient re-cycling. Palmquist et al. (1989), using hedonic analysis to study the effect of erosion control and drainage upon North Carolina land prices, estimated that a one ton soil loss reduced land price by US$6.19 per acre (US$ 15.29/ha); this would presumably also cover remedial costs.

This study adopted as a proxy for the whole study area the Brazilian estimate of Bastos Filho (1995) of US$36.50/ha/year (1991 value) for soil depreciation due to erosion under annual crops in SP State. Applying a net reduction of 76% in soil erosion under ZT versus CT. (De Maria, 1999) gives an adjusted value of US$27,74/ha/yr. This is much more conservative than the authors' estimate of US$48.91/ha/yr (prices for September 2000), based on comparative data from Paraná state (DERAL/SEAB, 2000) of a net annual margin advantage to ZT over Conventional Tillage in typical rotations. These figures also included significant yield increases under ZT, not related to erosion.

Water economy in overhead irrigation

The on-farm benefits of reduced consumptive use of water in overhead irrigation were estimated as equivalent to a water economy of 10%, applied to total pumping costs of US$200/ha, (800mm for two crops), giving on-farm benefits of US$20/ha/yr. To the 10% of water saved the value was imputed of a typical irrigation water charge of US$ 0.01/m^3 as the value of this benefit.

ESTIMATES OF OFF-FARM BENEFITS

Lower maintenance costs for rural roads

There are 1,265,907 km of earth roads in the study area (DNER, 2000). For earth road maintenance in PR, Carroll et al. (1997) estimated an annual economy of US$425/km/yr due to soil conservation practices; 50% of this is attributed to ZT adoption. It was also assumed that only 50% of the length of earth roads are located in arable areas.

Reduced treatment costs for municipal water

Surface water sources supply 96.5 million Brazilians (SUS-MS, 1991) with an estimated 40 l/day. In Paraná State, a reduction of 43% in turbidity reduced water treatment costs by US$ 5,77/10.000m^3 (Carroll et al., 1997). It is assumed that 50% of this benefit accrues to ZT adoption. Some 60% of the Brazilian population reside in the study area.

Increased reservoir life

To emphasise the magnitude of this factor, Carvalho et al. (2000) estimated an average annual loss of 0,5% in Brazils reservoir storage capacity, i.e., 2000 x 10^6 m^3, or about US$700.million/yr at reservoir replacement cost. For the state of São Paulo, Marques (1995 cited by López 1997) estimated annual losses of US$ 64 to US$ 74 million to hydroelectric generation due to the effects of erosion.. There is a net reduction of soil loss by erosion from annual crops of 17.2 ton/ha/yr due to adoption of ZT (De Maria, 2000). This soil loss is applied to arable areas in a typical medium-sized river basin area of 10,000 km^2 and a corresponding sediment yield of approximately 20% (Renfro, 1975), with an estimated 75% retention (Landers, 1996). A ratio of apparent density of 1.1 to 1.3 was estimated for eroded soil to sediment (J.E.F.W.Lima, Agência Nacional de Energia Elétrica, ANEEL – personal communication, 2001). The relatively low value shown in Table 1 is partially due to the dilution of arable area in the whole watershed, but the disparity appears too great when compared to the high figures cited above and requires an in-depth study.

Reduced dredging of rivers and ports

From the above, 25% of sediment from eroded soil ends up in rivers with catchments greater than 10,000 km^2, ports or the sea. It was assumed that only 10% of this requires future dredging at a cost of US\$3.90/m^3, (Landers, 1996). Dredging cost.

Improved aquifer recharge

Under ZT conditions, when compared with CT, the average reduction in rainfall runoff in a résumé of 30 experiments was of 95,3mm/ano (De Maria, 2000). The value of US\$ 0.01/m^3 was imputed as for on-farm water economy.

Water economy in irrigation

Under overhead irrigation, Stone and Moreira (1998) showed economies in consumptive use of water as high as 30%, with a very thick mulch; farmers report economies of 5–20%, depending on soil cover. Using an average of 10% water economy on annual consumptive use of 800mm gives an annual economy of 800 m^3/ha, which remains unused. Valuing alternative uses as equal to the irrigation water charge of US\$0,01/ m^3 gives a value of US\$8/ha/yr.

Carbon credits for fuel economy

Two estimates of fuel economy from adoption of ZT (Gentil, 1993 & Landers, 1999) averaged a reduction of 31 lit/ha/yr. The value of carbon credits was estimated by multiplying this value by the specific gravity and the carbon content of diesel oil (0.84 and 16%) and the above price of US\$10.91/ton of carbon.

Carbon sequestration in soil under ZT

Potential carbon credits were calculated at a value of US\$ 10.91/ton of carbon for reduced emissions of CO_2 under ZT through carbon sequestered in incremental soil organic matter (equivalent to US\$3/ton of carbon dioxide). Separate estimates were made for the four southern states, averaging the results for the 30cm depth of soil of 1.44Mg/ha/yr (Bayer et al. 2000) and 0,94Mg/ha/yr (Sá, 2000), Averaging these figures gives US\$12.98/ha/yr. For the remaining (tropical) states the value of Resck (1998) of 0.994 Mg/ha/yr for the 40 cm depth was taken, giving benefits of US\$10.84/ha/yr. Methane and nitrous oxide emissions were not quantified. Evidence from USA

(Robertson et al. 2000) shows a significant increase in mitigation of Global Warming Potential (GWP) when these emissions are computed.

Carbon sequestration in ZT surface residues

Average regional estimates of surface residue dry matter over the year were taken as 3 ton/ha in the South and 1 ton/ha in the tropics. Carbon equivalents were obtained by assuming an average carbon content of 250kg of carbon/ton of residue, giving on-farm credits of US$6.54 and US$ 2.18 once only, respectively, which is pro-rated over20 years to estimate the average annual contribution. This estimate ignores the slow accumulation of surface residues over time

Valuation of native vegetation preserved

Salati (1994) gives a value of 380 tons of carbon in the biomass of Amazon forest. If only 50% of this be converted to CO^2 on clearing, at a world market price ofUS$10.91/ ton of carbon, a proxy value of US$2122/ha can be established. For Cerrado vegetation half of this value is assumed. Since it is not possible to predict the ratios of these areas in the Amazon and Cerrado biomas which would be saved by incentives to the adoption of ZT in degraded pastures, half of each is assumed. An average value for native vegetation saved was taken as US$1568/ha, and an *ad hoc* estimate was made of 0,5 million hectares already under integrated ZT x livestock systems and an average annual increment was assumed of this same area was assumed over the next 20 years. The other benefits corresponding to this incremental area were not calculated.

RESULTS

Table 1 summarises the estimated benefits at the present adoption rate of 35% (12 million ha of summer planted crops) in the study area and at 80% of this area adopted (equivalent to an increment of 15,4 million ha under ZT).

Of the US$1,4 million in total current benefits for ZT adoption 57% were due the value of native vegetation saved due to ZT intensification on degraded pastures, 26% to on-farm benefits, only 4% in reductions in public spending and 13% to environmental impacts. Those impacts not yet quantified are summarised in Table 2.

DISCUSSION

To date, the expansion of Zero Tillage in Brazil has been driven by the comparatively higher margins it has generated, but these will tend to fall, as the market adjusts to the new efficiency levels of farmers. Also, the higher management level plus initial investments and incremental working capital required to convert to ZT mean that the more capitalised and larger farmers adopted first. This process may now be slowing down, requiring incentives to farmers and training both for technicians and for new adopters, plus a revision of teaching curricula.

Table 1. Annual economic impacts of Zero Tillage adoption in the area studied (Million US dollars)

Categories of Impacts	35% ZT adoption	80% ZT adoption
A. On-farm-benefits	***356,1***	***791,4***
Incremental net benefits ZT versus CT	332,9	739,7
Irrigation pumping economy	23,2	51,7
B. Off-farm reductions in public spending	**62,1**	**138,0**
Maintenance of rural roads	48,4	107,6
Municipal water treatment	0,5	1,1
Incremental reservoir life	9,2	20,4
Reduced dredging costs in ports and rivers	4,0	8,9
C. Off-farm environmental impacts	**184,1**	**409,1**
Greater aquifer recharge	114,4	254,1
Carbon credits for diesel economy	0,6	1,4
Irrigation water economy	6,6	14,8
Carbon sequestration in soil	59,5	132,2
Carbon sequestration in surface residues	3,0	6,6
D. Benefits to integrated ZT x livestock systems	**784,0**	**1742,2**
Total Benefits	**1386,3**	**3080,7**

A government policy of rewarding farmers for sustainable stewardship of natural resources is more than justified by the levels of public and environmental benefits indicated in Table 1. When the benefits to mitigating deforestation through recuperating degraded pastures with integrated ZT x livestock (ley) systems are taken into account, such a policy would appear to be unavoidable. By implicitly or explicitly valuing the natural resources so preserved, this would constitute a non-market transfer of part of the

Table 2. A list of un-quantified impacts of Zero Tillage

- Positive balance for terrestrial, soil and aquatic biodiversity.
- Lower chemical pollution levels in surface and coastal waters;
- Reduced emissions of methane and nitrous oxide;
- Value of environmental services in preserved natural resources (biodiversity, aquifer recharge, oxygen balance, scenic value and sustainable extraction).
- Longer concentration times for flood runoff and less flood damage;
- Possible mitigati0on of methane emissions by ruminants due to improved feed;
- Better air quality from less dust generation and elimination of burning;
- Greater food security; due to higher, more stable yields and lower food prices;
- Lower food prices per se *reducing living costs*;
- Greater rural incomes leading to less rural-urban migration;
- Better quality of rural life;
- Lower imports of petroleum and fertilisers and higher agricultural exports.

benefits already generated and not a subsidy. This practice is widespread in the USA and Europe for «Best Management Practices». (BMP´s). For instance, the principal of paying farmers for aquifer recharge at the same rates they pay for irrigation water extraction appears to be equitable.

The total carbon sequestration in soil and surface residue of US$63 million (present) and US$139 million (future) justify addressing policy stimuli to further ZT adoption, which has both immediate and ongoing impacts. The granting of carbon credits is subject to verification of «leakage». In the case of reduced diesel consumption this can be verified from fiscal notes of sale, however with a reversion to CT, sequestration of carbon in soil and residues would be negative and there is also danger from accidental burning of surface residues. These risks are considered low, due to the continued growth of the ZT area and the environmental conscience of the ZT farmer.

The estimated total value of US$ 9,2 million/yr for reservoir storage capacity spared by reduced silting at the current 35% adoption of ZT would rise to US$20,4 million/yr at 80% adoption. This value would appear to be low when compared to the equivalent total national loss of US$350 million/yr. An alternative calculation might use the reduction in costs of electricity generation. Economies in water treatment and carbon credits for lower diesel consumption are relatively minor impacts.

The majority of city dwellers, whose own sewage pollution levels and GHG emissions are played down, ignores the considerable achievements of the farm sector towards fulfilment of Agenda 21. The fact that the farmers' efforts have been overwhelmingly self-financed emphasises this merit. Recently, Brazil has been receiving many foreign visitors to see this technology, but Brazilian society and the world at large need to hear this positive news.

CONCLUSIONS

The economies and un-quantified impacts indicated above justify policy incentives to compensate farmers for these benefits to society. Such incentives are essential to the encouragement of vertical growth of Brazil's agricultural sector and consequent mitigation of deforestation. Beyond the overwhelming economic evidence presented to justify such incentives, they need to be homologated by society as a whole, accepting the principle of co-responsibility, past, present and future, for the preservation of natural resources. This must include the farmers' past actions as an economic agent of that society (world-wide, or national), which consumed the products of agriculture. The principles.of Brazilian ZT technology are universal; it has many for solutions for sustainable farming systems, especially in countries with similar soil/climate situations. The long list of additional impacts that were identified means that there is much scope for further research, especially in developing the basic data needed for economic analysis.

REFERENCES

Bassi, L. (1999) Better Environment, Better Water Management, Better Income and Better Quality of Life in Micro-catchments. *The Land Management II Project/World Bank. World Bank Rural Week*, 1999. World Bank, Washington, D.C.

Bastos Filho, G.S. (1995) Contabilizando a erosão do solo: um ajuste ambiental para o produto bruto agropecuario paulista. Dissertação (Mestrado). Escola Superior de Agricultura Luiz de Queiroz.

Bayer, C., Mielniczuk, J., Amado, T.J.C, Martin-Neto, L. and Fernandes, S.V. (2000) Organic Matter Storage in a Sandy Clay Loam Acrisol Affected by Tillage and Cropping Systems in Southern Brazil. In: *Soil & Tillage Research* **54**.

Broch, D.L., Pitol, C., Borges, E.P. (1997) Integração Agricultura-Pecuária: plantio direto da soja sobre pastagem na integração agropecuária. Maracaju-MS, FUNDAÇÃO MS para pesquisa e Difusão de Tecnologias Agropecuárias. *Informativo técnico*, **01**: 24

Carroll M. (1997) Paraná Rural – Projeto de Manejo e Conservação do Solo do Paraná. BIRD – Unidade de Gerenciamento – Brasil, Brasília, DF, Brasil. 29pp.

Carvalho, N.O., Filizola Júnior, N.P., Santos, P.M.C., Lima, J.E.F.W. (2000) Ocorrência de reservatórios assoreados no país. In *Guia de avaliação de assoreamento de reservatórios*. ANEL. Dupligráfica Editora. Ltda. Pp. 13-18

Colaccio, D., Osborn, T. and Alt, K. (1989) Economic damage from soil erosion.. In: Journal of Soil and Water Conservation, January/February. Washington, D.C. USA.

De Maria , J.C. (1999) Erosão e Terraços em Plantio Direto. Boletim Informativo da Sociedade Brasileira de Ciência de Solo, V.**24**:3 julho/Setembro. Viçosa, MG, Brazil. Pp. 17-21

FAO (1999) State of the World's Forests. Food and Agriculture Organization of the United Nations, Rome, Italy, 154 pp.

Gentil, L.V., Gonçalves, A.L.D, da Silva, K.B. (1993) Comparação econômica, operaciona e agronômica entre o Plantio Direto e o Convencional no Cerrado brasileiro. Universidade de Brasília, Brasília, DF, Brazil. 21pp. Xerox.

Hernani, L.C., Salton, J.C., Fabrício, A,C., Dedecek, R., Alves Júnior, M. (1996) Perdas por erosão e rendimentos de soja e de trigo sob diferentes sistemas de preparo de um latossolo roxo de Dourados (MS). *Congresso Latino-Americano de Ciência do Solo, 13o.* Águas de Lindóia, SBCS. [Proceedings on CD ROM].

IFPRI/UNEP report

IPCC (2000) Special Report on Land Use, Land-Use Change And Forestry

Landers, J.N. (1996) O Plantio Direto na agricultura: O caso do Cerrado. In: Lopes I V. et al.- *Gestão Ambiental no Brasil: experiência e sucesso.* Rio de Janeiro: Fundação Getúlio Vargas. pp. 03-33

Landers, J.N. (1999) How and why the Brazilian Zero Tillage explosion occurred. Paper presented to ISCO **(in press)** Lopes I V. et al.

López, A.O. (1997) Análise dos custos privados e sociais da erosão do solo – o caso da bacia do Rio Corumbataí. *Tese de Doutorado – Escola Superior de Agricultura Luiz de Queiroz.*

Mello J.F. (1995) Plantio Direto de Pastagens em Campo Nativo e Rotação de Culturas. *In: I Seminário Internacional do Sistema Plantio Direto.* Resumos. EMBRAPA- CNPT, Ponta Grossa, RS. Agosto. Pp. 119-120.

Muzilli, O., Vieira, M.J., Oliveira, E.L. (1994) Avaliação dos sistemas de plantio direto e convencional em diferentes sucessões de culturas. IAPAR-Programa Recursos Naturais. Relatório Técnico Anual. Londrina.

Palmquist, R.B., Danielson, L.E. (1989) A hedonic Study of the effects of erosion control and drainage on farmland values. Journal Series of the North Caroline Agricultural State University, no. 11633. USA.

Renfro, G.W. (1975) Use of erosion equations and sediment delivery ratios for producing sediment yield. In: *Present and prospective technology for producing sediment yield and sources.* USDA-ERS (ed.). Oxford, MS, pp. 33-45.

Resck, D.V.S. (1998) Agricultural intensification systems and their impact on soil and water quality in the Cerrados of Brazil. In: R. Lal (ed.) Soil quality and agricultural sustainability. Ann Arbor Press. Chelsea, MI. cap. 20. p.288-300.

Robertson, G.P., Eldor, A.P., Richard, R.H. (2000) *Greenhouse Gases in Intensive Agriculture: Contributions of Individual Gases to the Radiative Forcing of the Atmosphere.* 15 September. Vol **289** *Science . www.sciencemag.org*

Sá, J.C.M. (2000) Ph.D.Thesis. The Ohio State University, USA

Salati, E. (1994) Emissão versus sequestro de CO_2 – uma nova oportunidade de negócios para o Brasil. In *Anais do seminário : emissão versus sequestro de CO_2.* Rio de Janeiro, 1994.

Stobehn, R. (ed.) (1986) An economic appraisal of USDA erosion control programs: a new perspective. Agr. Econ. Report, no. 560, Ec. Res. Serv., USDA, Washington, DC.

Stone, L.F., Moreira, J.AA. (1998) A irrigação no plantio direto. In: Jornal Direto no Cerrado, Ano 3, No.8, Maio. APDC, Goiânia, GO, Brazil.

Tengberg, A., Stocking, M. and Dechen, S.C.F. (1997). The impact of erosion on soil productivity – an experimental design applied in São Paulo State, Brazil, *Geografiska Annaler,* vol. **79** A, p. 95 – 107.

Torres, E., Saraiva, O.F., Farias, J.R.B., Galerani, P.R. (2000) Avaliação de Sistemas de Preparo do solo, Rotação de Culturas e Semeadura da Soja. In: EMBRAPA. Centro Nacional de Pesquisa de Soja (Londrina, PR). Resultados de pesquisa da Embrapa Soja, 1999. Londrina. p. 104-109. (Embrapa Soja. Documentos, 142).

N. SOLORZANO

WATERSHED CONSERVATION FARMING «A FRIENDLY SOLUTION TO SOILS DEGRADATION»

FAO, Ministry of Agriculture and Husbandry; Apdo. 8198-1000, San José, Costa Rica

Strategically, the problem and solution to soils degradation in Costa Rica should begin by establishing the current correlation between watersheds and activities conducted in them. To associate the activities carried out at the watershed such as housing, commerce, agro-industry and production, among others, motivates people and awakens their interest. There are three elements common to the extension work conducted in the watershed, sub-watershed and micro-watershed: the aquifer recharge area, the spring and the water border division. The aquifer recharge zone is emphasized as the watershed area intervened by people. It is important for people to relate technology currently employed for production at the watershed with its potential to produce water, oxygen and clean air, as well as carbon capture, biodiversity, food production, landscape improvement and recreation. In this way, the interest to make a change towards conservation farming, as a way of producing without degrading natural resources, mainly soil and water takes place. Several examples of conservation farming practices in Costa Rica are also presented.
Key words: Watershed, Management, Degradation, Conservation Farming, National Plan.

INTRODUCTION

Traditionally, it has been very difficult to attract politicians, technicians and farmers attention and interest to develop, manage, use and preserve natural resources under an integrated watershed approach. It is estimated that a convenient strategy would be to formulate the problem going from the particular to the general. People at all levels talk about the importance of watershed conservation but they visualize it faraway from its real location. For this reason, it is necessary to create an extension-communication methodology to bring people nearer the watershed. An aspect to consider is that many activities develop daily at the watershed, such as: housing, commerce, agro-industry, roads, agriculture and livestock production, among others. It is important to make people aware of the watershed's capacity to produce water, oxygen, clear air, carbon capture, biodiversity, food, landscape betterment and recreation to improve life quality.

L. García Torres et al. (eds.). Conservation Agriculture, 351–354.

In Costa Rica, the execution of the project «Development and Application of Land Management Conservation Practices MAG-FAO/GCP/COS/012/NET», has generated experiences in pilot areas on conservation farming, showing that it is possible to produce without degrading natural resource. Since nowadays, there is a higher awareness level regarding the importance of watershed conservation, people should also think about conservation farming. Because of that, this paper presents the basic aspects involved in the future national agriculture conservation program.

MATERIALS AND METHODS

To define a conservation farming strategy, it is highly relevant to take into account the watershed. In this regards, the first aspect that will be defined is the watershed, then the watershed's management and last the technology currently employed for agriculture production.

Definition and characterization of a micro-watershed

The micro-watershed is defined as «A land area naturally delimited by higher zones (water division border) where surface and ground waters go to a common drain, namely: a river, creek, brook, spring» Viera and Cubero 1997.

There are three elements common to the watershed, sub-watershed and micro-watershed.

- The spring or common drain (river, brook, creek, etc.)
- The water division border (highest part, separating a micro-watershed from another)
- The aquifer recharge area (where people intervene, located between the water division border and the spring).

The watershed and its inhabitants

There is an ongoing effort to develop the concept of watershed in a more clearer way, to help people associate the watershed directly with the daily activities conducted within it, and to relate the technology they employ with events occurring in it. The purpose is to develop a disposition to change their attitude toward adopting conservation farming alternatives.

Human beings carry out farming, livestock, forest and industrial labors within the watershed; there, he develops housing projects together with agriculture activities and other commercial and socioeconomic transactions. All these activities generate products that generally mean economic earnings. However, without appropriate precaution measures regarding infrastructure planning, harvest residues management, garbage management and land management and utilization, they can become watershed degrading products and contaminants. Costa Rica's physical-geographical nature divides the country into countless watersheds, sub-watersheds and micro-watersheds.

Technology employed and consequences

Costa Rica has based its economy on agricultural activities, taking advantage of available natural resources such as water, soil, and forests, among others; but based on a strong colonization of different potential available areas. This situation has caused depletion of the agriculture frontier, causing serious surface and ground water contamination problems, watersheds deterioration and soils erosion. All this, make it possible to visualize difficult times to supply population food needs which, as a matter of fact, is growing at an accelerating rhythm. Guido, 2000.

Another aspect that strongly affects natural resources deterioration is the application of technology which aim is to maximize production and productivity, allowing high agrochemicals utilization, machinery and farming practices which favor erosion and soil fertility losses. In Costa Rica, there is an erosion rate of 28.3 t/ha/yr at the Birrís sub-watershed, located in the Reventazón watershed. The Reventado sub-watershed shows an erosion rate of 23 t/ha/yr, and the annual sediment average during the last three years for the Birris watershed is 1.200.000 t/ha/yr. Gomez and Associates 2000. As a result, water's infiltration capacity decreases and sedimentation rates increase at hydroelectric dams, endangering population's food security. Thus, it can be stated that the country's environmental, social and economical development will depend highly on how natural resources (mainly soil and water) are managed and utilized.

RESULTS

In Costa Rica, based on experiences generated mainly by the project «Development and Application of Land Management Conservation Practices MAG-FAO/GCP/COS/012/NET», conducted at national level under the figure of pilot areas with participation of political and technical levels of the agriculture sector, it has been possible to create conscience of the need to produce in a sustainable way, mainly of soil and water resources. Technological changes are one of the most important aspects to consider in times of strong competitiveness and currently, there is conscience that technology should help to attain sustainable production, mainly of soil resources. In this sense, efforts conducted to produce a change, involve the adoption of conservation farming practices through a practical and simple approach of the extensionists work to implement conservation farming. It is important that people think of the watershed and the micro-watershed as the action places where daily activities occur. They should be familiar with it in order to attain the watershed management necessary for a better future.

In this regards, there is State interest to include a conservation farming policy in the national agriculture development plan that will involve the following aspects:

Micro-watershed utilization as a management and planning unit

- There is an order defined by nature in the micro-watershed for natural resources performance; for this reason, it will be used as a basic planning unit for socio-productive activities.

- To develop technologies suitable to the country's conditions, involving agrochemicals reduction and implementation of minimum tillage systems. This strategy will seek to develop a more environmentally friendly agriculture, allowing food production in a more profitable way, thus contributing to preserve natural resources. Solórzano and Dercksen, 2000.
- It will consider diversified production systems, allowing a wide range of income, taking advantage of products and by-products. This will also reduce contamination and will favor a densely covered surface.

Participation of civil society local bases

The success of this plan depends on an active and conscious participation of civil society; namely, institutions, organizations, universities and municipalities; but above all, of farm owners at the micro-watersheds. This involvement will be attained by privileging the development of crop systems following natural resources preservation techniques which main objective will be to attain higher production using innocuousness and quality criteria.

Training addressed to the producer and his family

The central axis of the training processes will be the producer and his family. Training will be addressed mainly to young local leaders that will promote conservation farming.

Technical bases to execute the Plan

The following specific activities will be considered for the plan's execution: to maintain soil with coverage, to increase soil water infiltration, to reduce erosion and environment contamination, to increase organic matter and in general, to increase productivity.

REFERENCES

Viera, M. y Cubero, D. (1997) Agricultura Conservacionista. Ambitos de planificación participativa. Editorial Master Lito, MAG, FAO, San José, Costa Rica 56 p.

Guido, E. (2000) Informe de Costa Rica para la implementación de la convención de las Naciones Unidas para combatir la desertificación. I Informe de la Comisión para la Desertificación de Tierras (CADETI). San José, Costa Rica. 146p.

Gomez, C. y Asociados (2000) Plan de Manejo Integral de la Cuenca del río Reventazón. Instituto Costarricense de Electricidad. San José, Costa Rica. 175 p.

Solórzano, N. y Dercksen, P. (2000) Agricultura Conservacionista para productores y productoras en cuencas y microcuencas hidrográficas: Una herramienta para la producción sostenible. 1 Edición. Gestión Documentación e información, Gerencia General ICE. San José, Costa Rica. 48 p.

V.O. LOPES DE GERENYU, I.N. KURGANOVA, & D.V. SAPRONOV

CARBON DIOXIDE FLUXES FROM ARABLE SOILS AS AFFECTED BY TEMPERATURE AND MOISTURE

Institute of Physicochemical and Biological Problems of Soil Science, RAS
Pushchino, Moscow region, 142290 - Russia

Fluxes of carbon dioxide from arable gray forest soils (clay Humic Luvisol) were studied under field conditions during 36 months. The mean daily CO_2 evolution rates varied - from 0.9 to 246 mg C·m^{-2}·hour^{-1}. The annual flux of CO_2 from arable gray forest soils averaged 4.2 t C·ha^{-1}. Contribution of cold period (November - April) ranged from 12 to 16% of average annual flux. The exponential Q_{10} function over whole the period of investigations and over the different classes of soil temperature (Ts) and moisture (Ms) were calculated. Q_{10} value for all the data obtained was equal to 2.64 (n=104) and increased from approx. 0.8 at Ms<10% to approx. 5.4 at Ms > 30%. Significant differences of Q_{10} function were observed between different classes of Ts. The highest Q_{10} values were 13.3 at Ts = 5-10°C (Ms = 20-30%) and 8.8 at Ts = 0-5°C (Ms > 30%).

Key words: CO_2 emission, annual flux, agricultural soil, temperature dependence, Q_{10}.

INTRODUCTION

The carbon dioxide flux from soils is an important indicator of their microbial activity and intensity of organic matter decomposition. The processes of organic matter destruction are intensified by ploughing-up of virgin soils (De Jong & Schappert, 1972). As a result, CO_2 fluxes to atmosphere from just ploghed-up soils increase and can exceed the CO_2 evolution rate from virgin soils. Contribution of agricultural soils to global CO_2 flux is very significant. The current estimates of CO_2 fluxes are mainly based on studies conducted during growing season (Kudeyarov, Kurganova, 1998). It is suggested that carbon dioxide emission beyond the growing season is negligibly small. Our previous investigations (Lopes de Gerenyu et al, 2001) have demonstrated that neglect of CO_2 emission in cold season leads to incorrect assessments of annual CO_2 fluxes and carbon balance as a whole. The objective of this study is to determine the annual and seasonal carbon dioxide fluxes and to estimate the influence of soil temperature and moisture on CO_2 evolution rates.

L. García Torres et al. (eds.). Conservation Agriculture, 355–360.

MATERIALS AND METHODS

Site description

The investigations were conducted under field conditions during 3 years. The experimental plots (winter wheat) were located 4 km west of Pushchino (Moscow region, Russia, 54°50'N, 37°35'E) on moderately eroded clay gray forest soil (Humic Luvisols C_{total} 1.09 %, $\_\_{H2O}$ 6.0). The soils studied had been intensively used for agriculture since the 19th century.

CO_2 emission measurements

CO_2 emission by soils was measured by close chamber method over the period from November, 1997 to October, 2000 at 7-10 days intervals. The total number of measurements was 104. Measurements were conducted between 9 a.m. and 11 a.m. The number of replicates was 3-5. Soil moisture and temperature in the upper soil layer (0-5 cm) were also determined at each sampling date. The air samples (20 cm^3) were collected by syringe, transported to the laboratory in hermetically sealed flasks and analysed by gas chromatograph. The dynamics of CO_2 concentrations in the chamber was measured in 15, 30 and 45 min during warm period (May-October) and in 45, 90 and 135 min during cold season (November- April). The CO_2 flux was calculated according to following equation:

$$F=(C-C_0)\cdot H\cdot t^{-1},$$

where F is the CO_2 – C flux, mg $C\cdot m^{-2}\cdot h^{-1}$;$C_0$ - are the initial head-space concentrations of CO_2-C, mg $C\cdot m^{-3;}$ C is the head-space concentration of CO_2-C in time t (hour); H is the height of the head-space layer in the chamber, m.

Results

The average daily CO_2 evolution rate significantly varied during the study period, both temporally and spatially (Fig.1). The coefficient of spatial variation for CO_2 emission ranged from 5 to 173% with a mean 29%.The soil temperature changed from -2 to 31°C and soil moisture - from 2 to 61 %. Maximal carbon dioxide fluxes of 164-236 mg C $m^{-2}h^{-1}$ were observed in July-August, 1998, when hydrothermal conditions were optimal.

The annual flux of CO_2 from agricultural gray forest soils depended on weather conditions and averaged 4.2 t $C\cdot ha^{-1}$ (CV = 32 %). Contribution of cold period was 12.4% of average annual flux (Table 1). The CO_2 emissions from agriculture soils during winter, spring, summer and autumn were 0.17, 0.51, 2.15 and 1.36 t $C\cdot ha^{-1}$ $season^{-1}$ respectively. The contributions of seasons to annual emission were 4, 11, 52 and 33 %, accordingly.

Correlation between CO_2 emission and soil temperature was significant, but not very close (R= 0.50, F<0.01). The exponential equation described the dynamic of CO_2 emission more adequately than the simple linear equation. The correlation between soil respiration

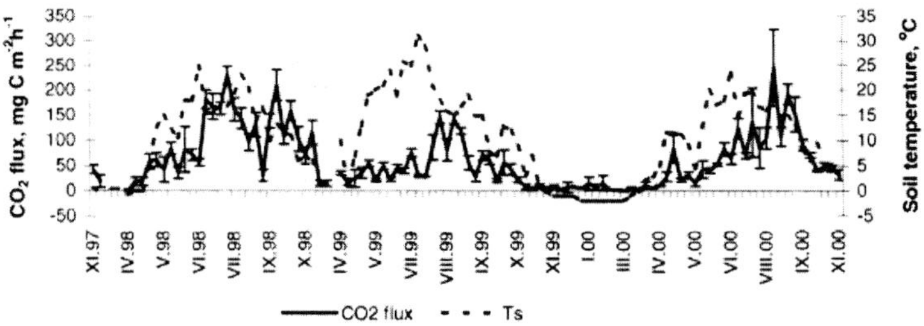

Fig. 1. Dynamic of CO_2 flux from arable gray forest soils and soil temperature at 5 cm.

rate and soil evolution rate more accurately, the average daily values of CO_2 fluxes were divided into classes of soil temperature (7 intervals, step 5°C) and furthermore subdivided into classes of soil moisture (4 intervals, step 10%). The highest CO_2 emission from agriculture soils (138-141 mg C·m^{-2}·hour^{-1}) was revealed at intervals Ts = 10-20°C and Ms = 20-30% (Table 2).

Table 1. Total CO_2 emission and contribution of different periods to annual CO_2 flux from agricultural gray forest soils

Period	Df's	CO_2 flux, t C·ha^{-1} Mean	SE's	CV, %	Contribution, % Mean	SE's	CV, %
Cold season (November-April)	2	0.52	0.10	19.5	12.4	2.56	20.6
Warm season (May-October)	2	3.66	1.26	34.3	87.6	2.56	2.9
Winter (December-February)	2	0.17	0.00	0.0	4.1	1.64	40.4
Spring (March-May)	2	0.50	0.13	24.8	12.0	1.50	12.5
Summer (June-August)	2	2.15	0.60	28.0	51.4	2.64	5.1
Autumn (September-November)	2	1.36	0.63	46.2	32.6	5.59	17.1
Total	2	4.18	1.35	32.3	100		

Table 2. Emission rates of CO_2 (±df's, mg C·m^{-2}·h^{-1}) from arable forest gray soils of different classes of soil temperature and moisture

Soil moisture range	Index	Soil temperature classes, °C 1 < 0	2 0-5	3 5-10	4 10-15	5 15-20	6 20-25	7 25-30	2:+30
< 10%	mean					49±23	52±22	34±6	46±20
	df's					4	3	2	11
10-20%	mean		14±5	26±6	53±27	79±38	102±40		58±38
	df's		2	3	9	7	2		27
20-30%	mean		19±14	68±39	137±46	141±62			97±66
	df's		8	9	9	11			40
> 30%	mean	7±5	22±18	41±19					14±15
	df's	15	2	2					21
1-30%	Mean	7±6	19±2	54±38	95±56	103±63	73±36	34±6	63±57
	df's	15	14	16	19	24	6	2	103

The exponential Q_{10} function (Katterer et al., 1998) over whole the period of investigations and over the different classes of soil temperature and moisture were calculated (Table 3). Q_{10} value averaged for all the data obtained was equal to 2.64 (n=104) and increased from about 0.8 at Ms<10% to about 5.4 at Ms > 30% (Table 3). Significant differences of Q_{10} values were observed between different classes of Ts. The highest Q_{10} values were: 13.3 at Ts = 5-10°C (Ms = 20-30%) and 8.8 at Ts = 0-5°C (Ms > 30%).

Discussion

CO_2 emission from arable gray forest soil estimated for growing season was at 1.5-1.8 t C·ha^{-1} (Pomazkina et al., 1999), and the annual CO_2 flux estimated was 1.8 t C·ha^{-1} (Larionova et al, 2001). The difference between our estimates (3.7 and 4.2 t C·ha^{-1} accodingly) and those in literature can be explained by different weather conditions and the lack of data on cold emission of CO_2. To obtain real estimations of annual and season CO_2 fluxes it is necessary to carry out long-term all-the -year-round measurements of carbon dioxide emission.

Q_{10} values depended on temperature and varied from about 13.3 to 0.4, decreasing with the increase of Ts. Possibly, this variation in Q_{10} values is due to temperature optima which can differ in time and space between organism communities (Katterer et al., 1998). So, the high respiration response to temperature increase in low temperature intervals can be explained by activation of soil psychrophyllic microflora. The high Q_{10}-values at temperatures below 5°C were found by T. Katterer et al. (1998). To use Q_{10} function for description of the dependence of CO_2 evolution on temperature, it is necessary to range the data with regard to Ts and Ms values.

ACKNOWLEDGEMENTS

This study was supported by Russian Foundation for Basic Researches (grant “Russian scientific schools” _ 00-15-97917, projects _ 01-04-48468) and Russian Ministry of Industry,Science and Technology (project «Biogenic sinks, sources and reservoirs of greenhouse gases»).

REFERENCES

De Jong, T. and Schappert, H.J.V. (1972) Calculation of soil respiration and activity from CO_2 profiles in the soil. Soil Science ***113(5)****: 328-333.*

Katterer, T., Reichstein, Andren, O. and Lomander, A. (1998) Temperature dependence of organic matter decomposition: a critical review using literature data analyzed with different models. Biology and Fertility of Soils *27: 258-262.*

Kudeyarov, V.N. and Kurganova, I.N. (1998) Carbon dioxide emissions and net primary production of Russian terrestrial ecosystems. *Biology and Fertility of Soils* **27**: 246-250.

Larionova, A.A., Yermolaev, A.M., Blagodatsky, S.A., Rozanova, L.N., Yevdokimov, I.V. and Orlinsky, D.B. (1998) Soil respiration and carbon balance of gray forest soils as affected by land Biology and Fertility of Soils *27: 251-257.*

Lopes de Gerenyu, V., Kurganova, I., Rozanova, L. and Kudeyarov, V. (2001) Annual fluxes of carbon dioxide from some Russian south-taiga soils. Eurasian Soil Science. *9: 1215-1226.*

Pomazkina, L.V., Lubnina, E.V., Zorina, S.Yu., Kotova, L.G. and Khortolomey I.V. (1996) Dymamic of CO_2 Emission from gray forest soil in the forest-steepe of Transbaikal region. Eurasian Soil Science ***12**: 1454-1459.*

B.G. SIMS & D.H. O'NEILL

THE ROLE OF DRAFT ANIMAL POWER IN SOIL AND WATER CONSERVATION

Silsoe Research Institute, Wrest Park, Silsoe Bedford, MK45 4HS, UK

Draft animal power (DAP) is fundamental to many small-holder farming systems in developing countries. Participatory research projects in Bolivia and Zimbabwe are investigating the use of this power source to develop soil and water conservation technologies for marginal farmers in semi-arid conditions. Work in Bolivia combines contour barriers of forage species with the use of single work animals on the forming terraces. High-lift harnesses and lightweight implements diversify the use of equids. Other conservation innovations in the area include winged chisel plows, tied ridgers and direct seeders. Work in Zimbabwe has also developed light-weight implements and contributed to the promotion of donkey power. Conservation tillage practices (no-till, ripping, mulching) have been evaluated with respect to their requirements and benefits and have been found suitable for adoption by smallholders provided they have adequate DAP. More on-farm trials in different conditions are needed to demonstrate the benefits more widely.

Key words: Bolivia, Zimbabwe, participatory research, draft animals, conservation

INTRODUCTION

Draft animal power (DAP) is of fundamental importance in developing country agriculture with about a quarter of the total energy requirement emanating from this source (compared with 70% from human power).

In Latin America, draft animals were introduced by the Spanish conquistadors over 500 years ago, and in many places the original technology (comprising draft oxen and wooden plows) is still in use. In hillside areas of Central and South America, smallholder farmers are frequently forced to farm fragile marginal areas (with these technologies originally designed for flat terrain) with the consequent degradation of the soil resource. The resulting decreases in farm productivity lead to reduced income and impoverished livelihoods.

The situation in Sub-Saharan Africa is similar although with a somewhat different history. DAP has been established for no more than about 100 years but, equally,

L. García Torres et al. (eds.). Conservation Agriculture, 361–366.

smallholders usually have access to only the poorer agro-ecological zones. Soil fertility has steadily decreased and increasingly erratic weather patterns have exacerbated the difficulties of crop production, mainly through drought. Furthermore, the droughts have had a significant negative impact on smallholder cattle ownership and, hence, access to their traditional source of draft power.

Farmer-participatory research projects in Bolivia and Zimbabwe are producing technologies that are responding to farm-family demands for an end to soil fertility depletion and for increased farm productivity. This paper provides, by means of two case studies, a brief summary of the DAP innovations which are contributing to the battle against rural poverty through the application of soil and water conservation (SWC) practices.

Bolivia case study

Since 1996, the Department for International Development (UK) has been implementing participatory research projects with smallholder farmers in the mid-Andean valleys of Bolivia. One of the projects (PROMETA) has been investigating the management, diversification, nutrition and health of working animals (bovines and equids). Sister projects have introduced farmer-demanded hillside conservation measures that include vegetative contour strips which have the additional advantage of producing dry-season forage (Sims and Rodríguez, 2001). This paper presents a small part of the achievements of these projects that relate to SWC.

MATERIALS AND METHODS

The mid Andean valleys of Bolivia are characterised by being high (2000 – 4000 masl), dry (<600 mm yr^{-1}) and cool (average temperatures of 8 C). Steep slopes are farmed using the draft oxen and ard plow introduced by the Spanish in the 16th Century. Soil erosion is rife resulting in falling crop yields.

The PROMETA methodology involves an iterative process of visits to its six participating communities located in contrasting areas (Sims, Dijkman and Zambrana, 2001). Farm communities prioritise technical problems that are channelled through locally elected research committees. PROMETA technical staff discuss these issues and produce technical solutions in their home-based workshop. Prototypes are tested on-farm and modified until the farmer customers are satisfied. The associated CIFEMA factory then produces a batch of about 30 units that are further tested on-farm in the PROMETA communities and further a field via the CIFEMA extensionists. Finally larger scale batch production (200 units) follows as farmer demand drives the manufacturing process.

The vegetative contour strips are also the product of participatory research and the species selected for the different agro-ecological niches found in the valleys region, have been selected by farming communities after on-farm evaluation. The practice is now being widely disseminated (Sims and Bentley, 1999).

RESULTS AND DISCUSSION

Contour-planted live-barriers produce two, very important, results. First they stabilise hillside plots and allow terraces to be formed slowly and naturally. Secondly they can produce palatable and nutritious forage at times of critical scarcity in the dry-season. The judicious mix of grass and legume species (phalaris grass - *Phalaris tuberoarundinacea* and woolly pod vetch - *Vicia villosa* ssp. *dasycarpa* have proved to be particularly prolific and adaptable) can secure forage supplies and avoid the necessity of selling-off working animals. As terraces form so they produce narrower, flatter surfaces for field work. This work can be performed by lighter, single animals if the right harnesses and tillage tools are available. PROMETA has produced a high-lift harness (Inns and Sims, 2000) and light-weight implements for horses and donkeys that include: reversible and conventional moldboard plows, harrows, ridgers, seeders and weeders. Diversification of equid use has meant a more efficient use of scarce fodder and animals.

Rain-water harvesting has a high potential for improving crop yields and reducing risk in this semi-arid environment. PROMETA has worked on the possibilities on two fronts; chisel plowing at the end of the dry season, and tied ridging for *in situ* collection of rain water. The chisel plow design (Inns, 2000) has inclined wings to increase the volume of soil disturbed per unit of draft force (Spoor and Godwin, 1978). Chisel plowing on the contour with this tool permits deep loosening (20-25 cm) of the soil that captures surface run-off at the first rains.

Direct seeding can conserve soil and soil moisture and so allow crops to be produced in lower rainfall conditions. PROMETA is collaborating with the International Maize and Wheat Improvement Center (CIMMYT) on the design, evaluation and manufacture of an animal-drawn direct seeder (Protrigo, 2000).

Zimbabwe case study

The small-scale sector in Zimbabwe comprises almost one million households practising mixed (i.e. crop / livestock) farming in agro-ecological zones with limited potential. They are characterised by low (unimodal) annual rainfall (c. 600 - 800 mm) and sandy soils with low water retention. Such soils need careful management if the productivity levels to support household needs are to be achieved and sustained. There is a perceived shortage of cattle within the smallholder sector (although there is considerable variability between households) which results in inadequate supplies of manure and limited availability of DAP, especially in the poorer households. A major need for smallholder farmers, therefore, in these circumstances is for technology to provide improved soil and water management practices at a reduced draft power demand.

MATERIALS AND METHODS

A portfolio of projects funded by DFID and other donors (e.g. see Nyagumbo, 1999) has enabled participatory on-farm research, in the communal areas (State land held in trust) of Zimbabwe, to be undertaken to address soil and water conservation issues particularly in the context of limited DAP for land preparation. The participating farmers are selected by their communities and run trials to investigate the characteristics and effectiveness of different tillage implements and practices with varying degrees of autonomy. The conservation tillage treatments evaluated in this way included no-till with hand planting, ripping and green manures, with or without pre-season plowing, and these were compared with conventional moldboard plow based practices. The inputs, draft requirements, crop responses (in all cases) and the soil responses (in most cases) were assessed both scientifically and by the participating households.

RESULTS AND DISCUSSION

The use of donkeys as a source of draft power for tillage operations has been promoted through the successful development of a light-weight plow, in conjunction with a major manufacturer (Mbanje *et al.*, 1997). The acceptance of donkey power increases smallholders' draft power resource and increases the potential uptake of conservation tillage implements requiring relatively low draft (e.g. <750 N, for a span of four donkeys or two oxen). Minimum tillage operations are widely held to increase the weed burden and, hence, the labour requirement at this critical period. However, research has shown that pre-season land preparation («winter plowing») inhibits this increase and the use of DAP for weeding, which often may be effected by a single donkey, alleviates the labour bottleneck. Whilst research continues to evaluate the requirements for, and benefits of, conservation tillage practices (e.g. ripping - see Mbanje *et al.*, 2001), it has already been established that, if sufficient DAP is available, the options for smallholders to adopt them are increased.

ACKNOWLEDGEMENT

This publication is an output from research projects funded by the United Kingdom Department for International Development (DFID) for the benefit of developing countries. The views expressed are not necessarily those of DFID. (Projects R6970, R5926 and R7352, Livestock Production Programme).

REFERENCES

Inns, F. (2000) Reporte de una visita a CIFEMA-PROMETA, Cochabamba. Silsoe, UK. Silsoe Research Institute report IDG/00/10. 48 p.

Inns, F. y Sims, B.G. (2000) Avances recientes en la gama de implementos de tracción animal de PROMETA-CIFEMA. En: Proyecto Mejoramiento Tracción Animal. II Seminario Taller Nacional sobre Tracción Animal. Cochabamba, Bolivia. Universidad Mayor de San Simón, Facultad de Agronomía. pp. 1-12.

Mbanje, E., Chatizwa, I., Koza, T., Machiwana, A., O'Neill, D. and Ellis-Jones, J. (1997) Implements for use with smaller/weaker draught animals. In: *Improving the Productivity of Draught Animals in Sub-Saharan Africa*, pp 156-168, eds J Ellis-Jones *et al.* Proceedings of a Technical Workshop, Harare, 25-27 February 1997. Report IDG/97/7, Silsoe Research Institute, Silsoe, UK.

Mbanje, E., Twomlow, S. and O'Neill, D. (2001) The potential for conservation tillage practices to improve smallholder maize production in Zimbabwe. These Proceedings.

Nyagumbo, I. (1999) Conservation tillage for sustainable crop production systems: experiences from on-station and on-farm research in Zimbabwe (1988-1997). In: *Conservation Tillage with Animal Traction*, pp 107-114, eds P G Kaumbutho, T E Simalenga. Animal Traction Network for Eastern and Southern Africa (ATNESA) Resource Book, ATNESA, Harare.

Protrigo (2000) Memoria Protrigo 1999. Ministerio de Agricultura, Ganadería y Desarrollo Rural, Bolivia. 99 pp.

Sims, B.G. y Bentley, J. (1999) Investigación participativa: Un juego de herramientas, pero no la clave del universo. *PROCAMPO* (Bolivia) **85**: 16-21.

Sims, B.G., Dijkman, J.D. y Zambrana, L. (2001) Vista panorámica del proyecto de mejoramiento de tracción animal. In: Memorias II Taller Nacional de Planificación de PROMETA. 17-18 May 2001. Cochabamba, Bolivia. Universidad Mayor de San Simón, Facultad de Agronomía. pp 4-29.

Sims, B.G. y Rodríguez, F. (2001) El proyecto PROFOCE: Producción de forraje y conservación de suelo en sistemas de laderas. En: La compatibilidad entre el manejo sostenible de malezas y la producción de forraje en sistemas agropecuarios de laderas. Memoria Taller Nacional del Proyecto Producción de Forraje y Control de Erosión (PROFOCE). Cochabamba, Bolivia. Universidad Mayor de San Simón, Facultad de Agronomía. pp. 1-8.

Spoor, G. and Godwin, R.J. (1978) An experimental investigation into the deep loosening of soil by rigid tines. *Journal of Agricultural Engineering Research* **23**: 243-258.

*M. WEISSBAC, *E. ISENSEE, **J. BRUNOTTE & **C. SOMMER

THE USE OF POWERFUL MACHINES IN DIFFERENT SOIL TILLAGE SYSTEMS

* *Institute for Agricultural Engineering, University of Kiel, Germany*

** *Production Engineering and Building Research, Federal Agricultural Research Centre, Braunschweig, Germany*

Soil is the most important production basis of the farmer. To provide a correct and gentle treatment, several improvements in technique and production systems were developed. Especially the heavy loads in harvesting represent a hazard to soil compaction. Modern terra tires and adequate chassis facilitate a gentle transmission of heavy loads to the soil. Conservation tillage improves the trafficability of the soil especially under humid conditions. A high loaded single passing of the soil must not lead to compaction in the subsoil. This is reflected by the soil physical parameters and the yield of wheat.

INTRODUCTION

Machines and vehicles in agriculture became more efficient over the time.

With the increase in efficiency also the static and total mass rises. A critical aspect for the soil is the increase in wheel loads.

While harvesting the risk is given, that the soil is subjected to heavy strain and stress. This effect is induced when the soil is passed over by heavy load harvesters and transport vehicles, e. g. as it occurs in the harvest of sugar beet at high soil moisture. Nowadays in Germany about 75% of the sugar beet areas are harvested with 6-row harvesters (Sommer & Brunotte 2000).

New terra tires and adequate chassis were developed to meet the demands of those high wheel loads.

Large volume tires with low air pressure transmit the heavy load on a large contact area to the soil.

Conservation tillage increases the load-bearing capacity of soil structure and improves trafficability. The effects of varying strains while sugar beet harvesting were examined for several years. Different measurements of the soil and the harvesters were carried out.

L. García Torres et al. (eds.). Conservation Agriculture, 367–374.

MATERIAL AND METHODS

Different soil tillage systems were examined on their effects of a high soil strain on different sites. Total mass of the applied 6-row self propelled sugar beet harvester (harvester type HOLMER) amounted to 44,6 t. More technical data is referred to in table 1.

Table 1. Technical data of the sugar beet harvester

	Frontwheel	**Rearwheel**
Tire	800/65R32	73x44.00-32
Wheel load [t]	11,3	11,0
Air pressure [kg/cm2]	2,5	2,3

Soil factors of influence are very complex in their interaction with the tire. Therefore different measuring systems must be applied.

Track depth

The track depth is the first indicator of the soil after driving across. To measure the track depth a sensor has to be installed at the sugar beet harvester. While driving the sensor measures online the track depth at the front axle of the harvester. Data is stored electronically (Brunotte et. al. 2000).

Soil pressure

Soil pressure is measured in the original soil by means of flexible sensors. The measurment system has been developed by Bolling (1987) and has continually been improved since then

(Weißbach 1994, Kath-Petersen 1994). While passing the measurement system records the soil pressure in four different depths (10, 15, 25 and 40 cm). The measurements are repeated fourfold. Analysis is performed with the respective maximum pressure data. Results from the first three depths allow expressions of the pressure pattern in the top soil. In the depth of 40 cm pressure in the subsoil is analyzed.

Porosity

Changes in soil physical parameters provide information about the compactive effects of different chassis. Soil samples are taken in similar depths according to soil pressure measurements.

Porosity is analyzed according to the methods described by Hartge & Horn (1989). Optimal porosity values for different soil types were derived by Czeratzki (1972), Maidl & Fischbeck (1995). Porosity highly interacts with the moisture content of the soil (Sommer & Hartge 1991).

Penetration resistance

Penetration resistance is a measure for soil density. The measured value shows a high correlation to bulk density and water content (Dumbeck 1986, Ermisch & Landmann 1982). The measuring method is suited to compare directly track and no traffic zones.

Penetration resistance is measured with the horizontal penetrograph (Weißbach & Wilde 1997, Weißbach 1998), whereby a cone is pulled in a defined depth through the soil. Measuring values are recorded in short intervals, e. g. an area of 1 hectar resolves to 17000 singular values. The measuring system is suited for big scale mapping.

Plant growth and yield

Plant development and yield of the follower crop – wheat – is well suited to quantify the effects of heavy load harvesting technique. They summarize all the factors that affect plant growth. No traffic zones were compared to areas that were driven on.

RESULTS

Track depth

Track depth has been recorded online while harvesting. Measurement values are depicted over the wheel load in order to quantify the bunker filling (fig. 1).

Conservation tillage results in a high level of soil firmness. Track depth remained quite constant at about 3 cm despite increasing wheel loads. The conventional tillage

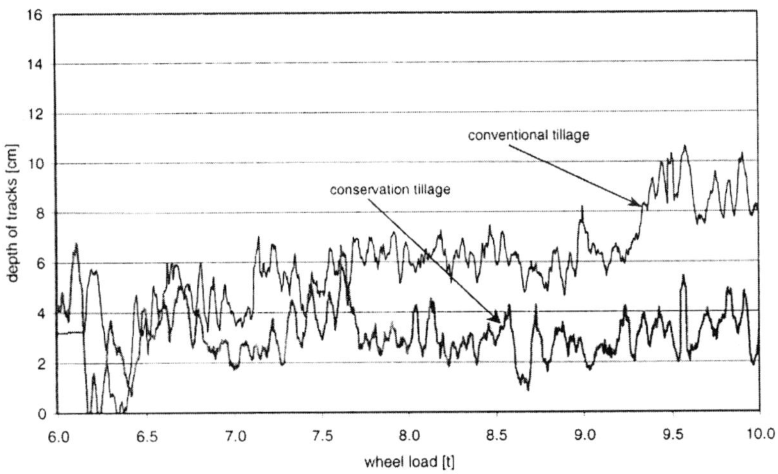

Fig. 1: Track depths of a 6-row self-propelled sugar beet harvester following different tillage

led to an increase in track depth, from 3 to 9 cm. Since this soil showed a lower load-bearing capacity the wheel sunk deeper into the soil with higher wheel loads.

Soil pressure

The pressure applied from the tire is the decisive parameter that leads to soil-strain. Transmission of the weight load from the tire to the soil occurs mainly by means of its air volume. The tire itself possesses only little bearing capacity. A strong interaction between tire air pressure and soil pressure in a depth of 10 cm was found in several studies with different types of radial tires (fig. 2).

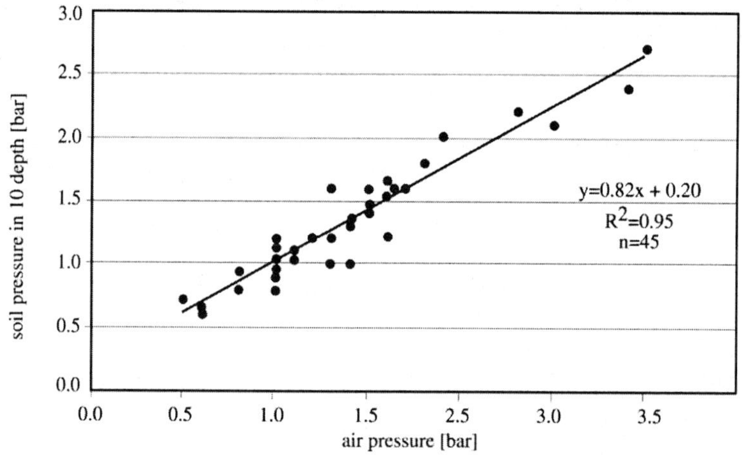

Fig. 2. Effect of tire air pressure on soil pressure

Measured data of the soil pressure in a depth of 10 cm indicate no difference between the tillage systems (fig. 3). In the following depths the two variants differ. The higher load-bearing capacity of the soil withconservation tillage is reflected by the soil pressure. The soil pressure decreases linearly with an increase in depth. Pressure is already very low in the subsoil, averaging at 0,4 bar.

The conventionally tilled soil shows a different pattern in soil pressure. In the top soil the pressure decreases very slowly according to the lower load-bearing capacity. It diminished from 15 to 25 cm only by a mean of 0,25 bar. Accordingly soil pressure in the subsoil is higher with 1,0 bar.

Porosity

Porosity and soil pressure are highly correlated. Already little pressure leads to a reduction of the pores in a soil with high porosity values (50 %), which represents the state of the soil at tillage. At this point in time soil pressure should not exceed 1 bar.

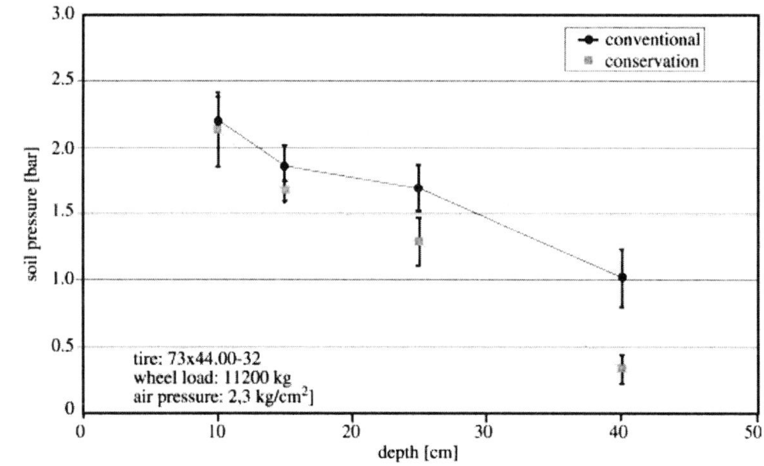

Fig. 3. Soil pressure underneath a sugar beet harvester following different tillage

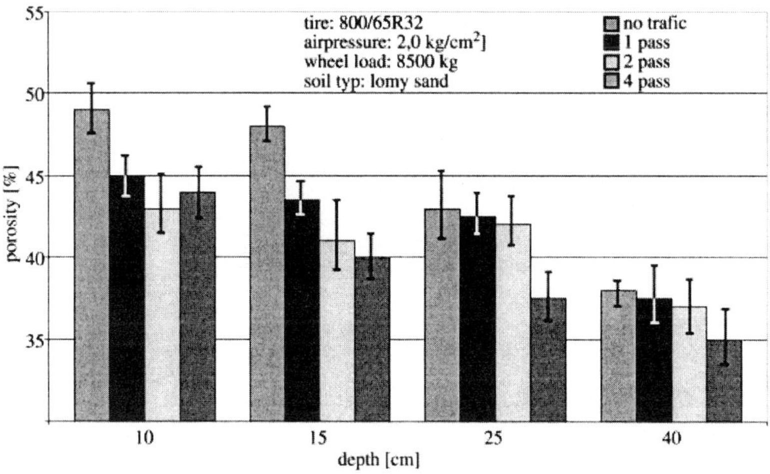

Fig. 4. Reduction in porosity multiple passages

At the harvesting time soil porosity is lower, resulting in a higher load-bearing capacity. At such soil conditions a soil pressure up to 2 bar does not impair the soil structure.

Effects on soil structure are exerted not only by the level of soil pressure, but also by the passing frequency. Under very humid conditions porosity is reduced step by step with each passing, despite constant soil pressure (fig. 4).

Soil compaction is not only limited to the top soil but can also be observed in the subsoil. To prevent such negative effects, agricultural industry has developed different chassis systems, with the aim to pass the soil only once, while harvesting.

Penetration resistance

To determine eventual soil compaction resulting from previous year harvesting, the examination area was analyzed with the horizontal penetrograph. Measurements were carried out in two depths, 20 cm representative for the topsoil and 40 cm representative for the subsoil. For each measuring depth 20000 readings were recorded. Soil density distribution was calculated from those values. In the topsoil (depth 20 cm) only tramlines with higher penetration resistance values were detected. The subsoil shows a very homogenous soil density distribution in all cases (fig. 5).

Effects of passing were not detected. Measured values show a general high level, resulting from a low soil moisture. Higher values were measured in the area of the sugar beet intermediate depot at the turnland, resulting from more frequent passing of the harvester. In this area the subsoil has been persistently compacted.

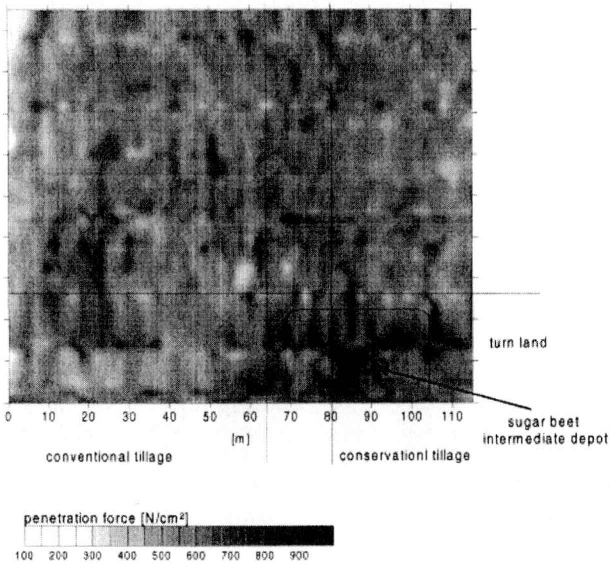

Fig. 5. Soil density distribution in the subsoil (depth 40 cm)

Yield

Crop yield results from the interaction of all factors affecting growth. After the sugar beet harvest the subsequent tillage was divided into three variants: conventional, conservational with loosening and conservational without loosening.

Yield of wheat was recorded on no traffic zones and zones passed by the sugar beet harvester (Tab. 2).

In general the yield level of the years investigated was very high, averaging about 105,9 dt/ha. Similar yield levels of the no traffic and passed zones do not indicate a soil compaction.

Cancelling the deep loosening for wheat tillage did not result in negative effects on yield. Repeated passing in the area of the sugar beet intermediate depot led to a yield depression of about 30 dt/ha caused by persistent soil compaction in the subsoil.

REFERENCES

Bolling, I. (1987) Bodenverdichtung und Triebkraftverhalten von Reifen –Neue Meß– und Rechenverfahren- MEG-Schrift 133, Dissertation München 1987

Brunotte, J. und C. Sommer (2000) Gute fachliche Praxis beim Einsatz leistungsfähiger Erntetechnik – Ergebnisse zum technikeinsatz bei Bodenbearbeitung und Ernte von Zuckerrüben– Entscheidungshilfe für die gute fachliche Praxis. Landtechnik, 55.jJg. H. 1, S. 14-16

Brunotte, J. C. Sommer, H. Rogasik, E. Isensee u. M. Weißbach (2000) Zur guten fachlichen Praxis beim Einsatz moderner Zuckerrüben-Erntetechnik. Zuckerrübe, 49. Jg. H. 1, S. 34-40

Czeratzki, W. (1972) Die Ansprüche der Pflanze an den physikalischen Bodenzusrtand. Landbauforschung Völkenrode 22 (1), S. 29-36

Dumbeck, G. (1986) Bodenphysikalische und funktionelle Aspekte der Packungsdichte von Böden. Dissertation Gießen

Ermich, D. u. Landmann, R. (1982): Beziehung zwischen Durchdringungswiderstand und Trockenrohdichte in Abhängigkeit vom Bodenwassergehalt auf ausgewählten Bodensubstraten. Wiss. Z. Univ. Halle XXXI '82 M, H. 4: 15-25

Hartge, K.H. u. Horn, R. (1989) Die physikalische Untersuchung von Ackerböden. 2. Auflage, Enke Verlag, Stuttgart

Kath-Petersen, W. (1994) Leisztungsfähige und bodenschonende Erntetechnik im Miscanthus. MEG-Schrift 260, Dissertation Kiel

Maidl. F.X. u. G. Fischbeck (1985) Wenn dem Boden die Luft ausgeht. DLG-Mitteilungen 23, S. 1354-1356

Sommer, C. u. K.-H. Hartge (1991) Verdichtung und Befahrbarkeit von Böden. Berichte über Landwirtschaft, 204. Sonderheft: Bodennutzung und Bodenfruchtbarkeit, Bd. 2: Bodengefüge, S. 104-117

Weißbach, M. (1994) Wirkung von Fahrwerken auf den Boden, insbesondere im Grenzbereich Boden/ Pflanze. MEG-Schrift 259, Dissertation Kiel 1994

Weißbach, M. and Wilde, Th. (1997) The Horizontal Penetrograph Big scale mapping device for soil compaction. Proceeding of the 3rd International Conference on Soil Dynamics (ICSD-III) Tiberias, Israel, 3.-7.August 1997 S. 244-250

Weißbach, M. (1998) Flächenbonitur mit dem Horizontalpenetrometer. Landtechnik2, 53. Jahrgang, S. 68-69

D. MC GARRY & G. SHARP

A RAPID, IMMEDIATE, FARMER-USABLE METHOD OF ASSESSING SOIL STRUCTURE CONDITION TO SUPPORT CONSERVATION AGRICULTURE

Natural Resource Sciences, Meiers Road, Indooroopilly, Queensland 4068, Australia

Sinclair Knight Mertz, 49 Annand Street, Toowoomba, Queensland 4350, Australia

Though soil structure degradation (SSD), often called soil compaction, is ranked as one of the most serious forms of land degradation, worldwide, it is regarded as the most difficult type of degradation to locate and rationalise, principally as it is a sub-surface phenomenon. Unlike erosion and salting that give strong surface evidence of the presence of land degradation; SSD requires physical input before it is uncovered and its extent, nature and cause resolved. The hidden nature of SSD leads to specific problems: poor crop growth or water infiltration due to SSD is blamed on other causes, SSD is often blamed for poor crop performance when it is actually not present, farmers rarely link their land management practices with causing SSD and remain unaware that many deep-ripping exercises worsen SSD.

This paper advocates the excavation of small spade-dug inspection pits to 50 cm as a rapid, repeatable, low cost method of immediately demonstrating, then rationalising, SSD in farmers' fields. Soil structure is recorded using five semi-quantified, globally accepted descriptors. Simplicity of the technique facilitates both digging many small holes in one field to check field variability, as well as teaching farmers the technique, so empowering them to routinely inspect and monitor the physical health of their soils toward more sustainable farming systems. Examples are drawn from the authors' published works in Argentina, Brazil, England and Australia to demonstrate the practicalities of the approach.

Key words: Tillage, compaction, field observations, farmer empowerment

INTRODUCTION

Soil structure degradation (SSD), often called soil compaction, is commonly ranked as one of the most serious forms of land degradation, worldwide. In one agricultural area of Australia, the Murray - Darling basin, SSD alone has cost $Aus144 million damage (Fray, 1991). Lal (1994) states a figure of one billion dollars, annually, due to yield losses caused by SSD in the USA, alone. Adding significantly to the problem,

L. García Torres et al. (eds.). Conservation Agriculture, 375–380.

SSD is commonly regarded as the most difficult type of degradation to locate and rationalise, principally as it is a sub-surface phenomenon. Unlike erosion and salting that give strong surface evidence of the presence of land degradation, SSD requires physical input before it is uncovered and its extent, nature and cause resolved. The composite effect makes SSD a major risk in the global «food security challenge».

The aim of this paper is to present a rapid, repeatable, low cost method of immediately demonstrating, then rationalising, SSD in farmers' fields, viz. the excavation of small «spade-dug inspection pits». One tool is required – a garden spade. The technique is a take-anywhere, readily teachable method, useable to empower farmers to routinely inspect and monitor the physical health of their soils. Farmers are then able to express and record the physical impact of their land management as a basis for establishing new, more sustainable conservation agriculture systems. As such, the method addresses many of the stated requirements of «improved indicators of resource condition» in less favoured land as specified in Theme-3 of IFPRI (The International Food Policy Research Institute, FAO). They specify the need for «developing and testing indicators and resource monitoring systems that not only flag emerging problems (or track improvements in resource condition when these occur), but also shed light on the causes of observed changes in resource conditions and the kinds of policy interventions needed to correct any serious problems» (IFPRI, 2001).

BACKGROUND

The hidden nature of SSD leads to specific problems. The most common is that poor crop growth or water infiltration due to SSD is blamed on other causes. As an example, McGarry (1990) presented the case of a measured 50% decrease in an irrigated cotton field. The farmer had been misinformed that the yield loss was due to root disease and he was considering expensive chemical-amelioration. However, spade dug soil pits immediately showed SSD beneath the cotton plants and a root system that stopped at 20 cm depth. Discussion with the grower permitted interpretation of the visible SSD; they were old wheel tracks from previous cotton crops, not completely removed before the current crop was (unknowingly) sown on top of them. Another scenario is where SSD is blamed for poor crop performance when it is actually not present as is the case with many hardsetting soils – naturally «compacted» with no human input. More generally, farmers rarely link their land management practices with causing SSD. If a farmer cannot see the SSD that was formed, it is assumed that the cultivation was successful. In the same way, farmers often deep-rip to rid themselves of SSD. If there was no prior soil examination to ensure SSD was present then the cultivation was wasted time and money, and if conducted in moist-wet soil the ripping would have caused further SSD. This is a prime example of an expensive «double negative» in conventional agriculture (McGarry, 2001).

RELATED STUDIES AND USE OF THE TECHNIQUE IN THE FIELD

Commonly, SSD has a strong, visible presence and most often is restricted to the top 50 cm of soil; easily within digging depth. The most common visible sign of SSD is massiveness

(Robertson & Erickson, 1980) where soil aggregates have been compressed into large and dense blocks that equate to reduced air space and increased soil strength. The second most common, visible form is «platiness» where the soil forms plate-like structure, horizontal to the soil surface; commonly 0.5 to 2 cm thick (McKenzie, 1998). Both massive and platy soil structures are strong barriers to plant root proliferation and water infiltration into subsoils.

There are five features that may be recorded in the field to describe and define soil structure (Table 1). The ones presented here are derived from the Australian soil description (McDonald *et al.,* 1984). However, the terms are common to many systems of soil profile description and descriptive/illustrative diagrams are commonly provided in the accompanying manual.

Several authors have used soil structure description to assess and present the nature and effects of SSD, worldwide. For example, Brammer (2000) and Douglas et al. (1999) described cultivation pans in Bangladesh and Malawi, respectively, using soil descriptions from hand-dug soil pits. Batey (2000) reviews previous work and compiles a wide range of visual and tactile methods of field soil assessment, strongly emphasising that such descriptors «provide a unique, sensitive, detailed and flexible means of assessing the physical condition of soil in the field». McKenzie (1998) presents a more complex system of scoring soil structure and other related soil profile features, usable when quantified information is carried-over between seasons, fields and operators. Application of this technique covers both cotton (McKenzie & McGarry, 1999) and vineyards (McKenzie, 2000). Additionally, there are several «farmer training and empowerment» manuals based on observations in soil pits to assess current practices toward more sustainable systems of conservation agriculture (eg FARM Programme, 1998; McKenzie, 1998; Adepetu *et al.*, 2000). In all of these works, evaluation was most commonly on a paired-site basis: untouched treeline vs field, wheel mark vs unwheeled, pre and post harvest, zero till vs conventional, etc. Simplicity of the technique facilitates the digging many small holes in one field to check field variability.

Table 1. Five features, recordable in the field to describe and define soil structure

1. **Type** of pedality provides a description of ped shape, eg. Platy, granular, lenticular, polyhedral, etc.
2. **Size** is the average least dimension of peds, used to define class intervals, eg small (0-2 mm), medium (2-5 mm), etc.
3. **Grade** is the degree of development and distinctness of peds, used to express the relative difference between the strength of cohesion within peds and the adhesion between adjacent peds. This is highly dependent on current water content. So, commonality of water contents between descriptions is to be aimed for.
4. **Fabric** is commonly restricted toVertisols (cracking clays). It records the lustre of ped faces, eg earthy, sandy, rough, smooth.
5. **Orientation** is most commonly used for Vertisols where peds in the subsoil commonly lie at 45° to each other. However, there is an inherent link between platy (type) structure and horizontal (close to 0°) orientation in all soils.

EXAMPLES FROM ARGENTINA AND BRAZIL, ENGLAND AND AUSTRALIA

Examples will be drawn from the authors' published works to demonstrate the application and the rapidity of the technique, as well as its use to resolve impacts of current land management systems, towards developing more sustainable practices.

Inspection was made of soil structure associated with zero tillage (ZT) and flotation tyres in cropping lands of Argentina and Brazil (McGarry, 2000a). As with each of the three examples presented here, a participatory approach was adopted as the landholders selected sites, then provided cropping histories towards rationalising the soil structure uncovered. In a 12-day period (six in each country) a total of 10 farms were visited, approx. 25 sites (fields) were inspected *via* 250 spade-dug pits. Soil structure was recorded at all sites and photographs taken of soil profiles, exposed on the spade blade. Evident was moderate to severe SSD associated with current and past farming practices on all farms visited in both countries. There was both soil surface SSD from traffic in the current ZT system, and platy layers below 20 cm associated with past deep cultivation. SSD was restricted to the top 30 cm of soil in Argentina and 40 cm in Brazil. The surface SSD was often associated with severe root restrictions. Fortunately, in the sites inspected in Argentina, the large earthworm population associated with ZT gave many SSD-bypass channels for plant root penetration to the subsoil. The cause of the SSD in both countries is seen as the random wheel traffic associated with current ZT practices. Discussion centred on the requirement for either controlled traffic or flotation tyres to remove the damaging effect of traffic wheels from the crop zone.

Soil structure associated with minimum tillage (MT) on farms to the east of Nottingham, England was investigated by McGarry (2000b). In this form of MT, mouldboard ploughing has been replaced by a one-pass, straight-tine cultivation each season. The system has reduced the number of tractors on this farm from 6 to 3, passes of farm machinery from 6 to 3 per season, and decreased horsepower requirements from 2 hp/ha to 0.5 hp/ha. In a five-hour visit, soil structure condition was assessed on two farms, at four field sites *via* 30 spade-dug pits. Evident was moderate to severe SSD in all fields visited on both farms; in strong contrast to the excellent, natural soil structure in an adjoining, untouched treeline: weak, subangular blocky, 3-4 cm diameter, breaking to fine, weakly structured crumbs. There was both surface SSD associated with narrow wheeled harvesters and haulouts, and strong platy layers below 5 cm depth, more likely associated with the one-pass cultivation. SSD at one site extended to 40 cm. Platy structure was associated with root restrictions, evident in both wheat and sugarbeet crops. There was a noticeable lack of earthworms present in almost all the cultivated soils inspected. In contrast, one set-aside area, under pasture, had large earthworm numbers. Discussions covered the need for eliminating the one-pass cultivation each year and the possibilities of introducing controlled traffic to remove random traffic from plant zones.

McGarry and Sharp (work in progress) are investigating soil structure condition associated with controlled traffic (CT) farming practices across the Northern grains area of New South Wales and Queensland, Australia. In a four-week period, soil structure condition to 50 cm was assessed on 35 CT/grain-growing farms, across a 2500 km north-south axis, *via* 2500 spade-dug pits. Results clearly showed better soil structure condition with CT in combination with reduced tillage practices and rotation crops on all soils and climatic regions visited. As expected, tramlines (permanent wheel tracks) were compacted but only to a maximum of 35 cm, with natural soil structure and actively growing roots below. Between tramlines the soil structure was in excellent condition, though some sites had been in CT for 10 years, growing a wide variety of crops. The excellent structure condition, evident in all CT fields, links well with farmers' anecdotal evidence of greatly increased water infiltration, root proliferation, and fertiliser and seed efficiencies with CT.

DISCUSSION

The quality and rapidity of observations collected *via* small «spade-dug inspection pits». demonstrate the power of this simple and readily taught technique. Many pits can rapidly be located over a field and the soil structure state ascertained. Initially, local farmers direct the location of the soil pits and are present at all sites. In this way they can interpret the soil condition uncovered, in terms of current and past cropping and tillage practices. In each of the three examples presented, above, the farmers thoroughly approved of and responded well to being able to immediately «feel» and see the SSD problems under their land. Many examples were uncovered of good soil structure, linked to conservation agriculture; ZT, CT and rotation/cover crops.

The examples presented demonstrate the wide range of applications of the spade-technique. Inspections of soil structure condition have been conducted under both conventional and conservation agriculture, including zero till, direct drilling, conventional tillage, harvester and haul-out impact, false seedbed formation, controlled traffic and increased earthworm dynamics across a wide range of soils in temperate to tropical locations. In this way strong support has been provided, directly at the farm level for important advances in conservation agriculture in several important, world-supply agricultural system.

Currently, follow-up studies to the rapid spade-technique are being conducted. These involve measures of bulk density, water infiltration and shear strength at the sites where soil inspections are recorded and photographed in spade-dug holes. The aim is to link the rapid spade-technique with more complex measures of soil physical condition.

REFERENCES

Adepetu, J.A., Nabhan, H. and Osinubi, A. (2000) *Simple soil, water and plant testing techniques for soil resource management.* Proceedings of a training course, Ibadan, Nigeria, 16-27 September 1996. Rome: FAO.

Batey, T. (2000) Soil profile description and evaluation. In *Soil and environmental analysis.* pp. 595-628. Eds K A Smith and C E Mullins. Ne York: Marcel Dekker.

Brammer, H. (2000) Ploughpans and tillage problems. In *Agroecological aspects of agricultural research in Bangladesh.* pp.151-158. Dhaka: UPL.

MATERIALS & METHODS

A statistical analysis was made of national olive area and yield in eight countries in the Mediterranean area. In order to rule out the alternate bearing behaviour of the olive tree and the effect of very dry years, area and yield increase over the last decade were defined as the difference between two five-year average periods (1987-1991 and 1996-2000). Subsequently, for Spain, Tunisia, Italy and Greece the regional importance of olive growing and soil erosion risk was assessed. Olive area statistics were available at provincial level (Spain, Italy) or at regional level (Greece, Tunisia). Potential erosion statistics were available from the CORINE soil erosion risk assessment (CEC, 1992) except for Tunisia, for which regional data on actual erosion are used (Projet PNUD-FAO TUN/86/020, 1992).

RESULTS

Olive production in Mediterranean countries

A general characterisation of the importance and trends in olive production was obtained by comparison of cultivated area, yield and production variability of the eight main olive producing countries (Table 1). The largest olive area can be found in Spain, yields are highest in Greece and Italy and variation in production is lowest in Greece and highest in Tunisia. Area increase over the last decade has been fastest in Spain, Morocco and Syria, while yield increase has been most evident in Spain, Greece and Syria. Thus, in Spain both important area and yield increases have occurred, although yields are still relatively low. Tunisia has an extensive olive area, but yields are both very low and very inconsistent, mainly due to the unfavourable (dry) climate. Olive area and yield have witnessed slight progression during the last decade. Italy's olive area and yield have shown little change over the last decade. Production variability is relatively low. Greece has a substantially lower area under olive, but it is increasing at a steady pace. Moreover, average yield is very high and still increases. Greek olive production demonstrates a very stable pattern.

Table 1. State and trends of area and yield of the main olive producer countries (1991-2000)

Country	*Average area (1000 ha)*	*Area increase* (1000 ha)*	*Average yield (kg/ha)*	*Yield increase* (kg/ha)*	*Variation in production (ratio lowest/highest)*
Spain	*2136*	*129*	*1665*	*540*	*0.29*
Tunisia	*1393*	*8*	*618*	*193*	*0.19*
Italy	*1123*	*-7*	*2675*	*225*	*0.54*
Greece	*728*	*45*	*2719*	*505*	*0.76*
Turkey	*558*	*38*	*1617*	*341*	*0.28*
Syria	*432*	*92*	*1148*	*473*	*0.29*
Morocco	*428*	*116*	*1209*	*-149*	*0.42*
Portugal	*324*	*-20*	*898*	*118*	*0.35*

*Source: after FAOSTAT (2001); * calculated on basis 5-year averages (1987-1991) and (1996-2000).*

Soil erosion and olive growing areas in Mediterranean countries

Potential soil erosion risk is high in many regions of the Mediterranean area (Fig. 1). In the following areas: southern Spain, Calabria, Toscana and the Italian islands and south and central Greece, soil erodibility and steep slopes are the main risk determining factors. The same factors account for actual erosion in central and southern Tunisia, along with sparse vegetation cover. Areas with a high degree of olive farming (Fig. 2) and areas of high potential erosion risk partially coincide: the Spanish provinces of Jáen, Málaga, Granada and Cáceres, the Italian region of Calabria and Lucca and Genova provinces, the Greek regions of Crete, Peloponnese, Aegean islands, Epirus and most of Central Greece and the Central and southern regions of Tunisia. With the exception of the Central East region in Tunisia, all of these areas are characterised by steep slopes.

Without soil conservation measures, erosion from olive orchards on sloping land is severe. For Andalucía, it is estimated at a general 80 t. ha^{-1} yr^{-1} (Pastor & Castro, 1995). Laguna & Giráldez (1990) found values ranging from 60-100 t. ha^{-1} yr^{-1} in

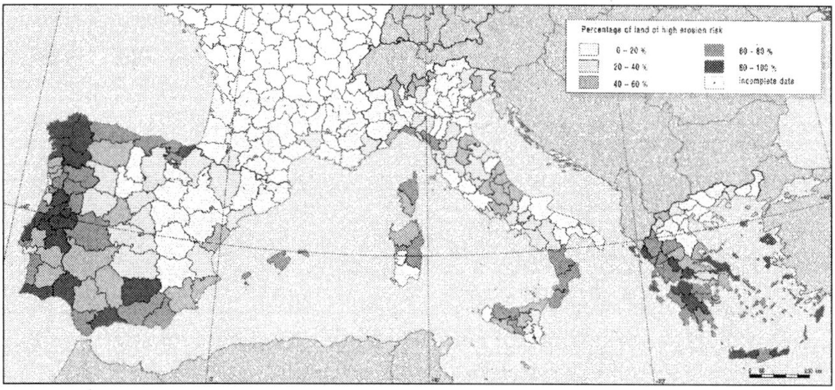

Fig. 1. High potential soil erosion risk by administrative regions (CEC, 1992).

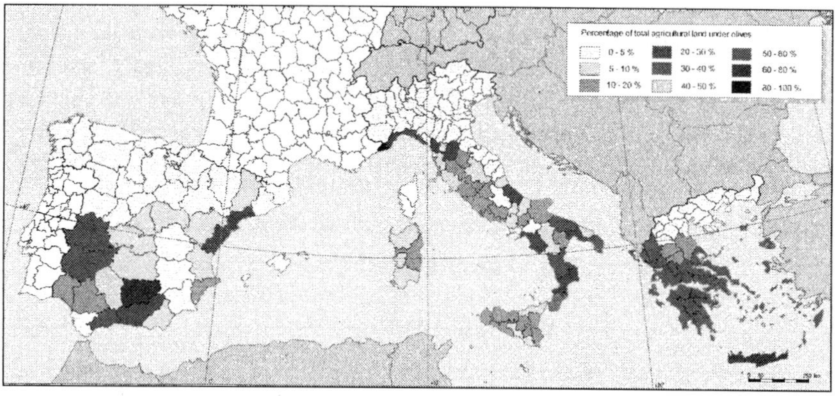

Fig. 2. Percentage of total agricultural land (excl. pastures) under olive trees by administrative regions (after M.A.P.A., 1999; ISTAT, 2001; Papageorgiou, 1987).

conventionally tilled olive orchards in Córdoba. In contrast, Kosmas *et al.* (1997) reported very low erosion (maximum 0.03 t. ha^{-1} yr^{-1}) for an olive grove with a dense soil coverage by annuals and plant residues in Spata, Central Greece. Arhonditsis, Giourga & Loumou (2000) also mention negligible sediment losses from terraced olive groves on a 50% slope with annual cultivation and undergrowth of annuals on Lesvos.

Soil conservation options for olive orchards on sloping land

Several soil conservation options, both physical and agronomical, have been applied or investigated in olive orchards on sloping lands:

- *Terraces;* although terraces have historically been established in olive orchards, they have not been well maintained or have even been removed in order to facilitate mechanisation. Constructing new ones is not feasible in economic terms.
- *Reduced and no tillage;* several tillage systems have been developed in advocacy of reducing erosion while simultaneously curbing production costs. Few of these have proved sufficient effective in soil conservation, crust formation and deep gullies being the main erosion problems.
- *Cultivation with soil cover;* inert and live plant cover provide good soil protection. Live plant cover (weeds or crops) should have an autumn/winter cycle with an early start of growth and it should be turned into mulch in early spring, either by mowing or application of contact herbicides to prevent competition for water (Tombesi, Michelakis & Pastor, 1996). Soil cover is especially important in traditional olive orchards with low tree densities. However, in dry areas cover crops fail frequently (Pastor and Castro, 1995).
- *Vegetated strips;* Martínez Raya, Francia, Martínez Vilela & Ruiz (2001) have found that soil loss from an orchard soil covered by cover crop strips was less than 10% of that of conventional tillage and about 5% of the no-tillage treatment. In a subsequent study, Martínez Raya *et al.* (2001) found that perennial shrubs (*Thymus baeticus*) performed much better as a strip cover crop than cereals and legumes, although it is not yet investigated whether the benefits of increased infiltration outweigh the increased competition for water.
- Other options include permanent pasture cover and strips of chopped pruning residues.

REFERENCES

Arhonditsis, G., Giourga, C. and Loumou, A. (2000) Ecological patterns and comparative nutrient dynamics of natural and agricultural Mediterranean-type ecosystems. *Environmental Management* **26-5**: 527-537.

Commission of the European Communities. (1992) *CORINE soil erosion risk and important land resources in the southern regions of the* European *Community: an assessment to evaluate and map the distribution of land quality and soil erosion risk.* Luxembourg: Office for Official Publications of the European Communities. 97 pp.

De Graaff, J. and Eppink, L.A.A.J. (1999) Olive oil production and soil conservation in southern Spain in relation to EU subsidy policies. *Land Use Policy* **16**: 259-267.

FAOSTAT. (2001) *FAO Statistical Databases*. http://apps.fao.org/

ISTAT. (2001) *Dati congiunturali sulle coltivazioni*. http://www.istat.it/Primpag/datiagri/ anno2000.htm

Kosmas, C. et al. (1997) The effect of land use on runoff and soil erosion rates under Mediterranean conditions. *Catena* **29**: 45-59.

Laguna, A. and Giráldez, J.V. (1990) Soil erosion under conventional management systems of olive tree culture. In *Proceedings of the seminar on interaction between agricultural systems and soil conservation in the Mediterranean belt, September 4-8, 1990,* pp.94-101. Eds A G Ferreira, M A Coutinho and P P Tomás. Oeiras: European Society for Soil Conservation..

M.A.P.A. (1999) *Anuario de Estadística Agroalimentaria 1999.* http://www.mapya.es/

Martínez-Raya, A., Francia Martínez, J.R., Martínez-Vilela, A. and Ruiz, S. (2001) Soil conservation techniques in the cultivation of perennials on steep slopes in semi-arid ecosystems. In *Man and soil at the third millenium (ESSC Third International Congress).* Eds J L Rubio *et al.* Valencia: European Society for Soil Conservation (in press).

Papageorgiou, C.L. (1987) The role of the olive tree in Greece. *Olivae* **19**: 7-11.

Pastor, M. and Castro, J. (1995) Soil management systems and erosion. *Olivae* **59**: 64-74.

Projet PNUD-FAO TUN/86/020. (1992) *Journée sur la mise en oeuvre de la strategie nationale de la conservation des eaux et du sol: la conservation des eaux et du sol en Tunesie – bilan et perspectives.* Sidi Thabet: Ministère de l'agriculture, Direction de la C.E.S.

Tombesi, A., Michelakis, N. and Pastor, M. (1996) Recommendations of the working group on olive farming production techniques and productivity. *Olivae* **63**: 38-51.

J. EPPERLEIN

DEVELOPMENT OF THE BIOLOGICAL ACTIVITY IN DIFFERENT TILLAGE SYSTEMS

German Association for Conservation Agriculture
(Gesellschaft für konservierende Bodenbearbeitung GKB-Germany)

Positive effects with regard to high-stability soil structure, higher load bearing capacity, improved infiltration and increase in biological activity were recorded from an experimental plot which had been exposed to ploughless cultivation for approximately seven years. Reduced tillage intensity led to loosening of the soil in the sole-shoe of the plough.
Remaining of organic matter in the topsoil promotes biological activity and improves the aggregate stability. Deep digging earthworm species take profit from the extensive nutrient supply at the soil surface in the conservation tillage system. Their share increased about 50 %.
Weed infestation increases with decreasing tillage depth. Though, an adequate weed- and crop rotation-management is able to control the weed.

INTRODUCTION

A research project on effects of non-invasive tillage procedures and their application to loam-sand sites in the *Uckermark* north-east of Germany was initiated 1995 under the heading of «Comperative Studies into Effects of Conservation and Conventional Cultivation on Selected Parameters of Tillage Systems in the *Schorfheide-Chorin* Biosphere Reserve». This research belongs to the integrated project of the Federal Ministry of Education, Science Research and Technology (BMBF) and the German Foundation for the Environment (DBU) on «Conservation of Nature in an Open Agriculturally Used Man-Controlled Landscape by the Example of the *Schorfheide Chorin* Biosphere Reserve».

Tillage intensity has crucial implications for the general conditions of farm soil, its structure as well as its flora and fauna. No-Tillage techniques offer many benefits for crop production and Environmental protection (TEBRÜGGE et al. ,1994). These benefits include reduced labour, time and energy inputs into crop production, improved soil physical conditions, aggregate stability and water conservation. Plough less tillage and direct sowing may provide for a number of advantages.

L. García Torres et al. (eds.). Conservation Agriculture, 387–394.

However, many practitioners are not experienced or not even familiar with the new approach and, therefore, have to meet the challenge of a complex of new demands in terms of cultivation and crop production.

Several authors, in recent decades, have dealt with biological processes in the soil, in particular on loamy and sandy sites, and adverse impact on them resulting from cultivation techniques.

METHODS AND SITES

Experiments were run on exemplary sites of two farms in the area of the *Schorfheide-Chorin* Biosphere Reserve. Tillage for seven years had been plough less (direct sowing or grubber down to 15 cm maximum) or conventional, with plough (autumn ploughing or seed furrow down to 30 cm). Crop rotation were arranged according to the farmers' specific programmes.

The following common conditions of departure were defined for comparability and comparison of either approach: Orthic Luvisol, soil type - Ls3 to Ls4 (moderate to strong sandy loam); valuation index of field 44-46; texture 50-55% silt, 18-19% clay; long-time average precipitins - 540 mm; crop rotation - winter barley, mustard as catch crop, sugar beet, spring barley. In 1996, sugar beet was directly sown into frozen mustard stubble in the conservation tillage variant. In 1997, a third site was added to the experimental program, a plot of land that had been uncultivated for 20 years an in witch parallel studies were to be conducted into the same parameters. Five measuring points were selected in cache of the test plots, about five hectares in size. Soil samples were collected from various depth levels at each of the measuring points (0-15 cm; 15-30 cm and 30-40 cm).

A catalogue of subject-related requirements was compiled together with an order of parameters that had to be investigated for the overall project. In setting up the catalogue, emphasis was laid on the need for an integrated approach to the greatest possible number of factors with relevance to the given land use system. They were to be recorded and evaluated in an interrelated manner rather than in isolation from each other.

RESULTS

Soil density and aggregate stability

Soil density and penetration resistance were measured on the two farm sites and the untouched natural plot at two different dates.

Investigations on penetration resistance in spring 1996 and 1997 with field capacity showed unambiguous differences between cultivation variants. The conservation tillage variant exhibited high strength of topsoil strata to 20-25 cm in depth but failed to show plough pan formation. Plough pan formation, on the other hand, was clearly recordable from the conventional variant together with subsoil compaction.

The curves recorded from the untouched natural plot were identical with those obtained from the plough less variant, which seemed to be indicative of stabilising, non destructive transition to the subsoil after 20 years of no plough cultivation (Fig. 1)

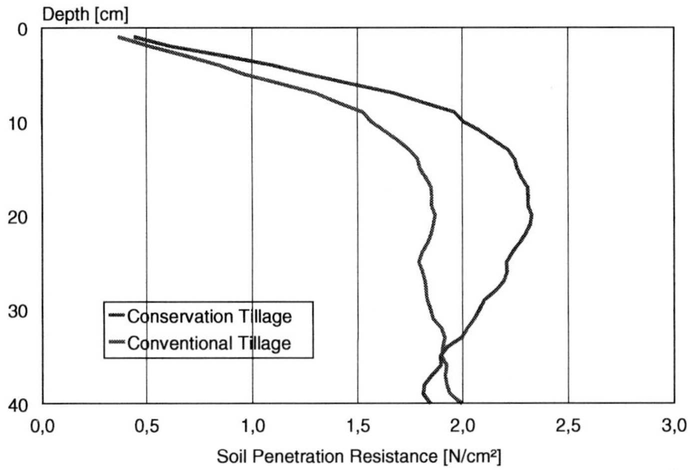

Fig. 1. Measurement of penetration resistance in different tillage systems

This primarily means to the practitioner clearly improved loadbearing and transport carrying capacity of soil in response to plough less cultivation, absence of plough pan and subsoil compaction and, consequently, improved water supply from the subsoil. Better infiltration may prove helpful in avoiding stagnant moisture and surface upsilting and is additionally supported by deep, well-draining earthworm and root capillaries.

Aggregate stability was determined (MURER et al.; 1993) for assessment of soil structures in response to different variants of cultivation. Measurements gave high soil strength and density valuce in response to conservation tillage as well as higher aggregate stability at all the three levels of measurement (Fig. 2).

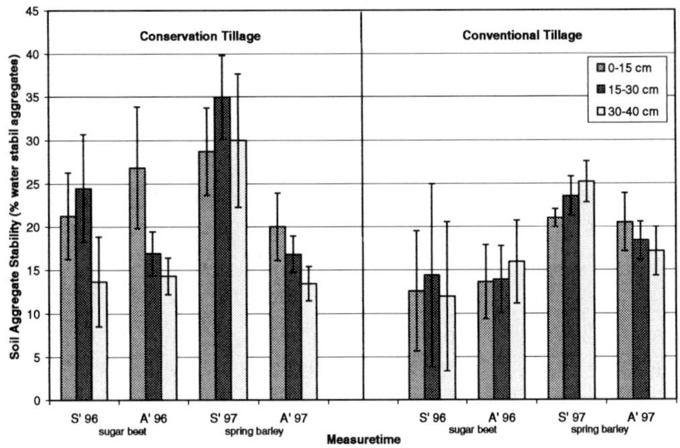

Fig. 2. Soil Aggregate Stability in different tillage systems

In other words, ploughing pulled to the surface instable, easily silting soil material and burried high-stability soil aggregates and soil-protecting plant residues (straw, mulch matter ect.). Pore system of higher stability were built up by ploughless cultivation and provided for an increase in water infiltration owing to a continnous macropore system and rise of soilborn oxygen levels. The mulching layer (mustard) clearly helped to reduce water loss (increasing the amount of water available to plant). The mulching layer and a compact soil structure also provided active soil protection against wind and water erosion during spring storms earlier this year.

Air permeability in different tillage systems

Important for the water infiltration is a high continuity of the pore space. The measurement of the pore space is carried out by recording the air permeability.

Decreasing of the intensity in soil tillage leads only to little change in the air permeability. In the transition horizons the pore space continuity is often disturbed.

At the sole-shoe of the plough - when tilled conventional - the air permeability is clearly reduced (Fig. 3).

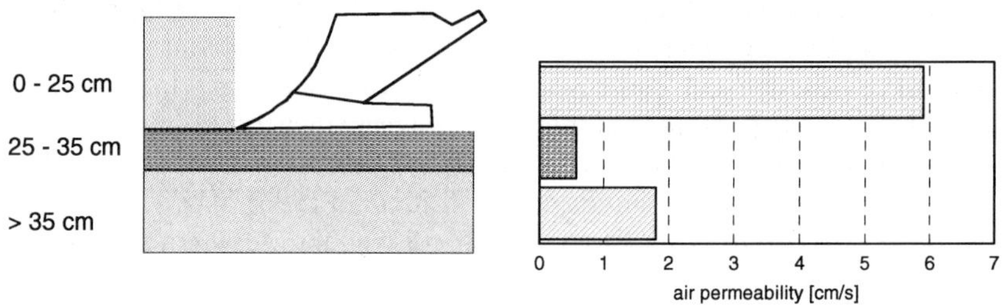

Fig. 3. Air permeability in conventional tillage

The problem of arising sole-shoes of the plough is well known in the literature (DUMBEK 1996, HOFMANN 1993, 1994). Such horizons may also occur with flat conservation tillage or after many years of constant depth tillage.

In the transition area from the tilled to the native soil (depth 15 – 20 cm) long lasting cultivator tillage leads to a decrease of the air permeability (Fig.4). The reduction is comparable to ploughing tillage.

Biological activity in different tillage systems

Soil samples were collected in spring and autumn, and CO_2 respiration as well as cellulose degradation were determined in laboratory experiments for assessment of biological soil activity in response to the two variants of cultivation.

In the conservation tillage variant, higher biological activity was recorded from the 0-15 cm soil stratum, which was attributable to increased concentrations of organic matter at topsoil level.

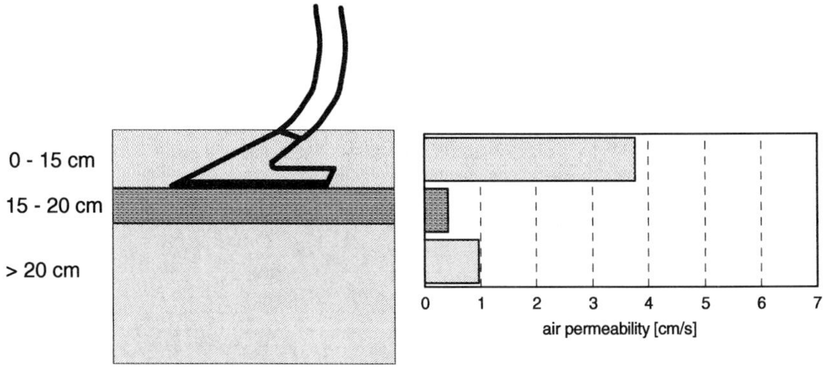

Fig. 4. Air permeability in conservation tillage

Higher biological activity also added to stability of soil aggregates and their resistance to upsilting and erosion. Shallow cultivation below 15 cm entailed reduction in microbial activity, as compared to the conventional variant, but did not impair aggregate stability.

The amount of organic matter on the surface is highest in the ploughless variant but than decreases more intensively.

Stability of soil aggregates is related directly to soil respiration, which increases with a rise in CO_2-respiration (Fig. 5).

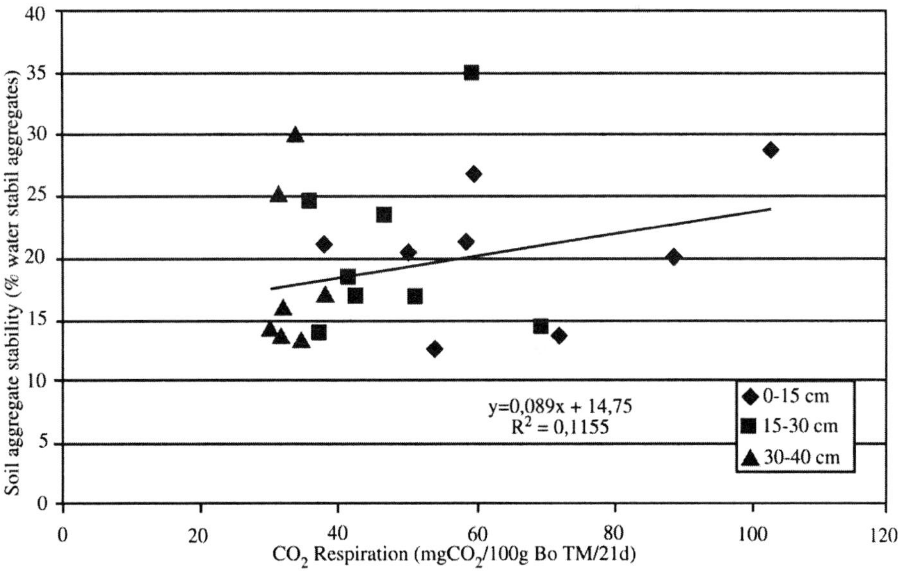

Fig. 5. Connection between Soil Aggregat Stability and CO_2-Respiration in different depth

Additionnally, gentle loosening in conservation tillage improves the development of earthworm population, especially *L. terrestris*. Latter increased in comparison to conventional tillage by 200 %. Furthermore the higher level of activity of L. *terrestris* leads to a higher pore continuity even in the subsoil (fig. 6 und 7).

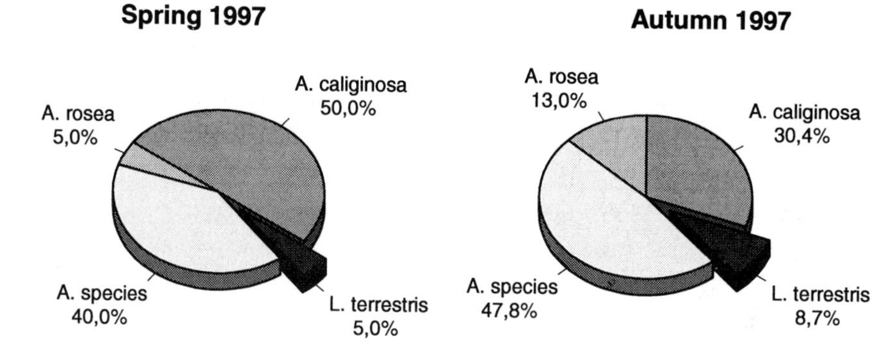

Fig.6. Earthworm-Population under Conventional Tillage

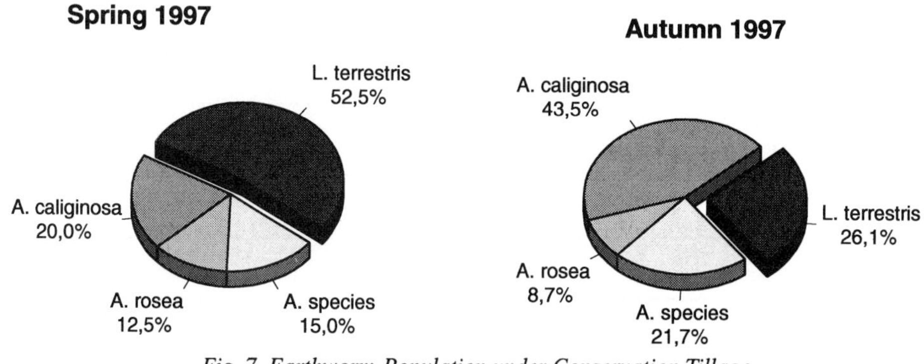

Fig. 7. Earthworm-Population under Conservation Tillage

Weed

More weed growth has often been considered as a problem in conservation tillage systems. This finding was confirmed by appraisal of potential germination under greenhouse conditions (plastic dishes of 20x14 cm; sampling in autumn 1995 and spring 1996) as well as by soil quality rating. The number of germinating seeds, according to expectation, was higher at unploughed topsoil levels (0-15 cm), wich was attributable to absence of the ploughing effect. The number of germinating seedlings, 260/280 cm2 , amounted to 9285 plants/m2, though the actual soil seed potential was much higher. In spring 1996, Chenopodium album, Veronica ssp. As well as Polygonum aviculare were the most common weed species in the ploughless variant, while Chenopodium album and Matricaria inodora were the most common species in the conventional variant. A weed control startegy should be drawn up on the basis of target and threshold valucs and should be adapted to the cultivation system used.

REFERENCES

Dumbek, G. (1986): Bodenphysikalische und funktionelle Aspekte der Packungsdichte von Böden. Dissertation Gießen

Tebrügge, F., Böhrnsen; A., Groß; U. and Düring, R: (1994): Advantages and disadvantages of no-tillage compared to conventional plough tillage. International Soil and Tillage Research Organisation (ISTRO), Proceedings of 13th International Congress; Denmark 1994 Aalborg July 24 to July 29. Volume II: 737-744.

Hofmann, B. & Sorge, B. (1993): Offset-Pflügen schont den Boden. DLG-Mitteilungen 108, H. 9, S. 36-38.

Hofmann, B. (1994): Möglichkeiten zur Vermeidung von Schadverdichtungen. Bodenbearbeitung und Bestellung von Großflächen, KTBL-Arbeitspapier 215, S. 18-28.

*P. GRANVAL, **M.B. BOUCHE, ***D. LECONTE, & ***B. MUYS

ADAPTATION OF SOWING TECHNIQUES TO MAINTAIN IMPORTANT LUMBRICID BIOMASSES IN THE HYDROMORPHIC SOILS OF NORMANDY

** Direction de l'Evaluation et de la Prospective (ONCFS), Domaine de Saint-Benoist F-78610 Auffargis.*

*** INRA, Laboratoire de Zooécologie du Sol, 2, Place Pierre Viala, F-34060 Montpellier cedex 1*

**** Domaine Expérimental Fourrager du Vieux Pin, F-61310 Le Pin au Haras*

Permanent pastures are occupying 10.5 million hectares (IFEN, 1996) and are only moderately productive. Sustained fertilization and judicious exploitation methods will allow to increase their output. In the Vieux Pin INRA domain (Le Pin-au-Haras, Orne), the production of a pastured prairie which had been monitored for fifteen years increased from 4.5 t DM/ha to 8 t DM/ha after they had been moderately fertilized.
Resowing a degraded prairie or destroying one, or even establishing a maize crop, means that the soil must be ploughed to make a good seedbed. If the ploughing works of the soil have a negative effect on the lumbricid populations, is it possible to conceive sowing techniques which would preserve this food source which is so important to the fauna? Is it possible to obtain a high forage production without degradation of the soils while maintaining the soils' bearing capacity and their feeding potential for game species in winter? What are the environmental impacts of the various technical crop management sequences on nitrate losses, on erosion and on the risks of diffuse pollutions? What is the influence of intensive forage practices on lumbricid populations?
Another resowing technique of the meadows, which is more respectful of the lumbricid populations and the properties of meadow soils, was tested in the Vieux Pin estate; it consisted in weeding in autumn and resowing in the next spring (Laissus, 1985; Leconte et al., 1998). The lumbricids which are very active in the period of the year when the days are short (autumn, winter) will take the organs of the plants which are destroyed by the herbicides, take them underground and then feed on them; thus they have an important role as_«cleaners» of the soil surface before sowing. They use the energy of the plant litters of dead vegetation to do a tilling job which creates porosity, aggregates of stable soils and and anti-erosives; they are recycling the fertilizing elements which are contained in the organic matter.
Key words: Hydromorphic soils, leached soils, plowing, direct seeding, autumn weeding, spring sowing, lumbricid biomass, *Lumbricidae*.

L. García Torres et al. (eds.). Conservation Agriculture, 395–400.

METHODS

Presentation of the experiment carried out in the Vieux-Pin estate

The objective of the tests which were carried out on the experimental domain of Vieux Pin, was to improve quantatively as well as qualitatively the production of the permanent pasture *and notably to obtain:* (besides quality and quantity):

- a greater durability of the resown prairie,
- a longer grazing period (earlier putting at pasture, later stalling which, a priori, is possible because of the improved bearing capacity of the soil),
- lesser number of hours worked and energy savings at the moment of resowing,
- a greater *soil bearing capacity* in the year following sowing,
- the possibility to finish sowing in one day with ordinary seeders.

Comparison of sowing techniques

Three sowing techniques (tables 1 et 2) were compared to those applied in permanent-meadow control plots:

- «Ploughing» process: mechanical shredding of the sward followed by ploughing, the run of a rotary harrow and sowing with a cereal seeder.
- «Simplified» process: destruction with herbicides in autumn (Aminotriazole + ammonium Thiocyanate at 3,600 g + 3,225 g AI/ha or Glyphosate at 1,000 g AI/ha, according to the vegetation that should be destroyed) then, in spring, rotary hoe before sowing, followed by sowing with a classical seeder, rolling after sowing.
- «Direct drillingprocess»: destruction by herbicides in autumn (Glyphosate at 1,000 g AI/ha) direct drilling with a special sowing drill *next spring.*

Measurements taken

The following parameters were compared:

- the cost of planting of the crop,
- the yields of two fodder crops (grass and maize),
- lumbricid biomasses,
- nitrogen mineralization,
- soil bearing capacity.

The expenditures for planting included the costs of pesticides, a reserve for the depreciation and maintenance of farm equipment at CUMA rates (Cran, *2000*), and the costs of labour for which the payment of the hours of work was charged to another account.

Lumbricids were collected by the etho-physical sampling method defined by Bouché and Beugnot (1972) whereby: the ground is wetted with formol, then soil samples are digged up with a spade at a depth of 20 cm, washed and sieved.

Net nitrogen mineralization was estimated by determining the amounts of nitrates in the soil (measurements of nitrates within a 0-30 cm horizon in the laboratory) and in crops in the field in late winter.

The penetrometer used was an artisanal device. It consisted of a 150-cm long rod, with a 3.0-kg weight sliding along the rod and a 4-cm diameter cap at the extreme end of a tube with a check at 12 cm from the cap.

RESULTS

Case of a resown prairie

During the first year, a significantly greater amount of grass will grow in plowed plots (table 1) because of the important mineralizable flux of a great amount of dead tissue (necromasses), obtained after soil aeration and tillage. In the second year, there is no longer any difference in productivity between the ploughed plot and those left without ploughing.

The classical technique of sowing after minimum tillage allows to plant very evenly at a lesser cost (1.711 F *all taxes included in 2000*) than with the traditional method of ploughing (2.554 F/ha in *2000*), i.e. at a lesser cost of 843 F/ha (33%). Total weed control before the winter followed by resowing in the following one is a reliable and cheap technique, which has been applied at a regular basis for more than 15 years in the Vieux Pin estate. The lower costs of this technique may in the first place be explained by the lesser time invested in farm labour (Leconte et al. 1998, table 1) .

The technique of direct drilling of seedlings is even less costly (a 43% reduction in the cost of sowing after ploughing). However, the results obtained by this method of total weed control followed by resowing later on in the season are very dependent of the preservation of the soil's humidity in the first two months after sowing and of the presence of bent grasses the exsudats of which are hindring the sown seedlings' growth (phenomenon of allelopathy described by Delabays et al., 1998).

Destruction of permanent pastures to plant maize

Thanks to the perfection of this technique of sowing without plouging through the adoption of simpler soil cultivation techniques, it has been possible to introduce the cultivation of maize in the Vieux Pin domain (table 2). No significant difference in yield has been observed whatever the adopted technical method. However, in the second year an (unsignificant) drop in yield has been noted for the plot that had been ploughed.

With regard to the method of direct drilling, the planting expenses of the crop had gone down by 30% in 1999 when direct drilling was applied; and moreover, the improved soil bearing capacity makes harvesting easier when it is raining in autumn, like in 1999 et 2000, the year when the many maize seedlings which had been planted after ploughing had incurred a great deal of additional expenses at the moment of harvesting (and to restore the state of the soil after harvesting), or had even been lost.

Effect of habitat and various sowing techniques on lumbricid populations

Meadows with hydromorphic soils, contain very large lumbicid populations (on average 2.8 t/ha), which is higher than than in leached brown soils (2.0 t/ha).

Earthworm populations are best preserved when meadows are weeded in autumn and are seeded directly in spring (74 à 83% of the control meadows with permanent grass, table 3); the effect of runs with a machine for superficial tillage with a rotary hoe, or of plowing, is that the lumbricid biomass will greatly be destroyed: respectivement 66-74% et 57% of the biomass when the direct drilling technique is applied.

Reduction of nitrate losses in the absence of deep soil ploughing

When soils which are rich in organic matter are cultivated, the risks are great that nitrates will be released. On the contrary, after weeding in autumn, there is only limited mineralizaton of the organic matter because no oxigenation of the soil takes place (table 4). In these cold soils, nitrate winter losses in the first 30 cm beneath ground level are very low compared to those which are deeply ploughed before autumn (between depths of 30 to 60 cm, no analyses were made of the clayey subsoil with glauconites which is very impermeable).

Improvement of the soil's bearing capacity and the absence of deep-soil tilling

The soil's bearing capacity before the meadows are grazed for the first time, is reduced by half when the «ploughing», option is applied, by 30% with superficial ploughing, and by 10% (NS) after after direct drilling of seeds (tableau 5). The effect is even more pronounced at the end of the winter. The greater resistance to trampling may contribute to the lengthening of the grazing season, while the pastures' perennity also increases.

Table 1. Cost-benefit analysis of the establishment of a prairie according to its technical specifications.

	Cost (F) *2000*	**Number** *of interventions*	**Hours of work** *(hour/ha)*	**Production* (t DM/ha)** *Year 1*	*Year 2*
Classical sowing after plowing	*2 554*	*5 – 6*	*10 - 11*	*13.1 a*	*10.9 b*
Classical sowing after simplified till	*1711*	*4*	*4-5*	*11.5 b*	*10.9 b*
Direct drilling	*1460*	*2*	*2-3*	*10.4 b*	*11.0 b*
Reference plot (permanent pasture)				*10.1 b*	*10.6 b*

** Test T: the different vertical letters indicate the significant differences between treatments at the threshold of P = 0.05%*

Table 2. Cost-benefit analysis of maize after prairie plantations according to their technical specifications.

	1st year		***2nd year***	
	Cost (F) 2000	***Yield (t DM/ha)****	***Cost (F) 2000***	***Yield (t DM/ha)****
Classical sowing after ploughing	*2 136*	*14.00 a*	*1 962***	*11.72 a*
Classical sowing after simplified till	*1962*	*14.12 a*	*1 962***	*13.2 a*
Direct drilling	*1570*	*13.29 a*	*1 962***	*13.32 a*

* *Test T: no significant differences between treatments at the threshold of P=0,05%*
** *For the 2e* year, same tilling method: 2 alternating passages of the cultivator in spring at 15-20 cm, 1 passage of rotary hoe and conventional sowing with a precision seeder.

Table 3. Influence of tilling on lumbricid biomasses (in t/ha)

	Plot1 (hydromorphic soil)	***Plot 2 (brown leached soil)***	***Plot 3 (hydromorphic soil)***
		1.32 ±0.40 D	
Classical sowing after simplified till	*1.43 ±0.29 D*	*1.72 ±0.40 CD*	
Direct drilling	*2.15 ±0.48 B*	*2.33 ±0.30 BC*	*2.73 ±0.30 AB*
Reference plot «permanent pasture»	*2.91 ±0.62 B*		*3.28 ±0.62 A*

A, B, C, D: 5% Newman-Keuls range test

Table 4. Effect of ploughing on the liberation of nitrogen during winter (kg N/ha, March 1990)

	Use of soil in winter	***Nitric nitrogen in late winter (kg N/ha)***	***Nitrogen exported by the crop (kg N/ha)***	***Liberated nitrogen (kg N/ha)***
Glyphosate in autumn	*Bare soil*	*21*	*0*	*21*
Glyphosate in autumn and direct drilling	*Triticale*	*7*	*20*	*27*
Glyphosate in summer, ploughing and sowing	*Triticale*	*69*	*81*	*150*

Table 5. Incidence of ploughing on the soils'penetrometry.

*Penetrometry**	***Classical sowing after ploughing***	***Classical sowing after superficial ploughing***	***Direct drilling***	***Control: permanent pasture***
end of winter	*1.8 (± 0.37)*	*4.0 (± 0.69)*	*6.6 (± 0.62)*	*7.5 (± 0.93)*
before 1st grazing	*4.5 (± 0.62)*	*6.9 (± 0.92)*	*8.9 (± 0.89)*	*9.8(± 1.06)*

* *Reference* specifications:

1 to 2: soil with low bearing capacity on which very large low-pressure tires should be used
5 to 6 : from this level on grazing is possible without degradation of the prairie
8 to 10: soil with a good bearing capacity for grazing as well as for cropping
12 to 16 : norm at the end of spring, beginning of summer

S. REUTER & R. KUBIAK

SOIL MANAGEMENT SYSTEMS TO SUPPORT SOIL MICROBIAL BIOMASS IN VINEYARDS

State Education and Research Center Neustadt, Department Ecology Breitenweg 71, Neustadt an der Weinstraße, 67435-DE

Vineyards at the Mosel and in the Palatinate (Germany) with different variants of soil management (mechanical or post-emergence chemical weed control, green cover, manuring with grape marc compost) were investigated for their soil microbial biomass and organic carbon content. Therefore, samples were taken from the topsoil (0-15 cm) of each variant. At all the vineyard sites the green cover had a stimulating effect on soil microbial biomass. Post-emergence chemical weed control leaded to higher biomass if the variants were established since several years. In contrast, manuring with compost showed pronounced short-term effects (one year). The quality of the soil organic matter to support microbial biomass was best under green cover and lowest in a soil with compost manuring that dated back two years.
Key words: viticulture, soil management, soil microbial biomass

INTRODUCTION

Soil biological activity has influence on a lot of soil functions like structure, water infiltration, nutrient supply and cycling, and organic matter. The support of these soil functions is important for the protection of the soil, the vigour of the vine, and the oenological potential of the grapes (Van Leeuwen *et al.*, 1998). Soil management and weed control systems are manmade factors interfering with theses functions. Soil management systems had often been regarded only in point of view of their efficacy for weed control, the security of vintage being the key factor. With the discussion about sustainable agriculture, also in viticulture effects of cultivation on soil ecological parameters have to be placed into the foreground.

In viticulture the permanent single crop cultivation and the rigid trellis systems require different soil cultivation techniques for the row between the vine-plants and for the line of vine-plants. Weed control is mainly performed by ploughing or herbicide application. The use of herbicides in vineyards is widely reduced to post-emergence application two times a year in the line of vine-plants because of difficulties for a

L. García Torres et al. (eds.). Conservation Agriculture, 401–406.

proper mechanical weed control under the vine-plants. Further, permanent green cover in the rows with several cuttings per year can be regarded as weed control. The combination of green cover in the rows and post-emergence herbicide treatment in the line is widely used in regions with sufficient rainfall and only moderate slopes like in the Palatinate (Germany). In contrast, green cover is avoided in dry areas (like in the Mediterranean) and at steep slopes with shallow and skeletal soils (like at the Mosel) because of water concurrence between vine- and cover-plants.

The aim of this work is to investigate vineyards by soil biological parameters to assess the influence of different soil management systems on the quality of vineyard soils. For this, soil samples were taken from vineyard sites with different soil characteristics. Different soil management systems were compared: ploughing (mechanical weed control), green cover, post-emergence herbicide application, and manuring with compost. The samples were analysed for soil microbial biomass and soil organic matter.

MATERIALS AND METHODS

The investigations were carried out at vineyard sites at the Mosel (MK and MS) and in the Palatinate (PD and PU) in Germany. The following soil cultivation systems were used.

In the row between the vine-plants:

- Mechanical weed control by ploughing two to four times a year.
- Permanent green cover composed by different grass and perennial weed species.

In the line of vine-plants:

- Chemical weed control with post-emergence herbicides two times a year.
- Mechanical weed control two to four times a year.

Further, manuring in the row of mechanical weed control was done with different organic matter: MS with grape marc compost one variant in 1996 and one variant in 1997 (same type of compost), PD with cacao peel in 1998, PU with waste compost in 1997. The different soil management systems at each site were established at MK since 1990 and at MS since 1995. At PD and PU the green cover was established since minimum 1995, but weed control in the line changed until 1998 from only mechanical to an additional chemical variant. Additionally to the sampling of vineyard soils, at MS soil samples were taken from an adjacent meadow, which can be regarded as a control soil. The soil physical and chemical properties are shown in table 1.

The sampling dates were 17 September 1998 at the Mosel and 27 April 1999 in the Palatinate. All soil samples were taken some days after a rainfall period from the top 15 cm of the soil. The vineyards in Palatinate were divided in 4 blocks, at the Mosel the vineyards were treated as one block. From each block 16 sampling points were taken and mixed to achieve one representative sample per block. After sewing at

2 mm the soil samples were adjusted to a water content of 60% of maximum water holding capacity and stored at 4°C prior to further analysis. For the analysis of microbial parameters the soil samples were allowed to adopt to room temperature during a minimum of two days. Soil microbial biomass was analysed either by the **F**umigation-**E**xtraction-**M**ethod (FEM) after Vance *et al.* (1987) or by the **S**ubstrate-**I**nduced-**R**espiration method (SIR) after Anderson & Domsch (1978) with a Wösthoff Ultragas US4 analyser.

Table 1. Soil physical and chemical properties of vineyard sites

	Mosel MK	Mosel MS	Meadow	Palatinate PD	Palatinate PU
Clay [%]	*20.1*	*21.7*	*19.6*	*11.1*	*18.0*
Silt [%]	*30.4*	*36.3*	*35.0*	*31.2*	*33.6*
Sand [%]	*49.5*	*42.0*	*45.4*	*57.7*	*48.4*
pH ($CaCl_2$)	*5.8*	*5.7*	*4.9*	*7.3*	*7.6*

RESULTS

The two vineyards at the Mosel (Tabs. 2 and 3) show an increasing effect of green cover on microbial biomass and organic carbon in the topsoil. Weed control by ploughing resulted in poorer living conditions for microbial biomass than application of post-emergence herbicides (Tab. 2).

Table 2. Soil microbial biomass (C_{mic} by FEM) and organic carbon (C_{org}) from the topsoil (0-15 cm) of a vineyard from the Mosel under different soil management systems. Sampling date: 17 Sept 1998. Number of blocks: n=1

Mosel MK	***Line, chemical weed control***	***Row, ploughing***	***Row, green cover***
C_{mic} [mg kg^{-1}]	*198*	*122*	*297*
C_{org} [%]	*1.8*	*1.3*	*2.3*
C_{mic}/C_{org} [%]	*1.08*	*0.92*	*1.29*

Manuring with compost from grape marc (Tab. 3) favours the soil microbial biomass in particular when compost application was recently (in 1997). High soil microbial biomass and highest ratio of C_{mic}/C_{org} (indicating the quality of organic carbon as nutritional source for the microorganisms) was found in the soil under meadow.

Table 3. Soil microbial biomass (C_{mic} by FEM) and organic carbon (C_{org}) from the topsoil (0-15 cm) of a vineyard from the Mosel under different soil management systems. Sampling date: 17 Sept 1998. Number of blocks: n=1

Mosel MS	**Row, ploughing, compost (1996)**	**Row, ploughing compost (1997)**	**Row, green cover**	**Meadow**
C_{mic} [mg kg^{-1}]	181	435	290	326
C_{org} [%]	2.3	3.6	2.1	2.1
C_{mic}/C_{org} [%]	0.80	1.22	1.37	1.55

The soil samples from the Palatinate (Tab. 4) also show the stimulating effect of green cover on soil microbial biomass. The row with ploughing has as a tendency towards higher values than both variants in the line, which show no clear differentiation in microbial biomass between chemical and mechanical weed control.

Table 4. Soil microbial biomass (C_{mic} by SIR) and organic carbon (C_{org}) from the topsoil (0-15 cm) of vineyards from the Palatinate under different soil management systems. Sampling date: 27 Apr 1999. Number of blocks: line n=4 ± SD, row n=8 ± SD

	Line, chemical weed control	**Line, mechanical weed control**	**Row, ploughing**	**Row, green cover**
Palatinate PD				
C_{mic} [mg kg^{-1}] 27 Apr	322 ± 84	324 ± 83	597 ± 78	601 ± 107
C_{org} [%]	1.2	1.4	2.1	2.1
C_{mic}/C_{org} [%]	2.68	2.31	2.84	2.86
Palatinate PU				
C_{mic} [mg kg^{-1}] 27 Apr	541 ± 71	599 ± 63	752 ± 73	928 ± 155
C_{org} [%]	1.3	1.3	1.9	2.1
C_{mic}/C_{org} [%]	4.16	4.60	3.96	4.42

DISCUSSION

The comparison of different soil management systems in vineyards demonstrated the positive effect of green cover and manuring on soil organic parameters. In particular the compost application one year before soil sampling had a strong effect (MS, compost 1997), whereas this effect was not observed, or not any longer observable with the compost applied already two years before soil sampling (MS, compost 1996). The comparison of the C_{mic}/C_{org} ratio shows, that the quality of organic carbon as nutritional source for the microorganisms is poor in the older compost variant, and that this ratio is best under green cover and the meadow. By analysing different C-fractions, Arshad *et al.* (1990) demonstrated, that a no-till system improves the quality of soil organic matter enabling enhanced biochemical activities in the soil. A high proportion of microorganisms in the organic fraction is important to stabilise the organic matter. In viticulture, there is only a low input of crop residues. Therefore green cover or organic manuring can serve as an important source of organic matter for the soil (Reuter *et al.*, 2000). Beside the function as nutrient source for soil organisms the soil organic matter improves the structure of the soil and favours water infiltration. This is important to reduce surface run-off in particular at steep sites (Meyer & Martínez-Casasnovas, 1999). At MK weed control by post-emergence herbicides had a positive effect on the soil in comparison to ploughing. The use of these herbicides transformed the developed weed coverage in a type of natural mulch of plant residues, which protect the soil surface and serve as easily degradable carbon sources. The comparison of pre- and post-emergence herbicide application in a vineyard in Burgundy/France resulted in an impoverishment of organic nutrient sources and lower soil microbial biomass under a long term use of pre-emergence herbicides (Reuter *et al.*, 2000). The alteration of soil properties by a differentiation of soil management became most obvious in the vineyard with longest history like MK. In the Palatinate the differentiation into mechanical and chemical weed control in the line of vine-plants was recently before the soil sampling. The long-term aspects of adaptation of soil organisms, soil organic carbon, and vine-plants to new soil management systems became also obvious in the work of Illmer (1999).

ACKNOWLEDGEMENTS

We thank the Ministry of Economy, Traffic, Agriculture and Viticulture of Rhineland-Palatinate for financial support.

REFERENCES

Anderson, J.P.E. and Domsch, K.H. (1978) A physiological method for the quantitative measurement of microbial biomass in soils. *Soil Biol. Biochem.* **10**: 215-221.

Arshad, M.A., Schnitzer, M., Angers, D.A. and Ripmeester, J.A. (1990) Effects of till vs no-till on the quality of soil organic matter. *Soil Biol. Biochem.* **22**: 595-599.

Illmer, P. (1999) Beeinflussung bodenmikrobiologischer Kennwerte und weinbaulicher Parameter durch unterschiedliche Bodenbedeckungen im Weinbau. *Mitteil. Deut. Bodenk. Ges.* **89**: 157-160.

Leeuwen van, C., Renard, R., Leriche, O., Molot, C., Soyer, J.P. (1998) Le fonctionnement de trois sols viticoles du Bordelais: conséquences sur la croissance de la vigne et sur le potentiel œnologique du raisin en 1997. *Revue Française d'Œnologie* **170**: 28-32.

Meyer, A. and Martínez-Casasnovas, J.A. (1999) Prediction of existing gully erosion in vineyard parcels of the NE Spain: a logistic modelling approach. *Soil & Tillage Research* **50**: 319-331.

Reuter, S., Chaussod, R., Kubiak, R. and Andreux, F. (2000) Soil microbial biomass in vineyard soils with respect to different weed control systems. In *XXVème Congrès Mondial de la Vigne et du Vin, section I viticulture*, pp. 155-161. Eds Ministère d'Agriculture et de la Pèche and OIV.

Vance, E.D., Brookes, P.C. and Jenkinson, D.S. (1987) An extraction method for measuring soil microbial biomass. *Soil Biol. Biochem.* **19**: 703-707.

A.J. BOT, T.J.C. AMADO, J. MIELNICZUK & J. BENITES

CONSERVATION AGRICULTURE AS A TOOL TO REDUCE EMISSION OF GREENHOUSE GASSES. A CASE FROM SOUTHERN BRAZIL

Freelance consultant FAO.

Professor Federal University of Santa Maria, RS, Brazil. CNPq Researcher.

Professor Federal University Rio Grande do Sul , Porto Alegre, RS, Brazil.

Senior Officer Land and Water Development Division, FAO, Rome.

Emissions of carbon dioxide (CO2), methane (CH4) and nitrous oxide (N2O), resulting from human activities are substantially enhancing the greenhouse effect, which is alleged to increase the average temperature of the earth's surface.
Conventional agricultural practices, like ploughing, mixing crop residue and other biomass into the soil surface and burning of residues, have contributed to the emission of carbon dioxide to the atmosphere as this is related to the mineralization and decomposition processes of soil organic matter by micro-organisms. A clear indicator of this is the decline in soil organic matter, which was estimated to be reduced in average with 50% in the soils of Rio Grande do Sul in hardly 15 years of conventional tillage.
However, the emergence, development and improvement of new systems of land preparation and land management, in short conservation agriculture, have changed this balance. Systems, based on high crop residue addition and no tillage, tend to accumulate more carbon in the soil, compared to the loss into the atmosphere. This turns the soil into a net sink of carbon. The figures presented confirm the potential of conservation agriculture for carbon sequestration, or at least reduce the amount of carbon dioxide to the atmosphere. Assuming an average accumulation of 1.0 t C ha-1 year-1, an area like southern Brazil (Rio Grande do Sul, Santa Catarina and Paraná) under cultivation, applying the three principles of conservation agriculture (no mechanical soil disturbance, permanent soil cover and crop rotations) would have the potential to sequester 8 million tonnes of C annually, which corresponds with 29 million tonnes atmospheric carbon dioxide.

INTRODUCTION

Emissions of carbon dioxide (CO2), methane (CH4) and nitrous oxide (N2O), resulting from human activities, are substantially increasing the average temperature of the earth's surface. Fifty percent of the increase in global warming, since the industrial revolution, is considered to be the consequence of an increased level of carbon dioxide and other gasses in the atmosphere (Lal, 1999). Sources of gas emissions include

L. García Torres et al. (eds.). Conservation Agriculture, 407–416.

burning of fossil fuels, industrial production, deforestation and agriculture. Although estimates of the total emissions vary widely, the contribution of agricultural activities and forestry products extraction to the emission of especially carbon dioxide, is estimated at only five percent of the global total (Benites, *et al.*, 1999). Conversely, the potential of agriculture and forestry for sequestering carbon (the absorption of carbon in biomass) is significant.

The effect of tillage systems versus no soil inversion, different cover crops and crop rotations on the accumulation of carbon in the soil, compared to the loss into the atmosphere, is being evaluated in long term soil management experiments in southern Brazil. The following case illustrates how certain water and soil conservation-effective practices contribute to the reduction of carbon dioxide into the atmosphere and to sequestering carbon into the soil.

BACKGROUND

In many areas of southern Brazil erosion problems started at the time of colonialization, when large areas were taken into cultivation for the production of forage crops and crops like coffee and sugarcane (Castro Filho *et al.*, 1993). Conventional agricultural practices, such as ploughing, monocropping of demanding crops, planting along the slopes, burning of crop residues, and excessive grazing, provoked little soil protection, causing an accelerated degradation (compaction, runoff and erosion) and led to the deterioration of soil physical, chemical and biological characteristics. A clear indicator of this is the decline in soil organic matter. Pottker (1977) estimated that in hardly 15 years of conventional tillage the organic matter content of the soils in Rio Grande do Sul was reduced with 50 percent (Figure 1) (Burle, *et al.*, 1997), which was a consequence of high erosion rate and high biologic oxidation of organic matter.

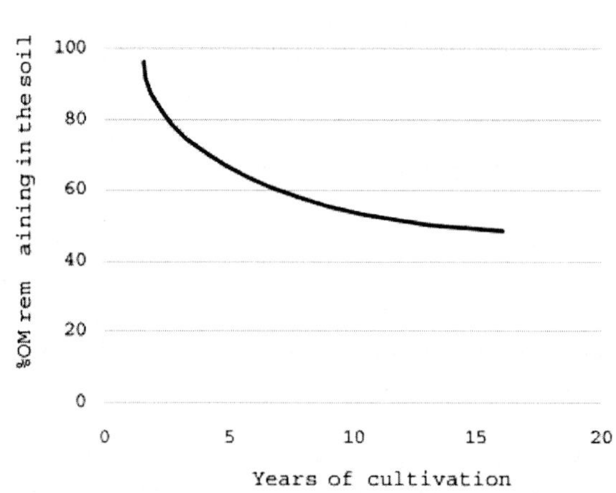

Figure 1. Reduction in organic matter content of soils of Rio Grande do Sul with conventional tillage (Pottker 1977).

Consequently, these practices have contributed to the emission of carbon dioxide to the atmosphere as this is related to the mineralization and decomposition processes of soil organic matter by micro-organisms (Lal, 1999). The CO2 emission from the soil is accentuated by ploughing, mixing crop residue and other biomass into the soil surface or burning of biomass.

A PROCESS OF CHANGE

Mainly due to relevant research, together with rural extension work the outlined situation began to change with awareness raising activities on natural resource conservation (Calegari *et al.*, 1998). Studies related to soil conservation research started in 1942 with plots to measure soil losses and then evolved to terracing, fertilisation and tillage systems. Reduced tillage systems were introduced in the 1960's and 1970's (Castro Filho *et al.*, 1993). It had become clear that in order to maintain soil losses within tolerable limits, similar to those which occur in nature, as for example in a forest, a farmer must cause minimum soil disturbance regardless of the situation and the agricultural activity (Eltz *et al.*, 1977; Wünshe *et al.*, 1980).

This resulted in the revival of the use of the ancient practice of *green manuring*, with a focus firstly to protect the soil from compaction and erosion. Research showed that, more important than using physical barriers to control runoff, the ideal solution is to maintain soils covered as much of the time as possible with cover crops and crop residues. By avoiding the detachment of soil particles by raindrop impact, which accounts for 95 percent of erosion, soil losses can be avoided and at the same time the soil can be cultivated in conditions similar to those found in forests (Freitas, 2000).

Nevertheless, the problem was not entirely resolved. Initially, green manure options were few, and knowledge about green manure was limited especially for the conditions of southern Brazil. Moreover, until a short time ago the prevailing concept was that green manures were to be incorporated into the soil to improve soil fertility by providing nutrients, especially nitrogen in the case of legumes. As the use of green manures increased, so did the use of conventional methods of land preparation which consisted of incorporating the green manure or crop residues by cultivating the whole soil surface by one or more ploughings and two or more passes with a harrow (Calegari and Alexander, 1998).

The present concept of green manuring or the use of cover crops consists of the practice to maintain the soil covered with the living or dead biomass of these crops for as long as possible, with the objective to improve soil quality by protecting the soil from the direct impact of rain drops, excessive insolation and wind action and in order to maintain and improve soil physical, biological and chemical characteristics (Calegari *et al.*, 1998).

This was accompanied by the emergence of new systems of land preparation as alternatives to the conventional practices introduced from temperate climates, like minimum tillage and direct sowing techniques, which appear more appropriate for tropical and subtropical climate conditions. Under tropical conditions, with high temperatures and humidity levels, crop residues tend to decompose rapidly, even when maintained on the soil surface, which allows for a good synchrony with the demands of the successive crop (Amado *et al.*, 2001).

Traditionally, both leguminous crops, like lupine (*Lupinus angustifolius*), vetch (*Vicia sativa*), clover (*Trifolium subterraneum*) and cowpea (*Vigna unguiculata*), and nonleguminous crops, like black oats (*Avena strigosa*), radish oil (*Raphanus sativus*) and wheat (*Triticum aestivum*), are being used as a cover crop during winter time in the production of maize (*Zea mais*) in southern Brazil. New cover crops include castor bean (*Ricinus communis*), mucuna (*Mucuna pruriens*), canavalia (*Canavalia* sp.) and lablab bean (*Dolichos lablab*).

EFFECT OF COVER CROPS AND REDUCED TILLAGE SYSTEMS ON SOIL PROPERTIES

The greater production of foliage in a system with cover crops and reduced tillage compared to monocrop cultures with conventional tillage, leaves a protective blanket of leaves, stems and stalks from the previous crops on the surface. In this way organic matter can be built up in the soil, which has great influence on the activity and the population size of the microorganisms.

The protective blanket, the greater biological activity and higher humus formation reduce the impact of raindrops on soil surface, and thus the runoff of rainwater. With this soil erosion is reduced close to the regeneration rate of the soil or even adding to the system as was found by Debarba and Amado (1997) in the cropping system oats+vetch/maize (Figure2).

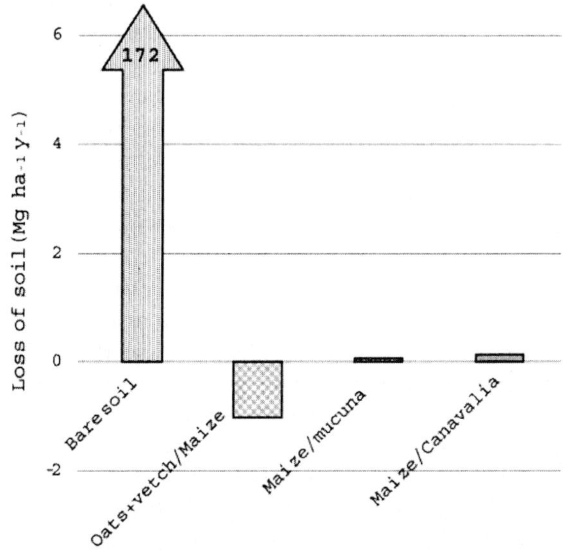

Figure 2. Soil loss due to water erosion (corrected with soil regeneration = 1.7 t ha-1 y-1) for different maize production systems (Debarba and Amado, 1997)

Infiltration of rainwater is increased under no-tillage because of the higher number of large pores due to biologic activity and roots growth (Roth, 1985). In southern Brazil rainwater infiltration increased from 20mm h-1 under conventional tillage to 45mm h-1 under no-tillage (Calegari *et al.*, 1998). In an experiment under natural rainfall conditions, Debarba and Amado (1997) found an increase in rainwater infiltration in maize-cover crop systems under no-tillage, with prominent results in oats+vetch/maize and maize/mucuna (Figure 3).

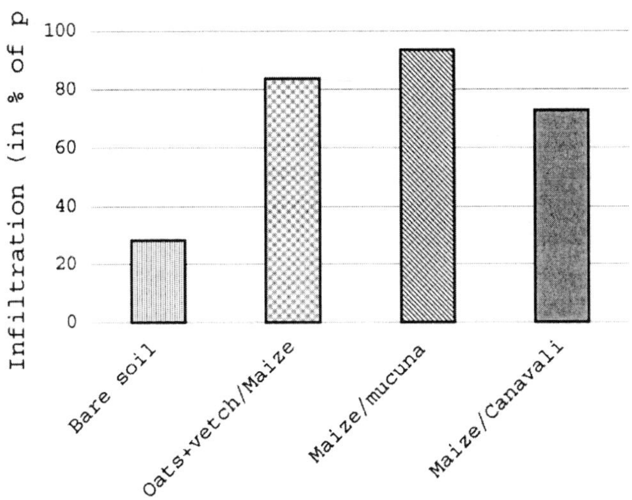

Figure 3. Infiltration of rainwater under different maize production systems (Debarba and Amado, 1997)

ACCUMULATION OF SOIL CARBON

According to Greenland and Adams (1992), systems, based on high crop residue addition and no-tillage, tend to accumulate more carbon in the soil, compared to the loss into the atmosphere. This turns the soil into a net sink of carbon (Reicosky *et al.*, 1995). The studies executed in southern Brazil show indeed an increase in organic carbon in the soil.

The initial incorporation of lime and fertilisers in an experiment in Rio Grande do Sul resulted in a decrease of the organic carbon content of the soil during the first four years, compared to the soil under natural vegetation, probably because of a reduction in physical structure of the organic matter, through breaking the aggregates and an increase in soil oxygen (Reicosky and Lindstrom, 1993; Amado *et al.*, 2001). From the fourth year the different maize/cover crops systems added carbon to the soil (Figure 4).

Bayer and Mielniczuk (1997) found that five years after the introduction of intensive cropping systems, containing leguminous crops (especially the cropping systems oats+clover/maize and oats+clover/maize+cowpea), the level of soil organic carbon was restored, at least in the superficial layers of the soil at the level as it was, before previous cropping systems were responsible for the loss of 8.3 t C ha-1.

For less intensive cropping systems this takes more time as shown in figure 4. However, also the cropping system containing maize/mucuna restored the carbon content after about 5∫ years. From 4th to 8th year this system had added an average of 1.6 t C ha-1 y-1 to the soil (Amado *et al.*, 2001). Other research indicate values of 0.9 to 1.6 t C ha-1 y-1 (Bayer *et al.*, 2000; Amado *et al.*, 2001). This high increase, compared to the other systems, is related to the high biomass production of which the mucuna contributed 44%, with an annual average biologic nitrogen addition/recycling of 140 kg ha-1.

Another experiment shows that 89% of the original soil carbon level of the soil layer 0-17.5 cm was restored thirteen years after the introduction of intensive cropping systems.Fifteen years of conventional agriculture were responsible for the loss of 12.2 t C ha-1 in relation to natural grass vegetation, in the same soil layer before the initiation of the experiment (Lovato, 2001).

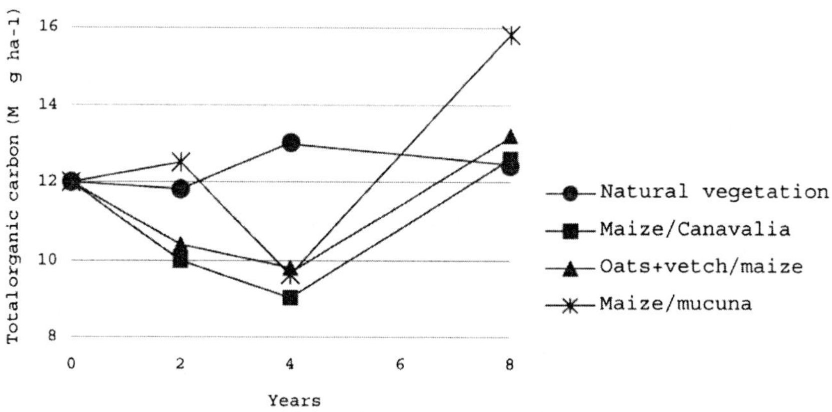

Figure 4. Dynamics of organic carbon (0-10cm) under different maize cropping systems (Amado et al., *2001)*

Another experiment carried out in Rio Grande do Sul, showed that during the initial years until establishment of the cropping system and the increase in soil aggregation, the increase in total organic carbon content was restricted to only the surface layers of the soil (0-2.5 cm) (Testa *et al.*, 1992). With time (fifth year), this effect reached also deeper soil layers (2.5-7.5 cm) (Bayer and Mielniczuk, 1997). At thirteen years this effect reached 7.5 to 12.5 cm (Lovato, 2001). Castro Filho *et al.* (1998) found a 29 percent increase of soil organic carbon in no-tillage compared to conventional tillage in the surface layer 0-10 cm of the soil, irrespective of the cropping system.

Compared to the cropping system winter fallow/maize, which was taken as a reference, soil carbon content increased with 47 percent in the system maize/lablab and with 116 percent in maize/castorbean cropping system. In systems were nitrogen was applied as a fertiliser the carbon content increased even more (Testa *et al.*, 1992).

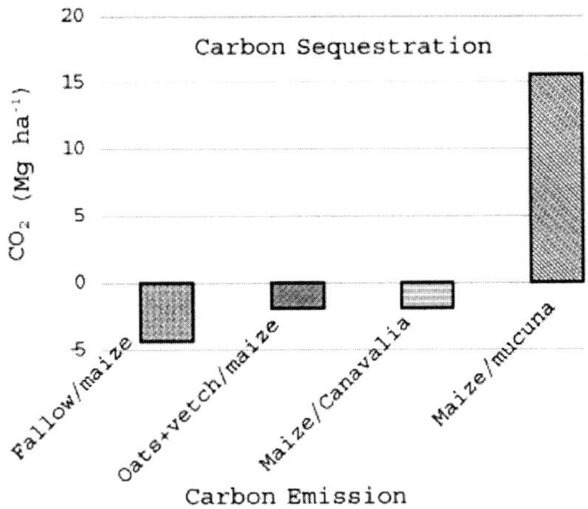

Figure 5. Estimation of emission and sequestration of CO2 under different maize production systems with cover crops under direct sowing compared to natural vegetation (Amado et al., *2001)*

The fact that agriculture can act as a sink for CO2 is shown in figure 5, in which the carbon stock in soils under natural vegetation is used as a reference (steady state C=0). During 8 years, the fallow/maize system liberated 4.32 Mg CO2 ha-1. Compared to fallow/maize system, all systems containing cover crops showed less CO2 emission. The maize/mucuna system showed a positive balance of almost 20 Mg CO2 ha-1 compared to fallow/maize (Amado *et al.*, 2001). Compared to soils under natural vegetation this means a sequestration of atmospheric carbon of more than 15 Mg CO2 ha-1 in eight years.

RECENT IMPROVEMENTS TO THE SYSTEM

In the majority of farms producing grains during summer and extensive grazing during winter in the same area, the preference for a particular species of green manure or cover crop, such as black oats, often associated with a specific subsequent crop such as soybean without rotation, has created some serious problems. These problems, like soil compaction, nutrient concentration in the surface soil, and certain pests, diseases and invading weed species, have resulted in an increased use of toxic pesticides.

For this reason it was not sufficient to maintain a minimum amount of residue on soil surface and to use tillage systems that cause minimum soil disturbance. The search for high biomass producing species, has motivated a lot of farmers to use a mixture of 3-5 different cover crops in winter season and rotate maize or sorghum with soybean in summer season. Some progressive farmers use radish oil during a small window between summer cash crop and winter cover crop, resulting in an extra addition of biomass to the system. The results are promising, especially with respect to nutrient recycling, reduction of weeds, pests and diseases and dry matter production.

In the integrated grain/livestock system, the best results for improvement were obtained through division of grazing areas, rotational grazing to allow the regrowth of pastures during 3-4 weeks, use of nitrogen fertiliser in grasslands, and especially the retraction of animals 30-45 days before planting of the summer crop. This interval permits an adequate biomass production before direct sowing of the next crop. Two critical points in livestock management are the retraction of animals during long periods of rainfall, when the soil is highly susceptible for compaction, and the maintenance of a permanent soil cover to reduce the effects of trampling.

Direct sowing has come to be considered as a system and not just a method of land preparation (Freitas, 2000). Direct sowing, with high addition of residues, through rotation of crops and cover crops has reduced the dependency on external inputs, such as mineral fertilisers, insecticides, fungicides and herbicides. The system has been optimised and become very efficient in small, medium as well as large farms. The adaptation of animal drawn and human operated machines in smallholder farms has brought a halt to the rural exodus in the southern part of Brazil.

CONCLUSIONS

With time, in systems with reduced tillage and high residue addition, soil life takes over the functions of traditional soil tillage, which is loosening the soil and mixing the soil components. In addition to that the increased biological soil activity creates a stable soil structure through accumulation of organic matter, and thus increases the levels of some soil nutrients and soil organic carbon.

The increase of soil organic matter transforms agricultural soils under the absence of soil tillage into sinks for carbon, reducing the atmospheric charge with greenhouse gasses like CO2. The figures presented confirm the potential of conservation agriculture for carbon sequestration, or at least reduction of the amount of carbon dioxide to the atmosphere. Assuming an average accumulation of 1.0 t C ha-1 year-1, an area like southern Brazil (Rio Grande do Sul, Santa Catarina and Paraná) under cultivation, applying the principles of conservation agriculture would have the potential to sequester 8 million tonnes of C annually, which corresponds with 29 million tonnes atmospheric carbon dioxide. Based on the figures of Landers *et al.* (2001) this would mean an estimated annual benefit of U$ 87 million.

A production system, which includes green manure, crop and cover crop rotation and no-tillage, can be regionally adapted and therefore can contribute to the sustainability of soil management in the region.

REFERENCES

Amado, T.J.C., C. Bayer, F.L.F. Eltz & A.C.R. de Brum. 2001. Potencial de culturas de cobertura em acumular carbono e nitrogênio no solo no sistema plantio direto e a conseqüente melhoria da qualidade ambiental. Revista Brasileira de Ciência do Solo, 25:189-197, 2001. Bayer, C. and J. Mielniczuk, 1997. Características químicas do solo afetadas por métodos de preparo e sistemas de cultura. Revista Brasileira de Ciência do Solo, 21: 105-112.

Bayer, C., J. Mielniczuk, T.J.C. Amado, L. Martin-Neto and S.V. Fernandes. 2000. Organic matter storage in a sandy clay loam Acrisol affected by tillage and cropping systems in southern Brazil. Soil & Tillage Research 54:101-109.

Benites, J., R. Dudal and P. Koohafkan, 1999. Land, the platform for local food security and global environmental protection. In: FAO. Prevention of land degradation, enhancement of carbon sequestration and conservation of biodiversity through land use change and sustainable land management with a focus on Latin America and the Caribbean. Proceedings of the IFAD/FAO Expert Consultation. Rome 15 April 1999. p. 37-42.

Burle, M.L., J. Mielniczuk and S. Focchi, 1997. Effect of cropping systems on soil chemical characteristics, with emphasis on soil acidification. Plant and Soil 190: 309-317.

Castro Filho, C., P.C. Corsini, D. Soares and W. Politano, 1993. Acceptance of soil and water conservation strategies and technologies in southern Brazil. Topics in applied resource management, Vol. 3: 341-361.

Castro Filho, C., O. Muzilli and A.L. Podanoschi. 1998. Estabilidade dos agregados e sua relação com o teor de carbono orgânico num Latossolo roxo distrófico, em função de sistemas de plantio, rotações de culturas e métodos de preparo das amostras. Revista Brasileira de Ciência do Solo, 22: 527-538.

Calegari, A. and I. Alexander, 1998. The effects of tillage and cover crops on some chemical properties of an oxisol and summer crop yields in southwestern Paraná, Brazil.

Advances in GeoEcology 31: 1239-1246.

Calegari, A., M.R. Darolt and M. Ferro, 1998. Towards sustainable agriculture with a notillage system. Advances in GeoEcology 31: 1205-1209.

Debarba, L. and T.J.C. Amado. 1997. Desenvolvimento de sistemas de produção de milho no sul do brasil com características de sustentabilidade. Revista Brasileira de Ciência do Solo 21, p. 473-480.

Eltz, F.L.F., N.P. Cogo and J. Mielniczuk, 1977. Perdas por erosão em diferentes manejos de solo e coberturas vegetais em solo Laterítico Bruno Avermelhado-distrófico (São Jerônimo). I. Resultados do primeiro ano. Revista Brasileira de Ciência do Solo, 1:123-127.

Freitas, V.H. de, 2000. Manejo de suelo en pequeñas fincas. Estrategiasy métodos de introducción, tecnologías y equipos. FAO Soils Bulletin 77, Rome. 66 pp.

Greenland, D.J. and Adams, 1992. Organic matter in the soils of the Tropics - From myth to complex reality. In: Myths and science of soils in the tropics. SSSA Special publication no. 29.

Lal, R., 1999. Global carbon pools and fluxes and the impact of agricultural intensification and judious land use. In: FAO. Prevention of land degradation, enhancement of carbon sequestration and conservation of biodiversity through land use change and sustainable land management with a focus on Latin America and the Caribbean. Proceedings of the IFAD/FAO Expert Consultation. Rome 15 April 1999. p. 45-52.

Landers, J.N., G. Sant'anna de C Barros, M. Theoto Rocha, W. A. Manfrinato, J. Weiss. 2001. Environmental impacts of Zero Tillage in Brazil – a first approximation. I World Congress on Conservation Agriculture, Madrid, 1–5 October, 2001

Lovato, T. 2001. Dinâmica do carbono e nitrogênio do solo afetada por preparos do solo, sistemas de cultura e adubo nitrogenado. Porto Alegre: UFRGS. 130p. Tese de Doutorado.

Pottker, D. 1977. Efeito do tipo de solo, tempo de cultivo e da calagem sobre a mineralização da matéria orgânica em solos do Rio Grande do Sul. Porto Alegre: UFRGS. 128p. Dissertação de Mestrado.

Reicosky, D.C. & Lindstrom, M.J. 1993. Effect of fall tillage method on short-term carbon dioxide flux from soil. Agron. J., 85:1237-1243.

Reicosky, D.C., W.D. Kemper, G.W. Langdale, C.L. Douglas and P.E. Rasmunssen. 1995. Soil organic matter changes resulting from tillage and biomass production. Journal of soil and water conservation 50(3):253-261.

Roth, C.H. 1985. Infiltrabilität von Latosolo-Roxo-Böden in Nordparaná, Brasilien, in Feldversuchen zur Erosionskontrolle mit verschiedenen Bodenbearbeitungssystemen und Rotationen. Göttinger Bodenkundliche Berichte, 83, 1-104.

Testa, V.M., L.A.J. Teixeira and J. Mielniczuk, 1992. Características químicas de um Podzólico vermelho-escuro afetadas por sistemas de culturas. Revista Brasileira de Ciência do Solo, 16: 107-114.

Wünshe, W.A., J.E. Denardin. J. Mielniczuk, J. Scopel, P. Schneider and E.A. Cassol, 1980. Projeto integrado de uso e conservação do solo - Um esforço conjunto para a conservação do solo no Rio Grande do Sul. III Encontro Nacional de Pesquisa em Conservação do Solo. Recife, Julho de 1980. In: Trigo/Soja 51: 20-25, Porto Alegra.

S.C. ROSE & A.D. CARTER

AGROCHEMICAL LEACHING AND WATER CONTAMINATION

ADAS Consulting Limited, Gleadthorpe, Meden Vale, Mansfield, Notts NG20 9PF, UK

ADAS Consulting Limited, Rosemaund, Preston Wynne, Hereford HR 1 3PG, UK

Agrochemical or pesticide leaching to surface waters or shallow groundwater takes place via a number of transport mechanisms and can be influenced by a range of hydrological and soil related factors. In addition, the movement of any pesticide away from the point of application to the target crop is greatly affected by the physico-chemical properties of the individual pesticide.

Cultivation of the soil greatly affects its potential to retain and degrade pesticides *in situ* or allow them to preferentially leach to water resources. Conventional cultivation causes a repeated disturbance of the soil profile, thereby affecting its structure, organic matter content, moisture content, biomass and pore connectivity. Reduced cultivation permits greater soil stability and soil pore connectivity, and increased opportunities for continuous cracks to develop in macroporous soils. Reduced cultivation does however increase the organic matter and hence the biomass which are important to the biodegradation of pesticides. These soil properties, together with the local hydrological conditions, generally dictate the leaching of pesticides within the near surface layers.

The research indicates that when comparing conventionally tilled soils to reduced cultivation soils the pesticides losses via the leaching mechanism may increase or decrease. Field hydrology, as affected by cultivation, greatly influences the routing, quantity and quality of agricultural drainage. Management systems for the crop, soil and pesticide applications require careful manipulation for a particular site to minimise the risk of pesticides leaching to water resources.

Key words: Tillage, agrochemicals, pesticides, surface water and groundwater contamination, leaching

INTRODUCTION

In recent years there has been increasing concerns in the UK about the number and frequency of detections of various pesticides in groundwater and surface water. For groundwater the immediate concerns are due to the possible human health effects when the water is subsequently used as a potable supply. For surface water the concerns are both for the human health issues at potable water abstraction points and for

L. García Torres et al. (eds.). Conservation Agriculture, 417–424.

potentially detrimental impacts on the aquatic ecosystems. The European Community has set a limit of 0.1mg/l for a single pesticide and 0.5mg/l for total pesticides in potable water (EC Drinking Water Directive: 80/778/EEC, revised as 98/83/EEC). At the time this limit was imposed it represented the known limit of detection, whether or not any harmful effects to the human population could be determined. As these limits exist in European legislation there is currently little desire by Member States to change them to a more toxicological risk based system.

Pesticides can reach groundwaters and surface waters from two sources and by a number of routes (Carter, 2000a). Diffuse sources are defined as those connected to the normal application of pesticides to agricultural fields. Given the right conditions pesticides have the opportunity to move through the soil in solution or attached to soil particles and enter water via underfield drainage systems or as surface or sub-surface water, leaching or bypass flow. Diffuse sources also include spray drift, precipitation and baseflow seepage. Point sources are defined as those derived from a localised situation where the water and associated pollution enters a water body at a specific or restricted number of locations. Typical point sources include spillages during handling operations in the farmyard, washings and waste disposal, sumps and soakaways and consented discharges.

The most common pesticides found in UK raw waters sampled by Water Companies that are intended for human consumption are the cereal herbicides, particularly mecoprop, isoproturon and MCPA and the non-agricultural herbicide diuron (Court *et al.* 1995). These agricultural herbicides are very extensively used in UK agriculture and so the probability of the conditions being favourable for them to be transported to water resources is quite high compared to other less widely used pesticides.

One of the main mechanisms by which pesticides can reach water resources, both surface water and groundwater, is leaching through the soil profile. Clearly, many characteristics of the soil in question (i.e. physical, chemical and biological) and its drainage potential (natural or artificial via underdrainage systems) will affect the amount and rate at which any particular pesticide enters water in solution or adsorbed onto soil particles, rather than being retained and degraded *in situ* (Carter, 2000b). In addition, the amount of pesticide applied, its persistence, solubility, degree of soil adsorption, vapour pressure over the treated area and the location of the chemical in the soil profile can all influence concentrations in sediments and water (Baker, 1992; Fawcett *et al.*, 1994). Soil characteristics can be greatly influenced by cultivation operations. Conservation tillage, reduced tillage and even zero tillage (e.g. direct drilling) have been widely promoted in the USA as a Best Management Practice (BMP) to reduce pesticide pollution. It has gained wide acceptance primarily because it reduces soil erosion by reducing surface runoff and minimises field operations, thereby generating energy savings (Isensee *et al.*, 1990). By increasing the rate of water infiltration into the soil has raised the question whether it might accelerate the leaching of pesticides to groundwater in particular. Data from leaching studies both in the UK and abroad can be conflicting in terms of the effects of cultivation on pesticide leaching. This paper reviews how cultivation affects soil characteristics and hence the potential for pesticides to leach to water. It also identifies pesticide properties and environmental factors which can influence the availability of a particular pesticide to leach and the conditions which may promote or limit the hydrological potential for the pesticide to leach.

EFFECTS OF CULTIVATION ON SOIL CHARACTERISTICS

According to Addiscott and Dexter (1994) there are three types of cultivation operation: primary, secondary and subsoiling. Primary cultivation loosens soil and involves total or partial soil inversion for weed control or crop residue incorporation. Secondary cultivation produces further breakdown of soil aggregates to prepare, for example, a seedbed. Subsoiling may be undertaken every few years below the depth of other cultivation operations to loosen compacted or dense subsoils or to create mole drainage channels.

Cultivation causes a disruption in the physical structure of the soil and can result in the breakdown of soil aggregate size and stability (Terbrügge, 1993; Silgram and Shepherd, 1999). Primary cultivation increases the porosity of the ploughed layer of the soil (Addiscott and Dexter, 1994). As a result of increased porosity in the topsoil the level of aeration is also elevated, leading to a generally warmer and drier soil. Ploughing does however disrupt the soil drainage characteristics, particularly the connectivity and continuity of the soil macropores which transport water from the topsoil to the deeper layers in the profile (Harris and Catt, 1999). This reduction in connectivity of the soil pores is made worse if the cultivation operation has created a smeared plough pan which restricts drainage or soil hydraulic conductivity below the cultivation depth. Due to the instability created in the soil aggregates by ploughing cultivated soils tend to be more susceptible to compaction problems caused by repeated vehicular trafficking when compared to reduced cultivation systems (Addiscott and Dexter, 1994). Compaction, which can extend down to at least 0.5m in the profile, has the effect of closing down the larger pores initially, which again reduces the amount of water that can be transmitted to the lower soil layers. The soil moisture conditions at time of cultivations dictates the severity of any compaction effects.

Subsoiling and mole drainage, if undertaken when the soil moisture is not too great, greatly disturbs the soil profile by creating a large network of cracks in the topsoil which permit the rapid flow of water down towards the lower layers. Subsoiling is generally undertaken to remove problems of compaction, either at the soil surface or lower down the profile. If mole drains have been produced (usually at a depth of 0.4-0.6m) these will connect into permanent underfield drainage pipe systems which discharge directly into open watercourses. If undertaken in the right conditions mole drains can remain effectively for several years.

The presence of organic matter in the soil greatly affects the stability of soil aggregates. Any cultivation operation that disturbs the soil increases the rate of organic matter mineralisation and hence decreases the organic matter content of the soil (Silgram and Shepherd, 1999). In reduced cultivation or zero cultivation systems the amount of soil disturbance is minimised therefore providing opportunities for the soil organic matter content and the soil aggregate stability to increase. In addition, an increased organic matter content could also cause the topsoil to become slightly more acidic.

Shepherd and May (1992) reported that a sandy loam soil dried out significantly more with ploughing than with direct drilling. Direct drilled soil often retains a lower temperature than the ploughed soil due to the presence of insulating crop residues at the soil surface. The differences between cultivated and non-cultivated soils, particularly in terms of the moisture and temperature conditions, will affect the amount of microbial activity in the soil. In turn, the microbial activity affects both the mineralisation of the soil organic matter and the biological degradation of organic pollutants, such as pesticides.

PESTICIDE PROPERTIES THAT AFFECT LEACHING POTENTIAL

The physico-chemical properties of the pesticides are very important in determining their fate in the environment once they have been applied to the target crop (Baker, 1992, Addiscott and Dexter, 1994). These physico-chemical properties are very different for different pesticides. The main pesticide properties that influence their fate are: soil persistence, soil adsorption, solubility and volatility.

The persistence of a pesticide, particularly near the soil surface will define how much of it is potentially available to be leached into the soil or lost with surface runoff at the time of the first significant rainfall event after application. Microbial degradation and volatilisation affect the pesticide persistence at the soil surface. Persistence in the generally unsaturated soil layers and in the underlying saturated layers will affect the amounts of pesticides reaching water resources.

Soil adsorption and solubility determine how much of the pesticide is adsorbed to soil particles and organic matter and how much is dissolved into the soil water. In general, pesticides with low water solubilities are strongly adsorbed, though the converse does not necessarily follow (Baker, 1992). Strongly adsorbed pesticides are lost mainly with sediment (Fawcett *et al.*, 1994). Moderately adsorbed pesticides tend to be lost in solution in surface runoff water. However, if preferential flowpaths exist then these pesticides can move rapidly to depth or to subsurface drainage systems. Weakly adsorbed pesticides are lost in the water phase both in surface runoff and in leachate.

Volatilisation of pesticides to the atmosphere is affected by the pesticide vapour pressure, the atmospheric conditions at and following the time of application and the location of application (on soil surface, within the soil, on crop surface). The more volatile pesticides tend to be applied within the topsoil, rather than on the soil surface or on crop, otherwise their effectiveness is severely restricted.

ENVIRONMENTAL EFFECTS THAT INFLUENCE PESTICIDE LEACHING

A number of environmental factors affect also pesticides losses from the point of application in the field. The condition of the soil surface at the time of application of the pesticide and afterwards (e.g. presence of crop or crop residue, degree of compaction, tilth) will affect the rate of infiltration into the soil surface. The duration of time and weather conditions between the date of application and the first significant rainfall event that follows will dictate how much pesticide might be lost through degradation processes and therefore not be available to leach. The characteristics of the first significant rainfall event (e.g. amount, intensity and duration) and the physical condition of the soil profile (moisture status, presence of cracks etc.) when the rain falls will all affect how much water and associated pesticide infiltrates through the soil surface or runs off. Once the pesticides have infiltrated through the soil surface the leaching of them to water resources may subsequently be relatively slow through the soil micropores or much more rapid through macropores which can effectively by-pass the soil matrix. Cultivation will influence capacity of the soil to retain and degrade pesticides *in situ* and the amount that can be lost via the leaching process.

EFFECTS OF CULTIVATION ON THE LEACHING OF PESTICIDES

All the effects of cultivation on soil characteristics will also play a part in the amount and rate at which surface or near surface applied pesticide move down the soil profile and into underlying strata or via drainage systems to open watercourses. Studies, particularly from the USA, provide sometimes conflicting views on the possible environmental benefits of conservation or reduced cultivation, particularly with regard to the loss of pesticides to water resources (Baker, 1987). It has been suggested that conservation tillage could increase the contamination of groundwater by pesticides (Logan *et al.*, 1987, Zacharias *et al.* 1991). The reason for this is the suggestion that crops grown within conservation tillage systems often require an increase usage of herbicides. In addition, due to the reduction in the disturbance of the topsoil, the connectivity and continuity of the soil macropores to depth are maintained thereby increasing the risk of the rapid transport of pesticides down the soil profile and into underlying strata (Harris *et al.*, 1993; Kamau *et al.*, 1996; Ogden *et al.*, 1999). However, on the other hand, conservation tillage increases the organic matter content of the topsoil, therefore increasing the possibility of pesticide adsorption and enhanced biodegradation. The disrupted soil matrix which results from repeated cultivation, together with the reduced organic matter content or lack of any crop residue associated with ploughed soils, tend to increase the risk of pesticides being lost in surface runoff (Watts and Hall, 1996). It has only been in the last ten years or so that longer term studies have been providing the necessary to data to effectively investigate the effects of different cultivation systems on pesticide leaching.

Fawcett *et al.* (1994) in their review of conservation tillage research aimed at surface water quality cited a number of US paired catchment studies in which zero tillage systems produced dramatic decreases in surface runoff and increases in infiltration. The decrease in runoff and increased in infiltration resulted from the development and maintenance of macropores, including cracks, root channels and earthworm burrows. The conventionally cultivated fields had significant surface runoff, soil erosion and pesticide loss. Edwards *et al.* (1993) cited a study at the North Appalachian Experimental Watershed in Ohio in the USA where 709mm of runoff was measured in a ploughed wheat watershed on a 6% slope over a four year period. Over the same 4 year period a nearby no-till wheat watershed on a 9% slope only produced 10mm of runoff. Therefore the possibility of pesticide losses in surface water in the ploughed system were substantially higher than in the no-till system. However, Baker (1992) points out that for the first significant runoff event severely tilled soils usually produce the least runoff, as infiltration of water into the disturbed soil is rapid. Therefore, if pesticide application follows on shortly after cultivation then conservation tillage may produce more pesticide losses in surface runoff, as the first event tends to result in the largest losses. This concurs with the work of Zaharias *et al.*, (1991) who found that the leaching losses of the herbicides atrazine and metolachlor in the first two weeks after application were greater in non-tilled soils than in tilled soil. After this period the differences between the two cultivation practices disappeared.

Isensee *et al.* (1990) and Smith (1993) found greater pesticide leaching through the soil profile under no-till wheat production than under conventional cultivation. Preferential flow pathways transporting the water and pesticides rapidly to the groundwater was suggested as the reason for this observation. Watts and Hall (2000) reported a similar result for mulch tillage versus conventional tillage trials, though the period of time between the pesticide application and the first significant leaching event was also seen to be very important. In contrast, Gish *et al.* (1995) working on deep, well drained soils found more of the herbicide atrazine leached from tilled fields than from adjacent no-till fields. Stable preferential flow paths in no-till soils supporting areas of enhanced biological activity was cited as the probable cause of this observation.

Work in the UK (Brown *et al.*, 1999) on comparing effect of a standard agricultural tilth with a finer deeper topsoil tilth on the leaching of the soluble herbicide isoproturon found total losses of the pesticide to be three times larger with the standard tilth. The reason given for this was a significant reduction in bypass flow through macropores created by the finer tilth, particularly during early drainage season events.

Logan *et al.* (1994) in a four year field study in NW Ohio, USA on a poorly drained, fine textured soil found that cultivation had little effect on surface runoff, underdrainage flow or pesticide losses. A similar result was reported by Kanwar *et al.* (1997). A study by Gaynor *et al.* (2000) in Canada looking at residues of the herbicides atrazine, metolachlor and metribuzin in clay loam soils under wheat subjected to different cultivation treatments found that environmental factors, such as rainfall, affected herbicide residues and their availability for leaching, rather than cultivation effects.

A number of studies have looked at the effect of cultivation on earthworm and microbial populations and have reported that higher earthworm numbers exist under conservation tillage systems than under repeatedly ploughed systems (Edwards *et al.*, 1992; Levanon *et al.*, 1994). Under direct drilled fields, for example, the residue-covered soil with its increased organic matter content provide favourable environmental conditions for earthworm populations and hence a much greater occurrence of burrows that could potentially preferentially carry water and pesticides (Edwards *et al.* 1993).

Under direct-drilling earthworms are the main agents for moving organic mater through the soil and increasing soil porosity. Levanon *et al.* (1994) working with two insecticides (carbofuran and diazinon) and two herbicides (atrazine and metolachlor) on 0-5 depth soil samples from long-term tilled and non-tilled fields found very different microbial populations and activity taking place. For the non-tilled soil the higher organic matter content permitted the enhanced adsorption of the pesticides and increased microbial populations. The active microbial biomass could in turn increase the biodegradation of pesticides in these soils. Laboratory studies by Edwards *et al.* (1992) has also shown that the lining of earthworm burrows are highly organic and contain large microbial populations. These structures can therefore very effectively remove pesticides from water travelling through them.

CONCLUSIONS

This review has shown that cultivation operations can directly affect pesticide leaching losses by altering the physical, chemical and biological characteristics of the soil. However, the research indicates that conservation tillage as widely practiced in the USA may, or may not, reduce pesticide leaching losses. The physico-chemical properties of the pesticide, together with a number of environmental factors at the time of application and in the weeks afterwards also exert strong influences on the availability of a particular pesticide to leach and on the rate and route of loss.

Good agricultural practices designed to minimise the risk of pesticides leaching to water resources need to manipulate the soil, crop and pesticide management at a particular site to produce the most beneficial system.

REFERENCES

Addiscott T M and Dexter A R. 1994. Tillage and crop residue management effects on losses of chemicals from soils. *Soil & Tillage Research*, **30**:125-168.

Baker J L. 1987. Hydrological effects of conservation tillage and their importance relative to water quality. Chapter 6 in: *Effects of Conservation Tillage on Groundwater Quality: Nitrates and Pesticides.* Eds – T.J.Logan, J.M.Davidson, J.L.Baker, and M.R.Overcash. Lewis Publishers Inc., Chelsea, Michigan, USA.

Baker J L. 1992. Effects of tillage and crop residue on field losses of soil-applied pesticides. Chapter 11 in: *Fate of Pesticides and Chemicals in the Environment.* Ed - J.L.Schnoor. John Wiley & Sons, Inc. 175-187.

Brown C D, Marshall V L, Carter A D, Walker A, Arnold D and Jones R L. 1999. Investigation into the effects of tillage on solute movement to drains through a heavy clay soil: I lysimeter experiment. *Soil Use and Management*, **15**:84-93.

Carter A D. 2000a. How pesticides get into water – and proposed reduction measures. *Pesticide Outlook*, **11(4)**:149-156.

Carter A D. 2000b. Herbicide movement in soils: principles, pathways and processes. *Weed Research*, **40**:113-122.

Court A C, Breach R A and Porter M J. 1995. Pesticides in drinking water – catchment protection or water treatment? BCPC Monograph No.62: Pesticide movement in water. Proceedings of a Symposium held at the University of Warwick, Coventry on 3-5 April 1995. British Crop Protection Council. 381-388.

Edwards C A, Edwards W M and Shipitalo M J. 1992. Earthworm populations under conservation tillage and their effects on transport of pesticides into groundwater. Paper No. 7C-16 in Proceedings of the Brighton Crop Protection Conference – Pests and Diseases. 859-864.

Edwards W M, Shipitalo M J, Owens L B and Dick W A. 1993. Factors affecting the preferential flow of water and atrazine through earthworm burrows under continuous no-till corn. *Journal of Environmental Quality*, **22**:453-457.

Fawcett R S, Christensen B R and Tierney D P. 1994. The impact of conservation tillage on pesticide runoff into surface waters: a review and analysis. *Journal of the Soil and Water Conservation*, **49(2)**:126-135.

Gaynor J D, Tan C S, Ng H Y F, Drury C F, Welacky T W and van Wesenbeeck I J. 2000. Tillage and controlled drainage-subirrigated management effects on soil persistence of atrazine, metolachlor and metribuzin in corn. *Journal of Environmental Quality*, **29**:936-947.

Gish T J, Shirmohammadi A, Vyravipillai R and Wienhold B J. 1995. Herbicide leaching under tilled and no-tillage fields. *Journal of the American Soil Science Society*, **59**:895-901.

Harris G L, Howse K R and Pepper T J 1993. Effects of moling on soil-water and runoff from a drained clay soil. *Agricultural Water Management*, **23**:161-180.

Harris G L and Catt J A 1999. Overview of studies on the cracking clay soil at Brimstone Farm, UK. *Soil Use and Management*, **15**:233-239.

Isensee A R, Nash R G and Helling C S. 1990. Effect of conventional vs. no-tillage on pesticide leaching to shallow groundwater. *Journal of Environmental Quality*, **19**:434-440.

Kamau P A, Ellsworth T R, Boast C W and Simmons FW. 1996. Tillage and cropping effects on preferential flow and solute transport. *Soil Science*, **161(9)**:549-561.

Kanwar R S, Colvin T S and Karlen D L. 1997. Ridge, moldboard, chisel and no-till effects on tile water quality beneath two cropping systems. *Journal of Production Agriculture*, **10(2)**:227-234.

Levanon D, Codling E E, Meisinger J J and Starr J L. 1993. Mobility of agrochemicals through soil from two tillage systems. *Journal of Environmental Quality*, **22**:155-161.

Levanon D, Meisinger J J, Codling E E and Starr J L. 1994. Impact of tillage on microbial activity and the fate of pesticides in the upper soil. *Water, Air and Soil Pollution*, **72**:179-189.

Logan T J, Davidson J M, Baker J L and Overcash M R. 1987. *Effects of conservation tillage on groundwater quality.* Lewis Publications, Chelsea, MI, USA, 292pp.

Logan T J, Eckert D J and Beak D G. 1994. Tillage, crop and climatic effects on runoff and tile drainage losses of nitrate and four herbicides. *Soil & Tillage Research*, **30**:75-103.

Ogden C B, van Es H M, Wagenet R J and Steenhuis T S. 1999. Spatial-temporal variability of preferential flow in a clay soil under no-till and plow-till. *Journal of Environmental Quality*, **28**:1264-1273.

Shepherd M A and May M J. 1992. Practical aspects of crop cover establishment, choice of species and cultivations. *Aspects of Applied Biology*, **30**:331-334.

Silgram M and Shepherd M A. 1999. The effects of cultivation on soil nitrogen mineralisation. *Advances in Agronomy*, **65**:267-311.

Smith Jr S. 1993. Effects of tillage pesticide concentrations in shallow ground water and surface runoff from fragipan soil in north Mississippi, USA. (1993). Proceedings of the Conference on Environmental Impact of Agricultural Activities. Washington DC, May 1993. American Institute of Hydrology. 79-89.

Tebrügge F. 1993. The environmental implications of tillage systems. Proceedings of the 5th Conference on Mechanization and Energy in Agriculture, 11-14 October 1993, Ku_adasi, Turkey. 55-65.

Watts D W and Hall J K. 1996. Tillage and application effects on herbicide leaching and runoff. *Soil & Tillage Research*, **39**:241-257.

Watts D W and Hall J K. 2000. Effects of conventional and mulch tillage on dicamba transport. *Weed Technology*, **14**:94-99.

Zacharias G, Heatwole C D, Mostaghimi S and Dillaha T A. 1991. Tillage effects on fate and transport of pesticides in a coastal plain soil: II leaching. Paper No. 91-2544 in Proceedings of the International Winter Meeting of the American Society of Agricultural Engineers, Chicago, Illinois, 17-20 December 1991.

VII. SOCIO-ECONOMIC PERSPECTIVES AND POLICY IMPLICATIONS FOR DEVELOPMENT

R.A. PEIRETTI

ECONOMIC GLOBALISATION AND CONSERVATION AGRICULTURE THE CASE OF MERCOSUR

AAPRESID: Argentinean No Till Farmers Association
CAAPAS: American Confederation of Farmers Organisations for a Sustainable Agriculture
Paraguay St. 777. 8th Floor. Office 4. 2000 Rosario. Argentina

Globalisation and some of the effects derived from it are analysed. The pressures that this phenomenon generates on top of the agricultural production of the Mercosur countries was reviewed. The reaction in front of this new reality and the changes generated, while adapting to the new scenarios, are described. The No Till farming system adoption and the utilisation of «last generation» technologies, as a valid –and to some extend– generalised answer to the globalisation forces, is discussed. The achievement of a higher and more stable productivity level within a more sustainable situation is described.
Key words: Globalisation, no till, holistic, modern technologies, sustainability.

INTRODUCTION

The Globalisation process can be considered as a spatial projection of the action of a human community to a different one. Mainly due to a combined effect between a global population growth –and its economic evolution– with the tremendous communication capabilities already developed, globalisation reaches all the world's corners with an ever-increasing speed. In many cases, globalisation impacts could be assimilated to the concept of an externality: «an externality is any action that affects the welfare of or opportunities available to an individual or group without direct payment or compensation, and may be positive or negative». (Pretty J. 2000).

The globalisation concept applies to almost all-human activities however it is frequently referred to economic issues. While impacting the economic variables, and within them the agricultural production, globalisation interacts with the present and future welfare and with the nutritional status of a given human group, country or continent. This happens through the modification of their ability to locally produce food or to the capacity to buy it.

L. García Torres et al. (eds.). Conservation Agriculture, 427–434.

Due to the existence of an extremely wide range of intensities and directions of the globalisation forces, they generate an extremely wide range of impacts or interaction products. In some cases, the acting globalisation forces are somehow balanced, and hence produce only a kind of almost neutral cross effect. However, most frequently, the equilibrium is not achieved so they create impacts that lead to different degrees of scenario changes. These scenario changes could be positive or negative according to their nature and way they are assimilated. Depending on the way they are interpreted by the impacted human group, they may be seen as a threat or as an opportunity. According to this behaviour, and in front of an equal scenario, a human group may not react properly and be negatively impacted while other may see and interpret it as an opportunity and act changing things for better. In this last case, the globalisation pressures could represent «the incentive» needed to apply the unavoidable efforts demanded to be able to change things in the right way. This mechanism is based on the idea that the challenge stimulate a response that when properly guided could be the source of the necessary will to generate changes that finally leads to some kind of benefits.

Going deeper in the analysis, when not only the short-term issues but also the medium and long range are taken into consideration, the concept of sustainability appears. To simultaneously achieve a good efficiency and competitiveness level at the present time, and also a proper degree of sustainability, we have to adequately satisfy a kind of combined constraint. Taking into consideration the actual and future human food needs, the approach that tend to satisfy both constraints, appears as an unavoidable obligation either from the socio-economic and ethic standpoint. When this situation is analysed across the world, in many cases we find that this multiple constraint or condition is not properly satisfied. However when focus on the CAAPAS (American Confederation of Sustainable Farmers Associations) countries, the opposite case is frequently found. Even the globalisation impacts can vary from light to intense; they should not be ignored in any case if a proper management of the present, and future of any human community, is been aimed.

MATERIALS AND METHODS

The methodology utilised in this study was based in the observation and interpretation of reality as well as on the consideration of several authors' findings, general publications and books referenced on this paper. Considering the empirical and observational part of the analysis herein presented, two main sources of information were utilised. While analysing the No Till system in countries other than Argentina, most of the information was either personally collected or obtained from different studies and publications. Sources like that provided by CAAPAS annual meetings reports and personal communications were rather important. Also the observation of the No Till Farming system utilisation across many different agro-ecosystems constitutes a very important part of the empirical validation of the system functioning. (Crabtree, 2000). While referring to Argentina, I was also utilising different sources. From personal data as well as information collected within AAPRESID (Argentinean No Till Farmers Association) to a variety of different small and large scale field trials.

Also, a wide range of information and data derived from Argentinean Agricultural Experimental Station, Universities and other Official Institution studies were utilised.

RESULTS

In most of the Mercosur countries, as well as in some other Latin American ones, like Chile, Argentina, Uruguay, Paraguay, Brazil, and to some extent Venezuela, Colombia, Costa Rica Mexico and others, the presence of different globalisation forces acting on top of the agricultural production systems were recognised.

Among other socio-economic issues the subsidisation applied to the agricultural activities, mostly within the developed countries, results in an «artificially» depressed commodity price which level is normally falling below that expected under a free economy and free market environment. This fact, along with the imposition of artificial trade barriers, represents some –but not all– the causes that generate many of the globalisation pressures. Also the general widespread idea that the environment is a prerequisite and not a consequence of the human action is derived from an echo-centric view rather than anthropocentric interpretation of the word and of our existence. This echo-centric point of view, that seems not to prioritise the proper satisfaction of the human basic needs, constitute one of the most flagrant contradictions of some environmental movements originated primarily within the developed world. They tend to promote the regression to a pristine natural stage of things that is not compatible with a proper level of satisfaction of the basic human needs generated by the six billion people that today exist on earth. This fact along with the application of the precautionary principle concept (that demands the absence of risk while acting on nature and hence an impossible full and absolutely precise prediction of the future), negatively interacts with the possibilities of real human and environmental progress. These concepts heavily impact on the Mercosur countries as well as in other developing countries of the world. (Sorman, Guy 2001). The consideration of the environment without including the human being and their actions as «natural» part of it constitutes an erroneous posture while looking for an improvement of the present and future satisfaction of the human needs. This facts, combined with the above mentioned effects originated on prevention of free economy rules, very frequently erodes the possibilities of the developing countries to evolve their economies, mainly supported by the production of basic goods. While referring to the agricultural production and human welfare of the Latin American countries, this phenomenon can be considered as one of the main source of negative globalisation pressures. In front of these pressures, several Latin Americans reacted trying to find a way to counteract their impacts. The acceptance of the new scenario imposed by the globalisation forces, the recognition of a higher level of complexity (Solbrig, 1997) and a more holistic and systemic approach was needed to be able to find a proper answer to the challenge. It was discovered that the real challenge was located in the necessity to be innovative and to undergo deep changes in the way of doing things. With these convictions in mind, an entirely new farming system named «The No Till System», was developed and applied in an ever increasing number of hectares either on the relative large commercial farming operations as well as on the case of the small farmers that mainly practice a subsistence agriculture. In regard to the small farmer's category, it is worth to mention the extraordinary success achieved when the No Till farming system was adequately promoted among them. While doing so, a better level of efficiency, productivity and sustainability was achieved. There are plenty of examples of this small farmer's successful stories in Brazil, Paraguay and to some extent in Argentina.

While referring to extensive –market oriented– agriculture the fitness of this new approach is clearly reflected by the important no till system adoption rate. On next figure number 1, we can see the results of the successive annual AAPRESID surveys in regard to the No Till farming system adoption in Argentina. In this country, no till started with only a couple of thousand hectares – seventeen years ago –, reaching around ten million hectares nowadays. These number of hectares that are been farmed under the no till system accounts for more than 40 % of the commercially (mostly cereals, oilseeds, fyber and to some extent forages) cropped of Argentina.

During the same period, mainly due to the same reasons that promoted the Argentinean adoption process, a similar adoption pattern can be found when we study the phenomenon in some other countries like Brazil, Bolivia, Chile, Uruguay and Paraguay; all of them members of the CAAPAS organisation.(Derpsch, 1998).

Farmers attitude and a proper reaction in front of the globalisation pressures, in many cases constitutes a good part of the reasons that partially explain the system adoption. In all these cases, globalisation forces were mainly interpreted as an opportunity to change for better rather than as a threat. The system constitutes an adequate framework within which all modern technologies are applied and the results were satisfactory from the beginning of the adoption process. Applying a great amount of innovation, willingness and wisdom to face the unknown –that always underline the changes–, the newest technological developments were incorporated into the system

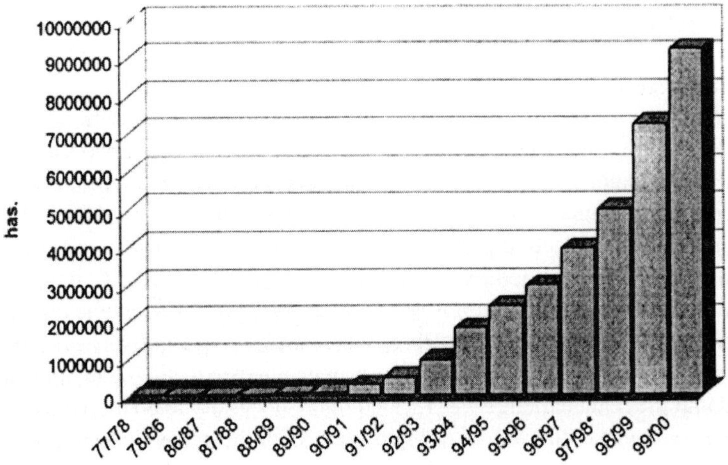

Figure 1. Argentinean No Till adoption process. Source AAPRESID

under a systemic and holistic approach. (Peiretti, R. 2001). Biotechnology, soil and plant nutrition, crops and inputs rotations, integrated pest management, as well as other technologies like precision farming constitute a representative example of them.

In all these countries, where the no till system expansion took place, a positive correlation between the achieved level of adoption and sustainability can be found. This new reality allowed to simultaneously reach a higher –and growing– productivity and a relative better economic result that normally ends on a higher competitiveness. Also, the «among years» yield variability was lowered and the output/input relationship of the production system were clearly increased. Some other very beneficial phenomenon can be identified along the No Till adoption process. Examples of them are: a noticeable control over the soil erosion and deterioration process of both types (water and wind)–, an improvement of the amount of water captured into the soil (which negatively correlates with the water run-off and water contamination), an important increase of the soil organic matter (which means carbon sequestration), a noticeable increase of the soil biotic load (Crovetto, 1992 and Crovetto, 1996); are all indicators of a better level of sustainability. (Derpsch R. 1997).

The no till farming system is based on a better understanding of nature. The full acceptance and assumption of complexity and the attempt to better understand and emulate the natural processes but now enhanced by the application of all the modern scientific knowledge and technological developments, constitutes its basic strategy. While compared with the other farming production system approaches, is that no till appears as unique from the standpoint that it is based on the complete avoidance of soil tillage and on the achievement of a soil permanently covered with crop residues. These two apparently simple facts create an entirely new and favourable agronomic and agro-ecological environment within which it became possible the achievement of a better management of central issues like soil erosion control and the water management.

On previous figure, N° 2, I include the results of a study that shows a very strong negative correlation between the percent of soil covered by crops residues and the soil losses by means of water erosion. In this study was found that only four tons per hectare of crop stoover on top of the soil – which are relatively easy to achieve- 60% of the soil surface is covered and a reduction of 90 % of soil losses is reached. (Papendick, 1995 and Papendick, 1996). Practical evidence at the farm level strongly supports these findings as well as a very effective control of the wind erosion.

While looking at the water availability on soils and according to (Dardanelli, 1998), soils under no till has a greater water infiltration capacity than conventionally tilled ones.Figure 2. This enlarged water capturing ability, not only provides more available water to crops, but also strongly diminishes water run off and water soil erosion. These two phenomena combined, normally ends in a significant reduction of water contamination which represents a very important environmental benefit.

Also the enlarged water availability enhance the entire agroecosystem functioning. When this facts is properly managed, it allows to achieve an increase of the organic matter soil content which at the bottom line means carbon fixation. There are several evidences that the No till farming system can sequester carbon from the atmosphere

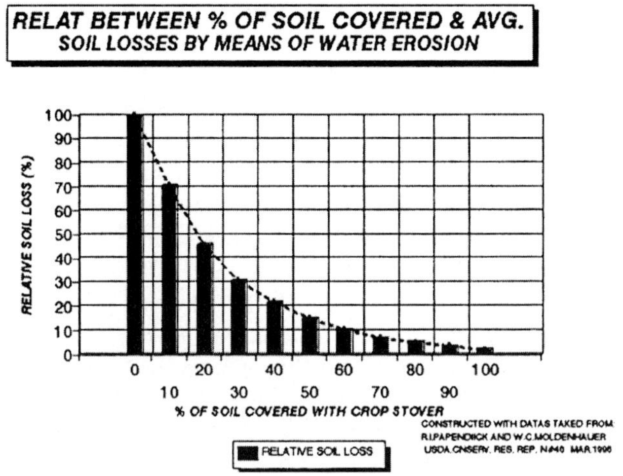

Figure 2. Relationship between the % of covered soil and the soil losses by means of water erosion. Source: Papendick, 1995

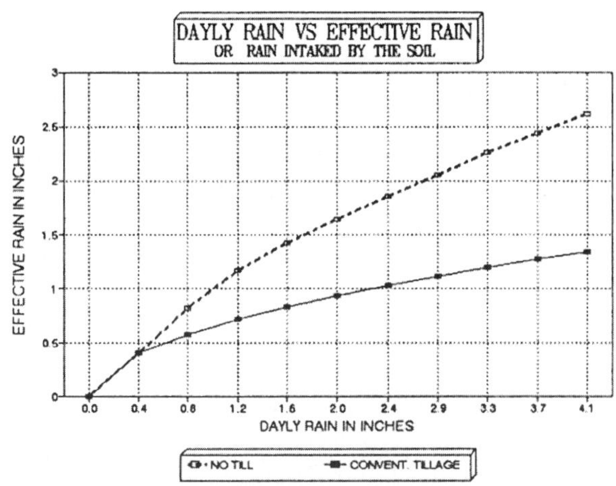

Figure 3. Water infiltration comparison between No Till and Conventional Till - Plough - Source: Dardanelly 1998.

and incorporate it into soils as part of the organic matter. (Aguilera, 1995) (Lal, 1998 (a) and Lal, 1998 (b)). This fact, also constitutes an important environmental asset.

It is worth to strongly notice that all these achievements, as well as several others that are not discussed on this paper, are reached within a sustainable frame and even allows us to go beyond sustainability accessing to an improvement stage. These positive and desirable trends that were detected since the beginning of the adoption process clearly keep growing at the present time.

DISCUSSION

To be able to find proper ways to keep developing the world human community, we must be prepared to face and accept the challenges imposed by a globalisation process that quickly speeds up. It would be highly desirable that those countries or international communities, that to some extent could be considered as highly responsible for the enhancement of several of the globalisation pressures, change their approaches while looking for solutions to solve their problems. It appears as a clear fact that none lasting solution to any problem could be found within the application of mechanism mainly based on the incentive of inequalities and inefficiencies which normally only leads to the achievement of a fragile, temporary and not sustainable shelters. Examples of these attitudes are those derived from protectionism, subsidisation, trade barriers and close economies and some other mechanism that are not based on the free economy and free market philosophy. Instead we would have to be prepared and accept the existence of continuous and permanent changes that will demand of us the best efforts to be able to find long lasting and hence valid solutions. The recognition and adequate characterisation of the globalisation forces, constitutes the first step to find a proper way to counteract them –minimising their potential negative impacts– an even taking advantage of the situation to change for better. The encouragement of the full utilisation and preservation of the comparative advantages –as a way to achieve and keep the highest possible level of efficiency– appears a much better way to stimulate a more healthier relationship among countries. This position appears as a more suitable way to face the globalisation process from both sides of the road or from a global point of view. This new scenario would eventually minimise the generation of artificially created asymmetric and negative impacts. If the humanity all prioritise these mechanisms, the achievement of a higher level of satisfaction of its basic needs for the present future generations, may became a reality. The achievement of this goal is not an easy task but constitute an unavoidable appointment that the whole humanity is demanding from us.

ACKNOWLEDGEMENTS

I specially want to thank my wife Cintia and all my family for the generous time they offered to me to be able to write this paper. Also to Dr. Otto T. Solbrig from the DRCLASS of Harvard University for his invaluable teachings. To the meeting organisers also my special thanks for inviting my to participate of it.

REFERENCES

Aguilera, S.M. et.al. (1999) Dinamica del carbono en suelos con distintos sistemas de labranza. In Frontera Agrícola, Vol. 5, N° 1 y 2. Ed Universidad de la Frontera : Temuco, Chile. Pp.33-38.

Crabtree, B. (2000) Incredible No Till Adoption in Western Australia. *In newsletter of the Western Australian No Till Farmers Association (WANTFA)*, January 2000, Vol **8** ,N° 1, pp.308.

Crovetto, C. (1992) Rastrojos Sobre el Suelo. Ed. Editorial Universitaria: Santiago de Chile, República de Chile.

Crovetto, C. (1996) El desarrollo de la Cero Labranza en el Fundo Chequen y su influencia en algunos parametros físicos químicos y biologicos. *In proceedings of the fourth AAPRESID annual No Till conference. AAPRESID* : Rosario. pp. 87-103.

Dardanelli, J. (1998) Eficiencia del uso del agua según sistemas de labranza. In *Siembra Directa,* J.L.Panigatti et. al. Eds. Hemisferio Sur : Buenos Aires. pp.107-115.

Derpsch, R. (1997) Importancia de la Siembra Directa, para obtener la sustentabilidad de la producció pn agrícola. In Proceedings of the fifth AAPRESID annual No Till Conference. AAPRESID: Rosario, pp. 153-176.

Deprsch, R. (1998) Historical Review of No-Tillage Cultivation of Crops. MAG-GTZ Soil Conservatio Project, Casilla de Corre 1859, Asunción, Paraguay.

Lal, R., et al. (1998) (a). Management of Carbon Sequestration in Soil. CRC Press LLC.

Lal, R., Kimble, J.M., Follet, R.F. and Cole, C.V. (1998-9) (b) The Potential of U.S. Cropland to Sequester Carbon and Mitigate the Greenhouse Effect. Chelsea, MI : Ann Arbor Press.

Papendick, R.I. and Moldenhauer, W.C. (1995) «Crop Residue Management to Reduce Erosion and Improve Soil Quality». USDA (United States Dpt. of Agriculture). *Agric. Research Service. Conservation Research Report* Nº **40**.

Papendick, R.I. (1996) No Tillage Impacts on Soils, twenty years of experience. *In proceedings of the fourth AAPRESID annual No Till conference.* AAPRESID: Rosario. pp. 59-85.

Peiretti, R.A. (2001) Direct Seed Cropping in Argentina: Economic, Agronomic, and Sustainability Benefits. In Otto T. Solbrig, Robert Paarlberg, and Francesco di Castri Eds. Globalization and the Rural Environment. Candbridge Massachusetts. Harvard University Press.

Pretty, J..N. et. al. (2000) An assessment of the total external costs of UK agriculture. *Agricultural Systems* **65** (2000) 113-136. Elsevier.

Solbrig, O.T. (1997) Sembrano el Futuro. Como desarrollar una agricultura para conservar la tierra y la comunidad. En AAPRESID. Manual de Conferencias ofrecidas durante el 5° Congreso Nacional. Paraguay 777. 8° piso of. 4. 2000 Rosario, Argentina.

Sorman, G. (2001) Prologue. In Otto T. Solbrig, Robert Paarlberg, and Francesco di Castri Eds. Globalization and the Rural Environment. Candbridge Massachusetts. Harvard University Press.

C. PIERI[1]

STRATEGIES FOR INTERNATIONAL COOPERATION

Senior Agro-Ecologist, Rural Development Department, Room MC5-771
The World Bank, 1818 H Street, N.W., Washington , DC 20433, USA

INTRODUCTION/BACKGROUND

At the onset of the 21st Century the Rural Sector is facing daunting global challenges from persistent poverty, undernourishment, in spite low global commodity prices, a new framework for rural life influenced by globalization, accelerating urbanization at the expense of rural areas, and continued natural resource degradation. However, recent studies (World Bank, 2001, IFAD, 2001) show that the agricultural sector as a whole is subject to a discernable decrease in support by national and international investors, while the food frights such as «mad cow» disease, the foot and mouth epidemic, and revelations of cancer-causing dioxin in food have sorely undermined the confidence of civil society at large in supporting new science applications required to increase farm productivity.

This paper calls for more focused and proactive cooperation among the main stakeholders at national, regional and international levels to support movement towards «Conservation Agriculture» (CA) and improved land husbandry, including farming practices such as No-Till system (NT) or Direct-Sowing-Mulch-based cropping systems that can raise farm productivity and create the enabling environment to jointly address the problems of poverty alleviation, food security and environmental protection. Practitioners of CA know that the adoption of improved land husbandry[2] (LH) requires, above all, a change in mindset among the main stakeholders. It entails much more than a simple switch from one technical package to a supposed better one, to include a collaborative effort on social mobilization, education and training, marketing and diversification, and environmental awareness without losing track of the vital short-term improvement of agricultural producers' livelihoods.

Before elaborating on the content of some appropriate strategies of international cooperation, it may be necessary to review the basic assumption under which cooperation is deemed necessary. Several recognized specialists (McCalla, 1999, Mellor, 2000) have convincingly established that to meet the growing global demand

L. García Torres et al. (eds.). Conservation Agriculture, 435–444.

for food, increased crop and animal production is required where demand exists, such as in developing countries where the rate of population increase is high. International food trade can be at best only part of the solution. What are the alternatives? «One plausible alternative –the substantial expansion of cultivated land– is not feasible for most rural populations in Asia, and increasingly in most of Africa and Latin America «(IFAD, 2001). Increased land productivity remains the best alternative for reducing food shortage and poverty on a long-term basis, at local as well as global levels.

Strategies for International Cooperation in Conservation Agriculture are discussed under this basic assumption.

HOW CAN INTERNATIONAL COOPERATION BE ACHIEVED

In this context, what does international cooperation focus on to foster the adoption of CA seen as an opportunity and a practical instrument to achieve the goals of Sustainable Development, Food Security and Poverty Alleviation in rural areas?

By reviewing the positive and negative experiences gained over the past decade by several international and bilateral organizations, it appears that a general framework for developing successful strategies of cooperation in CA should at least include the following three lines of actions: developing new partnerships, sharing common principles in cooperation, acknowledging that the adoption of CA is a process requiring time, commitment and increased knowledge.

Developing new Partnerships

There is an urgent need to revisit traditional «institutional partnerships», such as those between funding agencies and central governmental institutions and/or public agricultural services, to move into more «programmatic partnerships» open to new partners. A lassitude among traditional partners in institutional partnerships, as well as among international organizations «specialized» in cooperation is commonly acknowledged. Moreover, partnership between international (and/or governmental) institutions and organized «civil society», such as consumers' groups of urban origin, is often critical of and environmentally concerned with agriculture. Farmer' voices need to be heard in their own capacity and also in their diversity, including women farmers, through their own-managed organizations and not only through specialized NGOs. In addition, in a decentralized administrative setting, local rural authorities and elected bodies (such as «communes rurales», mayoral assemblies, etc.) should play an increasing role in representing all categories of rural dwellers, from the productive and non-productive agricultural sectors. Similarly, fruitful partnerships are sought with national agribusiness, and also with international agro-inputs companies as long as all partners share some common guiding principles and objectives for fair cooperation. Most recently, promising partnerships are developed between private corporations and farmers and rural institutions (e.g. in Canada, Europe, Brazil) to address the challenging issues of greenhouse gases balance (emission and sequestration), including the acknowledgment of the valuable environmental services provided by CA/NT farmers groups.

Objectives and guiding principles for successful cooperation

Although different forms of international cooperation should reflect the diversity in national and sub-national agricultural situations, strategies should be based on some common objectives and guiding principles, which can be summarized as follows:

- support the *new paradigm* of combined agricultural intensification and environmental protection, specifically through the development of profitable local land and crop or animal management systems which draw on improved biological interactions and the empowerment of rural producers and their organizations. This new agro-ecological and people-centered approach to environmentally sensitive agricultural intensification should be the preferred approach and the foundation of international cooperation for CA. It is recognized that neither low input-low output, nor low input-high output approaches are sustainable options, and the high input-high output agriculture that characterized the green revolution is not a sustainable and affordable alternative to solve the productivity increase problem in most rainfed land and complex cropping systems.
- *shift the role of State and public services* to rural producers, including an administrative and financial decentralization process. This process leads to increased participation and gives specific responsibility to the producers and main stakeholders in selecting and implementing local strategies of sustainable land use and land management. It acknowledges that decentralization and producer empowerment can dramatically change relationships among stakeholders, address conflicts, favor partnerships between Producers Organizations (POs)-local authorities-Agribusiness-Central Government, generate social capital development and, finally, create the right environment ready for change toward CA.
- facilitate *agricultural growth and fair competitiveness*: local strategies of CA must stimulate agricultural growth and generate short-term improvement of the livelihood of rural producers and rural poor in particular, such as decrease in drudgery and/or income generation. While the social and economic status of producers is rising, a new role the private sector is also emerging, including the development of more equitable relationships between private POs and private agribusiness, for which public and private international cooperation may provide cross-regional experiences and world-wide lessons learned.
- contribute to and implement *international agreements* inspired by Agenda 21 like the United Nations Framework Convention for Climate Change (UNFCCC), the United Nations Convention to Combat Desertification (UNCCD), the Operational Programs (OP) of the Global Environment Fund (GEF) such as OP#12 «Integrated Ecosystem Management» and OP#13 «Agro Biodiversity», which are conducive to this change in 'mindset', to people-centered development strategies and institutional shifts. International partners in CA have the opportunity and probably a specific responsibility to show and demonstrate that farming

should not be pictured as an unavoidable environmental problem but, that CA farming may well be an important part of a solution for sustainable global environment management. People-centered and locally adapted land husbandry techniques to control land degradation, have the potential to contribute to mitigate the impacts of global warming, to maintain appropriate water cycles and safe water quality, to prevent of loss of biodiversity and to combat rural migration and desertion of rural space. In a nutshell, CA/NT offers a powerful combination of positive impacts on the environment.

By acknowledging that change toward Conservation Agriculture is a *process* which requires time and knowledge for customizing known agro-ecological principles and participatory methods to *specific sites* (in physical, economic, and socio-cultural terms), the international community recognizes that achieving such change requires, as mentioned at the onset of this paper, a major shift in mindset among the many actors concerning explicit rejection of the increasingly irrelevant «top-down transfer of technology» model. Making the needed progress in this field will, in turn, require a) *long-term commitment* from all partners, including international institutions and funding agencies, and b) a *major effort in education* for all the agents of change, including farmers, front line agents, scientists, profit and non-profit private sector.

SOME IDEAS TO SUPPORT THE COOPERATIVE EFFORTS OF INTERNATIONAL INSTITUTIONS

These ideas are consistent with the renewed Rural Development strategy of the World Bank (2001) drawing on lessons learned from Projects on Land and Natural Resources Management implemented over decades in many developing countries, particularly in South America and more recently in Africa. They fit the agenda of many bilateral aid agencies, as well as well the program of several public and private organizations. They are also inspired more specifically by the results of the successful cooperation developed between the Rural Development Department and the Brazilian No-Till Producers Federation (FEBRAPDP) in organizing annual Study Tours in South Brazil, beginning in 1998. These study tours «Producer-led Rural Organizations for Sustainable Land Management (PRO- SLM) «, have been attended by about one hundred participants including farmers, agents from research and extension services, government officials and World Bank officials, from Africa, Latin America and Asia, and have significantly contributed to a positive change in the way they view the potential of CA and NT in reaching the goals of sustainable rural development.

Creating Enabling Pro-Rural Environment

An initial distinction should be made between generic policy and institution-related actions which are broadly «pro-rural», and those actions which will specifically support

CA. Of course both categories are closely linked, and should fit into a comprehensive agenda of international cooperation for CA, which is the main focus of this paper.

- **Advocacy role**. International organizations must have a specific macro and micro economic advocacy role, at international and national levels, in support of rural development as the best option for economic development ;
- **Social and Legal favorable conditions**. For the most part, natural resources degradation can be attributed to either market or government failures. Examples of these are insecure property rights in land resources, and overall inability to produce sufficient public goods in the form of appropriate flow of information to producers and other decision makers in land use and land management. More specifically, external assistance should concentrate on creating an environment that facilitates the productivity of any attendant domestic investments. External resources should be used to support the development of the other necessary elements for growth and poverty reduction, like rural education, health, and legal aspects such as land use rights and access, land market etc
- **Long-term international and national financial commitment**. Too often long-term financial commitment seems to be more supported by private profit-oriented organizations rather than public international partners. However, the World Bank has developed new lending instruments to meet this goal of long-term commitment, such as Learning and Innovation Loans (LILs), to test and learn and facilitate replicability and scaling-up, and Adaptable Program Loans (APLs)[3].
- **Decentralization and development of new partnership**. At the institutional level there have been some promising developments. Recently much effort has been dedicated to the development of demand-driven agricultural public services, and the participation of multi-sectoral partners in the national policy dialogue between international and bilateral funding agencies and developing countries. In addition, international and bilateral aid organizations have developed more programmatic partnerships with producers organizations (POs), to allow farmers to voice their point of view and be active partners i) at national level, during all the phases of development of rural projects, from preparation to implementation and evaluation, and ii) at international level, in addressing global social, economic and/or environmental issues. Examples include the partnerships at national level between the World Bank, FAO, bilateral organizations such as GTZ, CIRAD-French aid, DFID, NGOs such as ABLH, Pro-Natura, EcoCarbone and farmers organizations (Brazil, Madagascar, Mexico, Kenya, Senegal, Cote d'Ivoire etc.), and at international level between IFAP (International Federation of Agricultural Producers) the World Bank and CGIAR.
- This new partnerships will require the development of new «**rules of the game**» Scientific public and private organizations are currently developing new approaches for better sharing of activities on the international scene, such as the

initiative taken by the Global Forum on Agricultural Research. Similarly, efforts are currently made by the World Bank and IFAP in developing their partnership to find ways and financial mechanisms which support national/international settings for better networking among farmers without cutting them from their rural bases and without them forgetting the diversity of the farming sector. In general terms, it can be stated that farmers organizations should find among their own membership the necessary resources to make the organization sustainable including the implementation of basic networking activities, while outside funding is focused on targeted activities such as i) support to regional or international networking, including the dissemination of lessons learned from farming practices linking short-term economic goals with long-term maintenance of «common pool goods» in rural landscapes, and ii) capacity building in managerial skills of POs, including conflict resolution techniques and negotiation skills with agricultural inputs providers.

Specific Actions to promote CA

- On a short-term basis, the international cooperation can already take specific actions to foster the adoption of CA/NT, such as follows: Funding agencies should prioritize their support to projects adopting the new paradigm and refrain from investing in programs diametrically opposed to the LH/NT philosophy.
- Government commitment to develop NT systems and land husbandry appropriate to the majority of the rural producers, including rural poor, would provide international cooperation with an indicator which would trigger support.
- Within the framework of administrative decentralization specific institutional development conducive to CA and NT can be initiated and implemented by producer-led or producer-controlled entities such as watershed committees and/or regional SLM committees i.e. institutions specific to geographic units going beyond grouping of farms, villages and/or administrative boundaries. In South Brazil for instance, micro watershed committees have proved to be a very efficient and powerful instrument in achieving the goals of sustainable land management leading to rural development, linking the objectives of increasing land productivity and environmental sensitivity, and facilitating the development of improved political and economic relationships between rural and urban populations. This approach addresses also the issue of «large versus small» since large output attracts business, infrastructure development, and availability of required intensification inputs, while better managed small farms have the potential to generate direct social and environmental benefits for all agricultural producers, and to enhance economic activities and fiscal revenues.
- Innovative financial mechanisms, including cost-sharing and a voucher system. There is still a need for developing innovative financial mechanisms, including

cost-sharing mechanisms, to support change in farming practices which have the potential to both protect common goods, such as land and forest and the ecological services that they provide, and the alleviation of poverty. In addition, the problem of access of smallholders to productivity enhancing inputs through normal market channels is important and persists. The World Bank is proposing an approach that might address this issue, while being consistent with longer term development of private markets. This approach utilizes a voucher scheme (McConnell Brooks, K., 1999) targeting a group of smallholders and providing them with the tradable vouchers redeemable at face value for an identified menu of agricultural inputs. The Government's obligation would be strictly budgetary (to redeem the vouchers), and the private sector would deliver inputs, accepting vouchers as a supplement to cash in transactions. The intent of the program would be to assure the private sector that the Government would not compete as a distributor. It would enhance demand enough to increase the reach of the private sector, and would transfer income to a targeted group of rural poor.

- Measurement of economic and environmental impacts of CA would be an important part of international cooperation. On-farm pilots programs, partially funded by international partnerships, would be a preferred target to develop appropriate indicators of short and long-term impacts of improved land husbandry such as NT. One of the products would be valuing environmental goods and services protected by NT farmers, such as the balance of greenhouse gases (GHGs) resulting from improved land husbandry.
- Farmers and producers organizations should be encouraged by international cooperation to participate in the on-going debate on biotechnology and present a rural-poor perspective.
- Education: The CA/LH approach is typically site-specific and knowledge intensive. It leads to revisiting support system for technology development, to sharing responsibility with farmers and producers organizations in extension and adaptive research, and, eventually, to realigning the educational curriculum to incorporate principles of CA and improved land husbandry. Reinventing rural education is becoming increasingly urgent (Maguire C., 2000), and CA is a perfect example of a new focus for agricultural and rural education. Among several initiatives, the World Bank is exploring new avenues of knowledge management, including a distance learning internet-based system, and partnerships with producers organizations (e.g. Federation of Brazilian No-Till Planters, FEBRAPDP) and bilateral organizations (CIRAD, GTZ) to favor knowledge dissemination through study tours and the development of centers of excellence for practitioners, farmer-to-farmer contacts, capacity-building and decision-makers' awareness (see annex for details).
- Finally, international organizations such as World Bank and FAO, and OECD countries in their own right- should encourage a vigorous international and regional media campaign emphasizing the importance and relevance of CA as an entry point to the process of rural poverty alleviation, food security, and environmental protection.

NEXT STEPS

The challenge before us is to implement the actions described earlier. Let us not leave the «Worldwide Challenge» of this Congress without resolving to put in place the principles and mechanisms needed to implement CA/LH in the rural landscapes of our planet.

The next international platform will be the «Rio+10» Conference in 2002. We must bring the message of this gathering in Madrid to Rio in a compelling form. We must form a working group to synthesize the salient messages from this Congress and turn them into an unambiguous and convincing communication for «Rio + 10», where the voices of CA and NT farmers have to be heard loud and clear.

We have also to develop a full strategy to counter the deep rooted belief that tillage is the only way to create soil fertility. Selling such a strategy will not be an easy task or a short-term undertaking. I was born in a country proud of its rural roots, of the quality of the agricultural products harvested by experienced farmers from painstakingly managed land, and proud of the human values attached to farming activities. The deep Man—Nature bonds have been recorded, illustrated, and glorified by many French and European painters, by showing the bounty of the land, and the strength and courage that man needs to exhibit to tame nature. The plow epitomized this relationship between «honest» man and fertile nature. In religious rural France, it was common to bless the plow at seedbed preparation time. I believe that there is a need to change this noble but mistaken image of honest men and women «wresting a living from the land». We need to begin the process of overcoming the European tradition of tilling the soils, a tradition that spread to many parts of the world in the wake of colonialization. NT is about producing food and making a living for men and women by working in harmony with Nature, as ably described during the 7th National Encounter of No-Till farmers in Brazil. It is my wish that our message will inspire a new generation of painters and poets who will better understand how to change this ancestral and subliminal concept of the man-plow-fertility model. While for the artists' creativity to reflect the new and more sustainable reality, we can speed the process by developing high quality media products - film, video, CD-ROM, exhibits, posters, press releases – for world-wide distribution. To begin we must develop our message with clarity, incorporate the experiences of practitioners, add strength by including scientific and economic research data, and stress the contribution of the NT approach to sustainability and environmental protection. With this in hand we can approach appropriate donors for financial support in order to turn the convincing message into media products. Let us resolve to begin the process to-day.

ANNEX
NO-TILL CENTERS OF EXCELLENCE

These Centers of Excellence, developed at regional and sub-continental level would be part of the strategy to reinvent rural education and to «professionalize» farmers in land husbandry in developing countries. Controlled by farmers organizations and managed by education organizations, they would include the participation of academics, scientists, producers and practitioners to develop applied training and to share learning on a fee basis with farmers and agricultural technicians and scientists, particularly for those who are the main stakeholders in the preparation and implementation of CA projects. Centers of Excellence would establish linkages with International funding agencies, Foundations, International Research Organizations such as CGIAR, advanced industrialized countries, and with private companies. The World Bank's 22 Distance Learning Centers (DLC)[4] would be a natural partner in this knowledge management enterprise. – All would adhere to the guiding principles of CA cooperation, and be willing to help these Centers to become the engine for promoting CA in their region. The programs of the Centers of Excellence could be supplemented by other activities such as international exchange programs, study tours and the management of farmers-led demonstration sites. An initial model of a No-Till Center of Excellence is currently under development in Ponta Grossa, Paraná, Brazil. A second Center could be quickly developed in Madagascar under the leadership of a national NGO (*Association Nationale d'Actions Environnementales*, ANAE) and expert contribution of international organizations such as CIRAD.

NOTES

[1] *The author is Senior Agro-Ecologist, Rural Development Department, World Bank, Washington D.C. This paper and any judgments made herein do not necessarily reflect the views of the World Bank.*

[2] *In this paper both concepts of Conservation Agriculture and of improved land husbandry will be used alternatively with a common meaning i.e. people-centered and agro ecological practices applied at rural landscape level to achieve the double goals of productivity increase and environmental protection.*

[3] *APLs are meant to provide through a series of loans, phased and sustained support for the implementation of a long-term development program that reflects economic priorities and contribute to poverty reduction. Provides for funding of a long-term development strategy, starting with a first sequence of activities, according to agreed milestones and objectives.*

[4] *The World Bank Institute is currently participating and supporting 22 Distance Learning Center, including linkages with 25 African Universities in 15 different countries, and with 500 schools.*

REFERENCES

IFAD (2001) The Challenge of Ending Rural Poverty. Rural Poverty Report 2001. Oxford, University Press. UK.

Maguire, C.J. (2001) From Agriculture to Rural Development: Critical Choices for Agricultural Education. Rural Development Department, the World Bank, Washington DC, USA.

McCalla, Al. (1994) Agriculture and Food Needs to 2025: Why We Should Be Concerned. CGIAR, Sir John Crawford Memorial Lecture, International Centers Week, October 27, 1994. Washington , DC, USA.

McConnell Brooks, K. (1999) A proposal to improve access of smallholders to agricultural inputs through introduction of a voucher scheme. Discussion Note, The World Bank, November 4, 1999. Washington DC, USA.

Mellor, J.W. (2000) Agricultural growth, rural employment, and poverty reduction – Non tradables, public expenditure, and balanced growth. World Bank Rural Week 2000, «Poverty or Prosperity: Rural People in Globalized Economy», March 28-31, 2000, Washington D.C., USA.
World Bank (2001) *Rural Strategy Development; Reaching the Rural Poor*. Rural Development Department, the World Bank, Washington DC, USA.

F. DAUPHIN

INVESTING IN CONSERVATION AGRICULTURE

Investment Centre, FAO, Via delle Terme di Caracalla, 00100-ROME-ITALY

Although investing in Conservation Agriculture (CA) has been shown profitable and wise, Financial Institutions (FIs) seem reluctant to promote it through their agricultural projects. Reasons why they should be more interested are discussed, and lessons on how this could be done are drawn from a review of case studies of successes and failures. Investment projects can successfully promote CA through temporary subsidies, support to R&D conducted in partnership with farmers and the private sector, awareness building to change the attitudes of farmers and their advisors, and policy reform.

INTRODUCTION

Investing in conservation agriculture is a sound decision at both farmer and national level. The benefits are even greater at national level due in particular to reduced erosion, so it clearly makes sense for countries to invest in conservation agriculture, and for international agencies and financing institutions (FIs) to support such investment if this can achieve faster and more extensive adoption. Even if farmers stand to gain from Conservation Agriculture (CA), they do incur increased costs, usually in the form of suitable seed drills and other specialised equipment. Moreover, the change from conventional farming to CA typically takes several years before a new biological balance is achieved. During that period, there are risks of reduced yields and incomes, particularly if the farmer has no access to sound advice, or the system is not fully developed for a particular farmer's eco-zone. Thus, financing agencies (FIs) have invested very little in the promotion of CA, and are still hesitant as to the strategy which they should follow. A review of past experience will help understand the difficulties involved and suggest principles to guide CA development strategies.

WHY INVEST IN CONSERVATION AGRICULTURE?

The benefits for farmers of adopting CA do not need further demonstration where a CA system well adapted to local conditions has been developed. Huge adoption in countries such as Brazil and Argentina is best proof that the financial justification is

L. García Torres et al. (eds.). Conservation Agriculture, 445–456.

sound. While sustainable improvement of farm incomes could justify public investment, environmental benefits of CA, including the large reduction in soil erosion and the pollution of water bodies, are further compelling reasons (see Landers 1998 p 252 for a list of benefits to the society). Back on the farm, CA can bring income stability, by reducing the impact of annual weather fluctuations, and improving the long-term sustainability of the farming system. Such benefits also have a public dimension. However, to convince FIs and borrowing governments, we must also show that public financial support can make a real difference in speeding up adoption and reaching more farmers. The case studies in the next section will in fact show that this is far from self-evident.

In spite of such benefits, there have been few projects financed by FIs that had as their main objective, the promotion of CA. Until a successful system has been developed for a given environment and shown to be viable by early adopters, it is risky to invest in massive adoption measures. Fine-tuning of the mix of different technologies may take years, and projects rarely have the flexibility to adjust to the pace of technology development. A prudent approach consists in piggy-backing pilot development onto larger projects which have other objectives. But the risk then is that the CA pilot component does not receive the necessary attention and fails to produce results. Under these circumstances, piloting may be counter-productive.

FIs should invest in CA not only because of its potential benefits, but also because there are few other avenues left open for agricultural development in many countries. International lending for agricultural development has fallen dramatically in recent years. Developing countries have become less interested in borrowing for agriculture, which may well be due to the urban bias of the elite. It is often true that agricultural projects are seen as offering more risk and less tangible economic or political benefits than investments in infrastructure, health or education, often concentrated in or around urban areas. This is a pity, since poverty is often more prevalent in rural areas. But the scepticism that greets traditional agricultural development projects is partly justified. Most aim at promoting crop intensification, directly or indirectly and, in the tropics, rainfed farms often respond poorly to intensification. Almost invariably, intensification means more fertilisers, more pesticides, more water and better (more) tillage. If better yields are obtained, this is often short term, as fertilisers may mask the increasing depletion of soil organic matter. If so, fertilisers only allow farmers to use the land for a few extra years, until intrinsic fertility declines even further. This is a vicious circle which leads to impoverishment, indebtedness, and eventual urban migration.

CA has proved to have the opposite effect: soil fertility is restored, erosion decreased and farm productivity increased. CA may represent one of the very few real opportunities for sustainable development in poor rural areas. That it has already developed across such a range of agro-ecological environments and farm structures in the Americas is a powerful testimony to its potential for other regions. If CA represents such an opportunity, it must be promoted in areas where it is still not widely adopted, i.e. where it needs more research and adaptation. But can CA research and development be successfully supported and if so, how?

LESSONS FROM PAST EXPERIENCE

Brazil

The expansion of CA in Brazil, particularly in the subtropics and the *Cerrado*, the central tropical savannah, is probably the most impressive CA success story. CA adoption increased from 1,000 ha in 1973 to 400,000 ha in 1984 (Derpsch, 1998) and to over 12 million ha in 2000. What makes this development even more remarkable is that: (i) there was no precedent of CA adoption in the tropics on any significant scale; (ii) the route chosen was no-tillage rather than reduced tillage; (iii) whereas the expansion was driven by large farms in the 80s, small farms followed in the 90s and have also shown fast rates of adoption.

If elements of the technology were initially borrowed from US farms, the development of the system was essentially endogenous and included such radical innovations as the introduction of a living mulch to produce adequate soil coverage, and for small farmers the design of hand tools and draft animal implements for direct seeding. From the Cerrado experience, the following lessons were drawn (Landers, 1998):

- Technology development was led by private sector, overseas-sponsored activities and farmers;
- Farmer innovations and private sector R&D were much more important than government research (over which farmers had no control);
- Partnerships with input and machinery suppliers were easily developed;
- The primary mechanism for technology transfer was farmer-to-farmer contact, involving individuals, farmers clubs and associations, and a national NGO;
- The change to zero-tillage was a mental U-turn and farmers typically took 4-5 years to adjust.

Land management projects financed by the World Bank contributed to the rapid development of CA in sub-tropical Brazil among small farmers (see last section below). Colleagues in the FAO Investment Centre[1] who worked on those projects underlined the social dimension of change:

- conservation projects are planned in a participative way at the scale of micro-watersheds (each of a few thousand hectares);
- financial support is made available to groups only, not to individual farmers.

Lessons from the antipodes:

The ley farming system development in Australia …

Starting in the early 1900s, farmers in South Australia developed a wheat - grazed legume rotation which allowed them to increase vastly the number of sheep they could keep on the farm, while at the same time halting the decline in wheat yields and stopping soil erosion (Chatterton, 1996). The legume is either a «medic», an annual *Medicago* species (*M. trunculata*, *M. rugosa*, *M. polymorpha*, etc.) or in higher rainfall areas, a «sub-clover» (*Trifolium subterraneum*, *T. yanninicum*). Initially sown after wheat, the legume is grazed until the next wheat crop but allowed to produce seeds which will germinate[2] with the autumn rains. The rotation is typically over two years, but can respond flexibly to market or weather conditions. The system was progressively enhanced with the addition of phosphate fertilisers, and then of better varieties and seeders, both developed by private companies, and is now dominant in South Australia. In recent years, many farmers have adopted the «combine seeder» which provides shallow tillage, sowing and fertilisation in one pass, thus ensuring that the soil remains protected virtually all year round by wheat, wheat stalks or the legume pasture.

Lynne and Brian Chatterton note that ley farming[3] was strongly resisted by scientists and the South Australia administration, who until the 1940s advocated deep ploughing and clean fallow to ensure better weed control and soil moisture conservation (Chatterton 1996). It was only in 1954 that the scientific establishment publicly questioned deep ploughing. «We have not been able to show that deep ploughing is necessary or economic for cereal growing, nor is there anything to be gained from nicely inverting the furrow slice (Herriot 1954). The scientists' extraordinary faith in the superiority of deep tillage with soil inversion is not unique to Australia, but common in countries where agronomists or their professors were trained in Europe, and this belief is a major impediment to the acceptance of CA by those who should promote it.

... and its failure in the Mediterranean region

Starting in 1974, the Australian government financed projects in Libya, Algeria, Jordan and Iraq to introduce ley farming. Other tests involving Australian expertise were also conducted in Tunisia and Morocco in the late 80s and early 90s. The Australian projects all used project-managed demonstration farms. One of the major initial proponents of this co-operation was Seedco, the seed company established by South Australia farmers in the 60s when medic farming expansion became constrained by seed scarcity. The first project (Libya), was the only one in which Seedco was closely involved and also the only one which had a measure of success, possibly as the company was directly interested in CA development and worked with local farmers to produced the seeds. Ten years after project closure, some 600 farmers had adopted medic farming. Most projects were stopped after a few years with inconclusive results, and it was thought that the system required more research, or was simply not adapted to the local socio-economic conditions. Further research was needed to identify better-adapted local ecotypes of medics, a well-meaning objective, but probably of limited relevance inasmuch as the medic seed bank in the soil is contaminated by local ecotypes within few years. Socio-economic aspects are important, but their role difficult to assess in the absence of any effort to involve local farmers[4].

The major technical problem reported by project staff was poor medic regeneration, which is attributable to inadequate tillage. Modified ploughs were often used instead of the Australian scarifier, a tined implement for shallow tillage (<10 cm). Sometimes, deep ploughing was ordered by local agriculture officials who thought it essential for the wheat crop. Overgrazing, particularly in mid winter and at flowering, also affected regeneration (mostly the demonstration farms had no animals and borrowed them for periodic and intensive grazing).

In the description by L&B Chatterton, projects suffered all classical project weaknesses, including poor planning and unclear objectives, procurement delays (seeds arriving after planting time, scarifiers years later if foreseen in the package), lack of communication between projects and poor learning from past experience. Except in Libya through Seedco, the projects had virtually no contact with local farmers. The projects met with constant scepticism from local technicians and scientists. Finally, the authors believe that the Australian project managers were in fact not well acquainted themselves with the system. This is perhaps why their faith in it was easily shaken as new problems arose, and why projects easily shifted from demonstration to research, quickly loosing the system perspective. Australian farmers employed by some of the projects were only there to operate the demonstration farms, with no part in the decisions. International organisations, such as FAO and ICARDA, have at times supported the Australian initiative, but interest faded, as demonstrations were not conclusive and shifted towards research for new varieties and comparisons of different cereal rotations where grain legumes were preferred over medics.

The lessons from the failed introduction of ley farming to the Middle-East North Africa region (MENA) do more than illustrate the difficulty of changing a farming system: they reinforce major lessons which have emerged from the Brazil experience. First, since the change is so fundamental and has psychological and cognitive dimensions, a sustained, large-scale and well thought-out effort is required. Second, the R&D needs are best served by the private sector working in partnership with farmers. Finally, and without pre-judging the future potential of ley farming in MENA, it has not been disproved. It is suggested that CA may need to take on rather different forms in different regions. CA in North America requires residue management; in tropical Brazil a cover crop is required for mulch purpose. The grazed medic phase could be the right answer to the issue of conserving crop residues on the soil in systems where they are traditionally fed to animals.

ELEMENTS OF A STRATEGY

Reversing the plough?

It took centuries in Western Europe for the mouldboard plough to replace the antique symmetrical plough, the *ard*. The transition may have started in Roman times but, in many parts of Europe, it was only completed in the 20th century (Haudricourt *et al.* 1955). Ensuring weed control as well as a range of other benefits, including better drainage in heavy soils, the mouldboard plough had become synonymous with modern, scientific agriculture, and still is in many places. Clean fields, obtained through repeated and heavy tillage operations, were the mark of good farmers. Abandoning the plough is perceived as returning to middle-age agriculture, and relinquishing hard-gained control over nature. This perception can be a formidable obstacle to introducing CA. In the regions of Europe where the mouldboard plough was initially developed, soil organic matter content was high and could be maintained at a high level because of relatively high rainfall that produce large amounts of biomass, while cool temperatures slow the rate of decomposition. Carried outside its «natural» domain, the mouldboard plough produced environmental disaster in the U.S. – the infamous «Dust Bowl» in the 1930s, in Africa where «modern» plantations used it carelessly in the fifties, in Australia until the 50s, and still now in large parts of MENA[5] and the frontier lands of Latin America. This is because intensive tillage in semi-arid and tropical climates exposes the top soil, accelerates the decomposition of crop residues and organic matter, and breaks the soil structure, resulting in loss of organic matter and nutrients as well as physical erosion. In the words of an Australian pioneering environmentalist (Tepper, 1895, quoted in Chan and Pratley, 1998):

We plough and fallow like our European forefathers. Moreover, before doing so, we feed off all the weeds and stubble and rake them together and then burn ... leaving the soil as bare as a floor. Under the mistaken idea that it needs a rest, that weeds exhaust it, that the soil requires to be aerated ...acting on such fallacious ideas, we only succeed in converting our fields into deserts and blasting our future, for the causes which operated in reducing the 50 bushels per acre in early years to 5 at present are still acting and will reduce the 5 to less still, till not even grass or weeds will be able to exist.

The history of agriculture has been that of a progressive domestication and artificialisation of nature. By and large, the mouldboard plough represents a culmination of that process. CA is an effort to co-operate with nature rather than to dominate it; to accept and adopt CA requires a different attitude, new representations and new knowledge.

Building the Knowledge Where it is Needed

CA is more knowledge intensive than input-based technologies, for the reason that there is no single solution to management issues. Managing weeds and pests depends on knowing the species, the local environment and how they interact; and on being able to form a diagnostic, to select among a range of possible remedies one or several, to monitor their impact against relevant criteria, and to adjust the course of action according to the results of observation, either during the same cropping season or during subsequent cycles. It is not sufficient to have knowledgeable scientists in a research centre and good mechanisms for disseminating recommendations: the knowledge has to be shared by land managers who take day-to-day decisions.

In a reaction against earlier simplistic views on technology development and transfer, the concept of agricultural knowledge and information systems (AKIS) recognises that the production and dissemination of knowledge involves multiple actors, including researchers, universities and advisory services, but also farmers, the private sector and NGOs. Ensuring that the whole AKIS is mobilised in the pursuit of technology development for CA is essential. National AKIS can be mobilised for CA through competitive research grants to finance priority research, a system pioneered in several countries by the World Bank. Such grants can finance awareness building and technology dissemination as well as pure research and are opened to any capable operator, public or private. They can also facilitate networking between science institutions, farmer groups, traders, input supply firms, farm organisations, and environmental movements.

CA research must be participative, planned and conducted by farmers with the support of scientists where required. Demonstrations are also important, as they provide a focal point where practitioners, input suppliers and scientists discuss problems and solutions. There is a role for international and national research organisations, but their impact on CA development has been minimal so far, and this should be of serious concern. We noted the difficulty that scientists have in working on systems or in a system perspective, a result of discipline training. An obvious remedy it seems would be for scientists to work with system specialists, i.e. farmers, but accepting them as full partners rather than as ignorant implementers. Research could also help in developing decision tools for land managers. Farmer decisions are based on indicators which allow them to take corrective action when needed. For example, simple soil structure indicators that farmers could use to monitor trends in soil structure would raise their awareness and allow them to develop a practical knowledge of how it can be improved and what can destroy it (Chan and Pratley 1998).

Achieving Public Awareness

Awareness is required at farmer level as well as at the general public level. Conservation practices will not be adopted unless the farmer perceives a problem – hence the role of crises already mentioned. If the symptoms are more subtle and the effects of degradation only very progressive, the problem may not be perceived, and even less the relationship between current farm practices and soil degradation. Farmers who participate in the identification of soil and water conservation problems on their own land and of solutions to those problems are more likely to apply the information to their own situation (Osgood 1992).

Osgood also notes that it is not enough for the individual farmer to perceive that a problem exists. When the community understands the problem, there are more opportunities for group efforts and support. If a conservation practice is consistent with community values, it probably will be acceptable to the farmer; if not, few farmers will adopt. These remarks are consistent with the earlier noted importance of groups in analysing and planning changes in land management practices. There are also ecological reasons for a community approach to improved land management and CA. Every farm is under the influence of the land surrounding it and the ecological processes operating over the landscape, and practices in one farm will affect others as well. This is at the root of the Landcare philosophy in Australia as well as of the micro-catchment approach in Brazil.

Incentives and Subsidies

Subsidies can be used to facilitate adoption. However, economists caution that subsidies may have unwanted, perverse effects. They are often inefficient, expensive, and tend to outlive their utility (Giger 96). If needed over the long term, this is evidence that the proposed technology is not viable. The main reason for envisaging subsidies is to overcome risk-aversion by farmers. Such subsidies should be phased out when farmers have had their first experience using the new technique. Another case justifying subsidies is when society stands to gain from farming activities which are not compensated by market prices (e.g. watershed protection). In that case, the costs of these subsidies should be borne by those who actually benefit from the change in practices. The need for temporary financial support to facilitate CA adoption can be argued on two accounts. Firstly, there is an investment cost linked with the change in machinery and equipment, which many farmers would be unable to pay. Secondly, there is a transition period during which yields may initially decline. In some situations, the soil is too compact for a direct move towards reduced cultivation, and deep tillage might be initially needed. Finally, the farmer himself undergoes a period of training and adjustment, during which he is bound to make mistakes that reduce his gains and increase the risks.

The principle of sharing between farmers and government the costs of the transition, as practised in Brazil for example, is therefore sound. The same principle is also used in the soil conservation programmes of the United States and in some of the Landcare programmes in Australia. In all those cases, the decision to grant a subsidy is decided on the basis of a farm plan or a catchment management programme and involves a contract between a public agency and the individual or community manager.

The Role of Policies

Large scale adoption of conservation tillage (CT) in the United States is generally attributed to aggressive policies to protect soil and water, linking price support to conservation measures taken by farmers, in particular through the 1985 and 1990 farm bills. In FY 1998, total USDA administered programmes for conservation exceeded US$ 2.8 billion (Vasavada and Warmerdam, 1999). Of this, US$ 1.8 billion went to the Conservation Reserve Program providing rental payments to farmers who retire environmentally sensitive land, and 0,54 billion for technical assistance. However, the main incentive for CT was almost certainly provided through the Conservation Compliance principle established by the Food Security Act of 1985. Under this principle, farmers who grow crops on highly erodible land are required to implement an approved soil conservation plan to remain eligible for an array of subsidies which provide a large portion of farmers' incomes.

A 1996 review of conservation programmes indicates that the area where CT is practised increased from 28 million ha in 1989 to 40 million ha in 1994 (Sandretto 1996). Although this is an impressive progression[6], it also suggests that adoption did not wait for the 1985/1990 farm bills. Faeth et al. (1991) argue that commodity price support programmes in place during the 80s have in fact encouraged chemical intensive monocultures, which deplete soil nutrients and adversely affect ground and surface water[7]. In this view, the 1985 farm bill must be seen as an effort to counteract the disincentives to sustainable agriculture created by agricultural support payments, amounting annually to US$ 12 billion in the period 1982-1988. The history of CT in the US remains to be written, but it is likely that without unfavourable policies CT might well have spread faster.

In Europe, the task of redesigning the CAP (Common Agricultural Policy) to rationalise support for environmentally sound agriculture has hardly begun. Environment is now high on the European political agenda, and farmers receive several payments related to environment protection. The system is costly both to farmers and governments. One of the difficulties in designing environmental contracts is the information asymmetry between farmers and regulators. Rabinowicz (99) notes that rural development groups have an information advantage with better local knowledge of the environmental impact of specific policies and opportunity costs. The penalty for defaulting on contracts with the local community is probably, she remarks, much higher than for cheating the bureaucracy.

Other examples of inappropriate policies abound. In Australia, perhaps the best examples are in the settlement schemes that established inappropriate property sizes, the subdivision of marginal land, government support for irrigation schemes, and vegetation clearing promoted by taxation incentives (Curtis and Lockwood, 1998). Reviewing Australian recent experience in conservation policies, these authors underline the need to:

- establish roles and linkages between local community groups and regional planning bodies;
- develop cost sharing principles for allocating public money for investment on private land where there are community benefits;
- effectively support volunteer groups.

Still, more radical measures than currently used are needed to halt land degradation in Australia[8].

As a first step to change policy, these cursory remarks into the complex field of conservation policies first point to the need to thoroughly analyse existing agricultural policies[9] and understand how they affect agricultural practices. They also confirm the essential role of local community groups in reversing land degradation.

Role of Financing Institutions

Promoting CA requires an integrated strategy taking into account not only technical but also cultural aspects and the dynamics of collective action. The strategy must be based on a critical review of policies, promote awareness building, support to decentralised research, training and extension, and include temporary incentives to facilitate adoption. FIs, International Agencies and donors can help in financing a package of measures. These organisations have also acquired expertise in the preparation and implementation of such projects[10]. Successful projects have incorporated all the above elements and provide useful references. This is in particular the case of the various Land Management Projects financed in Brazil with World Bank support, which are shortly described below. It is clear that CA development and adoption started in Brazil before these projects were implemented; however, only large-scale farms were benefiting. Different technologies and approaches were needed for smallholders.

The Land Management I project (Parаña State) had a total cost estimated at US$ 150 million for a total project area of 5 million ha, of which 1 million ha of farmland would directly benefit. The costs covered research (13 %), extension (including awareness building, training and education: 18%), incentives (18%), redesign of roads (a factor in soil erosion: 26%), soil liming (4%), studies and mapping (14 %), and project management (7%). Incentives were managed through a Soil Conservation Fund and made available to participating communities under a contract for implementing a micro-watershed management plan. The level of subsidy was linked to farm size: from 70% for small farms to 30% for large farms in year 1, and decreased with time, to be phased out after 6 years. The total amount of direct subsidies came to 25 US$/ha for the whole period. Subsequent World Bank funded projects in Santa Catarina and São Paulo States had essentially the same components, including the establishment and financing of a soil conservation fund, with a per ha cost of direct subsidies ranging from US$15 to US$40. While significant, those costs are small when compared to the expected returns to farmers and to the economy as a whole (see for example Sorrenson (1998) for a precise evaluation of costs and benefits in Paraguay).

In addition to direct investment in CA development through such projects, FIs can press for policy changes, and promote the dissemination of information and partnerships among the main actors of change, as advocated by C Pieri and J Benites.

CONCLUSION

This paper has attempted to identify some common marks of success stories in conservation agriculture adoption. The lessons can be used for planning a strategy for promoting CT in new regions. ***Financial support*** to farmers is crucial, but may represent only a fraction of what is needed in terms of investment, and is to be used cautiously. ***Techniques*** are obviously pivotal, but it is clear that they cannot be developed centrally by scientists alone. Often underestimated, ***social and cultural*** aspects are what make change possible – or on the contrary, virtually impossible. This should not be interpreted as simply implying the need for understanding farmers and their culture. The proposed change also implies a radical shift in the culture and attitudes of scientists, technicians and politicians, and a real will by scientists and agriculture officials to work with farmers and the private sector.

NOTES

[1] Simon Hocombe and Katia Medeiros, FAO, pers. comm.

[2] Self-regeneration is more difficult with sub-clovers, which may need to be re-sown regularly.

[3] The authors prefer to call it a medic or sub-clover system as a true ley is cut for hay.

[4] In fact, there would seem to be clear incentives for a ley farming as against a pure cereal or a cereal-grain legume system, as the price ratio of meat to wheat is high in the region. Small farms use tractors and could use scarifiers, once there is a sufficient market for them. Protecting the medic pasture from neighbours' or nomads' sheep may be a more serious issue, but no insurmountable if worked at the community level.
[5] Where it tends to be displaced by disk ploughs, easier to operate and even more destructive.
[6] Particularly spectacular for no-till systems which went from 5 percent to 15 percent of agricultural land from 1989 to 1996.
[7] See also Steiner, 1990 for a discussion of the unintended impact of agricultural policies in the U.S.
[8] Despite the success of the Landcare movement in creating global awareness and enrolling some 30% of all farmers in groups devoted to conservation activities, Goldney and Bauer (1998) argue that this is still «too little too late».
[9] Bromley (1991) provides pertinent recommendations on how the policy analyst should proceed, avoiding the false pretence of rigorous scientific objectivity which separates too radically objectives and instruments.
[10] Obviously, most of the expertise, technical and managerial, regarding implementation of CT programmes resides in those countries which have successfully adopted.

REFERENCES

Bromley, D.W. (1991) Environment and economy: property rights and public policies, Blackwell, Oxford.

Chan, K.Y. and Pratley, J. (1998) Soil Structural Decline - Can the Trend be Reversed? In Agriculture and the Environmental Imperative, Pratley J. and Robertson A. (ed.), CSIRO.

Chatterton, L. and Chatterton, B. (1996) Sustainable Dryland Farming, Cambridge U. P.

Curtis, A. and Lockwood, M. (1998) Natural Resource Policy for Rural Australia. In Agriculture and the Environment Imperative, Pratley J. and Robertson A. (ed.), CSIRO, Australia.

Donovan, G. And Casey, F. (1998) Soil Fertility Management in Sub-Saharan Africa, Technical Paper No 408, World Bank, Washington.

Faeth, P., et al., (1991) Paying the Farm Bill: U.S. Agricultural Policy and the Transition to Sustainable Agriculture, World Resources Institute, Washington.

Giger, M. (1996) Using incentives and subsidies for sustainable management of agricultural soils, Paper presented for the 9th ISCO Conference, Bonn.

Haudricourt, A.G., Delamarre, J.M.B. (1955) L'Homme et la Charrue à travers le Monde. Gallimard, Paris.

Herriot, R I. (1954) Tillage Journal, Department of Agriculture, South Australia.

Landers, J. (1998) Zero Tillage development in tropical Brazil. In *Conservation tillage for sustainable agriculture.* Proceedings from an International workshop, Harare, 22-27 June 1998, FAO Rome.

Osgood, B. (1992) Public Awareness in Soil and Water Conservation Programs. In Conservation Policies for Sustainable Hillslope Farming, FAO, Rome.

Rabinowicz, E. (1999) Redesigning the CAP to meet the challenges of EU enlargement and the WTO, *European Review of Ag. Economics,* **26** (3): 265-281.

Sandretto, C. (1996). Crop residue management and tillage system trends. http://www.econ.ag.gov/epubs/htmlsum/residue.htm, Economic Research Service, USDA.

Sorrenson, W.J. (1998) Economics of no-till compared to conventional cultivation systems on small farms in Paraguay, Soil Conservation Project MAG-GTZ.

Srivastava, J. and Meyer, E. (1998) Is Conservation Tillage a Viable Option in the CIS? Draft paper, World Bank, Washington.

Steiner, F.R. (1990) Soil Conservation in the United States - Policy and Planning, J. Hopkins, Baltimore.

Stewart, B.A., Jones O.R. and Unger P.W. (1993) Moisture Management in Semiarid Temperate Regions. World Bank 13th Agricultural Symposium, January 1993, Washington.

Tepper, J.G.O. (1895) Remarks on weed: the soil and its fertility, *Agriculture Gazette NSW* **6**: 385-90, Australia.

Vasavada, U. and Warmerdam, S. (1998) Green Box Policies and the Environment. http://www.econ.ag.gov.briefing/wto/issues/environm.htm, USDA and U. of California.

World Bank (1994) Integrated Soil Management for the Tropics, *Agricultural Technology Notes* No **7**, Nov. 1994, AGRTN, World Bank, Washington.

P. KOOHAFKAN

WIN-WIN OPTIONS FOR FOOD SECURITY: CONSERVATION AGRICULTURE SOIL FERTILITY, SOIL BIODIVERSITY AND CARBON SEQUESTRATION NEXUS

Chief Land and Soil Fertility Management Service, Land and water Development Division, FAO, Rome, Italy

There is general consensus that food security and sustainable agricultural development face some unprecedented challenges:

I. At the year 2000, there were 790 million people hungry in spite of the fact that over the past 40 years, per capita world food production has grown by 25%, and food prices in real terms have fallen by 40%. People in 33 countries still consume under 2200 kcal per day and over a billion people remain under poverty line(less than 1 US$ day). Although a combination of increased production and more imports will mean average per capita consumption will increase to about 3000 kcal per day by 2015, food insecurity and malnutrition is expected still to persist for long time(FAO, 2000, Pretty, 2001).
II. Changes to natural ecosystems are occurring on a larger scale than ever before, involving entire landscapes. Such large-scale landscape changes through deforestation, expansion of agricultural land, and urban and suburban growth will likely dictate the physical condition and extent of terrestrial ecosystems in the next several decades. Progressive fragmentation of the world's remaining forest blocks; build-up in coastal areas; and the spread of cities, suburbs, and attendant roads and infrastructure over once-rural tracts will do much to degrade the habitat and watershed values of these areas (WRI, 2000).
III. The very scale of these landscape-level changes, as well as the increasing intensity of industrial and agricultural processes, are inducing changes in the global systems and cycles such as the atmosphere and the nitrogen cycle that underpin the functioning of ecosystems. These changes represent long-term environmental threats of a profound and far-reaching nature. Global warming from the build-up of greenhouse gases is the best-known example, with the potential for large-scale disruption of natural ecosystems, agriculture, and human settlements due to changes in rainfall and temperature patterns and rising sea levels. Disruption of the global nitrogen cycle through extensive use of fertilizers, the burning of fossil fuels, and other activities also has the potential to change the structure and composition of terrestrial and aquatic ecosystems.
IV. Threats to biodiversity and genetic resources for food and agriculture from all sources are quickly reaching a critical level that may precipitate widespread changes in the number and distribution of species, intensity and diversity of food production, as well as the functioning of ecosystems. Current species extinction rates and projected losses of habitat from land conversion, as well as increasing competition from non-native species, will have significant impact on biodiversity and food security world-wide.

L. García Torres et al. (eds.). Conservation Agriculture, 457–458.

Even as these threats and trends indicate food security and environmental challenges ahead, it is important to remember that they can be modified with human resolve. Already, the transition to more environmentally benign ways of growing food, producing goods and services, managing watersheds, and accommodating urban growth has begun in many far-sighted communities and enterprises (WRI, 2000).

The increased focus on holistic and systemic approaches has improved knowledge and understanding of the effects of land use changes, consumption patterns and management practices on critical life cycles and knowledge about the interdependencies of human activities and natural processes. Advances have been made, in particular, in understanding the importance of sustaining ecosystem functions and life-support systems such as the nutrient, hydrological and carbon cycles, climatic regulation, and pest and disease management processes that operate at local as well as watershed/ landscape levels.

Recent years have seen also significant convergence in both concepts and implementation themes for Sustainable Agriculture and Rural Development (SARD) for which Conservation Agriculture (CA) is one of promising options both in developed and developing countries. Conservation agriculture seeks the integrated use of a wide range of crop, soil, water, nutrient and pest management technologies building on farmers' local knowledge and improved technologies through participatory diagnosis and actions, using Farmers Field School Approach. Such regenerative and conservation effective agriculture can be highly productive, provided farmers participate fully in all stages of technology development and extension.

By adopting multiple-use strategies, many local and indigenous resource users and farmers manage, *in situ*, a continuum of agricultural and natural systems, obtaining a variety of products as well as ecological benefits. Diversified systems, such as those based on inter-cropping and agroforestry, and crop/livestock or crop/fish combinations, and those that manage the "associated biodiversity" of soil biota, pest and disease modulating organisms and others, are proven to be more sustainable and have been the target of considerable research. The favorable attributes are related to the higher levels of "functional" biodiversity and effects on the stabilization of agro-ecosystem processes. The challenge is to manage the agro-ecosystem so as to maintain or enhance key ecological functions and services such as nutrient cycling, biological pest regulation, and water and soil conservation. Many traditional and local systems and management practices have developed over years and generations to exploit such relationships between species and inter-linkages between biological and land resources.

The use of livestock, for example, is essential to recycle nutrients and maintain ecosystem resilience in the traditional extensive agro-pastoral systems developed over generations in the drylands of Sahelian Africa, as well as in the modified intensive systems using stall-fed animals in Java, Indonesia and other parts of Asia, where population pressures are high. Moreover, modern agro-ecological technologies that build on increased scientific knowledge of such synergies are recently proving to be more productive, especially in marginal lands, and when the biological structuring of the farm is improved and labor and local resources are efficiently used. They are being adopted in diverse environments and by certain groups of farmers where the socio-economic and political environment proves propitious, as exemplified by organic agriculture, for the production of environmentally friendly but often high value produce, especially for urban communities. The best example of spontaneous adoption on a much wider scale is that of the uptake of conservation tillage in many countries.

J. ROCKSTROM, P. KAUMBUTHO, P. MWALLEY
& M. TEMESGEN

CONSERVATION FARMING AMONG SMALL-HOLDER FARMERS IN E. AFRICA: ADAPTING AND ADOPTING INNOVATIVE LAND MANAGEMENT OPTIONS

IHE-Delft, PO Box 3015, 2601 DA Delft, the Netherlands,
KENDAT, Kenya Network for Draught Animal Technology, PO Box 30197, Nairobi, Kenya,

SCAPA, Ministry of Agric, PO Box 3163, Arusha,, Tanzania.

National Agricultural Mechanisation Research Center, PO Box 954, Nazareth, Ethiopia.

Despite wide adoption in several parts of the World, conservation farming still remains a largely unknown opportunity among small-holder farmers in Eastern and Southern Africa. This paper investigates the causes behind low adoption among smallholder farmers and discusses approaches to improve adoption. The paper also presents field results and intervention approaches from farmer driven trials on conservation farming in Tanzania, Kenya and Ethiopia. Focus is on semi-arid agro-ecosystems where water is a major constraint for crop growth. Here conservation farming is promoted as a form of water harvesting. On highly degraded crop land, especially fields with plough-pans, yield levels of maize are increased with more than a factor 2 just by changing implement from a disc-plough to a sub-soiler (tilling at approximately 35-45 cm depth). On less degraded land (the farming systems studied in Kenya) the mere change of implement is not enough to improved yield levels. Successful adoption of conservation farming will depend on several external driving forces. There is a large need to build capacity among extension staff, teachers at professional schools, researchers and managers. This requires agricultural policies that specifically favour conservation farming (as, e.g., in Zambia). Local markets need to be developed, in order to assure the availability of affordable and good quality implements at local level.
Keywords: semi-arid, animal traction, conservation tillage, farming systems, East Africa

INTRODUCTION – UPGRADING RAINFED AGRICULTURE

Economic development and social stability in countries of Eastern and Southern Africa still depend strongly on rainfed small-holder farming. In general at least 80 % of

L. García Torres et al. (eds.). Conservation Agriculture, 459–470.

the population depend to a large extent on small-scale rainfed farming. Land-holdings are small, in general with less than 5 ha of cultivated land. Crop yields are low and reflect decades of water desiccation and soil fertility decline. Staple food crop yields such as maize, millet and sorghum oscillate around 1 t grain ha^{-1}.

A major cause behind land degradation is intensive soil preparation by hoe or plough, which together with the removal or burning of crop residues, leaves the soil exposed to climatic hazards such as rain, wind and sun (Benites et al., 1998). Conventional tillage using ox- or tractor drawn ploughs has over the years been perceived as the indicator of farm systems modernisation in developing countries. However, it is becoming more and more apparent, that the ploughing techniques developed in temperate regions with gentle rains and low wind and water erosion, can have serious adverse effects on the long-term productivity of erosion prone tropical soils.

There are many documented examples of successful conservation tillage practices in Eastern and Southern Africa, where crop yields have increased through the conservation of soil, water and nutrients and/or draught power needs (Elwell, 1993; Oldreive, 1993; Vogel et al., 1994).

Despite these successes, and the increased adoption of conservation tillage in other parts of the World (e.g., North and South America), the adoption among small-scale farmers in Eastern and Southern Africa has been very low. On the other hand, the potentials of conservation tillage as a means of building up soil productivity on the long run and thereby attaining higher yields at lower costs, has been fully acknowledged and picked up by commercial farmers in, e.g., Zimbabwe and Tanzania (Oldreive, 1993; Rockstrom and Jonsson, 1999).

This paper explores factors affecting adoption of conservation tillage among small-holder farmers in Eastern and Southern Africa. A systems approach is taken to conservation tillage, and the implications such an integrated approach on farm development is discussed. Field results from farmer-driven trials on conservation tillage are presented from Ethiopia, Kenya and Tanzania.

WATER PRODUCTIVITY AND CONSERVATION TILLAGE

Conservation tillage aims at reversing a persistent trend in many production systems of reduced infiltrability due to compaction and crust formation and reduced water holding capacity due to oxidation of organic materials (due to excessive turning of the soil). From this perspective conservation tillage is a form of water harvesting, where runoff is impeded and soil water is stored in the root zone of the crop. This means that conservation tillage constitutes a very interesting approach to achieve improvements in water productivity and «crop per drop» increases, in line with the newly launched global dialogue on water for food and environmental security (Anonymous, 2001).

The above suggests that conservation tillage in sub-Saharan Africa (SSA) can have the most beneficial impact in water scarcity prone semi-arid and dry sub-humid areas. Semi-arid regions constitute some 40 % of the arable lands in SSA, and host some 40 % of the population.

Water balance analyses can be useful in assessing the potential of increasing water productivity through conservation tillage. As shown in Fig. 1, from synthesised data on water flow partitioning on field scale in crop producing systems in semi-arid SSA, only some 15 – 30 % of rainfall on average is used for productive crop growth (Rockstrom, 1999).

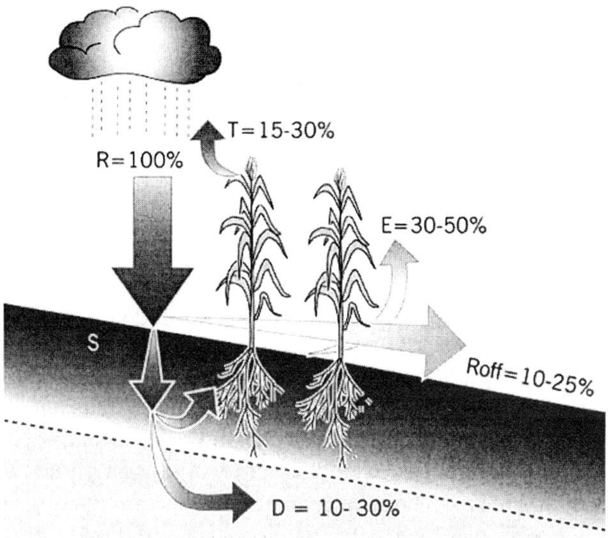

Fig. 1. Rainfall partitioning in the semi-arid tropics in Sub-Saharan Africa.

This analysis of rainfall partitioning indicates that for semi-arid tropical regions water scarcity (which often is the major constraint in crop growth) is not necessarily caused by to low cumulative volumes of rainwater, but rather a result of poor distribution of rainfall, large evaporative demand of the atmosphere, and large losses of water in the water balance.

One major goal with conservation tillage is to change the partitioning of rainfall in favour of infiltration and plant water uptake capacity. Rockstrom and Falkenmark (2000) has in a recent study shown that a doubling and in many cases even a quadrupling of crop yields in African drylands is feasible from a hydrological perspective if such measures are accomplished.

FARMING SYSTEMS APPROACH TO CONSERVATION TILLAGE

No-inversion of soil is the single most important step in the transition from a conventional to conservation tillage. But, isolated, the change in implements in the mixed farming systems practised by small-holder farmers in SSA, will not by itself turn a degraded farming system to a prosperous farming system *on the long term.*

LIVESTOCK LINKS

Mulching is a key component of conservation tillage. Free post harvest grazing is legion in most agro-pastoral communities, which means that no crop residues remain at the onset of the rainy seasons. Even if crop residues would remain, the lack of fuel wood and materials for construction is generally so severe, that residues are often used for these purposes. This means that a conservation tillage decision is also an animal husbandry and household management decision.

One of the most challenging initial steps in introducing CT, is training of draught animals to «walk in straight lines». The task is amplified by the wider yoke required for, e.g., ripping in maize.

TIMING LINKS

A major constraint in semi-arid farming systems of Eastern Africa is the poor timing of farm operations. A lack of strong animals for traction often means that ploughing can only be done after the onset of the rains when the topsoil is moist. Furthermore, many farmers do not have access to, primarily, oxen, making them dependent on borrowing or renting oxen from neighbours.

The effect is that primary tilling operations are seriously delayed in relation to the onset of the rains. In semi-arid areas, this can constitute the difference between getting a yield and total crop failure. For example, our own on-going farm trials in the dry semi-arid parts of Machakos district in Kenya, show that dry planting during the long rains of 1999 (between late March and late June), enabled a farmer to harvest between 600-700 kg DM grain ha^{-1} of maize, while the neighbouring farms planting 2-3 weeks after the onset of the rains experienced complete crop failure. These results are in line with long term findings by Stewart (1988), from the same region. A transition from conventional ploughing to ripping in permanent plant lines, enables resources poor farmers to carry out primary tillage during the dry season when draught animal power is easily available.

FERTILISATION LINKS

Small-scale farmers in the region generally do not have access to enough organic sources of soil nutrients to replenish soil fertility even on a short-term basis. For example the estimated annual average export of nitrogen (N) and phosphorus (P) from small-scale farms in drylands in Kenya amounts to 56 kg N ha^{-1} and 7 kg P ha^{-1}, respectively (Stoorvogel and Smaling, 1990). In-organic fertilisers are conceived by small-scale farmers as being too expensive. A problem with conventional fertiliser extension in the region is that organic and inorganic sources of soil nutrients are broad-casted over the entire crop fields, resulting in large losses (through weed nutrient uptake, and surface/sub-surface leaching). With conservation tillage fertilisation can more easily be concentrated to the sphere of soil surrounding the roots of the crop.

WEEDING LINKS

Farmers tend to identify weeding as their primary concern when discussing a transition from conventional mouldboard ploughing to a system of no-inversion of the soil. Their concern is well founded, and often herbicide use forms an integral part of promoted CT systems. Even though results generally show significant economic returns from minimum tillage systems using herbicides, the sustainability of promoting such a system among resource poor farmers has been questioned, due primarily to problems of access and affordability.

As pointed out by Brunner et al. (1998) the transition from a conventional tillage system to a conservation system has to be carried out progressively (they suggest a 5 year transition). A reason for such a transition is to progressively manage weed infestation. Decades of ploughing with no late-season weeding (thereby permitting weeds to flower and set seed), has resulted in a progressive and in many areas dramatic build-up of weeds. Farmers use ploughing to manage weeds, while simultaneously inducing more weeds due to lack of late season weeding. The soils get addicted to ploughing. It is common practice even in dry areas of e.g. Ethiopia to use the first rains only for weed germination. These weeds are then ploughed into the soil.

In our ongoing research on conservation tillage systems for small-holder farmers in E. and S. Africa we try to introduce an additional late season weeding (generally just before harvest of the crop) during the transitional phase in order to reduce the long-term weed load.

MARKET LINKS

Many semi-arid areas in Ethiopia, Kenya and Tanzania experience chronic food shortages not primarily due to a cumulative deficit of rainfall, but due to too small land-holdings in relation to the number of mouths to feed. In the Ethiopian highlands families try to survive on < 1 ha of arable land per household producing staple food yields of 0.5 - 1 t ha^{-1} (maize, tef, sorghum, millet). In such areas farming systems improvements have to focus on producing more - in terms of economic and nutritional returns - per unit soil and water. Introduction of conservation tillage must therefore be closely linked to market considerations, to enable a diversification of production, and if possible link conservation tillage to the production of high value crops.

Access to high quality and affordable conservation tillage implements is also a major challenge in remote communal farming areas.

MINIMUM TILLAGE IN ETHIOPIA

Tef (*Eragrostis tef* (Zucc.)) is the basic staple food crop in many semi-arid regions of Ethiopia. Tef is an annual grass with very small seed. This means that farmers traditionally till the soil extensively in order to establish a fine seedbed. The implement used for tilling is the Ethiopian *Maresha*, a wooden ard-plough, normally pulled by one pair of oxen. For Tef often up to five primary tillage operations with the *Maresha* are carried out before planting the Tef. The tillage is shallow (< 15 cm depth) and is generally carried out both along and perpendicular to the contour, which means that little or no ridging is produced. This tillage practice has two major side effects; (1) the excessive tilling at the same shallow depth leads to the creation of hardpans, and high risk of water and wind erosion, and (2) high traction needs over long time. The latter issue is perhaps the most critical. Many farmers do not have access to their own oxen, and depend on renting or borrowing oxen. The result is very late tilling and planting.

On-farm trials have been initiated to try and develop reduced tillage systems for Tef together with farmers. The trials have been carried out since 1999 in Nasret area (Wulinchity, South-West of Addis Abeba), and in Woldia (Amhara State) where reduced Maresha tillage is combined with the introduction of a locally developed wing-plough for seedbed preparation and weeding. The tillage practices are combined with tied ridging as a means of minimising runoff and with ripping (an implement attached to the Maresha ard-plough). As seen from Fig. 2a the results from the year 2000 rainy

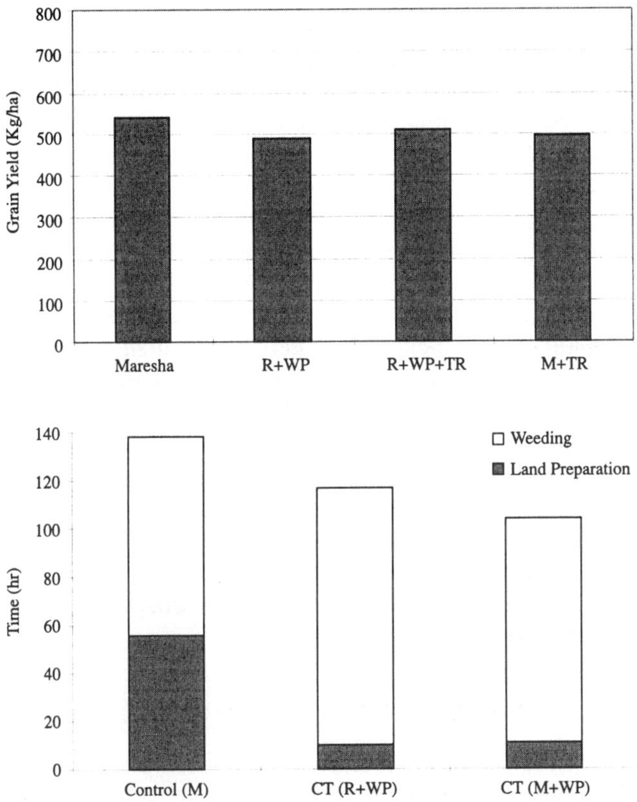

Fig. 2. (a) Tef grain yield (year 2000, Amhara State). Maresha = traditional practice, R+WP = ripper + wing plough, R+WP+TR = ripper + wing plough + tied ridging, M+TR = Maresha + tied ridging. (b) Labour time requirements for land preparation and weeding.

season (including 11 farmers in the Amhara State sites) suggest a very limited yield effect, with conventional Maresha ploughing giving an average yield level of 530 kg ha^{-1} compared with 490 – 500 kg ha^{-1} for conservation tillage practices combining ripping with wing ploughing and tied ridging.

The labour saving when abandoning the Maresha ploughing in favour of ripping at 80 cm spacing is high, with an 80 % reduction in traction time required (Fig. 2b). However, weeding requirements seem to increase, leading to a 30 % increase in weeding time.

Despite the limited increase in yields, farmers are very interested in reducing Maresha dependence, especially in combination with shallow scarification using the wing-plough.

SYSTEMS APPROACH TO CT IN KENYA

For decades land management efforts in Kenya have focused mainly on erosion control using structural soil conservation measures. The now famous *Fanya Juu* terraces within the context of the so-called catchment approach has resulted in a significant reversal of land degradation, e.g., in Machakos district (Tiffen et al., 1994). However, it is not until recently that the issue of «what do to between the *Fanya Juu* terraces» has been really highlighted, as a result of inert of stagnant yield levels despite soil conservation measures. Soil nutrient mining and tillage-induced soil moisture scarcity, caused by plough-pans as a result of continuous ploughing under wet soil conditions, are two important factors explaining the persistency of crop yields < 1 t ha^{-1}.

Farmer driven demonstration trials on CT systems for maize were initiated in 1999 in three locations; Kisii, Laikipia and Machakos. While the two latter are semi-arid, the Kisii site in Rachuonyo is located in the wetter Lake Victoria zone. The tested production systems include both animal drawn ripping and subsoiling and manual pitting with handhoe and pickaxe. In all three locations farmers interest in conservation tillage is strong, both for manual and animal drawn systems, based on primarily on productivity increase but on resource conservation and time saving. There are no statistically significant yield differences over the first 2 years of trials (three rainy seasons), even though CT systems in general result in a slight absolute yield increase (from e.g., 850 kg ha^{-1} grain yield for conventional ploughing in Rachuonyo short rains 1999/00 to 1000 kg ha^{-1} for ripping and 980 kg ha^{-1} for pitting with hand hoe).

The emphasis in the Kenyan CT initiative is not only on production enhancement but also to address the external system links, i.e., the external conditions that can make CT adoption possible. Focus here has been on capacity building and awareness creation, and the crucial task of involving the private sector for local implement manufacturing.

FARMER DRIVEN CONSERVATION TILLAGE IN TANZANIA

Farmer designed conservation tillage trials have been carried out since 1998 in semi-arid and dry-sub-humid (rainfall depth averaging 700 – 1000 mm yr^{-1}) parts of Arusha and Arumeru districts, North-western Tanzania. The basic tillage implements involved are; ox and tractor drawn sub-soilers, ox-drawn Magoye ripper (IMAG, 1999), and hand-hoe.

The trials included four principal production systems; (1) ripper/sub-soiler, (2) a ripped broad-bed system, (3) a manual pitting system and (4) the conventional ploughing system. These four systems were then combined and site-adapted regarding (i) inter-cropping (lab-lab or cowpea depending on location), (ii) fertilisation (manure, Mijingu rock-phosphate, and Urea), (iii) traction (oxen or tractor), (iv) crop rotation, (v) crop varieties. The main crop in all experiments is maize (*Zea mays*). Common to all sites was the use of a standard plant density of 80 x 30 cm, and the fertilisation (for all treatments expect the non-fertilised control). Cattle manure was applied at a rate of 3 t ha^{-1}. Locally produced Mijingu rock-phosphate was applied at a rate of 484 kg ha^{-1}, corresponding to some 15 kg of P ha^{-1}. Urea was applied twice during the growth season, at a cumulative rate of 30 kg of N ha^{-1}. The experiment was a randomised block design with two repetitions per farm-site (i.e., two blocks with 6 treatments each).

Sub-soiling was carried out either with tractor or with an ox-drawn sub-soiler (developed by IMAG-DLO in Zambia, IMAG, 1999). Sub-soiling was carried out during the dry season to a depth of 40-50 cm for tractor sub-soiling and to 25-35 cm with the ox-drawn sub-soiler. The sub-soiling was then followed by ripping, which was done to establish permanent planting lines, along the contour at 75 cm spacing. Sub-soiling prior to ripping was only carried out in two of the driest villages (Ngorobob and Mkonoo), where the farmers expressed serious problems with hard-pans. The sub-soiler/ripping treatment was divided in two plots, one with cover-crop (lab-lab) and one without cover-crop.

Planting was carried out manually. The manual pitting system was very similar to the *zai*-pitting found in the Sahel. A hand hoe was used to dig planting holes with a dimension of roughly 20x20x20 cm. Most important was that the depth exceeded the conventional ploughing depth (which in this region is 12-13 cm). The width and length of the hole is adapted to the hoe-size in order to enable highest labour efficiency in establishing the pits. The pits are then manually dry planted. A layer of fine soil is followed by a manure/soil mixture followed by the seed and fertiliser. The aim is to keep the soil level of the covered seed some 2-3 cm lower than the surrounding soil in order to enable the pits to function as micro-basins for rainwater harvesting.

The conventional system (control) was similar in all locations, based on the post-onset ploughing with mouldboard plough. All conservation tillage treatments were dry planted, and crop residue was left on the fields as mulch (except for maize leaves, which were taken for fodder). Weeding control was done manually, following the normal practices in the area. However, one additional weeding operation was carried out after harvest.

RESULTS

The trials started with the short rains 1998 (November), but due to a complete meteorological drought, farmers experienced total crop failure. Table 1 shows the

Table 1. Maize grain yield and statistical analysis for long rains 1999 and 2000, Tanzania.

			Comparative analysis			
			Control No-fertiliser		Control Fertiliser (FERT)	
Treatment	**Yield [kg ha^{-1}]**	**SD [kg ha^{-1}]**	**Sign.**	**Multiplyer**	**Sign.**	**Multiplyer**
Ripper + FERT	*3832*	*1472*	***	*2.9*	**	*1.4*
Ripper+CC	*3973*	*1698*	***	*3.0*	***	*1.5*
Ripper – FERT	*2429*	*1080*	**	*1.9*		*0.9*
Pitting	*3108*	*1197*	***	*2.3*	*	*1.1*
C+FERT	*2730*	*1460*	***	*2.0*		
C	*1334*	*672*				

average yield results from the long rains 1999 and 2000 for the 8 participating farmers. The average conventional ploughed maize yield is 1.3 t ha^{-1} which is a factor 3 lower than the ripper treatments (yielding on average 3.8 – 4.0 t ha^{-1}).

Compared with ploughed maize receiving fertiliser the conservation tillage treatment based on ripping yielded 40 – 50 % higher yield. Also the manual pitting system yield significantly higher yields compared to both fertilised and non-fertilised maize.

These results show very significant yield increases already after 1-2 years of CT introduction. This can be explained by the severe plough-pans generally experienced in the region. Interestingly, strong yield increases were experienced also on crop land with well developed soil conservation structures (terracing). This suggests that soil conservation alone will not enable improvements in water productivity in semi-arid rainfed farming systems. Also the results indicate that the water effect of CT can only give full synergetic effects on yield levels in combination with fertilisation.

Perhaps surprisingly, farmers found the manual pitting system most interesting. They perceived it to give the most stable yields, and despite being labour intensive, it was an effective measure to restore compacted and exhausted soils.

Similar results have been achieved in semi-arid Babati district, Tanzania, where tractor sub-soiling of maize resulted in an immediate (first season) 2.8 time increase in maize yields during the favourable rains 95/96 (from 1.7 to 4.8 t ha^{-1} for fertilised maize and 1.3 – 3.7 t ha^{-1} for non fertilised maize) and a 25 % increase in yields during a drought season (short rains 96/97) (Rockstrom and Jonsson, 1999).

CHALLENGES TO ADOPTION OF CT

Farming systems related challenges

Weeding, crop residue management, cover cropping and timing of operations are the most critical issues according to the local experience in the three countries when converting from mouldboard ploughing to no-inversion CT systems. Farmers are very concerned with the weed infestation resulting from no-inversion of soil. In Ethiopia the first 2-3 primary tilling operations are for weed control. Weed infestation may decrease over time (> 3-5 years) but farmers expect instantaneous productivity improvements. This may be a reason for the islands of success of promoting zero tillage systems combined with herbicide applications in Ethiopia, Tanzania and Kenya. In Ethiopia, Sasakawa-Global2000 have reported yield increases even in practically zero tilled tef cultivation (only one primary Maresha operation to incorporate fertilisers), but this is achieved with a conventional herbicide program (Ato Takele Gebre, pers. comm., 2000).

In Kenya our approach to weeding is to focus less on minimum tillage but rather only on non-inversion, and instead enable farmers to use animal traction for mechanical weeding. Suppressing weeds is also achieved by introducing cover crops and by addressing the critical issue of mulching.

A study carried out among the farmers participating in the on-farm trials in Tanzania showed that involvement of women in CT capacity building is critical for adoption success. Conservation farming is generally completely alien to the local rural community, and as women play a crucial role in many of the farm operations (weeding, planting, harvest and post harvest tasks, and at times also tillage), leaving them out of the capacity loop may result in poor CT performance or even non-adoption. Involving women in CT promotion can moreover improve mulching and cover cropping.

The free post-harvest grazing practises in Eastern and Southern Africa poses special challenges regarding mulching. Residue management thus has to be carried out at community level, with a clear insight that success of mulching depends on sustainable solutions for livestock grazing.

Timing of operations is crucial in semi-arid tropical farming. An opportunity with abandoning mouldboard ploughing is dry season land preparation and dry planting before onset of rains. This can increase the plant accessible proportion of rains with 25 %.

EXTERNAL FACTORS AFFECTING ADOPTION

Adoption of CT among smallholder farmers depends strongly on external factors related to extension services and the market. CT implements are not accessible on local markets in the studied countries. An important step is to involve private manufacturers in production of good quality and low cost implements.

Extension services are not trained to promote conservation farming. Instead, modernising of agriculture is still to be achieved through intensified adoption of conventional tillage implements. Grassroot level promotion of CT is important, but cannot succeed in isolation. Training and awareness creation is required also among mid and high level extension officers.

CONCLUSIONS AND DISCUSSION

The on-farm CT experiments from Ethiopia, Kenya and Tanzania clearly show that conservation tillage among smallholder farmers in drought prone environments has potential of not only improving farm productivity but also conserve resources and save time. A systems approach is required on the long term, addressing especially weeding, cover cropping, mulching and timing of operations. However, in highly compacted areas (e.g., the experience from Babati, Tanzania) immediate yield improvements can be achieved as a result of increased rainfall infiltration.

Capacity building among service giving institutions and involving the private sector in tillage manufacturing are two external factor that have to be addressed to enable successful adoption of site adapted CT systems for smallholder farmers in Eastern and Southern Africa.

ACKNOWLEDGEMENTS

The research in Ethiopia, Kenya and Tanzania is supported by the Regional Land Management Unit (RELMA) of Sida (Swedish International Development and Cooperation Agency).

REFERENCES

Anonymous (2001) Dialogue on Water for food and environment. Proposal, final version, May 2001. International Water Management Institute, IWMI, Colombo, Sri Lanka. p 38

Benites, J., Chuma, E., Fowler, R., Kienzle, J., Molapong, K., Manu, J., Nyagumbo, I., Steiner, K., and van Veenhuizen, R. (Eds.) (1998) Conservation Tillage for Sustainable Agriculture. Proceedings from an International Workshop, Harare, 22 – 27 June. Part 1 (Workshop Report). Deutsche Gesellschaft, GTZ, Eschborn, Germany. p 59.

Brunner, E., Mariki, W. and Elwell, H. (1998) Available technologies in conservation tillage. In: J., Benites, E., Chuma, R., Fowler, J., Kienzle, K., Molapong, J., Manu, I., Nyagumbo, K., Steiner, R., van Veenhuizen (Eds.). Conservation Tillage for Sustainable Agriculture. Proceedings from an International Workshop, Harare, 22 – 27 June. Part II, Annexes, pp 87 - 92. Deutsche Gesellschaft, GTZ, Eschborn, Germany.

Elwell, H.A. (1993) Development and adoption of conservation tillage practices in Zimbabwe. In: *FAO Soils Bulletin*, vol **69**: 129–164. FAO, Rome.

IMAG (1999) Smallholder agricultural mechanisation - Research, Development and Promotion. Department of International Development, Institute of Agricultural and Environmental Engineering (IMAG), the Netherlands. p 5

Oldreive, B. (1993) Conservation Farming for communal, small-scale, resettlement and co-operative farmers of Zimbabwe. A Farm management handbook. Mazongororo Paper Converters (Pvt) Ltd, Zimbabwe. p 76.

Rockström, J. and Jonsson, L.O. (1999) Conservation tillage systems for dryland farming: On-farm research and extension experiences. *E. Afr. agric. For. J.*, **65** (1): 101-114.

Rockstrom, J. and Falkenmark, M. (2000) Semiarid crop production from a hydrological perspective: Gap between potential and actual yields. *Critical Rev. Plant Sc.*, **19**(4): 319-346.

Stewart, J. I. (1988) Response farming in rainfed agriculture. The WHARF Foundation Press, Davis, California. p 103

Stoorvogel, J.J. and Smaling, E.M.A. (1990) Assessment of soil nutrient depletion in sub-Saharan Africa: 1983 – 2000, vol 1: Main Report. Report No. 28, Winand Staring Center, Wageningen, The Netherlands.

Tiffen, M., Mortimore, M. and Gichuki, F. (1994) More People, Less Erosion – Environmental recovery in Kenya. ACTSPRESS, African Centre for Technology Studies, Nairobi, Kenya. p 301

Vogel, H., Nyagumbo, I. and Olsen, K. (1994) Effects of tied ridging and mulch ripping on water conservation in maize production on sandveld soils. Der Tropenlandwirt, *J. of Agric. in the tropics and subtropics*, **3-4**: 33–44.

D. DE WRACHIEN

LAND USE PLANNING: A KEY TO SUSTAINABLE AGRICULTURE

EurAgEng Incoming President and Coordinator; of the Field of Interest on Soil and Water; Head of the Institute of Hydraulics Agricultural; State University of Milan, Italy

Sustainable use of the soil is a form of land management which retains the natural fertility of the soil and allows for the production of food and fiber supplies and renewable natural resources on a long-term basis. It implies that the natural environment should be treated and managed in such a way that the cycles and energy fluxes among the soil, bodies of water and atmosphere are considered, preserved or restored. To this respect, the term «sustainable land use» is more comprehensive than the term «sustainable soil use». Land, commonly, stands for a section of the earth's surface with all the physical, chemical and biological features that influence the use of the resource. It refers to soil, spatial variability of landscape, climate, hydrology, vegetation and fauna, and also includes improvements in land management, such as drainage schemes, terraces and other agrobiological and mechanical measures. The term «land use» encompasses not only land use for agricultural and forestry purposes, but also use of the land for settlements, industrial sites, roads and other human activities. Land use, in this meaning, can be termed sustainable only if is achieved such a spatial distribution or configuration of the different uses, as to guarantee biodiversity and preserve the eco-balance of the whole system. Rational land use planning is fundamental to this process. With reference to the aforestated issues, the paper describes the main physical, social and economic features of land use planning projects, along with their environmental impacts and constraints to sustainable development. The importance and role of institutional strengthening, sound financial and managerial frameworks, availability of human resources involved, research thrust, technology transfer and networking improvement are also analyzed.
Key-words: Land use planning, sustainable agriculture, networking system.

FOREWORD

The world's population is expected to grow from 6 billion today to at least 8 billion in the year 2025. It is, therefore, clear that achieving food security and improving the quality of life, while preserving the environment, will continue to pose major challenges to scientists, decision-makers and technicians in the years to come. The main activity of agriculture is the production of food, so increasing agricultural development in a sustainable manner will be crucial in responding to these challenges.

L. García Torres et al. (eds.). Conservation Agriculture, 471–484.

In the past, demand for growth in food has been met by expanding agricultural land. Nowadays, the availability of new land is limited; moreover, the more or less uncontrolled growth in agricultural production, during the past few decades, in industrial as well as developing countries, has pushed agricultural production to and, in many cases, over the edge of sustainability. This means that the traditional ways to increase production are facing a new challenge: to find a new balance between agricultural development and the conservation of the natural resources.

Agricultural engineering has been applying scientific principles for the optimal use and management of natural resources for centuries, and its role is increasing with the dawn of the new millennium. There are, at least, two reasons for this growing significance. First, it is well understood that the wise use of land resources will play a role of paramount importance in the provision of food for future generations. Second, the demand for different land uses is increasing tremendously, especially in the developed world. The land demands for cropping, grazing, forestry, wildlife, infrastructure, outdoor recreation, landscape and industrial and urban development are greater that the land resources available. To this end, rational land use planning will help to find a balance among these different demands and assure agricultural production, while conserving the natural environment.

With reference to the afore-mentioned issues, the paper, firstly, describes the main physical, social and economic features of the land use planning process, along with its environmental impacts and constraints to sustainable agricultural development. Finally, the importance and role of institutional strengthening, sound financial and managerial frameworks, availability of human resources involved, research thrust, technology transfer and networking improvement are analyzed.

THE CONCEPT OF SUSTAINABLE LAND USE

To meet future challenges of food security, further development of agriculture is necessary. This development has to guarantee both the growth in agricultural output and the conservation of natural resources. The conservation of the natural resources is important because of the dependence of agriculture on these resources. This means that the natural environment should be treated and managed in such a way that food production is secured now and in the future. So, food security is not only a matter of quantity, but also of continuity. Agriculture, thus, is forced to find a balance between development and conservation. In this process the responsible use of natural resources plays a role of paramount importance. Among the basic natural resources, upon which life depends, is the soil.

The responsible use of the soil can be described in terms of sustainability or sustainable development. Sustainability has been defined in many different ways and there is no single, universally accepted definition. According to the Brundtland Commission «sustainable development is a process of change in which the exploitation of resources, the direction of investments, the orientation of technological development and institutional changes are all in harmony and enhance both current and future potential, to meet human needs and aspiration». This process implies long-term perspective for planning and integrated policies for implementation. FAO has formulated its own definition of sustainability, specifically in the context of agriculture, forestry and fisheries: «sustainable development is the management and conservation of the natural resource base and the orientation of technological and institutional change in such a manner as to ensure the attainment and continued satisfaction of human needs for the present and future generations. Such sustainable development conserves land, water, plant and animal genetic resources, is environmentally non-degrading, technically appropriate, economically viable and socially acceptable».

Scarcity of suitable soil is a major constraint for further agricultural development in many countries of the world. Therefore, as the demand for soil continues to increase, it is imperative that this limited resource be used efficiently for agricultural and other uses.

Sustainable use of the soil is a form of land management which retains the natural fertility of the soil and allows for the production of high quality of foodstuffs and renewable natural resources on a long-term basis. This means that the natural environment should be treated and managed in such a way, as to preserve or restore the cycles and energy fluxes among soil, bodies of water and atmosphere.

The term «sustainable land use» is more comprehensive that the term «sustainable soil use». Land, commonly, stands for a section of the earth's surface, with all the physical, chemical and biological features that influence the use of the resource. It refers to soil, spatial variability of landscape, climate, hydrology, vegetation and fauna, and also includes improvements in land management, such as drainage schemes, terraces and other agrobiological and mechanical measures.The term «land use» encompasses not only land use for agricultural and forestry purposes, but also land uses for settlements, industrial sites, roads and so on. Land use, in this sense, can be termed sustainable if, and only if, is achieved such a spatial distribution or configuration of the different uses, as to guarantee biodiversity and preserve the eco-balance of the whole system. In other words, land use that limits the interactions among soil, water and atmosphere and degrades the habitat standards vital to biological diversity of flora and fauna cannot be defined sustainable. In this respect, the term «sustainable land use» combines technology, policies and activities aimed at integrating socio-economic principles with environmental concerns. The term bears more dimensions or meanings:

- Sustainable use in the meaning of husbandry. In this sense, it is related to concepts such as continuity, durability and equity in the exploitation of natural resources over long periods of time. The dimension refers to methods by which land is managed –crop rotation procedures, tillage systems and so on– all striving to preserve or restore the quality and fertility of the soil. This meaning is strongly related to the long-term physical and economic sustainability.
- Sustainable use in term of interdependence. This meaning is related to the spatial dimension of sustainability. It involves such aspects as fragmentation and relations among different land uses. On this facet of sustainability are, nowadays, focusing many land use planning studies, due to the fact that there is still a great lack of knowledge and uncertainty.

- Sustainable use in terms of ethical obligations to future generations. This refers to the losses and depletions of natural resources in combination with the expected increase in population. Land is not a simple commodity that can be stored and replaced, destroyed and remade, or even recycled in exactly the same way as manufactured goods. It is a complex and biological system, built up over long periods of time. The land could have lost its suitability for cropping or other uses by means of natural or anthropogenic causes. To restore its capacity for beneficial use, while protecting the environment, methods of reclamation have to be tailored to the specific problems at hand. In this field much needs to be done to ensure the future of mankind.

Any assessment of sustainability would be incomplete if it did not address all the dimensions previously described.

Clearly, there are conflicts among these goals. More equity may mean less efficiency. In the short term it may not be possible to meet the needs or demands of even the present generations, let alone the future ones, if these needs or demands are greater than what the environment can afford. Furthermore, degrading the natural resources will reduce their capacity to meet future needs, whatever those needs will be. So, demand management and degradation prevention play a basic role in the process of sustainable use and development of land. Decision – makers have to consider and agree upon a trade-off among different goals but, if the ecosystem as a whole is to survive, the use of natural assets must be compensated by the development of human or physical assets of equal or greater worth. In this regard, good and reliable information is essential, that is, information on the people's needs, land resources and on the economic, social and environmental consequences of alternative decisions. To this end, the task of the land-use planners is to ensure that decisions are made on the basis of consensus, to avoid disagreements on the ways and directions the natural resources should be exploited. Wise land use planning will help to reduce the trade-off costs and resolve conflicts by involving the community in the decision process.

LAND USE PLANNING: A TOOL TO ACHIEVE SUSTAINABILITY

Land use planning is the systematic assessment of land and water potential, alternatives for land use and economic and social conditions in order to select and adopt the best land use options. Its purpose is to select and put into practice those land uses that will best meet the needs of the people while safeguarding resources for the future. The driving force in planning is the need for change, the need for improved management or the need for a quite different pattern of land use dictated by changing circumstances. In the process all kinds of land use are involved: agriculture, forestry, wildlife conservation, urban and industrial expansions, tourism and amenities. Planning also provides guidance in case of conflicts among manifold alternatives, by indicating which areas are most valuable for any particular land use. Land use planning can be viewed as an iterative and continuous process, whose aim is to make the best use of land resources by:

- assessing present and future needs and evaluating the land's availability to meet them;
- identifying and resolving conflicts among competing uses and needs;
- devising alternative options and choosing those that best fit identified targets;
- learning from experience.

At every stage, as better information is available, the process may have to be changed to take account of it.

Goals are important elements in the planning process. They define what is meant by the best use of the land and they have to be specified at the outset of every planning project.Goals, normally, are divided into objects and targets.

Objectives are the general aims within the planning process. They allow the judging of different solutions of a concrete problem in the planning area, and lead to suitable propositions and projects for the use of the land.The targets are the most detailed aims of land use planning. They lead to the design of actual measures that have to be taken and carried out in an area to solve the problems at hand.

The objectives and targets identify the best use of the land. If two different forms of land use bring forth exactly the same profit (economically and socially), the objectives will determine which of the two land uses should be implemented, while the targets will indicate which procedures should be followed.

The goals, as a whole, may be grouped under three main headings: efficiency, equity and acceptability and sustainability.

- Efficiency refers to the economic viability of the land use plan.
  The plan should yield more than it costs. So one goal of planning development is to make efficient and productive use of the land. In general terms, for any particular land use, certain areas are better suited than others. Efficiency is achieved by matching different land uses with the areas that will yield the greatest benefit at the least cost. However, it is not always clear which land use is the most profitable one; this depends on the point of view. The point of view of individuals, for instance, focuses on the greatest return on capital and labour invested or on the greatest benefit from the area available. Government's point of view is more complex: it may include improving the foreign exchange situation by producing for export or for import substitution.
- Equity and acceptability represent the social features of land use planning.
  The plan must be accepted by the local population, otherwise the proposed changes will not take place. Equity refers to the levelling of the living standards of the residents. People living in the planning area are expected to gain from the land use plan, even if they do not own the land. Living standards may include levels of income, food security and housing. Planning to achieve these standards then involves the allocation of land for specific uses as well as the allocation of financial and other resources.
- Sustainability, as stated before, refers to a development in land use planning that meets the needs of the present while conserving resources for future generations.

This requires a combination of production and conservation: the production of the goods needed by the people now, combined with the conservation of the natural resources on which the production depends. So, land use to be sustainable, has to be planned for the community as a whole, because the conservation of soil, water and other land resources is often beyond the means of individual land users.

Other goals of the planning process could be:

- Livability After the land use plan is implemented, the area should still be a suitable place to live for the inhabitants;
- Flexibility The plan should be flexible and leave options for using the land in different ways, if needed, in the future;
- Public involvement Every group or individual with an interest in the plan should be allowed to participate in the process, to keep their land use from disappearing through the plan, or to be offered a new land use, as part of the plan.

On the whole, the land use planning, to be sustainable, should develop into an interdisciplinary, holistic approach that gives attention to all functions of the land and actively involves all land users through a participatory process of negotiation platform, be it at national or local levels. The aim of the process is to create the conditions to achieve an environmentally sound, socially desirable and economically appropriate form of land use.

RESEARCH AND DEVELOPMENT

International and national research, nowadays, needs to be focused more effectively than in the past on problems of land use planning and management. This is the only way to provide land users and planners with suitable and tested technologies for targeted measures to increase agricultural production while protecting the natural resources. The lack of research, application of research findings and access to new and advanced technology in this field is seen as one of the main reasons for the problems that plague the sector: poor land use efficiency, environmental degradation, high costs and lack of responsiveness to beneficiaries.

Successful research thrust on sustainable land use planning should include the following actions:

- Data base improvement;
- Adaptive research;
- Institutional strengthening;
- Socio-economic analysis;
- Environmental protection and conservation;
- Technology transfer and infrastructure.
- Data base improvement.
  Availability of reliable hydro-climatic and other associated natural resource data is an essential prerequisite for sustainable land use planning development.

As long as adequate and reliable data are not available, planning, design and management of land use programs will continue to remain guesswork, use of other natural resources haphazard and wasteful, and the development process unsustainable. Many land use projects were conceived and designed on a medium – to long-term basis, on the assumption that future climatic conditions will not be different from the past ones. This will not be so in the years to come, due to the global warming and greenhouse effect. Therefore, land use planning designers and managers should begin a systematic re-examination of engineering design criteria, operating rules, contingency plans and land allocation policies. Demand management and adaptation are essential components for increasing project flexibility to meet uncertainties of climate change. On the whole, land use planning programs can only be soundly formulated on the basis of adequate data on soil and its production capacity, potentially available water resources, performance of existing land use projects and other related factors.

- Adaptive research.
  A wide variety of techniques or methods are use in land use planning. They are taken from the natural sciences (climatology, hydrology, soil science, ecology), from technology (agriculture, forestry, irrigation and drainage engineering) and from the social science (economics, sociology). Research for land use planning requires enhanced field investigations and a large variety of tools such as: Information Management, System Analysis, Decision Support Systems, Multicriteria Analysis, Geographic Information Systems, Remote Sensing, Computer Image Analysis, Sensors, Modeling Technique, Neural Network Technology, Land Evaluation. All these tools have to be considered under a broad and integrated approach related to food and other agricultural commodity production, rational land use planning, water saving, resource conservation, environmental impacts and socio-economic effects. Current research thrust needs to be reoriented by recognizing the complex role of the land resources in agricultural development, and by following a broad-based holistic approach. To this end, adaptive research programs must be directed to investigate the actual and real problems associated with the planning, design, implementation and management of land use projects. It is important that the resulting methodology be technically feasible, environmentally and economically viable and socially acceptable.
- Institutional strengthening.
  The importance of a functional and coherent institutional framework aiming to promote, at both national and international levels, sustainable land use planning development, has been fully recognized at present. The solution may not always require the creation of new and enlarged institutions and establishment of larger governmental services. An important criterion in reorganizing and/or establishing new institutions should be the ability of such institutions to address successfully the multi-dimensional problems that are generally faced by the land users at both local and national levels. Such institutions should be capable of undertaking,

regulating, stimulating and facilitating the roles and the tasks carried out by the land users. These institutional frameworks need to be strengthened or restructured to meet more efficiently the land users' requirements and to promote sustainable land use planning development. Principal institutions should have effective linkages with all other related frameworks, so as to optimize the use of physical, financial and human resources involved.

The necessary actions are the following:
- review, strengthen and restructure, if required, existing institutions in order to enhance their capacity in land use planning activities;
- review, assess and revise, if needed, existing legislation on land management within the broader framework of legislation for the development, use and conservation of land resources.
- Human resource development.
  Successful technology and research thrust on land use planning depends on the number, orientation and quality of human resources (decision makers, professionals and research-related people) involved. They orient appropriate knowledge and skill to solution of priority issues and emphasize the adaptation of available techniques to solve local problems. These knowledge and skill will include the ability to:
- identify local hurdles and constraints;
- formulate research strategies;
- design suitable technologies for testing, monitoring and evaluating;
- assess the technical, economic, social and institutional aspects regarding the application and adaptation of modern and advanced technology.Moreover, this body of human resources will help national and international institutions, improve educational contents and training in land and other natural resources related topics, as well as scientific organizations identify subjects to be further analyzed and investigated.

The necessary actions can be summarized as follows:
- assess training needs for land use planning and management;
- develop practical training courses for improving the ability of extension services to disseminate technologies and strengthen land users' capabilities;
- enhance the capabilities of decision makers, administrators and officers at all levels, involved in land use planning programs.
- Social economic analysis.
  Social and economic analyses are important features of the land use planning process. A land use project, like many other projects, can be implemented only if the total benefits exceed the total costs. Therefore, sustainable land use planning should meet two basis considerations, namely economic viability and social acceptability. Comparisons of social with economic analyses can highlight the need for policy changes. A particular land use may be degrading and thus destroying other land resources. If the economic analysis shows the

use to be advantageous from a land user's point of view, it is likely to continue, whether the process is environmentally sound or not. Economic analysis should take account of damage to land resources and the consequent lowering of their productivity.

A great many land use planning projects in the past have failed due to the inadequate attention given to social and economic aspects in their design and implementation. Application of appropriate socio-economic analysis in all phases of the planning process is urgently required in the development of land use projects. In this regard it is recommended that:

- effort should be made to incorporate economic and social analyses in land use planning methodologies;
- governments, relevant international and national institutions and decision – makers should ensure that socio-economic analyses are adequatly applied in the formulation and selection of land use planning projects for implementation.
- Environmental protection.

Sustainable land use planning has to find a balance between agricultural development and conservation of natural resources. Thus, development and environment are two aspects of the same process. Much agricultural land is deteriorating due to inappropriate soil and water management. Soil erosion, nutrient depletion, salinization and waterlogging all reduce productivity and jeopardize long-term sustainability. Wise management of the environment requires ability to forecast, monitor, measure and analyze environmental trends and assess the potentials of the land resources at different levels, ranging from the farm to the watershed. Adopting suitable environmental impact assessments will enable decision-makers, professionals and institutions to plan land use without irreversible environmental damage and allow sustainable natural resource use. Environmental impact assessments should be followed by monitoring and appropriate actions in order to maximize positive impacts of development and minimize environmental hazards. In this regard, environmental protection and conservation of natural resources must be made an integral part of development.

The necessary actions have to:

- carry out objective environmental impact assessments in order to ensure the sustainability and environmental acceptability of land use projects and programs;
- establish environmental monitoring, evaluation and feedback systems on a long term basis;
- expand, improve and coordinate international assistance to enhance the capabilities of less developed countries to assess, manage and protect their environment and natural resources.
- Technology transfer and infrastructure.

The success of a land use planning project is strongly influenced by the availability of technology and whether or not appropriate choices have been made to suit the local conditions. So, a framework for information transfer which includes storing, disseminating, receiving feedback and updating information is urgently needed to support sustainable land use activities. As in all economic activities, agricultural development, particularly involving the land use sector, has infrastructural requirements to ensure its success. Farmers and other land users must have appropriate funds, food supplies must be delivered in time and in adequate quantities, and proper marketing facilities and pricing structures must be assured. In addition to physical infrastructure, services such as education and health are also necessary. The necessary actions have to:

- establish effective methods to facilitate the transfer of new and tested techniques and practices;
- encourage and provide required facilities for transfer of knowledge and experiences among developed and developing countries;
- enhance the development of a more effective production environment.

STRATEGIC ACTION PROGRAM

The above described themes and principles, strike at the root of the major problems encountered in the land use planning process. To be effective, they have to be translated into actions through the formulation of programs which have to take into account the actual conditions of the environment where they are expected to be implemented. These programs have to include:

- the adoption of a comprehensive approach that views land and water use and management and environment in an integrated manner;
- the promotion of regional cooperation to ensure that the concerns of all parties are factored into decisions;
- the recognition of the linkages among the different land uses;
- the encouragement of broad-based participation, including governments, professional and research institutions and non-government organizations;
- the endorsement of a phased program of action at the national and local levels.

This regional approach makes up and outlines the body of a Strategic Action Program, which is a critical measure for implementing priority actions at both national and local levels. The objectives of the Strategic Action Plan are to:

- evaluate trends;
- assess causes and implications;
- provide a cost estimate for investments;
- establish a framework for monitoring and evaluating;
- identify priority actions to address key issues.

Priority selection has to follow the criteria listed below:

- ensure selectivity, in order to concentrate resources on significant problems;
- avoid duplication and overlap;

- emphasize adaptive and cost effective solutions through adaptation and/or improvement of existing technology to specific tasks;
- select topics for investigation and research that are likely to realize the greater benefit, considering return on investment, response time, probability of success and impact on agricultural production.

This integrated approach is expected to bring forth clear benefits in environmental and economic terms, a more sustainable use of land resources in agriculture and higher yields and incomes.

CONCLUDING REMARKS

- Sustainable land use planning is a process that aims to integrate ecological with socio-economic, and political with ethical principles in the management of land, for productive and other functions, to achieve intra – and inter – generational equity.
- For formulating and implementing policies and strategies for land use planning it is essential to collect, process and disseminate timely and reliable information and utilize modern land assessment and evaluation technologies, to create sound scientific knowledge for proper decision support.
- The establishment of an effective networking system can greatly improve, enhance and speed up the process of collection, selection and exchange of information avoiding duplication and overlap.
- No detailed layout for sustainable land use planning can be drawn up for a region as a whole. A regional strategy can, at best, give a general idea of what needs to be achieved at the country level. Each country, then, will have to tailor its sustainable development strategy in view of its particular problems, constraints and comparative advantages.
- Regional strategies must set priorities and identify relevant projects, assess the environmental impacts of policies, investigate mechanisms to mobilize resources, enhance and encourage the participation of all concerned parties.
- The promotion and implementation of land use planning projects will not come free of cost. Major emphasis should, therefore, be paid on developing new sources of funds to supplement the national budgetary allocations. Chief among these approaches are measures that seek to mobilize local funds, in particular under the «user pays» principle.
- The challenging, but widely acceptable concept of sustainable land use planning calls for new approaches on development and, therefore, on land use and management. To this respect, new perspectives are required to manage the land and its associate resources. This is not only a question of allocating and controlling the use of the land, but of combining the knowledge of pressure influencing the resources themselves, with the relations among users and human and social objectives, the technologies available to improve and enhance the land use planning process, and the maintenance of both the biodiversity and the natural equilibrium.

- The lessons learned demostrate that it is necessary to make a decisive break from past policies to embrace a new holistic approach in land use planning and management, that is comprehensive, participatory and environmentally sustainable.
- There is an urgent need for adequately trained professionals who can work in the multisectorial environment of integrated natural resource management.
- Finally, to achieve a sustainable land use planning development, objectives and goals, policies and regulations should be grounded in local realities, traditions and natural resource management strategies. The environmental and socio-economic impacts of such policies and regulations should be assessed before they are implemented.

REFERENCES

Beek, J.K., Burrough, P.A. and Mc Cormack, E.D. (Eds) (1987) Quantitative land evaluation procedures. ITC Publication 6, Enschede, The Netherland.

Brouwer, F.M., Thomas, A.J. and Chadwich, M.J. (Eds) (1991) Land use changes in Europe. Processes of change, environmental transformations and future patterns. Kluwer Publisher, Dordrecht, The Netherlands.

Bunting, A.H. (1987) Agricultural environments. Characterization, classification and mapping. Wallingford, U.K.

Chisci, G. and De Wrachien, D. (1999) Soil regeneration and improvement. In: CIGR Handbook of Agricultural Engineering, Vol 1, H.N. van Lier (Ed). St, Joseph, US

Cloke, P.J. (Ed) (1988) Rural land use planning in developed nations.Unwin Hyman Publisher. London, UK.

Davidson, D.A. (1980) Soil and land use planning. Longman Publisher. London, UK.

De Wrachien, D. (2001) Irrigation and drainage. Trends and challenges for the 21st Century. Proceedings of the 19th ICID Regional Conference on Sustainable Use of Land and Water. Brno and Prague, Czech Republic.

De Wrachien, D. (2001) Sustainable land use. The role of Agricultural Engineering. Proceedings of the International Conference on Sustainable Soil management for Environmental Protection. Soil Physical Aspects. Florence, Italy.

De Wrachien, D. and Chisci, G. (1999) Soil reclamation. In: CIGR Handbook of Agricultural Engineering, Vol 1, N.H. van Lier (Ed). St. Joseph, US.

Dongen, M.W. and van Lier, H.N. (1999) Balancing agriculture between development and conservation. In: CIGR Handbook of Agricultural Engineering, Vol. 1, N.H. van Lier (Ed). St. Joseph, US.

Ervin, D., Batie, S. and Livingston, M. (1995) Developing indicators for environmental sustainability. The nuts and bolts. Proceedings of the Resource Policy Consortium Symposium. Washington D.C., US.

FAO (1990) An international action program on water and sustainable agricultural development. Rome, Italy.

FAO (1991) Guidelines on land evaluation for extensive grazing. FAO Soils Bulletin n. 58. Rome, Italy.

FAO (1991) Land use planning applications. Proceedings of the FAO Expert Consultation. Rome, Italy.

FAO (1993) Guidelines for land use planning. Rome, Italy.

FAO (1995) Planning for a sustainable use of land resources. Towards a new approach. Rome, Italy.

Fleischhauer, E. and Efer, H. (1998) Can sustainable land use be achieved? An introductory view on scientific and political issues. In: Towards Sustainable Land Use, Blume H.P., Eger H., Fleischhauer, Hebel H., Reij C. and Steiner K.G. (Eds). Reiskirchen, Germany.

Fleischhauer, H.E., Hebel, A. and Sombroek, W.G. (1998) Conclusion and Recommendations. In: Towards Sustainable Land Use, Blume H.P., Eger H., Fleischhauer, Hebel H., Reij C. and Steiner K.G. (Eds). Reiskirchen, Germany.

Forman R.T.T. and Godron M. – 1986 – Landscape ecology. Wiley Publisher. New York, US.

Fresco, L.O., Huizing, H.G.Y., van Keulen, H., Luning, H.A. and Schipper, R.A. (1992) Land evaluation and farming systems analysis for land use planning. FAO/ITC. Wageningen, The Netehrlands.

Hamdy, A. and De Wrachien, D. (1999) New policies and strategies on land and water development in the mediterranean region. Proceeding of the 2nd International Conference on Environment Water. Lausanne, Switzerland.

Herdt, R.W. and Steiner, R.A. (1995) Agricultural sustainability. Concepts and conundrums. In: Agricultural Sustainability. Economic, Enviromental and Statistical Consideration, Barnett V., Payne R., and Steiner R. (Eds). Chichester, UK.

Hes, M.B.A. and van Lier, H.N. (1999) Land and water use planning. In: CIGR Handbook of Agricultural Engineering, Vol 1, N.H. van Lier (Ed). St. Joseph, US.

Jaarsma, C.F. New approaches for planning of rural road networks within the frame of sustainable land use planning. Proceedings of the 2nd Workshop on Sustainable Land Use Planning with Special Regard to Central and Easter European Countries. Godollo, Hungary.

Jaarsma, C.F. (1999) Planning of rural road notworks. In: CIGR Handbook of Agricultural Engineering, Vol 1, N.H. van Lier (Ed). St. Joseph, US.

Jurgens, C. (1992) Tools for the spatial analysis of land and for the planning of infrastructures in multiple land use situation. Proefschrift. Wageningen, The Netherlands.

Kerkstra, K. and Vrijlandt, P. (1990) Landscape planning for industrial agriculture. A proposed framework for rural areas. Landscape and Urban Planning, n. 18.

Kneib, W.D. (1996) A landscape development plan as a tool for sustainable ecological planning. Proceedings of the 2nd Workshop on Sustainable Land Use Planning with Special Regard to Central and Eastern European Countries. Godollo, Hungary.

Laconte, P. and Haimes, Y.Y. (1985) Water resources and land use planning. A system approach. NATO Advanced Studies Institute Series. Nijhoff, The Netherland.

Lier, H.N. van (1996) Sustainable rural systems. A challenge also for land use planners. Proceedings of the 2nd Workshop on Sustainable Land Use Planning with Special Regard to Central and Eastern European Countries. Godollo, Hungary.

Lindgren, D.T. (1985) Land use planning and remote sensing. Principles and applications. Longman Publisher. Harlow, UK.

Loomis, R.S. and Connor, D.J. (1997) Crop ecology. Productivity and management in agricultural systems. Cambridge University Press Publisher. Cambridge, UK.

Mollett, J.A. (1984) Planning for agricultural development. Croom Helm Publisher. London, UK.

NCR (1991) Towards sustainability. A plan for collaborative research on agriculture and natural resource management.National Research Council (Ed). Washington DC, US.

OECD (1997) Environmental indicators for agriculture. Paris, France.

Rennings, K. and Wiggering, H. (1997) Steps towards indicators for sustainable development. Linking economic and ecological concepts. Ecological Economics. Vol. 20.

Scarascia Mugnozza, G.T., Tinelli, A. and Benedetti, A. (1999) Interdisciplinary studies for the management of a sensitive natural area. Proceedings of the International Workshop on Soil Vulnerability and Sensitivity. Florence, Italy.

Stigiani, W.M., Brouwer, F.M., Munn, R.E., Shaw, R.W. and Antonosky, M. (1989) Future environments for Europe Some implications of alternative developments paths. Executive Report 15, IIASA Publisher. Laxenburg, Austria.

UN (1992) Report of the United Nations Conference on Environment and Development United Nations (Ed). New York, US.

WCED (1987) Our common future. Oxford University Press. London, UK.

World Bank and UNDP (1990) Irrigation and drainage research. World Bank (Ed). Washington D.C., US.

Yokoi, Y. (2000) Measuring the environmental impacts of agriculture. OECD Agri-environmental indicators. Soil related issues. Proceeding of the International Workshop on Soil Vulnerability and Sensibility. Florence, Italy.

Young, A. (1986) Land evaluation and diagnosis and design. Towards a reconciliation of procedures. Soil Survey and Land Evaluation, n. 5.

W.R. RITCHIE, C.J. BAKER & M. HAMILTON-MANNS

THE DEVELOPMENT AND TRANSFER OF A NEW NO-TILLAGE TECHNOLOGY

Centre for International No-Tillage Research and Engineering, Feilding, NZ

Centre for International No-Tillage Research and Engineering, Feilding, NZ

Monsanto New Zealand Limited, Christchurch, New Zealand

A new no-tillage opener (Cross Slot™) is described. It creates a horizontal rather than vertical slot, returns residues over the slot, separates seed and fertilizer in the slot with a single opener and has exceptional seeding depth control resulting in reliable seedling emergence and superior crop yields. All design criteria were dictated by extensive peer-reviewed biological research, which also formed the basis of the technology transfer phase. Graduate students formed the basis of a knowledge team, which in conjunction with on-farm demonstrations, a comprehensive book summarizing the underlying science and a field manual outlining the management issues that compliment superior technology, ensured its successful uptake.
Key words: No-tillage, opener, Cross Slot™, seed slot, technology transfer, books.

INTRODUCTION

When the modern concept of no-tillage, which was based on double or triple disc openers, was first introduced to New Zealand in the 1960's, failures were common. A 28-year program at Massey University began in 1967 seeking to isolate the causes of failure and to design improved no-tillage openers that were able to tolerate sub-optimal soil, residue and climatic conditions. It was felt that only when farmers found no-tillage to be as fail-safe as tillage, would they embrace the new technique with enthusiasm, regardless of its environmental benefits.

Biological research

Extensive biological and soil physical research preceded engineering design. This research centred on identifying and isolating the causes of seedling emergence failures and impaired crop performances under no-tillage in dry and wet soil conditions. Important direct effects were identified, attributable to the nature, amount and positioning of surface residues relative to the seed; slot shape; slot cover; slot micro-

L. García Torres et al. (eds.). Conservation Agriculture, 485–488.

environment; slot wall compaction; slot smearing; oxygen diffusion rate within the soil surrounding the slot; slot temperature; soil-seed contact; consistency of seeding depth; pressing on the soil adjacent to the seed; and the positioning of fertilizer relative to the seed. Indirect effects were also studied involving interactions with soil fauna, weed seed germination, and soil-borne diseases (Baker, Saxton and Ritchie, 1996).

The studies, which later involved cooperative work with the USDA at Washington State University, USA and the New South Wales Department of Agriculture, Australia, involved more than 30 graduate students and numerous full time and visiting staff. Highlights of these studies were improvements in seedling emergence in dry soils of up to 1,400%, and in wet soils of up to 400% as a function of slot shape (Baker, 1976,1995; Baker, Chaudhry and Springett, 1988; Choudhary and Baker, 1981); and marked crop yield responses to horizontal banding of fertilizer in the slot (Saxton and Baker 1990). An experimental horizontal (or inverted «T») shaped slot promoted consistent seedling emergence responses compared with more traditional «V» or «U» shaped slots, especially in sub-optimal conditions.

Engineering development - the Cross Slot™ opener

The key aspects of no-tillage equipment design that affect risk and biological reliability are:

Shape of the seed slot (seed micro-environment). The pore space of untilled soils at all practical moisture levels, remains at 100% relative humidity (Scotter, 1976). Tillage reduces equilibrium humidity well below this level (Baker *et al*, 1996). The varying ability of different no-tillage drill opener designs to harvest this moisture vapour profoundly affects seedling emergence. Seed slots with surface residues returned over the slot (inverted T) trap more moisture vapour than those that are open at the top and/or difficult to cover (V- or U-shaped slots). Loose soil is only partially effective as a covering medium.

Control of seeding depth. No-tillage openers must cope with greater changes in soil strength and surface smoothness than tillage openers. Openers must faithfully follow the soil surface and ignore changes in surface residues, soil type and strength while still maintaining a constant depth, especially with small seeds. The downforces from mechanical springs vary as the springs elongate or contract. Mechanisms incorporating depth/press wheels and variable hydraulic downforce systems are essential for accurate depth-control and surface following.

Residue handling and micro-management. Residues on the soil surface provide moisture retention, erosion control, weed suppression, organic matter, nutrients, and encourage earthworm numbers and activity. No-tillage drills must «micro-manage» such residues, placing them back over the seed zone for maximum benefit whilst ensuring they avoid direct contact with seed and do not block equipment.

Fertiliser placement. Separate fertiliser placement beside and/or below the seed has been shown to create superior crop yields with no-tillage compared with surface broadcasting (Baker *et al*, 1996). This contrasts with tilled soils. No-tillage experts in the USA rated separate fertiliser placement as the single most desirable function of no-tillage openers (C J Baker, *unpublished. data,* 1981).

The above factors have three things in common: (1) their effects are biological, demonstrating that the design of no-tillage equipment should be driven mostly by biological parameters (factors influencing plant establishment and growth) rather than mechanical expediency. (2) the effects have all been measured scientifically, thus providing a scientific basis for comparing designs of opener and machines. (3) all are controllable using appropriate equipment designs.

The Cross Slot™ no-tillage opener is the product of research that defined the above biological parameters, and engineering design that captured the results. Its central disc and winged side blades create an inverted T-shaped slot that places seed on a horizontal shelf to one side of the vertical disc cut and fertiliser on a separate horizontal shelf on the other side. Following press/gauge wheels fold the soil and residue back over the seed zone in the same relative positions as they were prior to drilling, thus creating an ideal, and controlled, microenvironment. The press/gauge wheels also control seeding depth close to the point of seed release and, together with individual nitrogen-cushioned hydraulic rams on each opener, accommodate ground surface irregularities up to 50 cm with minimal change in down-force. An electronic downforce monitor automatically alters the downforce on-the-move in response to changes in soil strength across individual fields at drilling/ planting speeds up to 16 km/hr.

Technology transfer

An on-farm field-scale no-tillage programme was established to test developments with equipment (Ritchie and Baker, 1987). The programme was initiated to give researchers access to the wide range of crops, climates, soil types, topographies and management systems that characterise New Zealand agriculture. A self-contained unit comprising tractor, truck and no-tillage drill together with an expert operator undertook contract drilling for farmers throughout New Zealand on a partially self-funded basis. As the reliability of the equipment improved the programme developed additional functions. Technical and management problems were identified and expertise from other disciplines introduced. An overall consultancy package was developed to provide as much support as possible to farmers wishing to introduce no-tillage at an advanced level as a management tool on their properties.

When the drilling technology became commercially available, the need for the single-machine demonstration programme ceased but the support package was up-scaled. Although commercial hardware had become available to farmers and contractors, knowledge of the management of no-tillage systems was still limited. The Massey University post-graduate students associated with the no-tillage project throughout its 34 years (including the period after the project left the university) became powerful tools of technology transfer. They not only produced refereed scientific papers, conference lectures and popular press articles, their subsequent infiltration into professional employment within the wider agricultural sector served to raise the level of knowledge within the industry. Until then, general agricultural advisors in New Zealand were often poorly skilled in no-tillage matters and some provided a direct barrier to its uptake.

Coincidental with the release of Cross Slot™ no-tillage equipment onto the Pacific and North American markets in 1995 (Baker, Collins and Choudhary, 2001) two significant books were published. A comprehensive summary of the biological science underpinning the technology (*«No-tillage seeding: Science and practice»,* Baker *et al,* 1996) was published by CABI. This publication explained the science underpinning the development and evolution of the Cross Slot™ opener, while drawing attention to the agronomic and engineering factors responsible for impaired crop performance. Readership has included agricultural consultants, scientists and farmers and the text has been sold worldwide.

Shortly after, an on-farm management manual (*«Successful no-tillage in crop and pasture establishment»*, Ritchie *et al, 2000*) was published by Monsanto New Zealand. This provided a checklist of the input factors necessary for successful no-tillage. The text did not revisit the earlier scientific reports but drew on this fundamental information to relay advice on such topics as field selection and preparation, weed and insect control, and drill/planter selection, calibration and operation. Even no-tillage practised with less-than-advanced equipment has benefited from the guidelines outlined in Ritchie *et al,* (2000).

REFERENCES

Baker, C.J. (1976) Experiments relating to the techniques of direct drilling of seeds into undisturbed sprayed pasture. *Journal of Agricultural Engineering Research,* **21 (2)**: 133-145.

Baker, C.J. (1995) Is hi-tech no-tillage inevitable? *Proceedings National No-Tillage Conference, Indianapolis, IN, USA.* pp. 154-164.

Baker, C.J. and Afzal, M.A. (1986) Dry fertilizer placement in conservation tillage: seed damage in direct drilling. *Soil and Tillage Research.* **7**: 241-250.

Baker, C.J., Chaudhry, A.D. and Springett, J. (1988) Barley seedling establishment by direct drilling in a wet soil: Comparison of six sowing techniques. *Soil and Tillage Research.* **11**: 167-181.

Baker, C.J., Saxton, K.E. and Ritchie, W.R. (1996) *No-Tillage Seeding: Science and Practice.* 258 pages. Published by CABI, Oxford, England.

Baker, C.J., Collins, R.M. and Choudhary, M.A. (2000) Factors affecting the uptake of no -tillage in Australia, Asia and New Zealand. *Proceedings 1 World Congress on Conservation Agriculture, Madrid, Spain.* (in press).

Choudhary, M.A. and Baker, C.J. (1981) Physical effects of direct drilling equipment on undisturbed soils: III Wheat seedling performance and in-groove micro-environment in a dry soil. *New Zealand Journal of Agricultural Science,* **24**: 189-195.

Ritchie, W.R. and Baker, C.J. (1987) A field programme for evaluating direct drilling technology. *Proceedings Agronomy Society of New Zealand.* **17**: 77-82

Ritchie, W.R., Baker, C.J. and Hamilton-Manns, M. (2000) *Successful No-tillage in Crop and Pasture Establishment.* 102 pages. Published by Monsanto New Zealand Ltd.

Saxton, K.E. and Baker, C.J. (1990) The Cross Slot drill opener for conservation tillage.*Proceedings of the Great Plains Conservation Tillage Symposium, Bismarck, North Dakota, USA.* pp. 65-72.

Scotter, D.R. (1976) Liquid and vapour phase transport in soils. *Australian Journal of Soil Research.* **14**: 33-41.

C.A. MEISNER*, M. BODRUZZAMAN**, M.R. AMIN**, E. BAKSH**, A.B.S. HOSSAIN**, M. AHMED** & A. SADAT**

CONSERVATION TILLAGE OPTIONS FOR THE POOR, SMALL LANDHOLDERS IN SOUTH ASIA

** CIMMYT Bangladesh PO Box 6057 Gulshan Dhaka-1212 Bangladesh*

*** Wheat Research Centre, Bangladesh Agriculture Research Institute, Bangladesh*

Rice-wheat is the major cereal cropping system in Bangladesh and the eastern Gangetic Plains of India; however, due to the lateness of the monsoon rice harvest and the traditional tillage operations, sowing of wheat is often quite late, reducing the potential for wheat yields by 1.3% per day sown late. Many alternative tillage options have been tested in Bangladesh by participatory methods with growers, agricultural engineers, economists and agronomists. These include:

- Power operated tiller cum seeder behind the Bangladesh Hand Tractor (BHT): There are more than 600,000 BHTs in Bangladesh, and the use of this accessory that prepares a shallow but effective seedbed (~5 cm deep) and sows the wheat in a line and presses the seed in the soil in one pass is very effective.
- Surface Seeding
- Minimum tillage

All methodologies, experiments, and research findings with economic analyses will be discussed in detail.
Key words: Tillage, wheat, yield, machinery

INTRODUCTION

South Asia historically had depended upon its monsoon rice crop to feed its populations. However, as its populations have exponentially expanded, the need to use the dry season winter months for cropping has grown. While many parts of South Asia lack irrigation for the dry winter months, those having irrigation facilities are exercising options for diverse cropping patterns. These literally beginning just over one decade ago, growers are learning to shift cropping emphasis from the monsoon rice to the winter. Thus, tillage practice, fertilizer management and other agronomic practices for non-rice crops are being learned as well as adapted by the growers and researchers.

L. García Torres et al. (eds.). Conservation Agriculture, 489–494.

Traditionally, just 10 years ago, the tillage practice for wheat showed that there were over 6 plowings with the country plow behind the bullock and over 12 ladderings (plankings to level the soil) (Saunders, 1990 &1992, Meisner, 1992, 1996, 1999, & 2001). Rice stubble was excised and thrown into a pile at the side of the fields. Wheat researchers had demonstrated through well-placed minimum tillage demonstrations (over 1,500 since 1990) to growers the benefits of practicing minimum tillage. Currently, over 70% of the wheat fields are using less tillage than a decade ago. Rice stubble is being left in the fields and incorporated in the few tillage practices used.

Continuing to gain from conservation tillage practices, wheat researchers studied zero tillage wheat after rice in southwestern Bangladesh. Wheat is sown in existing monsoon rice 2 weeks before rice harvest. However, the soil must be saturated at the time of sowing for this 'relay' sowing to work. Since rice is hand harvested, feet press any seed on the surface not germinated into the moist soil. Growers adapted this practice and so far research has shown this to be a very economical practice with very little cropping system demerits. Only foliar diseases in wheat appear to be greater with this practice. There are no increases in insects or harmful pests in this cropping system after years of monitoring and data collection.

Bangladesh, having over 600,000 of these hand tractors imported from China, could benefit from the use of the accessories. Thus, a participatory stakeholder partnership was formed between the scientists of CIMMYT, the Bangladesh Agriculture Research Institute, the Bangladesh Rice Research Institute, the Wheat Research Centre, private manufacturers, NGOs, and other interested partners. We have been meeting as a group for over 3 years now at least twice each year. Our goal as a group is to introduce these power operated tiller/seeders to the growers in participatory ways so that all parties learn the benefits and problems of the use of such equipment under the Bangladesh context. This paper will describe the methodologies used, measurable benefits, and status of the ongoing work. Plans for up-scaling this to the Eastern Gangetic Plains will be detailed.

MATERIALS AND METHODS

Power operated tiller cum seeder behind the Bangladesh Hand Tractor (BHT)

Introduction of new technologies including conservation tillage concepts, has been difficult for the public sector often due to lack of resources to partner with the grower community. CIMMYT, in its catalytic working role in South Asia, invited partners to join a "Participatory Stakeholders Group' in Bangladesh for the purpose of introducing the power operated tiller/seeder to the grower community to supplement the growers' expanded use of the hand tractor into other resource conserving practices. The partners in the group included agricultural engineers and agronomists from the various research institutes, private manufacturers interested in manufacturing the equipment, growers, hand tractor operators, NGOs and other interested parties. We have been meeting together twice yearly since 1997 when only 10 seeders were introduced to various NGOs and public sector research institutes. Currently there are over 50 seeders.

Through these meetings, the equipment's benefits, problems, and suggested changes are discussed freely among the group. The physical setting of the meeting is very important. Everyone sits in a rough oval around a low-stature table that allows everyone to be seen and heard. Culturally, this is quite radically different to the accepted norm since usually the scientists sit up front with the growers usually relegated to the rear. Traditionally, speakers sat behind tables in the front with their microphones insulated themselves from the other partners. CIMMYT acts as the moderator for the group, only to ensure that everyone has a chance to speak and accurate notes taken. Anyone who remains quiet during the meeting is involved in the discussions by the moderator. Bangla, the language of Bangladesh, is the accepted and common language for the group.

Meeting notes are taken and distributed among the group. Training opportunities are provided to the stakeholders yearly, sponsored by public and private sector. The agriculture engineers have written manuals for the operation of the equipment in the local language.

Minimum Tillage

For the introduction of the concept of minimum tillage to the growers, use of well-placed demonstrations throughout Bangladesh over the years coupled with training and growers' rallies has shown that growers readily implemented minimum tillage in areas surveyed (data not shown). The Wheat Research Centre prepares a 'kit' with the correct fertilizers, improved wheat varieties, signboards, and instructions for successfully implementing the demonstration to the sponsoring agency. Since the success of minimum tillage is less time for turnaround from rice harvest to wheat sowing, there should be a difference in the sowing dates of the plots. One plot demonstrating minimum tillage uses fertilizer applied to the soil surface with one plow in parallel lines, followed by another plowing in perpendicular lines where seed is dropped in the furrow. This is followed by 2 ladderings (plankings) taking only 2-4 days. The other plot uses conventional tillage defined by the grower him/herself and is usually 6-8 plowings followed by over 11 ladderings, taking up to 15 days time. Usually a community will have up to 5 of these demonstrations, so a small training class is conducted among the growers to ensure their cooperation and understanding of the demonstration.

Surface Seeding

Demonstrations for zero tilled wheat or referred to 'surface-seeded' wheat were placed throughout Bangladesh with other scientists, extensionists with the Department of Agriculture Extension, and NGOs. The Wheat Research Centre prepares a 'kit' with the correct fertilizers, improved wheat varieties, signboards, and instructions for successfully implementing the demonstration to the sponsoring agency. The placement of this demonstration has proved difficult due to the perfect timing of soil moisture. If the soil is not saturated, then wheat establishment is hindered. Though similar grower training is conducted, many times, the soil was not fully saturated, and the demonstration was not successful.

RESULTS AND DISCUSSION

As of October 2001, there are over 50 seeders; all but 5 were manufactured locally with the original 5 imported from China. Because of the actions of the participatory stakeholders' group, the currently manufactured seeder in Bangladesh is quite different from the original seeder imported from China three years ago. The manufacturer has made many modifications based on the operators' and other stakeholders' suggestions, coupled with the agriculture engineers who have assisted in the redesign. The distribution of the seeder, while centered in the major rice-wheat growing areas of NW Bangladesh has been widespread throughout the country. During the first year where there were only 15 seeders, the seeder sowed over 100 hectares of wheat. During the second year where there were about 25 seeders, over 150 hectares of wheat and other crops were sown. By the third year, 2000/2001 season, there were over 35 seeders and together they had sown 200 hectares of wheat, 7 hectares of pulses, 20 and hectares of potatoes. Since the past season, many public sector institutions have ordered these and are implementing their proliferation under their programs. Thus, over 50 seeders will be operated during 2001/2002 season.

However, the ultimate test of the success of this equipment will be its purchase by and proliferation among the growers. A program between multi-partners, including CIMMYT, agriculture engineers and agronomists, one NGO located in western Bangladesh, and other local agencies will begin a program of a revolving loan for growers to purchase sets of this equipment. Each set will contain a seeder, reaper, pump set for irrigation via the hand tractor engine, and hose pipe. The reaper is another Chinese-manufactured accessory to the 2-wheel hand tractor that cuts 4 rows of a cereal crop, depositing it in a single line to the side. The use of the hosepipe, curiously, is a grower-defined technology becoming widely practiced among the growers—eliminating the need for canals and increasing the effectiveness of the irrigation. This NGO had been providing loans for the 2-wheel hand tractor; however, coupled with the successful completion of a training course for the operators in the tractor's maintenance and repair. Similar training will be conducted for the seeder, reaper, and irrigation pump maintenance and repair for the revolving loan to begin during the 2001/2002 season. If successful, the model of introduction will be up-scaled to other areas of the country, partnering with other similar NGOs who are providing loans for such hand tractor purchases. Currently, the banks in Bangladesh are not providing small loans for such purchases. However, if the revolving loan program proves successful, more public and private banks will be invited into the participatory stakeholders group.

Through the demonstration kits described previously, minimum tillage is being practiced now for wheat throughout the country. Informal survey results (data not shown) indicate acceptance by growers in the wheat producing areas by over 70%.

Retaining the rice straw in the system will increase the soil moisture holding capacity, the soil fertility, and decrease the crusting of the soils by the increase in organic matter. Rice straw is valuable to the rural communities as the major cattle feed and fuel for cooking. Rarely is burning of straw in the fields practiced due to the high value of the straw. Though the demonstrations were simple, their single focus was sufficient to being widely accepted by the grower community. Economic studies on minimum tillage versus conventional tillage either by the country plow or the power tiller indicate that there was significant net profit from using minimum tillage and/or the power tiller compared to the country plow if one factored in the feed and care of the animals.

Through the demonstration kits described in the Methodologies section, surface seeding has not been able to be implemented by the growers. Informal surveys indicate that often, these demonstration kits arrive too late for the saturated soil conditions to be met, creating the demonstration's failure to illustrate the benefits. Additionally, growers are weary of sowing directly on the surface due to the amount of seed open for consumption by birds as well as the weeds that will germinate even before the wheat seed can germinate. The growers have stated that in comparison with minimum tillage and the power operated seeder, the use of surface seeding will remain a low priority. However, increasing the growers' knowledge of this sowing method will continue through use of other methods for its demonstration as it is a useful option when soil moisture conditions are such that timely sowing of wheat by any other means will be delayed.

CONCLUSION

While there are many methods to disseminate knowledge or technologies on tillage options, our paper showed that forming participatory stakeholders' groups proved useful so far in the slow but steady introduction of new accessories to the 2-wheel hand tractors currently being using in Bangladesh. Use of demonstration kits through engaged partners proved sufficient to increase awareness of minimum tillage and for the growers to accept this tillage option. However, such demonstration kits were not successful in increasing awareness of surface seeding, probably due to the more complex nature of the option and the growers' requirement for some minimum tillage for control of weeds. Yet, during years of excess rainfall at rice harvest, knowledge of surface seeding may mean that growers can sow wheat still timely, ensuring their yield to be high.

Up-scaling plans are being implemented for using the participatory stakeholder concept for other areas in the Eastern Indo-Gangetic areas of India and Nepal. While the proliferation of 2-wheel hand tractors has not been as successful, lessons learned from Bangladesh can be introduced in those areas.

REFERENCES

Meisner, C.A. (1992) Report of an on-farm survey of the Mymensingh region and Tangail: Wheat growers' practices, perceptions, and their implications. *Monograph* No. **9**, Bangladesh Agriculture Research Institute. Wheat Research Centre, Nashipur Dinajpur. September 1992.

Meisner, C.A. (1996) Report of an on-farm survey of the Greater Comilla Region: Wheat growers' practices, perceptions, and their implications. *Monograph* No. **13**, Bangladesh Agriculture Research Institute. Wheat Research Centre, Nashipur Dinajpur. June 1996.

Meisner, C.A. (1999) Report of an on-farm survey of the Greater Faridpur Region: Wheat growers' practices, perceptions, and their implications. *Monograph* No. **14**, Bangladesh Agriculture Research Institute. Wheat Research Centre, Nashipur Dinajpur. June 1999.

Meisner, C.A. (2001) Report of an on-farm survey of the Greater Sylhet Region: Wheat growers' practices, perceptions, and their implications. *Monograph* No. **16**, Bangladesh Agriculture Research Institute. Wheat Research Centre, Nashipur Dinajpur. July 2001.

Saunders, D. (1990) Report of an on-farm survey of the Dinajpur District: Farmers' practices, and problems, and their implications. *Monograph* No. **6**, Bangladesh Agriculture Research Institute. Wheat Research Centre, Nashipur Dinajpur. June 1990.

Saunders, D. (1991) Report of an on-farm survey of the Jessore and Kushtia Districts: Wheat farmers' practices, and their perceptions. *Monograph* No. **8**, Bangladesh Agriculture Research Institute. Wheat Research Centre, Nashipur Dinajpur. June 1990.

T. GREGOIRE & S.H. CLANCY

AN INNOVATIVE APPROACH TO DEVELOP CONSERVATION AGRICULTURE PROGRAMS AND FUNDING

North Dakota State University Extension Service, P.O. Box 477, Devils Lake, North Dakota. 58301 USA

Conservation Agriculture, Natural Resources Trust, P.O. Box 31, Devils Lake, North Dakota. 58301 USA

The premise for the Conservation Agriculture farm demonstration is to develop actions and programs to increase profitability of agriculture by farming the best soils and establish alternatives on the rest. Communications among conservationists and farmers have been made more effective by breaching comfort zones through working in a team setting. The demonstration project improves communication on three levels: the farm family and their team, the thirteen-member advisory board of eight farmers and five agency representatives, and the partners which include non-profit organizations, federal, and state agencies. The resource analysis team, an interdisciplinary technical team, works directly with the farm family to develop a whole farm plan which addresses the trilogy of sustainability: ecology, sociology and economics. The Advisory Board oversees the Conservation Agriculture farm demonstration project, develops innovative programs, and finds funding to facilitate the whole farm plan implementation. Partners fund programs, offer advise on the process, discuss, revise, and forward the best agricultural policy ideas for debate to the state and federal levels.
Key Words: whole farm planning, resource planning teams, alternative land-use, sustainable agriculture.

INTRODUCTION

The Drift Prairie Region of the Red River Basin has been in a climatic wet cycle since the summer of 1993. The Red River Basin suffered monumental losses from flooding in 1997 and Devils Lake, a closed basin located in this region, has risen 25 feet and expanded from 43,000 surface acres in size to 122,000 surface acres over the past 10 years causing major economic damages. The area has had nine Presidential Disaster Declarations in the past eight years. Many organizations and agencies in the Red River Basin have been working to address the future management of water, the reduction of flooding in its watershed, and to revitalize and stabilize the area's economy (Hollevoet, 1999). During this time agricultural commodity prices have plummeted

L. García Torres et al. (eds.). Conservation Agriculture, 495–500.

and costs have continued to increase causing many producers to consider new uses for land with poor productivity.

All private farm landowners are dealing with farm income issues, salinization of soils, wetland designations and regulations, drainage issues, and private property rights (Hollevoet, 1999). Many landowners have an animosity toward agencies and environmental groups that sponsor conservation of natural resources, especially those issues dealing with wetlands. They have many concerns or conflicts regarding water storage to reduce flooding. Communication between farmers and conservation groups is often strained. There is a historical inclination among agency staff to be most comfortable talking to conservationists within their own discipline (Darling, 1941) such as bird specialists talking only to bird conservationists or soil technicians talking to those interested in land issues, etc. These conservation agencies offer a whole host of conservation incentives (Gregoire, 2001) that many farmers find out about after the funds are exhausted or the agency has withdrawn the offer.

The concept for Conservation Agriculture was born out of the numerous meetings among diverse stakeholders drawn together to discuss water issues (Hollevoet, 1999). After listening to the farmers attending these meetings Roger Hollevoet (District Director, U.S. FWS Devils Lake Wetland Management District) and Ray Horne (farmer and Board of Director for the Natural Resources Trust) articulated the concept for Conservation Agriculture. The premise is to develop actions and programs to increase profitability of agriculture by farming the best soils and putting alternatives on the rest. This strategy would lower input costs, make farming more efficient and lower the use of fuels, herbicides, and fertilizers. Alternatives on the landscape would provide a host of societal functions within the watershed.

Hollevoet and Horne presented this concept to the Natural Resources Trust, a private non profit agency, (formerly North Dakota Wetlands Trust) in June of 1999. The Trust board of directors responded by initiating the five year conservation agriculture demonstration program and dedicating $100,000 in funding for each year with a matching fund stipulation.

METHODS

Several techniques and objectives have been established to move the project to a successful conclusion.

1.Farmers and Resource team selected year 2000; team meets with farm family two times per year through year 2004. The project has engaged four Drift Prairie farmers and their farm operations in a five year demonstration project to improve natural resource protection while maintaining net profits. Each farm works under a whole farm plan developed by the producer and a six member resource team comprised of agriculture and conservation professionals. Prime soils are treated with best management practices while non-prime soils are set-aside to an alternative use such as hay-land and or wildlife habitat.

2. *Farm families enrolled in five year programs 2001; practices implemented years 2001 to 2004.* The project draws upon conservation and alternative income programs

offered by the partners. The advisory board for Conservation Agriculture also developed land treatment programs to respond to the needs identified by the farmers in a manner that fits the mission of the project. Dynamics of the board ensures a landowner majority voice throughout the life of the project.

3. *Farm families enrolled in Adult Farm Management classes years 2000 to 2004.* To conduct the farm analysis and determine the change in farm inputs and efficiency, each demonstration farmer has been enrolled in an Adult Farm Management Class. Through this program, an economic database was established and assistance with income and expense projections by enterprise were realized. An economist assembled an annual report including the year-end analysis, net income changes, and conservation investments.

4. *Baseline land-use data entered year 2000; updated years 2001 to 2005.* Baseline land-use data for each farm and sub-watershed was established and placed in a Geographic Information System (GIS). The GIS monitors annual change by acre across each year defining acres remaining in farmland and acres by type of alternatives uses.

Baseline inventories of wildlife response to land-use changes provide the benchmark to assess the success of the program relative to wildlife. Baseline counts were conducted to determine the avian species presence and density by primary habitat types. Two additional counts will be conducted to monitor the trends of these species to demonstrate changes in species richness and diversity from restored habitats and by restored-diversified habitat types (LSP 1998, Meeks 2001).

Baseline soil measurements for all prime and non prime soils using a minimum data set will be used to screen the condition, quality, and health of soil (Land Stewardship Project 'LSP' 1998) for all prime and non prime soils. Since soil quality factors change slowly, a minimum data set will be measured again at the end of the project.

Baseline plant and invertebrate densities and diversity are good indicators of wetland health. Zones in wetlands (both existing and restored) which contain suitable habitat were examined to determine plant/invertebrate composition and diversity. Because these densities are known to fluctuate across space and time due in part to changing water levels and the amount of organic litter, the habitat zones will be inventoried throughout the study period (Meyer 2001).

5. *Educational outreach provided through annual watershed workshops; begin autumn year 2000 and each year through 2004.* The workshops are intended to develop an understanding of watershed elements and to develop a better understanding of the relationship between land-use, watershed management, landscape function and societal benefits among stakeholders. Each demonstration farm was identified on a map within a watershed and current land-use was illustrated and will be monitored through the five year project term.

RESULTS

Accomplishments the first year of the five year Conservation Agriculture demonstration project include naming the thirteen member advisory board, and engaging 25 partners. Demonstration farmers were chosen based on landscape, watershed, enterprise diversity, landowner attitude, typical North Dakota farms, conservation-wetland management potential, and accessibility for farm tours. Resource teams were chosen to represent the following disciplines: adult farm management specialist, agronomist, soils, wildlife biologist, conservation planner, and quality of life specialist. A coordinator was selected by the Conservation Agriculture advisory board.

Farmers, it was determined, wanted a broad array of programs that were accessible, voluntary, and flexible. The advisory board responded with twelve new program ideas. In developing contracts with the farmers, the board found that a paradigm shift in program targets was required. The shift was away from landowners who were inclined to use conservation programs as a personal retirement program to renters/owners who would grow grass and provide landscape functions for the benefit of society within their whole farm plan. Societal benefits identified by the advisory board were improved water quality, reduced erosion and siltation, increased plant, mammal, avian and invertebrate diversity, sequestered carbon, recharged ground water, improved flood storage function, and increased recreational opportunities.

Initial work with the demonstration farmers has resulted in 181 acres of wetlands and 547 acres of uplands enhanced, 100 acres of cropland dedicated to pasture and 626 acres of pastures engaged in rotational grazing for a total program cost of $190,000 U.S. dollars over five years. To encourage more conservation practices the advisory board has drafted a sliding scale of incentive payments for conservation tillage starting at 30% residue remaining after seeding to 60% residue remaining after tillage with no acreage limitations. At the conclusion of the project demonstrations, the four farmers are projected to have 2000 acres of wetlands-streams enhanced, 3500 acres enrolled in conservation tillage, 2500 acres enrolled in alternative use, and 350 acres of prime farmland set aside for efficiency of operations.

There have been three partner meetings since the onset of Conservation Agriculture demonstration project. The U.S. Congress is working on a new farm bill and one North Dakota delegation aid has hand carried a draft of Conservation Agriculture policy recommendations to Washington.

DISCUSSION

The initiation of the Conservation Agriculture project seems is timely. Farmers who have been reluctant to work with conservation agencies on land management problems have been willing to approach Conservation Agriculture, though tentatively. The seriousness of the current farm crisis may be attributed as the motivating factor. Agencies, who understand the crisis and wanting to be helpful to farmers but lacked opportunity, have found a useful vehicle in Conservation Agriculture. In the process they have been intrigued to hear the technical exchange around the farm family table and have come to a new appreciation for other resource persons. Societal benefits have been defined for policy makers who need to gather non farm elected representatives to support farm policy. Conservation may become the «safety net» for farmers rather than commodity price supports.

Whole farm planning has been the centerpiece for sustainable agriculture organizations such as Holistic Resource Management®, the Land Stewardship Project,

Sustainable Farming Associations, and the states that have Institutes for Sustainable Agriculture. This technique has enabled the resource team to address rather difficult conservation issues in a less personal more productive manner.

REFERENCES

Darling, J.N. (1941) The need of balance. The Land Vol. 1(1) reprinted in The Land Report, A publication of the Land Institute. The Land Institute, 2440 E. Water Well Rd., Salina, KS 67401

Gregoire, T. (2001) Income alternatives for farmers and ranchers. NDSU Ext. Pub. WL-942(revised). North Dakota State University, Fargo, North Dakota 58105. 7 pages. March

Hollevoet, R. (1999) A comprehensive and Multipurpose approach to watershed management-a concept. *Proceedings of the North Dakota Academy of Science.* Vol **53**:29-32

Land Stewardship Project. (1998) The monitoring tool box. Land Stewardship Project, 2200 Fourth Street, White Bear Lake, MD 55110 USA

Meyer, G.W. (2001) Biological implications of surrounding land use on wetland health: a conservation agriculture perspective. A Proposal to the Wetlands Trust. Dept. of Biology, University of North Dakota, Grand Forks, ND 58202-9019. pages 40

Meeks, W.A. (2001) Breeding bird surveys and baseline wildlife inventories on lands participating in the conservation agriculture demonstration project within the Drift Prairie Region of North Dakota. U.S. Fish and Wildlife Service, Devils Lake WMD, P.O. Box 908, Devils Lake, ND 58301. April

J. EKBOIR

ADOPTION OF NO-TILL BY SMALL FARMERS: UNDERSTANDING THE GENERATION OF COMPLEX TECHNOLOGIES

CIMMYT, Apdo. Postal 6-641, 06600 Mexico, D.F., Mexico, e-mail: jekboir@cgiar.org

INTRODUCTION

Zero tillage (ZT) is a complex process technology that has had major positive impacts on poor farmers in many parts of the world. ZT reversed soil degradation, boosted farmers' profitability, improved life quality and increased the sustainability of agricultural systems.

A comparison of different ZT experiences can provide important insights for the design of research and extension policies for small farmers. Around the world ZT evolved through the interactions of a large number of agents. Many of these interactions were planned while others happened by chance; that is, the evolution of ZT was a random process in which many technological alternatives were tried and discarded. These alternatives were developed simultaneously by different networks of agents (e.g., researchers, input suppliers and farmers) around the world. Accessing information developed in other regions has been one of the most important tools of successful ZT programs; at the same time, it has been one of the most difficult to use because funding for this type of activities has been hard to obtain. Small farmers and manufacturers of small machinery usually lack the resources to search for or generate appropriate technologies. Extension programs aimed at these agents should minimize the costs of access to information and the technology. This requires, for example, a larger number of demonstration plots compared to a program for commercial farmers and the financing of visits to other countries by key farmers and manufacturers.

ZT for small farmers was developed through a system of participatory innovation, which emerged as a consequence of the weakness of public research and extension systems, the commercial interests of input suppliers, the community commitment of some farmers and the activities of international research institutions and cooperation agencies. In section 2, I

L. García Torres et al. (eds.). Conservation Agriculture, 501– 508.

review the concept of a *National Innovation System* and sketch an alternative framework for the analysis of research systems. In section 3, I present several features that define the economic properties of ZT. In section 4 I discuss three successful experiences of ZT for small farmers while in section 5 I analyze the particular problems of developing ZT packages for small farmers.

INNOVATION SYSTEMS AND TECHNICAL CHANGE

An innovation is defined as anything new introduced into a production or social process (OECD). Innovations, and consequently economic growth, depend on the ability of many social agents to form networks to engage in collective learning. These networks co-evolve with the technologies they help to develop (Rycroft and Kash). The transformation of knowledge flows into innovations depends on the singular characteristics of knowledge, formal and informal regulations (including laws), and each process's history. The agents involved in the innovation process, their actions and interactions, and the formal and informal rules that regulate this system constitute the National Innovation System - NIS (OECD).

In this framework, technical change has a new meaning. The traditional analysis of technical change is based on a linear vision of science: knowledge starts with basic science, continues with applied research, technology development, and ends with adoption. However, technological developments in general have preceded the scientific understanding of the underlying phenomena (e.g., thermodynamics and the steam engine), or have occurred in production lines by the transformation of known processes. In the systemic vision of the NIS, innovations result from complex interactions among agents, including several feedback loops (Rycroft and Kash). More than a linear process, the NIS resembles a spider web.

The NIS is larger than the national research system (NRS). A NIS can be strong even when the NRS is weak. This happens when public investments in research are small, but agents seek new technologies and adapt them to their local environment.

ECONOMIC ISSUES IN THE ANALYSIS OF ZERO TILLAGE

ZT with burning of residues has been practiced by small farmers in developing countries for centuries. This practice has usually been associated with shifting cultivation. Increasing population pressure is forcing a more intensive use of natural resources, resulting in severe degradation. ZT with mulch has solved these problems in those regions where a suitable package was developed.

Even though the name ZT refers only to one practice, it really is a farm management system that includes all agricultural practices (i.e., planting, residue management, weed and pest controls, and rotations). For this reason, the system has to be adapted to local conditions. The most complex ZT package, as the one used in South America, is a combination of special machinery (planters and sprayers), agrochemicals (particularly herbicides) and knowledge. Due to the traditional practices used by small farmers in some countries, some elements were not required for the initial adoption. For example, in Ghana farmers usually plant with a cutlass; consequently, the ZT package developed is basically a weed management system based on herbicides but does not use planters. In the irrigated areas of the Indo-Gangetic plains weeds are controlled mainly by water management and agricultural practices, consequently, herbicides were not used in ZT as intensively as in South America, but development of the package depended on the design of adequate drills.

The many benefits of ZT for commercial farmers have been well documented (Ekboir and Parellada). For small farmers, the main benefit is the reduction in labor requirements per unit of output thus freeing time for other income-generating activities while still covering the basic food needs. Other benefits are a reduction in the effort demanded by agricultural tasks, cost reductions (in spite of the need to buy more inputs) and the possibility of planting closer to the optimal date.

The main restrictions to the adoption of ZT by small farmers are 1) the move from conventional till to ZT involves learning the dynamics of a system out of equilibrium, which usually takes a long time to reach a steady state,[1] some farmers never succeed in mastering the package and revert to conventional tillage, 2) small farmers have different needs than commercial farmers, consequently they usually cannot use the packages developed for the latter, and 3) ZT requires substantial adaptation efforts to local conditions.

Research on ZT for small farmers is characterized as follows:

- ZT is a complex package of commercial inputs and knowledge. Complex technologies have the characteristic that they cannot be developed by a single agent but require a network that shares complementary assets. Participation in the network is defined by both formal and informal arrangements .
- The networks are always initiated and held together by one or a few agents who play a catalytic role. After the network becomes established, sub-networks may emerge and many agents may play catalytic roles. Every single agent probably becomes less important.
- Small farmers lack the resources to search for or develop their own technologies. Because they do not participate in markets as intensively as commercial farmers, private firms usually do not have the incentive to develop technologies nor to invest in technology dissemination for them. Public resources are required to set up these programs; in some cases, though, private agents have teamed up with public researchers to establish ZT research and extension programs for small farmers.

THREE ZT EXPERIENCES FOR SMALL FARMERS

In this paper I analyze three experiences of development of ZT packages for small farmers: Brazil, Ghana and the Indo-Gangetic Plains. Even though there are on-going programs in other countries, they are either too new to show an impact or the information available is too fragmented.

Brazil

Here I review two ZT programs aimed at small farmers: the Foundation ABC and the project METAS. Other programs were established later by cooperatives and the State of Parana.

The first efforts to develop a package for small farmers started in the late 1970s. After several experiences, two farmers from the state of Parana traveled in 1979 to the U.S. to gather information on ZT. Both were active in their local cooperatives and had very strong community commitments. Upon their return they invited other farmers to form an informal group to discuss ZT issues. Eventually this group evolved into the ABC Foundation, a research and extension institution funded by three cooperatives. Because many of the associates of these cooperatives were small farmers, the Foundation established an extension program for them. The Foundation played the catalytic role; other members of the network were ICI, public research institutions, foreign universities and machinery manufacturers.

In 1990, a researcher from EMBRAPA's Wheat Center and a Monsanto technician conducted a study to identify the causes of low adoption of ZT among small farmers in the state of Rio Grande Do Sul. They identified 3 factors: the package was ill-adapted to local conditions, adequate planters for small farmers had not been developed and extension agents did not command the package. Following this diagnosis, Monsanto promoted in1993, the project METAS which involved five public and private institutions.[2]

It was estimated that in 1999 almost 100% of grain production in southern Brazil was produced with ZT.

Ghana

Continuous research on ZT in Ghana started in the 1990s with a joint project between the Crops Research Institute (CRI) in Kumasi and the Ghana Grains Development Project (GGDP). The GGDP was a collaborative effort that involved the CRI, the Grains and Legumes Development Board, the Ministry of Food and Agriculture, CIMMYT and IITA with funding from the Government of Ghana and the Canadian International Development Agency. Roberto Soza, a CIMMYT agronomist working with the GGDP, organized ZT research in the region from 1990 until 1996. In 1991, the GGDP adopted ZT for planting maize and grain legumes in 5 research stations. The system was also tested extensively in farmers fields across the country. In 1993, Sasakawa Global 2000 (SG 2000) and Monsanto joined the GGDP and the public extension service to promote ZT. The program received strong political support from the government (Findlay and Hutchinson, 1999).

Monsanto also assisted the CRI to evaluate glyphosate on farmers fields with maize and beans. Trial protocols and guidelines were established by Monsanto and discussed with CRI researchers, who implemented the trials with support from SG 2000. The results obtained by the researchers were used to train the extension service field officers who carried out their own series of demonstrations in farmers plots.

For all practical purposes, only one CRI researcher has worked intensively on ZT in the 1990s. Today, extension agents and researchers at CRI work closely together with a participatory approach, to the point that farmers cannot distinguish the activities of each one. Some Rural Banks and District assemblies have also joined in the promotion of ZT by providing credit to selected farmers.

A second program is currently administered by the University of Development Studies at Tamale (Northern Ghana) and is implemented by the extension services supported by SG 2000 and Dizengoff Ghana, Ltd., a local distributor of fertilizers and pesticides (Findlay and Hutchinson, 1999).

The emphasis in Ghana was on weed management; this differed from other countries where a substantial effort was also put in the development of adequate machinery (i.e., planters and mechanized sprayers). The emphasis on weed management was the consequence of the weakness of the Ghanaian research effort and the fact that most farmers in the humid zones do not use planters. The lack of research on machinery prevented the development of an adequate package for large mechanized farmers.

It was estimated that in 2000 ZT was used by 100,000 small farmers on 45,000 ha.

The Indo-Gangetic Plains

In 1983, CIMMYT and the Pakistani national wheat program started a ZT research program with funding from USAID. CIMMYT imported a drill from New Zealand and made it available to local manufacturers; several demonstrations in farmers fields were made with these drills in the next years. By the time CIMMYT closed its office in Pakistan in 1988, about 30 drills have been made. Adoption was scant because ZT was strongly opposed by the research and extension authorities.

In 1988 CIMMYT started a similar program in India. Researchers in the Universities of Pant Nagar and Haryana joined the national wheat program in the ZT network. Engineers at Pant Nagar University modified the old rabi wheat seed drill by replacing the old seed coulters with the inverted T-openers tested in Pakistan. A manufacturer in Punjab became interested in the drills and started to produce them. The technology was immediately tested by farmers.

In 1994 the Director General of On Farm Water Management in Punjab, Pakistan, realized that ZT could save substantial amounts of irrigation water. He collected the decade old drills and started a large program of on-farm demonstrations. In 1995, this Director General visited India and saw the advances that had been made to the equipment; immediately a small exchange program of ZT researchers from both countries was started. The Rice-Wheat Consortium imported an Indian drill into Pakistan, which was subsequently modified to develop a local prototype.

In 2000 CIMMYT imported from its program in Bolivia an animal-pulled drill that could plant wheat through loose residues. The drill had to be modified because as agriculture in Bolivia is rain fed, there are less residues than in the irrigated areas of India and Pakistan. The original drill was improved by a local Indian manufacturer.

The wheat area planted with ZT in the rice-wheat region of the Indo-Gangetic Plains in 2000 was estimated at 100,000 ha; however, the technology for ZT rice has not yet been developed.

PROBLEMS IN DEVELOPING ZT PACKAGES FOR SMALL FARMERS

Market failures are the traditional justification for public involvement in research. The recent literature on innovations highlights a new area for public policy: system failures that arise from lack of interaction among agents (OECD). In the case of ZT for small farmers, these include:

a. Public research institutions were (and most still are) based on a linear concept of science. It is common for public institutions to plan their activities with little (and in many cases, ineffective) interaction with technology users. This lack of interaction is also evident in most national funding agencies. Public agents hesitate to recommend technologies that had not been developed by formal research procedures. Finally, the structure of incentives in most public research institutions do not favor innovation, and many employees try to avoid the problems of changing established working and research programs.
b. Public research institutions are going through a transformation process, but the objectives and instruments often conflict. The main features of these changes are 1) new priority-setting mechanisms (usually relying on more formal procedures); 2) emphasis on diversifying the sources of funding, which, in all cases include a substantial reduction in direct budgetary allocations; 3) greater pressure to generate resources through the sale of goods and services; and 4) reductions in the staff of researchers and support personnel. The pressure on public research institutions to generate their own resources has forced them to concentrate on the production of goods with market value –reducing the production of «public goods»– or in research that responds to political needs with very short-term objectives.
c. Insufficient interaction among agents in the innovation network. Every agent in the network develops specific areas of expertise. This is especially true for individual researchers in public institutions The efficiency of the whole NIS can be strengthened through greater interaction among the best teams. But these interactions are scarce due to a lack of a culture of collaboration and insufficient funds.
d. Micromanagement of research programs. Until recently, ZT has not been a priority for public research institutions. At the beginning, individual researches set ZT research projects opposed by the majority of their peers and authorities. As the trend to centralize priority setting strengthen, it becomes increasingly difficult for individual researchers to develop lines of research that are not officially recognized.
e. Manufacturers of planting equipment for small farmers have difficulties in accessing information. The planting equipment for small farmers is usually

manufactured by local blacksmiths or small shops. Because of their small scale, small scale manufacturers lack the resources to search for what has been developed for environments similar to those they work in. National and international research institutions have been instrumental in transferring prototypes across locations.

f. There is adverse selection among manufacturers for small farmers. When manufacturers for small farmers reach a sufficiently large scale, they target large farmers since they are a more profitable market. This means that only the less successful manufacturers or those limited by the market size remain in the small farmer market.

g. Small farmers lack resources to search for information and for developing their own technologies. Searching for information requires not only money but human capital to identify useful sources among an increasing number of alternatives. Commercial farmers have formed associations to take advantage of economies of scale in the generation of information. Small farmers have been able to do it only in a few cases where they had strong community ties (e.g., Brazil). In most cases, however, they were able to access information only when they joined research efforts organized by other agents.

CONCLUSIONS

The generation and adoption of ZT occurs in a complex innovation network. The system's efficiency is reduced by market failures that characterize the production of knowledge, and system failures that arise from the lack of interactions among agents (OECD).

The system failures are generally more important for small farmers than for commercial ones, because the latter have the human and financial resources to act as catalysts of the innovation network. Small farmers and small manufacturers of equipment, on the other hand, generally lack these resources to search for or produce information. Because of this weakness, other agents have to play the catalytic role of organizing the innovation network.

In this paper I reviewed three successful ZT programs aimed at small farmers. In the three cases, different agents organized the network of researchers, extension agents and private firms; the identities of the catalytic agents usually changed over time. In Brazil this role was initially played by two committed members of local cooperatives, or a public researcher together with a Monsanto employee. Later several programs were organized by farmers, cooperatives or state governments. In Ghana, a CIMMYT researcher was the first organizer of the network; this role was then taken by Monsanto, SG 2000 and a public researcher. In the Indo-Gangetic Plains the initial organizational role was played by a CIMMYT researcher and a few researchers of the Pakistani wheat program. Later, researchers at public universities (in India) or public water management officers (in Pakistan) played a major organizational role in the network.

In all three cases the key to the success of the network was the major effort of committed individuals. They gathered the financial and human resources required for developing the technical package. In some cases, these individuals worked within institutions that recognized the importance of participatory research programs; in other cases, these individuals had to work outside their institutions.

The failure to recognize technological and market opportunities was not a characteristic only of the public sector. Often private companies failed to see the market potential of small farmers while researchers from public or international institutions were key players in the dissemination of ZT.

NOTES

[1] It is not clear even if a steady state exists, and if the system ever reaches it.

[2] Monsanto (herbicides), EMBRAPA (research), Trevo (fertilizers), Agroceres (seeds), and Semeato (planters).

REFERENCES

Ekboir, J.M. and Parellada, G. (2001) Public-Private Interactions and Technology Policy in Zero-Tillage Innovation Processes - Argentina, in Byerlee, D. and Echeverría, R. (eds.), *Agricultural Research Policy in an Era of Privatization: Experiences From the Developing World.* Oxon:CABI.

Findaly, J.B.R. and Hutchinson, N.C., Development of Conservation Tillage in African Countries: A Partnership Approach. In: Breth, S. (ed.), *Partnerships for Rural Development in sub-Saharan Africa.* Geneva: Centre for Applied Studies in International Negotiations.

Nelson, R. and Rosenberg, N. (1993) Technical Innovation and National Systems, in National Innovation Systems, A Comparative Analysis, Nelson, R., ed., Oxford University Press, NY.

OECD (1999) Managing National Innovation Systems, France.

Rycroft, R.W. and Kash, D.E. (1999) *The Complexity Challenge: Technological Innovation for the 21st Century*, Printer A Casssell Imprint, UK.

M. HAMELL

DEVELOPING SOIL PROTECTION IN THE EU

Environment Quality and Natural Resources
Environment Directorate-General, European Commission
Rue de la Loi, 200 B-1049 Brussels – BELGIUM

This paper deals with the steps being taken towards a full soil protection policy in the EU. It presents several examples of legislation and other policy instruments, which provide partial protection but notes the lack, to date, of a comprehensive approach at EU level. It indicates steps being taken in that direction including in the Sixth Environmental Action Programme and outlines emerging thinking in relation to a draft Commission Communication on soil protection. Finally it indicates some possibilities for soil protection within the EU common agriculture policy.

INTRODUCTION

The broad issue of soil protection including within agriculture has been moving up the agenda of environmental issues to be tackled in Europe in recent years. This reflects not just a desire to place soil protection and conservation at the same level as that provided for air and water but also as part of a growing realisation that the features and functions of soil are under growing threat.

It is also being recognised that the protection of soil for the future will require measures and actions in many areas of policy development and implementation. The complexity of the challenges to soil is such that successful protection will depend not just on the actions of those most directly involved in working the soil such as farmers but also on areas which, at first sight, have little relationship to the soil. Together with the oceans, soil is the end receptor of so much human activity and we ignore the threats to it at our peril.

In this paper, it is intended to sketch briefly the current situation regarding soil protection, the emergence of soil as an important issue in the EU and to indicate in detail some background and substance for a communication which the Commission

L. García Torres et al. (eds.). Conservation Agriculture, 509–514

intends to bring forward in the very near future. Finally, I wish to outline current approaches to soil protection within the EU common agriculture policy.

CURRENT APPROACHES TO SOIL PROTECTION

Although a formal European Community policy on soil protection does not yet formally exist, a very broad range of Community environmental legislation is targeted at or has an influence on soil protection. This is especially true, for instance, for legislation dealing with nature, water and air the implementation of which is generally protective for soil. But it is also true for legislation in the area of planning, pollution prevention, waste and waste disposal. A number of examples provide concrete examples of an already rather comprehensive although unfocused approach.

Legislation to prevent nitrate water pollution from agriculture is designed with a general approach requiring good farming practice and particular actions aimed at resolving problems in areas where nitrate levels in water are high. These actions require the avoidance of run-off from land through the provision of winter cover thus linking water and soil protection in the same practical management task.

Recently introduced water framework legislation sets the objective of achieving and maintaining good status for all waters, rivers, lakes and coastal water and for groundwater. Wherever soil contamination contributes to water pollution, actions will have to be taken to eliminate this pollution on the basis of river basin management plans.

Diffuse air pollution is a serious long-term threat to soil and Community legislation requires Member States to monitor and limit pollutant concentrations in air. Pollutant lists and limit values are already in place for several pollutants and these lists will be extended in coming years.

In order to combat acidification, eutrophication and ozone depletion, all with serious consequences for soil, the EU is currently finalising legislation dealing with national emissions ceilings. Changes in agricultural, transport and industrial practice in coming years resulting from the implementation of this legislation will protect soils in many instances far distant from the original source of the air pollution.

Nature protection requires soil protection. In response to nature protection legislation, EU Member States are currently finalising lists of protected sites covering about 15% of the EU land area and will have to establish management plans for which will involve the protection of their soils.

In the area of waste, particular legislation dealing with soil governs the use of sewage sludge in agriculture. This legislation provides for maximum heavy metal concentrations in sewage sludge for re-use in agriculture. It also provides for maximum heavy metal concentrations in soil in order to avoid that the spreading of sludge on the same agricultural plots should increase heavy metal concentrations to such a level that could potentially cause adverse effects on the soil ecosystem.

Waste policy in the EU is based on guidelines of prevention, re-use, recycling, optimisation of disposal and a preference for recovery of material over energy recovery. The focus towards prevention and recycling is fostering an interesting in composting as an alternative to landfill. About 100 million tonnes of municipal solid waste in the EU is compostable annually. Properly composted, this material has potential to contribute to the restoration of soil quality. In addition to environmental legislation, EU policy in several areas has particular relevance to soil even if this is not the major thrust of the policy itself.

Within the context of the European Climate Change Programme, issues including the reduction of emissions of nitrous oxide are being addressed. The EU Bio-diversity Action Plans on both agriculture and natural resources involve the protection and sustainable use of soil and land.

Regional and agricultural policies have important implications for soil. As the agri-environmental aspects of policy development will be discussed in more detail in part 5, a few words on regional approaches will suffice here. Regional programmes are used, inter alia, to support the improvement and protection of soils through integrated land use planning, including the growing use of geographic information systems (GIS), the rehabilitation of derelict and polluted land and the encouragement of forestry as a means of combating or preventing erosion.

DEVELOPING A FOCUSED SOIL PROTECTION POLICY

Despite this considerable volume of legislation and policy with important implications for soil, it has been clear for some time that, in order to afford soil protection on an equal footing with air and water a more focused approach is necessary. At European level, there have been several initiatives recently to achieve this.

For the past three years, the European Commission, together with the Member States, the candidate countries and others have been discussing aspects of soil protection within the European Soil Forum specifically set up for this purpose with a view to raising soil's political profile. More recently, the European Commission published its sixth Environmental Action Programme, "*Environment 2010, Our future, our choice*" in which it laid out its approach to soil protection. In this strategic document, the Commission stressed the need for a systematic approach to soil protection covering:

- Erosion and desertification
- Pollution from landfill waste sites, industry and mining
- Pollution from air water, and from some agricultural practices and the application of sewage sludge contained by heavy metals, organic pollutants or pathogens
- Loss of land and therefore soil to development
- The role that soil plays in climate change as a carbon sink.

The European Council and Parliament have spent considerable time in the past half year debating the Commission's approach. They are currently deliberating on a thematic strategic approach to soil protection addressing the need for measures dealing with pollution, erosion, sealing, desertification, and hydrological risk while at the same time, respecting regional diversity. The Commission will have to present and the European Council and Parliament decide on the strategy within five years.

Finally, in a recent paper setting out its approach to sustainable development the Commission recognised as a major threat to sustainability that soil loss and declining fertility are eroding the viability of agricultural land. In effect, then, at European level, the phase of increasing the political profile of soil is drawing to a close and we are now entering a period of preparation for actions so that full soil protection can be achieved.

FORTHCOMING EUROPEAN COMMISSION COMMUNICATION ON SOIL

As part of the preparation for establishing the European soil protection policy, the Commission has begun work on a formal communication to the European Council and Parliament. In order to ensure a comprehensive approach, an initial draft of this document was placed on the Commission Internet web-site during early summer 2001 and an updated version of the draft will be available shortly for further comment. Therefore my comments with relation to its final shape and content should be seen as tentative and open to further development.

The communication places considerable emphasis on the functions of soil and the salient features relevant to policy makers. These include its local nature, its subjection to property rights (as a component of land), the competitive demands on it (how to protect the most productive agricultural soil) and the risks inherent in the degradation (sometimes unseen) of its considerable buffering filtering and transformation capacity.

The individual threats to soil, including erosion, point and diffuse contamination, sealing and compaction, declines in organic matter and bio-diversity and salinisation are described. It is recognised that the process of soil degradation usually takes place as a result of the combined action of several simultaneously occurring processes resulting from these threats.

Looking at these various processes, it can be broadly stated nevertheless that

- Soil degradation processes are often driven or exacerbated by human activity. By damaging the soil's capacity to perform all its functions, they constitute an insidious and most serious threat to the long-term development of society.
- There is solid scientific evidence that several soil degradation processes are occurring widely in Europe.
- It is likely that degradation processes are increasing rather than diminishing.
- More detailed information is required to determine the extent and significance of the degradation processes so as to identify the different pressures and the main factors involved with a view to their resolution.

However, while there is need for comparable and detailed policy relevant data, it is not at all necessary to await the arrival of such data before getting on with the business of protection. Policies, be they regional, national or international need to be examined to decide whether or not they themselves contribute to soil degradation or can contribute to protection.

As a first step, then, the Commission communication foresees the establishment of an efficient European soil database, which ideally will build on existing information and allow the flexibility to focus on particular or local degradation processes. Additionally, it will be valuable to examine within the various European policies and instruments the extent to which soil protection is or needs to be included or strengthened. Of course this does not exclude action at national and regional level where the essential action to protect or restore soils needs to be taken. It is both necessary and desirable to establish the real extent of the threats to soil and the ways of avoiding them. Action at both European and local level can then contribute to their resolution.

SOIL PROTECTION WITHIN AGRICULTURE

In the past, the main focus of agricultural policy in the EU tended to be market orientated. Over the past decade, this focus has been broadened considerably in recognition of the many functions of farming going far beyond food production itself. In terms of soil, the farmer himself has a vested interest in its maintenance and protection and, to a large extent, he was expected to farm in a manner as to achieve that protection.

Unfortunately, economic and social pressures have undermined the protective role of farmers with respect to soil. Modern farming methods and returns and labour availability and rewards combine to restrict the amount of care given to soil. It is difficult to imagine that farmers would return to the backbreaking process of returning soil from the bottom to the top of a slope after every few harvests as a means of countering erosion. Yet this practice was once common.

In the framework of a widening approach to agriculture and rural development, the EU introduced agri-environment measures as a part of general agricultural policy in the 1992 common agriculture policy reform. This followed a series of tentative steps recognising the greater environmental role of agriculture during the previous decade.

The basis for agri-environment is that the farmer agrees contractually to deliver environmental services to society, which go beyond what is considered to be good farming practice for the region in which he is farming including compliance with existing environmental legislation. He is paid for the service he delivers.

During the last decade, very considerable effort has gone into the design of agri-environmental measures, which of course can be very diverse in purpose and specificity particularly when adapted to local conditions. The overall purpose of a measure is generally to protect or enhance the quality of water, air, soil, bio-diversity or landscape and many measures address a number or all of these purposes.

Measures may be sharply focused such as the protection of a riverbank or the maintenance of field boundaries or directed towards drinking water sources or habitat protection. On the other hand, they may relate to a more general farm based approach such as organic farming or integrated crop management or an overall farm focus of the delivery of environmental services beyond good farm practice.

Agri-environment is proving popular with many farmers. At present, over 20% of the utilised agricultural area of the EU is farmed under agri-environment contracts.

There is, however, much greater uptake in some Member States than in others. Uptake in Sweden, Finland and Austria covers more than 60% of the farmed areas while in Spain, Greece, Denmark, Netherlands and Belgium uptake is less than 5% of the farmed area.

Many reasons can be put forward to explain this difference in uptake including the probable lack of attractiveness of the measures for farmers in intensive crop or animal producing areas or the lack of local or regional commitment to them. The agri-environment measures with a particular focus on soil protection include organic farming, input reduction, the conversion of arable land to grassland and rotations, undersowing and cover crops, strips to prevent erosion and landscape protection.

It is not intended to present an evaluation of measures of relevance to soil in this paper. An evaluation was carried out in 1998, which more or less coincided with the close of the first phase of widespread experience of the measures. It can be stated, however, that well targeted measures have brought benefits in terms of improved soil bio-diversity and in organic content as well as in the reduction of erosion. This reinforces the view that agri-environment measures have a very positive role to play in soil protection.

THE AGENDA 2000 CAP REFORM, A MORE SYSTEMATIC APPROACH TO ENVIRONMENTAL PROTECTION IN AGRICULTURE

The current shape of EU policy on agriculture was established via a reform due to apply for the period up to 2006. A central element to the policy was the requirement for Member States to establish rural development plans at the appropriate level. This necessitated an initial examination of the economic, social and environmental pressures on farming and the subsequent construction of plans to address these problems.

This process has allowed the development of measures (including agri-environment and forestry) more tailored to the particular needs of a region and is playing a significant role in pushing environmental issues to the foreground. The focus of the measures to tackle problems is largely as before but monitoring and evaluation to ensure the delivery of targets is now more central to the approach.

Of course, it is not reasonable to expect farmers taking up agri-environment measures to deliver soil protection while not requiring a reasonable standard of care through good farm practice from others. The reform now in place addresses this by requiring that Member States take the environmental measures they consider appropriate for each agricultural sector including for instance the arable, dairy, olive oil and wine sectors. In order to complete this task, each Member State will have to carry out an analysis of the environmental effects of each production sector. If, as a result of this analysis, they need to establish environmental protection requirements for producers, they can do so by making the fulfilment of these conditions conditional for any support measures. If Member States use this power in a constructive manner, they can begin the process of eliminating farm practices associated with soil problems.

AUTHOR INDEX